Knowing the Odds
An Introduction to Probability

Knowing the Odds
An Introduction to Probability

John B. Walsh

Graduate Studies
in Mathematics
Volume 139

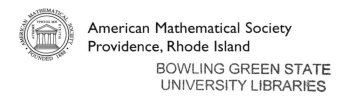

American Mathematical Society
Providence, Rhode Island

BOWLING GREEN STATE
UNIVERSITY LIBRARIES

EDITORIAL COMMITTEE

David Cox (Chair)
Daniel S. Freed
Rafe Mazzeo
Gigliola Staffilani

2010 *Mathematics Subject Classification.* Primary 60-01.

For additional information and updates on this book, visit
www.ams.org/bookpages/gsm-139

Library of Congress Cataloging-in-Publication Data
Walsh, John B.
  Knowing the odds : an introduction to probability / John B. Walsh.
    p. cm.– (Graduate studies in mathematics ; v. 139)
  Includes bibliographical references and index.
  ISBN 978–0–8218–8532–1 (alk. paper)
  1. Probabilities. I. title.

QA273.W24 2011
519.2–dc23
                                                                                          2012013119

Copying and reprinting. Individual readers of this publication, and nonprofit libraries acting for them, are permitted to make fair use of the material, such as to copy a chapter for use in teaching or research. Permission is granted to quote brief passages from this publication in reviews, provided the customary acknowledgment of the source is given.
  Republication, systematic copying, or multiple reproduction of any material in this publication is permitted only under license from the American Mathematical Society. Requests for such permission should be addressed to the Acquisitions Department, American Mathematical Society, 201 Charles Street, Providence, Rhode Island 02904-2294 USA. Requests can also be made by e-mail to reprint-permission@ams.org.

© 2012 by the American Mathematical Society. All rights reserved.
The American Mathematical Society retains all rights
except those granted to the United States Government.
Printed in the United States of America.

∞ The paper used in this book is acid-free and falls within the guidelines
established to ensure permanence and durability.
Visit the AMS home page at http://www.ams.org/

10 9 8 7 6 5 4 3 2 1   17 16 15 14 13 12

To my wife, Joke

Contents

Preface		xi
Introduction		xiii
Chapter 1.	Probability Spaces	1
§1.1.	Sets and Sigma-Fields	1
§1.2.	Elementary Properties of Probability Spaces	6
§1.3.	The Intuition	8
§1.4.	Conditional Probability	15
§1.5.	Independence	18
§1.6.	Counting: Permutations and Combinations	22
§1.7.	The Gambler's Ruin	30
Chapter 2.	Random Variables	39
§2.1.	Random Variables and Distributions	39
§2.2.	Existence of Random Variables	45
§2.3.	Independence of Random Variables	48
§2.4.	Types of Distributions	49
§2.5.	Expectations I: Discrete Random Variables	54
§2.6.	Moments, Means and Variances	60
§2.7.	Mean, Median, and Mode	63
§2.8.	Special Discrete Distributions	65
Chapter 3.	Expectations II: The General Case	75
§3.1.	From Discrete to Continuous	75

§3.2.	The Expectation as an Integral	81
§3.3.	Some Moment Inequalities	85
§3.4.	Convex Functions and Jensen's Inequality	86
§3.5.	Special Continuous Distributions	89
§3.6.	Joint Distributions and Joint Densities	96
§3.7.	Conditional Distributions, Densities, and Expectations	103
Chapter 4.	**Convergence**	**117**
§4.1.	Convergence of Random Variables	117
§4.2.	Convergence Theorems for Expectations	122
§4.3.	Applications	127
Chapter 5.	**Laws of Large Numbers**	**133**
§5.1.	The Weak and Strong Laws	134
§5.2.	Normal Numbers	137
§5.3.	Sequences of Random Variables: Existence*	140
§5.4.	Sigma Fields as Information	142
§5.5.	Another Look at Independence	144
§5.6.	Zero-one Laws	145
Chapter 6.	**Convergence in Distribution and the CLT**	**151**
§6.1.	Characteristic Functions	151
§6.2.	Convergence in Distribution	162
§6.3.	Lévy's Continuity Theorem	170
§6.4.	The Central Limit Theorem	176
§6.5.	Stable Laws*	182
Chapter 7.	**Markov Chains and Random Walks**	**191**
§7.1.	Stochastic Processes	191
§7.2.	Markov Chains	192
§7.3.	Classification of States	201
§7.4.	Stopping Times	204
§7.5.	The Strong Markov Property	208
§7.6.	Recurrence and Transience	211
§7.7.	Equilibrium and the Ergodic Theorem for Markov Chains	218
§7.8.	Finite State Markov Chains	226
§7.9.	Branching Processes	234
§7.10.	The Poisson Process	242

§7.11.	Birth and Death Processes*	250
Chapter 8.	Conditional Expectations	265
§8.1.	Conditional Expectations	265
§8.2.	Elementary Properties	268
§8.3.	Approximations and Projections	272
Chapter 9.	Discrete-Parameter Martingales	275
§9.1.	Martingales	275
§9.2.	System Theorems	282
§9.3.	Convergence	290
§9.4.	Uniform Integrability	295
§9.5.	Applications	304
§9.6.	Financial Mathematics I: The Martingale Connection*	315
Chapter 10.	Brownian Motion	335
§10.1.	Standard Brownian Motion	336
§10.2.	Stopping Times and the Strong Markov Property	344
§10.3.	The Zero Set of Brownian Motion	348
§10.4.	The Reflection Principle	351
§10.5.	Recurrence and Hitting Properties	352
§10.6.	Path Irregularity	354
§10.7.	The Brownian Infinitesimal Generator*	359
§10.8.	Related Processes	363
§10.9.	Higher Dimensional Brownian Motion	368
§10.10.	Financial Mathematics II: The Black-Scholes Model*	374
§10.11.	Skorokhod Embedding*	377
§10.12.	Lévy's Construction of Brownian Motion*	388
§10.13.	The Ornstein-Uhlenbeck Process*	390
§10.14.	White Noise and the Wiener Integral*	394
§10.15.	Physical Brownian Motion*	404
§10.16.	What Brownian Motion Really Does	410
Bibliography		413
Index		415

Preface

In the long-forgotten days of pre-history, people would color peach pits differently on the two sides, toss them in the air, and bet on the color that came up. We, with a more advanced technology, toss coins. We flip a coin into the air. There are only two possible outcomes, heads or tails, but until the coin falls, we have no way of knowing which. The result of the flip may decide a bet, it may decide which football team kicks off, which tennis player serves, who does the dishes, or it may decide a hero's fate.

The coin flip may be the most basic of all random experiments. If the coin is reasonably well-made, heads is as likely as tails to occur. But... what does that mean?

Suppose we flip a coin, and call "Heads" or "Tails" while it is in the air. Coins are subject to the laws of physics. If we could measure the exact position, velocity, and angular velocity of the coin as it left the hand—its initial conditions—we could use Newton's laws to predict exactly how it would land. Of course, that measurement is impractical, but not impossible. The point is that the result is actually determined as soon as the coin is in the air and, in particular, it is already determined when we call it; the result is (theoretically) known, but not to us. As far as we are concerned, it is just as unpredictable as it was before the flip. Let us look at the physics to see why.

The outcome is determined by the exact position, angular position, velocity, and angular velocity at the time of the flip. Physicists represent these all together as a point in what they call phase space. We can picture it as follows.

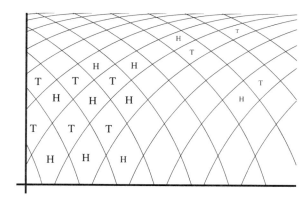

Figure 1. Phase space

This represents the initial condition of the coin in phase space. Some points lead to heads, some to tails. But a small difference in initial conditions completely changes the result. The conditions leading to heads are a union of very small regions, which are evenly mixed up with those leading to tails.

This means that no matter how we try to toss the coin, we cannot zero in on a particular result—our toss will be smeared out, so to speak, over the "Heads" and "Tails" regions, and this will happen no matter how carefully we toss it. This leads us to say things like: "Heads and tails are equally likely," or "Heads and tails each have probability one-half."

Philosophers ask deep questions about the meaning of randomness and probability. Is randomness something fundamental? Or is it just a measure of our ignorance? Gamblers just want to know the odds.

Mathematicians by and large prefer to duck the question. If pressed, they will admit that most probability deals with chaotic situations, like the flip of a coin, where the seeming randomness comes from our ignorance of the true situation. But they will then tell you that the *really* important thing about randomness is that it can be measured—for probabilities measure likelihood—and that we can construct a mathematical model which enables us to compute all of the probabilities, and that, finally, this model is the proper subject of study.

So you see, mathematicians side with the gamblers: they just want to know the odds.

From now on, probability is mathematics. We will be content just to note that it works—which is why so few casino owners go broke—and we will leave the deeper meanings of randomness to the philosophers.

Introduction

There is an order to chaos. Unpredictability is predictable. In fact, randomness itself is so regular that we can assign a number to a random occurrence which tells us in a precise way how likely it is. The number is called its probability.

That is not to say that we can predict the result of a single toss of a fair coin. We cannot. But we can predict that between forty and sixty out of a hundred tosses will be heads. We might—rarely—be wrong about that, but only once or twice in a hundred tries, and if we continue to toss: a thousand times, a million times, and so on, we can be sure that the proportion of heads will approach $1/2$.

So randomness has its own patterns. Our aim is to understand them.

Probability is a rather unusual part of mathematics. While its full birth as a mathematical subject can be traced to the correspondence between Fermat and Pascal[1] in the summer of 1654, the subject wasn't put on a rigorous footing until 1934, 270 years later, when A. N. Kolmogorov showed it was properly a part of measure theory[2]. But probability had been around for several centuries before measure theory existed, and it is quite possible to study the subject without it. In fact, probability is taught at many different

---

[1] Pascal and Fermat were by no means the first to study probabiity, but their work on the "problem of points" was so much deeper than what had gone before that it is properly considered the true beginning of the subject. See Keith Devlin's "The Unfinished Game" [13] for an account.

[2] See [22] for an English translation of Kolmogorov's landmark paper. It showed that all of probability theory could be regarded as a part measure theory, giving a general existence theorem for stochastic processes (not present, alas, in this book, but see [12] or [9]) and a rigorous definition of conditional expectations (see Chapter 8), which had previously been confined to special cases. This was quite a change from the more intuitive approach, and it took some time to replace "could be taken" by "is." That was completed by Doob, culminating in his seminal book Stochastic Processes [12].

levels, according to the mathematics the students know: in elementary and high school, first year college, third or fourth year college, as well as in graduate school. Certain things are common to all of these courses, but the the more mathematics the student knows, the deeper he or she can go. This particular text is drawn from a two-semester course taught over the years at the University of British Columbia, mainly to fourth-year mathematics honors students. It assumes the student is familiar with calculus and knows some analysis, but not measure theory. Many of the students, but by no means all, take a concurrent course in Lebesgue measure. It is not necessary, but it adds depth, and gives the student some "Aha!" moments, such as the sudden realization: "Aha! The expectation is nothing but a Lebesgue integral[3]!"

We begin with the basic axioms of probability, and the all-important ideas of conditional probability and independence. Then we quickly develop enough machinery to allow the students to solve some interesting problems and to analyze card games and lotteries. Just to show how quickly one can get into non-trivial questions, we work out the problem of the gambler's ruin.

The systematic study of classical probability begins in Chapter Two. Its aim is to prove two of the basic classical theorems of the subject: the law of large numbers and the central limit theorem. Far from being recondite, these theorems are practically part of Western folklore. Who has not heard of the law of averages? That is another name for the law of large numbers. What student has not been subject to "grading on a curve", a direct (and often mistaken) application of the central limit theorem? It is surprising how much of the curriculum is determined by the modest aim of understanding those two results: random variables, their expectations and variances, their distributions, the idea of independence, and the ideas of convergence are needed merely to state the theorems. A number of inequalities, the theory of convergence in distribution, and the machinery of characteristic functions, are necessary to prove them. This, along with enough examples to supply the intuition necessary to understanding, determines the first six chapters.

The second part of the book introduces stochastic processes, and changes the viewpoint. Stochastic processes evolve randomly in time. Instead of limit theorems at infinity, the emphasis is on what the processes actually do; we look at their sample paths, study their dynamics, and see that many interesting things happen between zero and infinity. There is a large selection of stochastic processes to study, and too little time to study them.

---

[3]On the other hand, students who take probability before measure theory have their "Aha!" moment later, when they realize that the Lebesgue integral is nothing but an expectation.

Introduction

We want to introduce processes which are major building blocks of the theory, and we aim the course towards Brownian motion and some of its weird and wonderful sample path properties. Once more, this determines much of the curriculum. We introduce the Markov property and stopping times with a study of discrete-parameter Markov chains and random walks, including special cases such as branching processes. Poisson and birth and death processes introduce continuous parameter processes, which prepares for Brownian motion and several related processes.

The one non-obvious choice is martingales. This deserves some explanation. The subject was once considered esoteric, but has since shown itself to be so useful[4] that it deserves inclusion early in the curriculum. There are two obstructions. The first is that its whole setting appears abstract, since it uses sigma-fields to describe information. Experience has shown that it is a mistake to try to work around this; it is better to spend the necessary time to make the abstract concrete by showing how sigma-fields encode information, and, hopefully, make them intuitive. The second obstruction is the lack of a general existence theorem for conditional expectations: that requires mathematics the students will not have seen, so that the only case in which we can actually construct conditional expectations is for discrete sigma-fields, where we can do it by hand. It would be a pity to restrict ourselves to this case, so we do some unashamed bootstrapping. Once we show that our hand-constructed version satisfies the defining properties of the general conditional expectation, we use only these properties to develop the theory. When we have proved the necessary martingale theorems, we can construct the conditional expectation with respect to a general sigma field as the limit of conditional expectations on discrete sigma fields. This gives us the desired existence theorem ... and shows that what we did was valid for general sigma-fields all along. We make free use of martingales in the sequel. In particular, we show how martingale theory connects with a certain part of mathematical finance, the option pricing, or Black-Scholes theory.

The final chapter on Brownian motion uses most of what we have learned to date, and could pull everything together, both mathematically and artistically. It would have done so, had we been able to resist the temptation to spoil any possible finality by showing—or at least hinting at—some of

---

[4] The tipping point was when engineers started using martingales to solve applied problems, and, in so doing, beat the mathematicians to some very nice theorems. The coup de grâce was struck by the surprising realization that the celebrated Black-Scholes theory of finance, used by all serious option-traders in financial markets was, deeply, martingale theory in disguise. See sections 9.6 and 10.10.

the large mathematical territory it opens up: white noise, stochastic integrals, diffusions, financial mathematics, and probabilistic potential theory, for example.

A last word. To teach a course with pleasure, one should learn at the same time. Fair is fair: the students should not be the only learners. This is automatic the first time one teaches a course, less so the third or fourth time. So we tried to include enough sidelights and interesting byways to allow the instructor some choice, a few topics which might be substituted at each repetition. Most of these are starred: *. In fact, we indulged ourselves somewhat, and included personal favorites that we seldom have time to cover in the course, such as the Wiener stochastic integral, the Langevin equation, and the physical model of Brownian motion.

Chapter 1

Probability Spaces

It is said that the best way to enter a cold swimming pool is to dive in head first, not to inch in one toe at a time. Let us take that advice and begin with a splash: the basic mathematical model of probability. We will explain the intuition behind it afterwards.

1.1. Sets and Sigma-Fields

Let Ω be a set.

Definition 1.1. A class $\mathcal{G}$ of subsets of Ω is a **field** if

  (i) $\emptyset \in \mathcal{G}$,
  (ii) $A \in \mathcal{G} \implies A^c \in \mathcal{G}$,
  (iii) $A, B \in \mathcal{G} \implies A \cup B \in \mathcal{G}$;

it is a **σ-field** if

  (i) $\emptyset \in \mathcal{G}$,
  (ii) $A \in \mathcal{G} \implies A^c \in \mathcal{G}$,
  (iii) $A_1, A_2, \cdots \in \mathcal{G} \implies \bigcup_{i=1}^{\infty} A_i \in \mathcal{G}$;

it is a **monotone class** if

  (i) $A_1 \subset A_2 \subset \cdots \in \mathcal{G} \implies \bigcup_{i=1}^{\infty} A_i \in \mathcal{G}$,
  (ii) $A_1 \supset A_2 \supset \cdots \in \mathcal{G} \implies \bigcap_{i=1}^{\infty} A_i \in \mathcal{G}$.

Thus a field[1] is closed under complementation and finite unions, a σ-field is closed under complementation and countable unions, and a monotone

---

[1] You might encounter these under other names. A field has also been called an algebra, a Boolean algebra, and a finitely additive class. A σ-field—pronounced sigma-field—has been called a σ-algebra, a Borel field, and a countably additive class.

class is closed under countable increasing unions and countable decreasing intersections.

For our purposes, the σ-field is by far the most useful. This is because it is closed under countable set operations. Finite unions and intersections are not enough. We will be dealing with limits, and limits involve infinite operations. However, in practice, the field is the easiest to come by. Once we have a field, we often ask, "Is it actually a σ-field? If not, can we extend it to be one?" The monotone class is an important tool for this: it is easier to show that something is a monotone class than to show it is a σ-field. Under certain conditions, this implies that it *is* a σ-field.

Remark 1.2. (*i*) A^c is the complement of A *relative to* Ω: $A^c = \Omega - A$.

(*ii*) Note that a field or σ-field contains Ω as well as $\emptyset$, since $\Omega = \emptyset^c$.

(*iii*) A field (resp. σ-field) $\mathcal{F}$ is closed under finite (resp. countable) intersections, too. For if $A_1, A_2, \cdots \in \mathcal{F}$, then $\bigcap_n A_n = \left(\bigcup_n A_n^c\right)^c \in \mathcal{F}$ by (*ii*) and (*iii*).

There are some obvious relations between the three:

(*i*) σ-field $\implies$ field and monotone class.

(*ii*) Field and monotone class $\implies$ σ-field.

(*iii*) Let $\{\mathcal{G}_\alpha, \alpha \in I\}$ be fields (resp. σ-fields, monotone classes). Then $\bigcap_\alpha \mathcal{G}_\alpha$ is a field (resp. σ-field, monotone class.)

(*iv*) If $\mathcal{C}$ is a collection of sets, there exists a minimal field (resp σ-field, monotone class) containing $\mathcal{C}$. This is called the **field** (resp. **σ-field, monotone class**) **generated by $\mathcal{C}$**.

Proof. (*i*) is obvious. For (*ii*), if $\mathcal{G}$ is a field, it is closed under complementation and finite unions, and if $A_1, A_2, \cdots \in \mathcal{G}$, then

$$\bigcup_n A_n = \bigcup_n \left(\bigcup_{j=1}^n A_j\right) \in \mathcal{G}.$$

Indeed, the finite unions are in $\mathcal{G}$ since $\mathcal{G}$ is a field, so a countable union can be written as a union of increasing sets, and hence is also in $\mathcal{G}$ because $\mathcal{G}$ is a monotone class, proving (*ii*). And (*iii*) is easy! Just verify the defining conditions.

To see (*iv*), note that if $\mathcal{C}$ is any class of subsets of Ω, there is at least one field (resp. σ-field, monotone class) containing it, namely the class of all subsets of Ω. Thus the intersection of all fields (resp. σ-field, monotone classes) containing $\mathcal{C}$ makes sense, and, by (*iii*), is itself a field (resp. σ-field, monotone class). It is necessarily minimal. □

Example 1.2.1. Here are some σ-fields.

- All subsets of Ω.
- $\{\Omega, \emptyset\}$. (This is called the **trivial σ-field**.)
- All countable subsets of $\mathbb{R}$ together with their complements.
- The Lebesgue-measurable subsets of $\mathbb{R}$.

The first two are clearly σ-fields, for (i), (ii) and (iii) are immediate. Note that every σ-field contains both $\emptyset$ and Ω, so the trivial σ-field is also the smallest possible σ-field.

The third example contains the empty set and is clearly closed under complements. In addition, it contains countable unions: if $A_1, A_2, \ldots$ are all countable, so is $\bigcup_n A_n$ since a countable union of countable sets is countable. If at least one of the A_n is the complement of a countable set, so is $\bigcup_n A_n$. In either case, it is closed under countable unions, and therefore a σ-field.

The last example—the Lebesgue-measurable sets—is for the benefit of those who are familiar with Lebesgue measure, where they are defined and shown to form a σ-field. It is also proved there that there are non-Lebesgue-measurable sets, so that the first and fourth examples are actually different[2].

Exercise 1.1. Show that the following two classes are fields, but not σ-fields.

(a) All finite subsets of $\mathbb{R}$ together with their complements.

(b) All finite unions of intervals in $\mathbb{R}$ of the form $(a, b]$, $(-\infty, a]$, and (b, ∞).

Let us take a close look at one σ-field.

Definition 1.3. The class $\mathcal{B}$ of **Borel sets** in $\mathbb{R}$ is the smallest σ-field which contains all open sets.

This is a natural class of subsets of $\mathbb{R}$. While it is smaller than the class of Lebesgue sets, it contains nearly any set we can think of.

We will only be concerned with the Borel sets of the line for the time being, but let us note for future reference that they can be defined for any topological space: if $\mathcal{Z}$ is a topological space, the class of Borel sets $\mathcal{B}(\mathcal{Z})$ of $\mathcal{Z}$ is the smallest σ-field of subsets of $\mathcal{Z}$ which contain all open sets. In particular, the Borel sets $\mathcal{B}(\mathbb{R}^n)$ can be defined for n-dimensional Euclidean space.

They can be generated by far smaller classes than the open sets. For instance, an open set is a countable union of open intervals, so that any σ-field containing the open intervals also contains all open sets, and therefore contains $\mathcal{B}$, which is the smallest σ-field containing the open sets. So the

---

[2] That is, they are different if you believe in the Axiom of Choice: the existence of non-Lebesgue measurable sets is equivalent to the Axiom of Choice.

open intervals generate the Borel sets. Here are a few more possibilities. Note that to prove that a class of sets generates the Borel sets, we need only show they generate the open intervals.

Exercise 1.2. Show that the Borel sets are generated by any one of the following classes:

(a) The closed sets.

(b) All closed intervals.

(c) All closed intervals with rational end points.

(d) All intervals of the form $(a, b]$.

(e) All intervals of the form $(-\infty, x]$.

(f) All intervals of the form $(-\infty, x]$, where x is rational.

We say two sets A and B are **disjoint** if $A \cap B = \emptyset$.

Definition 1.4. Let $\mathcal{F}$ be a σ-field. A **probability measure** P on $\mathcal{F}$ is a real-valued function defined on $\mathcal{F}$ such that

(i) if $A \in \mathcal{F}$, $P\{A\} \geq 0$;

(ii) $P\{\Omega\} = 1$;

(iii) if $A_1, A_2, \ldots$ is a finite or countably infinite sequence of disjoint elements of $\mathcal{F}$ (i.e., $i \neq j \implies A_i \cap A_j = \emptyset$), then

$$P\{\bigcup_n A_n\} = \sum_n P\{A_n\}.$$

$P\{A\}$ is called the **probability** of A.

Property (iii) is called **countable additivity**. Note that it holds for infinite unions, not just finite ones. We will see shortly that this is really a continuity property in disguise.

The Monotone Class Theorem★. This section states and proves the Monotone Class Theorem, and can be safely skipped for the moment. We will not use it until Section 2.3. It is one of those theorems which seems very technical... until it's needed to make some otherwise-painful proof easy.

Theorem 1.5 (Monotone Class Theorem). *Let $\mathcal{F}_0$ be a field and $\mathcal{G}$ a monotone class. Suppose $\mathcal{F}_0 \subset \mathcal{G}$. Then $\sigma\{\mathcal{F}_0\} \subset \mathcal{G}$. In particular, the monotone class and the σ-field generated by $\mathcal{F}$ are the same.*

Proof. This proof makes extensive use of minimality. Let $\mathcal{G}'$ be the smallest monotone class containing $\mathcal{F}_0$. Then $\mathcal{G}' \subset \mathcal{G}$. Since $\sigma(\mathcal{F}_0)$ is also a monotone class containing $\mathcal{F}_0$, $\mathcal{G}' \subset \sigma(\mathcal{F}_0)$. We will show $\sigma(\mathcal{F}_0) = \mathcal{G}'$.

Define two classes of subsets of Ω:

$$\mathcal{C}_1 = \{A \in \mathcal{G}' : A \cap B \in \mathcal{G}', \forall B \in \mathcal{F}_0\},$$

1.1. Sets and Sigma-Fields

$$\mathcal{C}_2 = \{A \in \mathcal{G}' : A \cap B \in \mathcal{G}', \forall B \in \mathcal{G}'\}.$$

Let $A_1, A_2, \cdots \in \mathcal{C}_1$, (resp. $\mathcal{C}_2$) and $B \in \mathcal{F}_0$ (resp. $B \in \mathcal{G}'$). If $A_1 \supset A_2 \supset \ldots$, then

$$A_1 \cap B \supset A_2 \cap B \supset \ldots \downarrow \left(\bigcap_n A_n\right) \cap B \in \mathcal{G}'$$

since $A_n \cap B \in \mathcal{G}'$ and $\mathcal{G}'$ is a monotone class. Similarly, if $A_1 \subset A_2 \subset \ldots$, then

$$A_1 \cap B \subset A_2 \cap B \subset \ldots \uparrow \left(\bigcup_n A_n\right) \cap B \in \mathcal{G}'$$

for the same reason. Thus $\mathcal{C}_1$ (resp. $\mathcal{C}_2$) is a monotone class. Moreover, $\mathcal{C}_1$ contains $\mathcal{F}_0$, since $\mathcal{F}_0$ is already closed under intersections, so that $\mathcal{F}_0 \subset \mathcal{C}_1 \subset \mathcal{G}'$. But $\mathcal{G}'$ is the minimal monotone class containing $\mathcal{F}_0$, hence $\mathcal{C}_1 = \mathcal{G}'$.

Now $\mathcal{C}_2 \subset \mathcal{G}'$ by definition, and $\mathcal{F}_0 \subset \mathcal{C}_2$. Indeed, if $B \in \mathcal{F}_0$ and $A \in \mathcal{G}' = \mathcal{C}_1$, then $A \cap B \in \mathcal{G}' \implies B \in \mathcal{C}_2$. By minimality again, $\mathcal{C}_2 = \mathcal{G}'$. Therefore, $\mathcal{G}'$ contains $\mathcal{F}_0$ and is closed under finite intersections.

Let us check complementation. Let $\mathcal{C}_3 = \{A \in \mathcal{G}' : A^c \in \mathcal{G}'\}$. Then $\mathcal{F}_0 \subset \mathcal{C}_3$. If $A_1 \subset A_2 \subset \cdots \in \mathcal{C}_3$, then

$$\left(\bigcup_{j=1}^\infty A_j\right)^c = \bigcap_j A_j^c.$$

But $A_1^c \supset A_2^c \supset \cdots \in \mathcal{G}'$, and $\mathcal{G}'$ is a monotone class, so that $\bigcap_{j=1}^\infty A_j \in \mathcal{G}' \implies \bigcup_{j=1}^\infty A_j \in \mathcal{C}_3$. Similarly, if $A_1 \supset A_2 \supset \cdots \in \mathcal{C}_3$, then

$$\left(\bigcap_{j=1}^\infty A_j\right)^c = \bigcup_j A_j^c.$$

But $A_1^c \subset A_2^c \subset \cdots \in \mathcal{G}'$ and $\mathcal{G}'$ is a monotone class, so that $\bigcap_{j=1}^\infty A_j \in \mathcal{G}' \implies \bigcap_{j=1}^\infty A_j \in \mathcal{C}_3$. Therefore $\mathcal{C}_3$ is a monotone class containing $\mathcal{F}_0$. Since $\mathcal{C}_3 \subset \mathcal{G}'$, and $\mathcal{G}'$ is minimal, $\mathcal{C}_3 = \mathcal{G}'$. Therefore, $\mathcal{G}'$ is closed under complementation.

Thus $\mathcal{G}'$ is closed under finite intersections and complementation and therefore it is also closed under finite unions, so it is a field. But it is a monotone class, so it is also a σ-field. This proves the first conclusion of the theorem.

The second conclusion is almost immediate. Since $\sigma(\mathcal{F}_0)$ is the smallest σ-field containing $\mathcal{F}_0$, $\sigma(\mathcal{F}_0) \subset \mathcal{G}'$. But $\sigma(\mathcal{F}_0)$ is also a monotone class containing $\mathcal{F}_0$, and $\mathcal{G}'$ is minimal, $\mathcal{G}' \subset \sigma(\mathcal{F}_0)$. Thus $\sigma(\mathcal{F}_0) = \mathcal{G}' \subset \mathcal{G}$. □

Here is a typical application of the monotone class theorem. Let $(\Omega, \mathcal{F}, P)$ be a probability space. Suppose that $\mathcal{F}_0$ is a field, but not a σ-field. Then it generates a σ-field, $\sigma(\mathcal{F}_0)$. The question is, what kind of sets do we add to make it a σ-field? Are the new sets close to the old? In fact they are, in the sense that every set in $\sigma(\mathcal{F}_0)$ can be approximated arbitrarily well by a set in $\mathcal{F}_0$. (We could borrow some terminology from topology and say that in a sense, $\mathcal{F}_0$ is dense in $\sigma(\mathcal{F}_0)$.)

Definition 1.6. The **symmetric difference** $A \bigtriangleup B$ of two sets A and B is
$$A \bigtriangleup B = (A - B) \cup (B - A).$$

Note that $P\{(\bigcup_{j=1}^\infty A_j) \bigtriangleup (\bigcup_{j=1}^\infty B_j)\} \leq \sum_{j=1}^\infty P\{A_j \bigtriangleup B_j\}$.

Theorem 1.7. *Let $\mathcal{F}_0 \subset \mathcal{F}$ be a field and $\Lambda \in \sigma(\mathcal{F}_0)$. For any $\varepsilon > 0$, there exists $\Gamma_\varepsilon \in \mathcal{F}_0$ such that $P\{\Lambda \bigtriangleup \Gamma_\varepsilon\} < \varepsilon$.*

Proof. This is typical of many proofs in measure theory: it defines a class of sets with the desired property, and then shows it is a monotone class.

Let $\mathcal{G} = \{\Lambda \in \mathcal{F} : \forall \varepsilon > 0 \; \exists \; \Gamma_\varepsilon \in \mathcal{F}_0 \text{ such that } P\{\Lambda \bigtriangleup \Gamma_\varepsilon\} < \varepsilon\}$. Clearly $\mathcal{F}_0 \subset \mathcal{G}$.

Let $\varepsilon > 0$ and let $\Lambda_1 \subset \Lambda_2 \subset \cdots \in \mathcal{G}$. Choose n such that $P\{\Lambda - \bigcup_{j=1}^n \Lambda_j\} < \varepsilon/2$.

For $j = 1, \ldots, n$ choose $\Gamma_j \in \mathcal{F}_0$ such that $P\{\Lambda_j \bigtriangleup \Gamma_j\} < \varepsilon/2^{j+1}$. Then set $\Gamma_\varepsilon = \bigcup_{j=1}^n \Gamma_j$. Note that $\Gamma_\varepsilon \in \mathcal{F}_0$ since $\mathcal{F}_0$ is a field, and

$$\begin{aligned} P\{\Gamma_\varepsilon \bigtriangleup \Lambda\} &\leq P\Big\{\Lambda - \bigcup_{j=1}^n \Lambda_j\Big\} + P\Big\{\Big(\bigcup_{j=1}^n \Lambda_j\Big) \bigtriangleup \Big(\bigcup_{j=1}^n \Gamma_j\Big)\Big\} \\ &\leq \varepsilon/2 + \sum_{j=1}^n P\{\Lambda_j \bigtriangleup \Gamma_j\} \\ &\leq \varepsilon/2 + \sum_{j=1}^n \frac{\varepsilon}{2^{j+1}} \leq \varepsilon. \end{aligned}$$

Thus $\Lambda \in \mathcal{G}$. A similar argument shows that if $\Lambda_1 \supset \Lambda_2 \supset \cdots \in \mathcal{G}$, then $\bigcap_j \Lambda_j \in \mathcal{G}$. Therefore, $\mathcal{G}$ is a monotone class. It contains the field $\mathcal{F}_0$, so by the monotone class theorem, $\mathcal{G}$ contains $\sigma(\mathcal{F}_0)$. □

This corollary follows immediately:

Corollary 1.8. *Let P and Q be probability measures on a sample space $(\Omega, \mathcal{F})$. Let $\mathcal{F}_0 \subset \mathcal{F}$ be a field. Suppose $Q\{A\} = P\{A\}$ for $A \in \mathcal{F}_0$. Then $Q\{A\} = P\{A\}$ for $A \in \sigma(\mathcal{F}_0)$.*

1.2. Elementary Properties of Probability Spaces

Let $(\Omega, \mathcal{F}, P)$ be a probability space.

1.2. Elementary Properties of Probability Spaces

Proposition 1.9. *Let A and B be events. Then*
  (i) $P\{\emptyset\} = 0$;
  (ii) $P\{A^c\} = 1 - P\{A\}$;
  (iii) $A \subset B \implies P\{A\} \le P\{B\}$;
  (iv) $P\{A \cup B\} = P\{A\} + P\{B\} - P\{A \cap B\}$.

Proof. (i) $\emptyset \in \mathcal{F}$ and $\emptyset \cap \emptyset = \emptyset$—the empty set is disjoint from itself (!)—so that $P\{\emptyset\} = P\{\emptyset\} + P\{\emptyset\} = 2P\{\emptyset\}$. Thus $P\{\emptyset\} = 0$.

(ii) $A \cap A^c = \emptyset$ and $A \cup A^c = \Omega$, so $P\{A\} + P\{A^c\} = P\{\Omega\} = 1$.

(iii) Since A and $B - A$ are disjoint, and $B = A \cup (B - A)$, $P\{B\} = P\{A\} + P\{B-A\}$. This is greater than or equal to $P\{A\}$, for $P\{B-A\} \ge 0$.

(iv) Write A, B, and $A \cup B$ as disjoint unions: $P\{A\} = P\{A \cap B\} + P\{A - B\}$ and $P\{B\} = P\{A \cap B\} + P\{B - A\}$. Thus $P\{A\} + P\{B\} = \bigl(P\{A-B\} + P\{B-A\} + P\{A \cap B\}\bigr) + P\{A \cap B\} = P\{A \cup B\} + P\{A \cap B\}$, and (iv) follows. $\square$

Countable additivity ((iii) in the definition) is important. It is not enough to have additivity just for finite unions. In fact, countable additivity is a continuity property in disguise. The following proposition shows that probability measures are continuous from above and from below: $\lim_n P\{A_n\} = P\{\lim_n A_n\}$ if the sequence (A_n) is either increasing or decreasing.

Proposition 1.10. (i) *If $A_1 \subset A_2 \subset A_3 \subset \ldots$ are events, then $\lim_{j \to \infty} P\{A_j\} = P\{\bigcup_j A_j\}$.*

(ii) *If $B_1 \supset B_2 \supset B_3 \supset \ldots$ are events, then $\lim_{j \to \infty} P\{B_j\} = P\{\bigcap_j B_j\}$.*

Proof. (i) $\bigcup_j A_j = A_1 \cup (A_2 - A_1) \cup (A_3 - A_2) \cup \ldots$, which is a disjoint union. Thus, by countable additivity and the definition of the sum of a series:

$$P\Bigl\{\bigcup_j A_j\Bigr\} = P\{A_1\} + \sum_{j=2}^{\infty} P\{A_j - A_{j-1}\}$$

$$= P\{A_1\} + \lim_{N \to \infty} \sum_{j=2}^{N} P\{A_j - A_{j-1}\}.$$

But $A_{j-1} \subset A_j$, so $P\{A_j - A_{j-1}\} = P\{A_j\} - P\{A_{j-1}\}$, and the series is

$$= P\{A_1\} + \lim_{N \to \infty} \sum_{j=2}^{N} \bigl(P\{A_j\} - P\{A_{j-1}\}\bigr)$$

$$= \lim_{N \to \infty} P\{A_N\}.$$

(ii) follows from (i) by taking complements. We leave the details as an exercise. □

Definition 1.11. A **probability space** is a triple $(\Omega, \mathcal{F}, P)$, where Ω is a set, called the **sample space**, $\mathcal{F}$ is a σ-field of subsets of Ω, and P is a probability measure on $\mathcal{F}$.

Definition 1.12. Elements of $\mathcal{F}$ are called **events**.

Remark 1.13. (i) In many cases, $\mathcal{F}$ is the class of all subsets of Ω. But sometimes we do not know the probabilities of all subsets, and sometimes—for technical reasons we need not go into here—it is simply not possible to define the probability of all sets. For those who have seen it, the primary example of this is Lebesgue measure. It cannot be defined on all subsets of the line, but only on a class of sets called the Lebesgue-measurable sets.

(ii) In any case, the class of sets on which we can define a probability is a σ-field. This is implicit in the very definition of a probability measure, since, for example, (iii) implies that the probability of a countable union of events can be defined, so the union must itself be an event.

(iii) There is another important reason for using σ-fields. It will not come up until Chapter 5.4, but it is worth mentioning here. It turns out that σ-fields are a very flexible way to represent information. One could say that they are information in the raw.

Problems 1.2

1.3. Let A, B, and C be sets. Prove or disprove:
  (a) $(A - B) - C = A \cap B^c \cap C^c$.
  (b) $(A \cap B) - C = (A - C) \cap (B - C)$.
  (c) $A \cup B = B \iff A \subset B$.

1.4. Let n be an integer, let $\Omega = \mathbb{R}$, let $\mathcal{F}_n$ be the smallest σ-field containing all the intervals $(\frac{i}{n}, \frac{i+1}{n}]$, and let P be any probability measure on $\mathbb{R}$. Describe the random variables on $(\Omega, \mathcal{F}_n, P)$.

1.5. Show that the following two classes are fields, but not sigma-fields.
  (a) All finite subsets of $\mathbb{R}$ together with their complements.
  (b) All finite unions of intervals in $\mathbb{R}$ of the form $(a, b]$, $(-\infty, a]$, and (b, ∞).

1.6. Let $\mathcal{C}$ be the class of all subsets of $\mathbb{R}$ which are either countable or are the complement of a countable set. Show that $\mathcal{C}$ is a σ-field.

1.7. Finish the proof of Proposition 1.10 by showing that if B and B_n are events, then $B_n \downarrow B \implies P(B_n) \downarrow P(B)$.

1.3. The Intuition

Think of a probability space as the mathematical description of an experiment. The experiment could be, for example, tossing a coin, rolling dice,

1.3. The Intuition

taking a number in a lottery, or getting up in the morning, opening a window, and checking the weather. In each case, there is a certain amount of randomness, or unpredictability in the experiment. The end result is called the *outcome* of the experiment.

To describe this mathematically, start with what we observe: the outcome. Think of Ω as the set of all possible outcomes of the experiment: each point of Ω represents an outcome. An event is a set of outcomes. The probability measure gives the probability of events. For instance, the probability that the outcome falls in a given event A is $P\{A\}$. The collection of all events is the σ-field $\mathcal{F}$.

In short, the probability space $(\Omega, \mathcal{F}, P)$ can be thought of as a mathematical description of an experiment: the elements of Ω are the possible outcomes, the elements of $\mathcal{F}$ are the events, and P gives us the probabilities of all the events—and, in particular, the probabilities of the outcomes.

We should emphasize that there is no need to tie a probability space to a particular experiment. That is just for the intuition. Indeed, Ω can be any set, $\mathcal{F}$ can be any σ-field of subsets of Ω, and P can be any probability measure on $\mathcal{F}$. Together, they form a probability space.

Example 1.13.1. Suppose the experiment is a roll of a single die. A die is a cube whose six faces each have a different number of spots, one through six. Take the the outcome to be a number from one to six. Thus $\Omega = \{1, 2, 3, 4, 5, 6\}$, and the σ-field $\mathcal{F}$ is all subsets of Ω. To get the probability measure, we note that if the die is well-made, the six sides are identical except for the number of spots on them. No one side can be more probable than another. Thus they must all have the same probability: $P\{1\} = P\{2\} = \cdots = P\{6\}$. The probabilities add up to one, so that $P\{1\} = P\{2\} = \cdots = P\{6\} = 1/6$.

Typical events would be $\{1, 3, 5\}$ = "odd"; $\{3, 4, 5, 6\}$ = "greater than two". Notice that each individual outcome is itself an event[3]. The probabilities are easily computed by simply counting the number of outcomes; for example $P\{2, 4, 6\} = P\{2\} + P\{4\} + P\{6\} = 1/2$.

Example 1.13.2. If the experiment is tossing a coin, the outcomes are heads and tails, and we can take $\Omega = \{H, T\}$. If the coin is well-balanced, there is no effective difference between the two sides, and the probabilities of heads and tails should be the same—$P\{H\} = P\{T\} = 1/2$. The σ-field $\mathcal{F}$ is the set of all subsets, which, in this case, is so simple that we can list it: $\emptyset$, $\{H\}$, $\{T\}$, and Ω, having probabilities zero, one-half, one-half, and one, respectively.

---

[3] Technically, the event is the set whose only element is the outcome.

Example 1.13.3. Equally Likely Outcomes. Many games of chance are designed to provide equally likely outcomes. In that case, the probability of each outcome is one over the total number of outcomes: if $|A|$ denotes the number of elements in A, the probability of each outcome is $1/|\Omega|$. This leads directly to the following formula for probabilities.

Proposition 1.14. *Suppose $|\Omega|$ is finite. If all outcomes are equally likely, then the probability of any event A is*

$$P\{A\} = \frac{|A|}{|\Omega|}.$$

Example 1.14.1. Roll two dice. In order to tell them apart, color one red and one green. Suppose they are well-made, essentially perfect cubes. Each die has a number of pips painted on each side, with the number of pips going from one to six. (Assume that the paint and the color of the dice make no difference to the outcome.) The outcome for each die is the number on the top face when the die comes to rest. We can represent the outcome by an ordered pair (m, n), where m is the number showing on the red die, and n the number showing on the green die.

Let the sample space Ω be the space of outcomes $\{(m,n), m, n = 1, \ldots, 6\}$. Then Ω has 36 points. To find the probability measure P, we note that since the dice are perfectly symmetric, no number can be more likely than any other. In fact, no pair of numbers can be more likely than any other pair. To see this, note that we can change any outcome into any other merely by repainting the faces: to replace $(2, 5)$ by $(1, 4)$, just interchange the numbers 1 and 2 on the red die, and 5 and 4 on the green. Repainting the dice doesn't change their probabilities—the paint is too light to make a difference. We conclude, purely from symmetry considerations, that all ordered pairs are equally likely. There are 36 outcomes, so that the probability of any event A is $|A|/36$. For instance, the event that the sum of the two dice is a four consists of ordered pairs $(1, 3), (2, 2), (3, 1)$ so that $P\{\text{ sum is 4 }\} = 3/36 = 1/12$.

Example 1.14.2. Choose a number at random from the interval $[0, 1]$. To choose at random implies, intuitively, that no number is more likely than any other. As there are uncountably many numbers in $[0, 1]$, the probability of each is zero, so we had better rephrase this: the probability that the number falls in any sub-interval $[a, a + h]$ is the same as the probability it falls in any other sub-interval of the same length. A moments reflection suggests that the probability must be proportional to the length of the interval, and, since the whole interval has length one, we conclude that the probability of falling in a sub-interval of $[0, 1]$ is actually equal to its length.

1.3. The Intuition

Therefore, we take the sample space Ω to be $[0,1]$ itself: the space of all outcomes. We will take the σ-field to be $\mathcal{B}(0,1)$, the σ-field of Borel[4] subsets of $[0,1]$. The probability measure is $P\{dx\} = dx$, $x \in [0,1]$, that is, Lebesgue measure. So for any event A, $P\{A\}$ is the length of A and—most importantly—$P\{A\}$ is the probability that our randomly-chosen number lies in A.

Thus $([0,1], \mathcal{B}(0,1), P)$ is a probability space, where $P\{dx\} = dx$ is Lebesgue measure.

Note. The previous examples concerned finite sample spaces, and the countable additivity of the probability measure was automatic. However, Example 1.14.2 is far from finite, and the fact that Lebesgue measure on $[0,1]$ is a probability measure on the Borel sets is far from trivial. While it certainly has total mass one, the fact that it is countably additive is deep. We will accept it here, and refer the reader to a text in measure theory, e.g. [**31**], §3.3, for its proof.

Remark 1.15. The probability space for an experiment is not unique. For example, the sample space for a single coin toss can be $\Omega = \{H, T\}$. But we could also use the phase-space representation mentioned in the Preface, which would make Ω a subset of twelve-dimensional Euclidean space! We indicated there that any probability measure sufficiently "smeared out" would work. Another possibility would be to take Ω to be the unit interval $(0,1)$, where any point $x \in (0, 1/2)$ corresponds to heads, and any $x \in [1/2, 1)$ corresponds to tails. In this case, $\mathcal{F}$ could be any σ-field of subsets of $(0,1)$ which contained $(0, 1/2)$ and $[1/2, 1)$, and P could be any probability for which $P\{(0, 1/2)\} = P\{[1/2, 1)\} = 1/2$.

We (perhaps too often) think mathematics is neat and tidy, and some might find this plurality of probability spaces off-putting. However, it is useful in practice: it allows us to choose the probability space which works best *in the particular circumstances,* or to not choose one at all; we do not always have to specify the probability space. It is usually enough to know that it is there if we need it.

1.3.1. Symmetry. The probabilities in the preceding examples were all derived from symmetry considerations. In each case, the possible outcomes were indistinguishable except by their labels. ("Label" just means "name" here. The two sides of a coin are labeled by the embossed figures we call heads and tails, dice are labeled by the number of pips on each side, cards are labeled by the pictures printed on their faces, and lottery tickets are labeled by the number printed on them.)

---

[4]We could equally well take the σ-field of Lebesgue sets, which is larger than $\mathcal{B}$, but it turns out that we cannot take the σ-field of *all* subsets.

This is no coincidence. In fact, it is about the only situation in which we can confidently assign probabilities by inspection. But luckily, while nature is not always obliging enough to divide itself into equally-likely pieces, one can, through art and science, start with the equally-likely case and then determine the probabilities in ever-more complex situations. Which is what the subject is about.

The idea of symmetry applies to events, not just outcomes. Consider a physical experiment with finitely or countably many outcomes, labeled in some convenient fashion.

Symmetry principle. *If two events are indistinguishable except for the way the outcomes are labeled, they are equally likely.*

"Indistinguishable" refers to the physical events in the experiment, not to the mathematical model. The principle says that if we can't tell the difference between one event and another except by asking for the names—the labels—of the individual outcomes, they must have the same probability. For example, roll a die and consider the events "even" and "odd", i.e., $\{2,4,6\}$ and $\{1,3,5\}$. If we physically renumber the faces of the die, so that we interchange n and $7-n$ on each face, so that $1 \leftrightarrow 6$, $2 \leftrightarrow 5$, and $3 \leftrightarrow 4$, then the events "even" and "odd" are interchanged. The symmetry principle says that the two events must have the same probability. (So does the very easy calculation which tells us that both have probability $1/2$.)

Let us illustrate this graphically. Recall the glasses used to view 3-D movies. They have one lens colored red, the other colored blue. One lens filters out blue colors, the other filters out red. Take a blank die. Put n red spots, and $7-n$ blue spots on each face, where n goes from one to six. Put on the 3-D glasses, and look at the die with one eye closed. If the left eye is closed, say, you look through the red lens—which filters out red—and you only see the blue spots. If the right eye is closed, you see the red. Roll the die and close one eye: suppose you see an even number of spots. Now open that eye and close the other: you now see an odd number. So the events "even" and "odd" must have the same probability, since, physically, they are the same event! They are just observed differently.

This is essentially J. M. Keynes "Principle of Indifference" [**18**]. To use it in a concrete case, number five identical marbles from one to five, put them in a sack, mix them up, close your eyes, and choose one. The numbers have no influence on the draw since we cannot see them. Then what is the probability that marble one is drawn first? What is the probability that marble five is drawn last? What is the probability that both happen: marble one is first and marble five is last?

Clearly, by symmetry all five marbles are equally likely to be the first choice, so the probability that number one is first is $1/5$. In the second

1.3. The Intuition

case, we could count the number of ways to choose all five marbles in order so that number five is last, but that is unnecessary, since it is again clear by symmetry that all five marbles are equally likely to be the last choice. Indeed, the problem is symmetric in the five marbles, so no one can be more likely than any other to be last, and the probability is 1/5 again.

The third probability might appear harder, but it is not. Simply look at the ordered pair (first marble, last marble). By the symmetry principle, all possible orders are equally likely. (To see that (1,5) has the same probability as, say, (2,3), just note that we can renumber the marbles so that marble 1 $\mapsto$ marble 2, and marble 5 $\mapsto$ marble 3, so that $(1,5) \mapsto (2,3)$. Thus $(1,5)$ must have the same probability as $(2,3)$.)

For a last example, let us take another look at the probabilities for a pair of dice. We argued above that we could distinguish the two dice by coloring them differently, and write the outcomes as 36 equally likely *ordered* pairs. So, for instance, the point 12 can only be made by $(6,6)$, so it has probability 1/36, while the point 11 can be made with $(5,6)$ or $(6,5)$, so it has probability 2/36.

But, in fact, a pair of dice are usually of the same color, and essentially indistinguishable. This led Gottfried Leibnitz, the co-inventor (with Newton) of calculus, to write in 1666 that the points 11 and 12 were equally likely. This was because each point can be made in only one way: the 12 by two sixes, the 11 by a six and a five[5].

This contradicts the previous calculation. Rather than laugh at Leibnitz, let us restate what he said. In effect, he thought that the 21 different *combinations* are equally likely, not the 36 different *ordered pairs*. (After some reflection, it may appear less obvious that Leibnitz was wrong. Indeed, we use equally-likely combinations to calculate the probabilities of card hands; why shouldn't we do it for dice too?)

Problem: Why are the 21 combinations not equally likely? (Hint: Is it possible to relabel the dice to change any given combination into any other? If it is, the symmetry principle would tell us Leibnitz was right, and we would have a genuine problem!)

Problems 1.3

1.8. (*i*) Throw 64 dice. What is the probability that their sum is odd? Even? [Hint: Try renumbering *one* of the dice.]

1.9. Construct probability spaces to describe the following experiments.

(a) A fair coin is tossed three times.

(b) A single coin is tossed and a single die is rolled.

---

[5] Do you think that Leibnitz gambled a lot?

1.10. Find the most ridiculous probability space you can think of to describe the roll of a single fair die. (But... you must be able to explain how your probability space actually describes the experiment!)

1.11. A ping pong ball is drawn at random from a sack containing 49 balls, numbered from one to 49.

(a) Find the probability that its number is divisible by 2 but not by 3.

(b) Find the probability that its number is divisible by both 2 and 3.

(c) Find the probability that its number is prime.

1.12. A pair of fair dice are rolled. Find the probability that the total number of spots showing is at least 11. Which number has the greatest probability?

1.13. What—if anything—would go wrong if we calculated the probabilities in the roll of a pair of fair dice by taking the *unordered* outcomes to be equally likely, instead of the ordered outcomes? That is, if we regarded, for example, $(3, 4)$ and $(4, 3)$ as the same outcome?

1.14. Roll six independent fair dice.

(a) Find the probability they all show different faces.

(b) Find the probability they form three pairs.

1.15. One card is lost from a standard bridge deck. Subsequently, a card is drawn from the remaining cards. What is the probability that the card drawn is an ace?

1.16. P.R. de Montmort, writing in 1708, posed a problem on the French game Jeu de Boules. The object of the game is to throw balls close to a target ball. The closest ball wins. Suppose two players, A and B have equal skill. Player A tosses one ball. Player B tosses two. What is the probability that player A wins?

1.17. Deal cards one-by-one from a well-shuffled bridge deck until only one suit is left. What is the probability that it is spades?

1.18. A number is chosen at random from the interval $[0, 1]$. It divides the interval into two parts. Find the probability that the longer part is at least twice as long as the shorter part.

1.19. Sicherman dice are like ordinary dice, except that they have different numbers of pips on their faces. One has 1,3,4,5,6,8 on its six faces, and the other has 1,2,2,3,3,4. A pair is thrown. Find the probabilities of rolling the numbers one thru twelve with these three dice. Compare with the usual probabilities.

1.20. 100 passengers wait to board an airplane which has 100 seats. Nassif is the first person to board. He had lost his boarding pass, but they allowed him to board anyway. He chose a seat at random. Each succeeding passenger enters and takes their assigned seat if it is not occupied. If it is occupied, they take an empty seat at random. When the final passenger enters, there is only one empty seat left. What is the probability that it is that passenger's correct seat?

1.21. Suppose A and B are disjoint events with probability $P\{A\} = .4$, $P\{B\} = .5$. Find

(a) $P\{A \cup B\}$

(b) $P\{A^c\}$.

(c) $P\{A^c \cap B\}$.

1.22. Let A and B be events. Show that $P\{A \cup B\} + P\{A \cap B\} = P\{A\} + P\{B\}$.

1.23. Let A, B and C be events. Show that $P\{A \cup B \cup C\}$ equals

$$P\{A\} + P\{B\} + P\{C\} - P\{A \cap B\} - P\{B \cap C\} - P\{C \cap A\} + P\{A \cap B \cap C\}.$$

1.4. Conditional Probability

Conditional probability is legendary for producing surprising results. Let us consider a particular—though unsurprising—example which will lead us to its definition.

Toss a dart at a circular dart board. Let the sample space Ω be the unit disk, which represents the target. The outcome, the point where the dart lands, is a point in the unit disk, and an event is just a subset of the unit disk. There is always a certain amount of randomness in where the dart lands, so the landing-point will be governed by a probability measure, say P: $P\{A\}$ is the probability that the dart lands in the set A.

Let C be the lower half of Ω. C is an event itself. Let A be another event. Suppose we know that C happens, i.e., that the dart lands in the lower half of the board. What is the probability of A given that this happens? In general, it is different from the probability of A. (For if A is in the upper half of Ω, we know the dart cannot have landed in A, so the new probability of A equals zero!)

Let us give this new probability a name and a notation: we call it the **conditional probability of A given C** and we write

$$P\{A \mid C\}.$$

Let us do a thought experiment to determine this. Throw a large number of darts at the target, independently, and use the law of averages[6]. Since this is a thought experiment, there is no limit to the number of darts we can throw. To find the probability of A, we could throw, say, N darts, count the number, say $n(A)$, which land in A, and say that $P\{A\}$ is the limit of $n(A)/N$ as $N \to \infty$.

Suppose we wish to find the probability the dart lands in A, *given that the dart lands in C*. Then we would simply count *only those darts that land in C*—there are $n(C)$ of them—and we would estimate this probability by the proportion of *those* darts which land in A. If we are only counting the darts that land in C, then the darts which land in A and also are counted are the darts which land in $A \cap C$. So we throw N darts, of which $n(C)$

---

[6]We all know the law of averages, and, wonder of wonders, it is actually true. We will prove it later.

land in C, and, of those, $n(A \cap C)$ land in A. So we estimate $P\{A \mid C\}$ by $n(A \cap C)/n(C)$. Now use the law of averages again.

$$P\{A \mid C\} = \lim_{N \to \infty} \frac{n(A \cap C)}{n(C)} = \lim_{N \to \infty} \frac{\frac{n(A \cap C)}{N}}{\frac{n(C)}{N}} = \frac{P\{A \cap C\}}{P\{C\}}.$$

Conclusion: The conditional probability must be defined as follows.

Definition 1.16. Let A and C be events. If $P\{C\} > 0$, the **conditional probability of A given C** is

$$P\{A \mid C\} = \frac{P\{A \cap C\}}{P\{C\}}.$$

Here is a useful consequence:

(1.1) $$P\{A \cap C\} = P\{A \mid C\} P\{C\}.$$

We said that conditional probability is legendary for producing surprising results. Let us look at some examples to support this.

Example 1.16.1. An ancient Chinese sandalwood chest has three drawers. The first contains two silver coins, the second contains a gold and a silver coin, and the third contains two gold coins. A drawer is picked at random and a coin is taken at random from the drawer. The coin is gold. What is the probability that the *other* coin in the drawer is also gold?

Think about this before reading further. Is the answer obvious?

Answer. There are two ways to choose the gold coin: choose the drawer with two gold coins or choose the drawer with one gold and one silver coin and choose the gold coin from it. Thus $P\{\text{1st gold}\} = P\{\text{choose drawer 3}\} + P\{\text{choose drawer 2}\}P\{\text{choose gold} \mid \text{choose drawer 2}\} = 1/3 + 1/3 \times 1/2 = 1/2$. The probability that both the first and second coins are gold is the probability of choosing drawer 3, namely $1/3$. Thus $P\{\text{2nd gold} \mid \text{1st gold}\} = P\{\text{both gold}\}/P\{\text{1st gold}\} = (1/3)/(1/2) = 2/3$.

So...the answer is *not* $1/2$. Why not? Because the method of drawing makes all six coins equally likely to be drawn first. So if the first coin is gold, it is twice as likely to have been chosen from drawer 1, which has two gold coins, as from drawer 2, which has only one.

Note that the information that the first coin drawn was gold has changed the probabilities. Initially, there were three possible drawers to choose, each having probability one third. Once the first coin is drawn, the probability of the drawers is no longer $1/3$. Indeed, given the information that the coin chosen was gold, the new (conditional) probabilities of choosing drawers 1, 2, and 3 are, respectively, 0, $1/3$, and $2/3$.

1.4. Conditional Probability

Example 1.16.2. False positives in HIV testing. The antibody test for HIV infection is thought to be over 99% accurate, and estimates range even higher[7]. This means that a given test will be correct with probability at least .99 and will be in error with probability no more than .01. (For reasons that will shortly be apparent, HIV tests are normally confirmed with one or more additional tests. We only consider a single test here.)

The test can actually make two kinds of errors: it can indicate that an uninfected person has HIV—this is called a *false positive*—and it can indicate that an infected person does not have HIV, which is called a *false negative*. Estimates of the conditional probability of these individual errors are less reliable than the estimate of the total error, so, for the purpose of this example, let us assume the total error is one percent, and that both error probabilities are equal, so that the conditional probability that a non-infected person receives a false positive and the conditional probability that an infected person receives a false negative are both one percent.

The prevalence of HIV infection in Canada, as of 2005, was approximately 2 per 1,000 in the population. A person is chosen at random from the population, is tested, and tests positive. What is the probability that this person is actually not infected?

Answer. We are asked for the conditional probability that the individual does not have the disease given that the test was positive, i.e., the probability of a false positive. This turns out to be somewhere between surprisingly and astoundingly large. Let us see why.

$$P\{\text{not infected} \mid \text{test positive}\} = \frac{P\{\text{not infected and test positive}\}}{P\{\text{test positive}\}}.$$

The probability of being infected is 2 per 1000, or 0.002, so the probability of not being infected is .998. The probability that the test is wrong on a non-infected person is .01, so $P\{\text{not infected and test positive}\}$ equals $P\{\text{not infected}\}P\{\text{test positive} \mid \text{not infected}\} = .998 \times .01 = .00998$. To find the denominator, note there are two ways of testing positive: either be infected and test positive or not be infected and still test positive. Thus the probability of testing positive is $P\{\text{infected}\}P\{\text{positive} \mid \text{infected}\} + P\{\text{not infected}\}P\{\text{positive} \mid \text{not infected}\} = 0.002 \times 0.99 + 0.998 \times 0.01 = .00698$, and

$$P\{\text{not infected} \mid \text{test positive}\} = \frac{.00998}{.01196} = 0.834\,.$$

This means that the odds are about five to one that the test is a false positive! This is in spite of the fact that the test is 99% effective.

---
[7]Screening for HIV: A Review of the Evidence for the U.S. Preventive Services Task Force, Chou et al., Annals of Internal Medicine, July 5, 2005, vol. 143, no. 1, 55-73.

Now that we have done the calculation, we can see why the number of false positives outnumbers the number of true disease detections, even with an efficient test. It is because the disease is so rare. There are so few who have the disease and so many who do not, that even one percent of the people without the disease outnumber all the people with it.[8]

Example 1.16.3. Bayes Rule. We did the previous examples from basic principles, but we could have applied a general formula, Bayes rule. (In fact, Bayes rule involves the same calculations, so it is no faster, and, at the beginning, at least, it is better to use basic principles than general formulas.) Suppose that there are a number of mutually exclusive possibilities, say $B_1, \ldots, B_n$ (the three drawers, in the first example) with known probabilities. An event, A, is observed (i.e., a gold coin is drawn.) How does this new information change the probabilities of the B_i? The answer is given by Bayes' Rule.

Proposition 1.17 (Bayes Rule). *Let $B_1, \ldots, B_n$ be a partition of Ω. Then for any event A,*

$$(1.2) \qquad P\{B_j \mid A\} = \frac{P\{A \mid B_j\}P\{B_j\}}{\sum_{i=1}^{n} P\{B_i\}P\{A \mid B_i\}}.$$

Proof. This is a straightforward calculation:

$$\begin{aligned} P\{B_j \mid A\} &= \frac{P\{B_j \cap A\}}{P\{A\}} \\ &= \frac{P\{B_j\}P\{A \mid B_j\}}{\sum_{i=1}^{n} P\{A \cap B_i\}} \\ &= \frac{P\{A \mid B_j\}P\{B_j\}}{\sum_{i=1}^{n} P\{B_i\}P\{A \mid B_i\}}. \end{aligned}$$

Remark 1.18. The formula can be derived so easily that there is little point in memorizing it: just remember that it exists, and rederive it from basic principles when needed.

1.5. Independence

Intuitively, events are independent if they have no influence on each other, or, more exactly, if they have no influence on each other's probability. Events A and B are independent if $P\{B \mid A\} = P\{B\}$ and $P\{A \mid B\} = P\{A\}$. That is, if events are independent, the conditional probability of one should not depend on whether or not the other occurs. Assume that $P\{A\}$ and

---

[8]The problem of false positives comes up in the diagnosis of rare diseases. When a disease is rare, a randomly-chosen person is very unlikely to have it, so it requires very strong evidence to confirm that the person is actually infected.

1.5. Independence

$P\{B\}$ are non-zero. Then we have $P\{B\} = P\{A \cap B\}/P\{A\}$ and $P\{A\} = P\{A \cap B\}/P\{B\}$. If we multiply either out, we get $P\{A \cap B\} = P\{A\}P\{B\}$. This leads us to the formal definition of independence.

Definition 1.19. Two events A and B are **independent** if

(1.3) $$P\{A \cap B\} = P\{A\}P\{B\}.$$

Remark 1.20. (i) This is symmetric in A and B, so there is no difference between "A is independent of B" and "B is independent of A"; both mean that A and B are independent.

(ii) The definition makes sense even if $P\{A\}$ or $P\{B\}$ vanishes. Indeed, if $P\{A\} = 0$ or 1, then A is independent of *all* events B. For if $P\{A\} = 0$, (1.3) is $0 = 0$, while if $P\{A\} = 1$, (1.3) is $P\{B\} = P\{B\}$.

Example 1.20.1. Toss two fair coins. There are four outcomes, according to whether the first and second coin fall heads or tails. Represent them as the ordered pairs (T,T), (T,H), (H,T), (T,T). By symmetry they are equally likely, so each has probability $1/4$. Let A be the event "tails on the first coin" and let B be the event "tails on the second coin". Then A and B are independent, since $P\{A\} = P\{(T,T) \text{ or } (T,H)\} = 1/4 + 1/4 = 1/2 = P\{B\}$, while $P\{A \cap B\} = P\{(T,T)\} = 1/4 = P\{A\}P\{B\}$. (The other pairs, A and B^c, A^c and B, A^c and B^c are also independent, so that the two coin tosses are independent.)

Exercise 1.24. Suppose that the events A and B are independent. Show that A^c and B^c are independent, as are A and B^c, and A^c and B.

We must be careful when we define independence for more than two events. Strangely enough, the obvious extension is not sufficient.

Definition 1.21. A finite family of events $A_1, \ldots, A_n$ is **independent** if, for each subsequence $i_1, \ldots, i_k$ of $1, \ldots, n$,

(1.4) $$P\{A_{i_1} \cap \cdots \cap A_{i_k}\} = P\{A_{i_1}\} \cdots P\{A_{i_k}\}.$$

Definition 1.22. Let I be an index set. A family $\{A_\alpha, \alpha \in I\}$ of events is *independent* if each finite subfamily is.

Remark 1.23. The family can be infinite, even uncountable. In all cases, the independence of the whole family is determined by the independence of finite subfamilies.

Exercise 1.25. Let $A_1, \ldots, A_n$ be events, and for each i, let B_i equal either A_i or A_i^c. Show that the A_i are independent if and only if for any choice of the B_i, $P\{B_1 \cap \cdots \cap B_n\} = \prod_{i=1}^n P\{B_i\}$.

Example 1.23.1. A family of events is **pairwise independent** if any pair of elements in the family are independent. For a family of two events, pairwise independence and independence are the same. But for larger families, they are not. Here is an example of three events A, B, and C which have the property that A and B are independent, A and C are independent, B and C are independent, but A, B, and C are not independent. In other words, they are pairwise independent, but not independent.

Toss two independent fair coins. Let $A = \{$1st coin is heads$\}$, $B = \{$2nd coin is heads$\}$ and $C = \{$1st coin and 2nd coin show different faces$\}$. All three events have probability one-half. A and B are independent since the tosses are. To see that A and C are independent, note that $A \cap C = (H, T)$ so $P\{A \cap C\} = 1/4 = P\{A\}P\{C\}$. Similarly, B and C are independent. But the three events are not independent since, as $A \cap B \cap C = \emptyset$, $P\{A \cap B \cap C\} = 0$ while $P\{A\}P\{B\}P\{C\} = 1/8$.

Likewise, the condition that $P\{A \cap B \cap C\} = P\{A\}P\{B\}P\{C\}$ is *not* sufficient to imply that A, B, and C are independent. For instance, let $\Omega = [0, 1]$ and $P = $ Lebesgue measure for the probability space, and let $A = B = [0, 1/2]$ and $C = [1/3, 1]$. These are clearly not independent—A and B are identical—but they satisfy the equation.

Problems 1.5

1.26. A pair of fair dice is rolled. What is the conditional probability that one of the dice is a four, given that the sum is seven?

1.27. Three students are chosen at random from a class of ten girls and twelve boys. What is the probability that at least one is a boy, given that at least one is a girl?

1.28. A fair coin is tossed. If it is heads, one fair die is rolled. If it is tails, two fair dice are rolled.

(a) What is the probability that the number of spots showing is 2? 5? 1? Less than 6?

(b) Suppose that there are five spots showing. What is the probability that the coin was heads?

Information is transmitted over the internet via communication channels. These transmit a series of bits, i.e., a series of 0's and 1's. A certain amount of noise—static—is unavoidable, so there is always a probability that a digit will be incorrectly transmitted: a 0 may be received as a 1 and vice-versa. There are various ways to combat this, involving multiple transmissions, error-correcting codes, and other schemes. The following two problems explore this.

1.29. To reduce the error in a very noisy communication channel, the channel transmits 000 instead of 0 and 111 instead of 1, and uses majority decoding. Let the probability that a given bit is incorrectly transmitted be $p > 0$. Suppose that

1.5. Independence

zeros and ones are equally likely in the original message. If a 1 is received, i.e., the digit is a 1 after decoding, what is the conditional probability that a 1 was sent?

1.30. A communication channel can increase the probability of successful transmission by using error-correcting codes. One of the simplest of these is called a "parity scheme". In such a scheme, the message is divided into blocks of a fixed number of digits. Then a single bit, called the "parity bit" is added to each block. The parity bit is 0 if there are an even number of 1's in the block, and 1 if there are an odd number of 1's. The receiver compares the block with the parity bit. If they do not agree, the block is retransmitted. If they do agree, the block is accepted as received.

The original block of, say, n bits is transmitted as $n+1$ bits. Suppose the blocks are three bits long—four, including the parity bit—and that the probability of mistransmitting a single bit is 0.1.

(a) Find the probability of having at least one error in a block of three bits.

(b) Given that there is at least one error in the bit, what is the probability that it will be retransmitted?

[Hint: This scheme only detects an odd number of errors. Don't forget that there may also be an error in transmitting the parity bit.]

1.31. Show that if B is an event of strictly positive probability, then $P\{\cdot \mid B)\}$ is a probability measure on $(\Omega, \mathcal{F})$.

1.32. Prove the "Sure-Thing Principle": if A, B, and C are events, and if $P\{A \mid C\} \geq P\{B \mid C\}$ and $P\{A \mid C^c\} \geq P\{B \mid C^c\}$, then $P\{A\} \geq P\{B\}$.

1.33. Independent events A and B have the same probability p. Find $P\{A \cup B\}$.

1.34. Suppose A and B are events and the probability of B is either zero or one. Show that A and B are independent.

1.35. According to statistics, an RAF pilot in World War II had a 5% chance of being shot down each mission. A pilot is assigned ten missions. Answer without calculating: is the probability of completing them safely greater than, less than, or equal to one-half? Now calculate the exact probability.

1.36. In an ESP experiment, one person chooses two different numbers between 1 and 10, inclusive. A second person in a different room then attempts to guess them. Suppose the second person correctly guesses the two numbers. (Order is not important.) What is the probability that this happens by pure chance?

1.37. Suppose a fair die is rolled twice, independently. Which of the following pairs of events are independent?

(a) The first roll is a two; the second roll is odd.

(b) The first roll is a two; the two rolls are equal.

(c) The first toss is a two; the sum of the two rolls is six.

1.38. In the baseball World Series, two teams play best-of-seven. The first team to win four games wins the series, so it must go at least four games, and may last seven. Suppose that the two teams are equal, so that each team has probability one-half of winning each game, and that the games are independent. Which is more probable: a six game series or a seven game series?

[Hint: Consider the situation after the fifth game has been played.]

[Note: Baseball records show that there are significantly more seven-game series than six-game series. Discuss this in light of your answer.]

1.39. Four fair coins are tossed independently. Find the probability they all have the same face up.

1.40. Three points are placed randomly and uniformly on a circle. What is the probability that all three are in the same semi-circle?

1.41. Three fair dice are rolled independently, and the number of spots showing on the three are multiplied. Find the probability that the product is two.

1.42. Assume that boys and girls are equally distributed in the population.

(a) Choose a family at random from those families in the population who have exactly two children. Suppose one of the children is a girl. What is the probability that the other child is a boy?

(b) Suppose the oldest child is a girl. What is the probability that the youngest child is a boy?

(c) Choose a girl at random from the children of all two-child families. What is the probability that her sibling is a boy?

1.43. What can you say about the probability of an event A which is independent of itself?

1.44. The inhabitants of an island tell the truth with probability 1/3 and lie with probability 2/3, independently of each other. One islander makes a statement. A second islander says that the first islander told the truth. What is the probability that the statement is true?

1.6. Counting: Permutations and Combinations

The mathematics of probability began when some gamblers asked mathematicians to explain why certain bets were profitable and others were not. The subject has never completely lost its raffish roots: gambling motivates it to this day. So it is time to develop the tools to analyze some games of chance, and to compute other interesting probabilities as well.

Most gambling games—cards, dice, and lotteries, for example—involve a set of equally likely outcomes. In that case, the computation of probabilities comes down to counting outcomes, and therefore to combinatorics.

Here is a basic counting principle: *if a task is done in two stages, and if the first stage can be done in m_1 ways and, for each way that the first stage can be done, the second stage can be done in m_2 ways, then the two together can be done in $m_1 m_2$ ways.*

An easy induction argument extends this to n-stage tasks.

1.6. Counting: Permutations and Combinations

> Suppose that a task is done in $n \geq 2$ stages, and suppose that
> - the first stage can be done in m_1 ways, and
> - if $1 < k \leq n$, then for each way that stages $1, \ldots, k-1$ can be done, there are m_k ways of doing the kth stage. Then
> - there are $\prod_{k=1}^{n} m_k$ ways to do the whole task.

Example 1.23.2. The native language of computers is binary: a string of zeros and ones. To represent text, computers must code each letter and punctuation mark into a string of zeros and ones. A common code for this is called ASCII, short for American Standard Code for Information Interchange. Each letter is represented by a sequence of seven zeros and ones, or seven bits. For instance, A is coded 1000001, and B is coded 1000010.

To see how many characters this code can represent, note that the representation involves the choice of seven digits. There are two ways to choose the first digit—either 0 or 1—and two ways to choose each of the remaining six. The product of these seven 2's is $2^7 = 128$, so there are 128 characters in ASCII code.

We will use the counting principle to develop two main ideas: permutations and combinations. A **permutation** of objects is simply a reordering. For two objects, say a and b, there are two possible permutations: ab and ba. For three objects, a, b, and c, there are six: abc, acb, bac, bca, cab and cba. But the number of possible permutations grows so rapidly with the number of objects that it soon gets beyond the point where listing them is an option. There are 24 permutations of $abcd$, for example, 720 permutations of $abcdef$, and the number of permutations of the entire alphabet is far larger than the entire past and present human population of the earth; the number is roughly a 4 followed by 26 zeros. We must think of a better way to count them.

Consider n distinct objects. We can rearrange them by picking one to be the first, one of the remaining ones to be second, and so on. In fact, a moments thought reveals that a permutation is a 1-to-1 correspondence between the n objects and the numbers $1, 2, \ldots, n$. So the number of permutations equals the number of such correspondences. Let us count them.

There are n possible objects which can correspond to 1. After choosing the first, there remain $n-1$ objects which can correspond to 2. Now, regardless of which two we have chosen first, there remain $n-2$ objects to correspond to 3. And so on: after we have chosen the first $n-1$ objects, there remains a single object to correspond to n. By the above principle, the number of possible ways this can be done is the product $n(n-1)(n-2)\cdots 2 \cdot 1 = n!$. Conclusion:

> There are $n!$ different permutations of n distinct objects.

We approached permutations as the task of choosing the n objects one at a time. But suppose we don't want to choose all n objects, but a smaller number, say r? How many ways can we choose r distinct objects in order from a collection of n distinct objects? Exactly the same reasoning works: we can choose the first one in n ways; since we can no longer choose that same object, there remain $n-1$ objects to choose for the second, and so on, until the rth object, which can be chosen in $n-r+1$ ways. This is called *choosing an ordered sample of r from n*. This leads to the conclusion that

> The number of possible ordered samples of r from n distinct objects is $n(n-1)\ldots(n-r+1) = \dfrac{n!}{(n-r)!}$

Notice that a permutation of n objects is just an ordered sample[9] of n from the original n objects.

The above method of choosing a sample is called **sampling without replacement.** The chosen items are necessarily distinct—none can be chosen twice. This is to distinguish it from **sampling with replacement** in which each item, after being chosen, is returned to the original set, so that it is available for choosing a second time, and the chosen items may not be distinct. In sampling with replacement, the number of possible ordered samples of r from n is n^r, since each of the r choices presents the same n possibilities.

The next question is, "What if the order is not important? How many ways are there to choose r distinct objects out of n then?" Note that if the order is not important, we are simply choosing a set of r objects, and this question can be reworded: how many r-element subsets does a set of n elements have?

Notice that each subset of size r can be rearranged in $r!$ ways to give $r!$ ordered samples. If there are N subsets in all, the total number of ordered samples of size r must be $N\,r!$. So $N\,r! = n!/(n-r)!$, hence $N = n!/r!(n-r)!$.

---

[9]Note the change: from "permutation" to "ordered sample." These two intimately related ideas come from different sources: "permutations" comes from combinatorics, "samples" from statistics. This mixing of metaphors is common in probability, that most syncretic of mathematical subjects.

1.6. Counting: Permutations and Combinations

In this context, a subset of size r is called a **combination**[10] **of n things taken r at a time**. To summarize:

> The number of combinations of n distinct objects taken r at a time is
> $$\binom{n}{r} \stackrel{\text{def}}{=} \frac{n!}{r!(n-r)!}$$

Another name for $\binom{n}{r}$ is **n choose r**. They are collectively called **binomial coefficients** because of their connection with the binomial theorem.

We now have enough machinery to calculate the probabilities in some popular games of chance.

Example 1.23.3. Let us see how many 5-card poker hands there are. A hand consists of five cards drawn from a standard deck of 52 cards. The order is unimportant, so a hand is just a combination of five cards drawn from 52. Therefore, there are $\binom{52}{5} = 2,598,960$ possible hands, and each has probability $1/2,598,960$.

Hands are named and ranked according to their contents: one pair, two of a kind, two pair, three of a kind, and so on. They are ranked in order of decreasing probability: the most valuable hands are those with the smallest probability.

Similarly, a bridge hand consists of thirteen cards, so there are $\binom{52}{13} = 635,013,559,600$ bridge hands.

Before the cards are dealt, the deck is well-shuffled. The aim of shuffling is to put the deck in totally random order, which means that, ideally, all $52!$ possible permutations of the 52 cards are equally likely. This implies that all hands—that is, all five-card combinations—are equally likely. (Why?)

Exercise 1.45. Suppose a deck of 52 cards is placed in a random order in which all $52!$ permutations have the same probability. Show that if r cards are dealt, then all $\binom{52}{r}$ combinations have the same probability.

Example 1.23.4. A standard deck is arranged in four suits, called hearts, diamonds, spades and clubs, each of 13 cards. Within each suit, the cards are called Ace, 2, 3, 4, ..., 10, Jack, Queen, King.

A *flush* is a hand with five cards from the same suit. Let us see how many combinations are flushes. The flush can be in one of four suits—so, four ways to choose the suit—and five cards can be chosen from the 13 cards in the suit in $\binom{13}{5}$ ways. There are $\binom{52}{5}$ possible hands, so by Proposition 1.14, the probability of a flush must be $4\binom{13}{5}/\binom{52}{5} = .00198$.

---

[10] Combinatorics preceded set theory. The word "combination" was well-established before Cantor used "set", and it stuck.

A *full house* has three cards of one denomination, two of another. Let us calculate the number of full houses. We first have to choose the denomination for the threesome, which can be done 13 ways, then choose a different denomination for the pair, which can be done 12 ways. Then we choose three cards of the four of that denomination, which can be done $\binom{4}{3} = 4$ ways. The pair can be chosen $\binom{4}{2} = 6$ ways. Together there are $13 \cdot 12 \cdot 4 \cdot 6 = 3744$ such hands. Thus the probability is $3744/\binom{52}{5} = .00144$. This is less than the probability of a flush, which accounts for the fact that a full house is ranked the higher of the two.

Example 1.23.5. Binomial theorem. Let a and b be real numbers and $n \geq 1$ an integer. Then

(1.5) $$(a+b)^n = \sum_{k=0}^{n} \binom{n}{k} a^k b^{n-k}$$

To prove this, write $(a+b)^n$ as the n-fold product: $(a+b)(a+b)\ldots(a+b)$. One way to do the multiplication is to choose either an a or a b from each of the n binomial terms and multiply them together. Do this in all possible ways, and add the products. The result is $(a+b)^n$. (It takes some thought to see this is true; it can be proved by induction, but to understand how it works, just check the cases $n = 2$ and $n = 3$.)

If we choose k a's from the n terms, the other $n - k$ terms must be b's, so the product will be $a^k b^{n-k}$. There are exactly $\binom{n}{k}$ ways to choose the k a's, so there are exactly that many terms of the form $a^k b^{n-k}$. Thus the term with a^k in it is $\binom{n}{k} a^k b^{n-k}$. Sum over k to get (1.5).

1.6.1. Pascal's Triangle. It is not always necessary to use factorials to calculate binomial coefficients. For small values of n, it is easier to use **Pascal's Triangle**[11]:

```
                    1
                 1     1
              1     2     1
           1     3     3     1
        1     4     6     4     1
     1     5    10    10     5     1
  1     6    15    20    15     6     1
1    . . . . . . . . . . . . .        1
     . . . . . . . . . . . . . . .
```

---

[11]The discovery of Pascal's triangle and its use in calculating probabilities was another result of the Pascal-Fermat correspondence of 1654. It was actually a rediscovery: Chinese mathematicians knew of it—but not its use for probability—some 300 years before Pascal.

1.6. Counting: Permutations and Combinations

There is a simple way to construct this. It only requires addition. With practice, it gives a relatively quick way of calculating the binomial coefficients $\binom{n}{r}$ for small values of n.

- There is a single 1 on top, and each line below that starts and ends with a one, and
- except for the 1's, each number is the sum of the two numbers above it in the triangle.

Number the rows from the top, calling the single 1 at the top "row zero". Similarly, number the elements in each row starting with zero: the 1 at the left of the row is the zeroth element. Then the nth row gives the binomial coefficients $\binom{n}{0}, \binom{n}{1}, \binom{n}{2}, \ldots, \binom{n}{n}$, and the kth element of the nth row is $\binom{n}{k}$.

This is not hard to see. The first row is 1, 1, which equals $\binom{1}{0}, \binom{1}{1}$ and the second row, 1, 2, 1 is just $\binom{2}{0}$, $\binom{2}{1}$, $\binom{2}{2}$. For any n, $\binom{n}{0} = \binom{n}{n} = 1$, which is where the 1's come from. The rest follows by induction on n and the formula (which we leave to the reader to verify):

$$(1.6) \qquad \binom{n-1}{r-1} + \binom{n-1}{r} = \binom{n}{r}.$$

Example 1.23.6. (Path Counting.) There is a congenial combinatorial interpretation of Pascal's triangle involving a stroll thru the triangle itself. It also gives (1.6) as a by-product.

If we stroll thru the triangle, we follow a path. The path begins at the apex, and goes down one step at a time, each step going from a number to one of the two numbers directly below it. So, for instance, if a path reaches the first 4 in the fourth row, it goes down to either the 5 or the 10 below it in the fifth row.

We claim that there are exactly $\binom{n}{k}$ such paths which go from the apex to the kth element in the nth row. This is exactly the number at that point in Pascal's triangle.

Note that a step in the path goes either diagonally down and to the left, (the reader's left, that is) or diagonally down and to the right. If it goes to the left write L. If it goes to the right, write R. Then the path can be represented by a string of L's and R's. Each possible string corresponds to a path. For instance, the illustrated path is LRLRRL. If the path is all L's, it stays on the zeroth diagonal (the one with all the 1's.) If there is a single R, the path ends up on the first diagonal (which has 1, 2, 3, 4,...). Two R's puts it on the second diagonal, and so on.

In order to reach the kth element of the nth row, the path must be n steps long, with exactly k R's and $n-k$ L's. But there are $\binom{n}{k}$ ways to place k R's in the n steps, each corresponding to a different path, so there are exactly $\binom{n}{k}$ paths to the kth element of the nth row.

We can also see (1.6) from this. Indeed, to reach the kth element of the nth row, the path has to pass through one of the two points above it, either the $k-1^{\text{st}}$ or the kth element of the $n-1^{\text{st}}$ row. No path passes through both of those points, so that the number of paths reaching that point—namely $\binom{n}{k}$—is the sum of the number of paths reaching the two points above, $\binom{n-1}{k-1}$ and $\binom{n-1}{k}$, respectively, giving (1.6).

Problems 1.6

1.46. Harry's Holonomic Pizza Emporium offers the following ten toppings to go on its cheese pizza: pepperoni, mushrooms, green peppers, Italian sausage, anchovies, shrimp, Canadian bacon, onions, olives, and tomato. The customer can have any combination of these toppings, including all or none of them. The shop claims that it offers more that 1000 different pizzas. Is this true? (Note that a pizza with no topping at all is a plain cheese pizza, which counts.)

1.47. The TV show "Who wants to be a Millionaire" often asks people to put 4 items in order, e.g., put 4 actors in order of birth, put 4 events in the order they happen. What is the probability that a random guess is correct?

1.48. Eric took a history test. One question asked him to put four historical events in the order they occurred: the Thirty Years War, the Hundred Years War, the War of the Roses, and the War of Jenkins' Ear. Eric knew that the Hundred Years War preceded the Thirty Years War, but had no idea of the rest, so he chose them at random. What was the probability that he was correct?

1.49. N people, including Barbie and Ken, sit at a round table with N seats. How many ways can they sit if Barbie and Ken cannot sit together. How many ways if they *must* sit together?

1.50. How many ways can a committee of three be chosen from a group of ten people? How many ways are there to choose a president, secretary, and treasurer?

1.51. Four cards are face-down on a table. Two are red and two are black. Knowing this, you must guess the colors of all four, so you guess the two you think are red, and call the other two black. If you choose them at random, what are the probabilities that you get 0, 2, 4 of the cards correct?

1.52. How large must an alphabet be if one million people can be uniquely distinguished by three initials?

1.53. Find the sum of the binomial coefficients $\binom{n}{0} + \binom{n}{1} + \cdots + \binom{n}{n-1} + \binom{n}{n}$.

1.54. Prove that an n-element set has 2^n subsets. (The empty set and the whole set are included.)
[Hint: In the language of combinatorics, a combination of k elements drawn from a set of n is just a subset of k elements.]

1.55. Four shoes are chosen at random from a closet containing five pairs of shoes. What is the probability that there is at least one pair among the four?

1.6. Counting: Permutations and Combinations

1.56. A five-card poker hand is drawn from a standard deck. What is the probability that it contains four of a kind, i.e., all four cards of the same denomination?

1.57. A five-card poker hand is drawn from a standard deck. What is the probability that it is three of a kind? (Remember to exclude the possibility that it is actually either four of a kind or a full house.)

1.58. A five-card poker hand is drawn from a standard deck. What is the probability it is two pair? That is, a pair of one denomination, a second pair of another denomination, and a fifth card of yet another denomination.

1.59. Five cards are dealt. What is the probability that they form a straight, i.e., that are all in sequence? (For example, A,2,3,4,5 or 8,9,10,J,Q. They can be of any suits. For this purpose the ace can be either high or low, so 10,J,Q,K,A is also a possibility.)

1.60. In the game of bridge, 13 cards are dealt to each player from a standard deck. A hand which contains no aces and no cards above a nine is called a *Yarborough*. That is, all 13 cards are either 2, 3, 4, 5, 6, 7, 8, or 9. This is named after the British nobleman Charles Worseley, the Second Earl of Yarborough, who is said to have bet a thousand to one against the dealing of such a hand. Compute the probability of a Yarborough. What do you think of his bet?

1.61. Texas Hold'em is a kind of poker in which each player is dealt two cards face-down, called the *hole cards*, and then five cards, called *the board*, are dealt face-up on the table. A player can choose a five card hand from his or her two hole cards and the five cards on the board—so each player has seven cards to choose from.

(a) What is the probability that a player has at least a pair, i.e., that there are at least two cards among the seven of the same denomination?

(b) Suppose a player's hole cards are of different denominations. What is the probability that one of them is paired, that is, what is the conditional probability that at least one of the five cards on the board has the same denomination as one of the hole cards? What is the conditional probability if the hole cards are of the same denomination?

1.62. Jorge Luis Borges, in the story "The Library of Babel" describes a library. The reader gradually realizes that it contains every possible book. (Literally every possible book, so the library would have one book consisting of nothing but the letter A, another whose every page is filled with "abcabcabcabc..." and so on.) The books each have 410 pages, each page has 40 lines and each line has 80 characters. There are exactly 25 symbols, including the space and the punctuation marks.

Assuming that every possible book is included in the library, and that there are no duplicates, how many books does the library contain?

Some astronomers estimate that there are up to 10^{87} sub-atomic particles in the universe. Are there more or fewer books in the library than particles in the universe? How does this number compare to a googol (10^{100}) and a googolplex ($10^{10^{100}}$)?

1.63. A certain lottery, called 6-49, draws six different numbers at random from 1 to 49, and then draws a seventh, called the *bonus* number. The player provides a list of six numbers. If the first six numbers drawn match the six numbers on the players list in any order, the player wins the grand prize. If five of the six drawn numbers match numbers on the players list, and the bonus number matches the remaining number, the player wins second prize. The third prize goes to those players whose list contains any five of the six lottery numbers, but not the bonus number.

Calculate the probability of getting the grand prize, the second prize, and the third prize.

1.64. (Continuation.) If more than one person wins the grand prize, it is split: each winner gets an equal proportion of the prize. Suppose you win the grand prize, and that five million other people also enter. What is the probability that you will have to split it?

1.65. (The Banach Matchbox Problem.) A certain absent-minded pipe-smoking professor[12] carries two matchboxes, one in his left-hand coat pocket, and the other in his right. Each time he lights his pipe he chooses one of the two at random and takes a match from it. Eventually, he finds that the matchbox he chooses is empty. If both boxes initially contain the same number, n, of matches, what is the probability that there are k matches in the other box at that time?

1.66. The following algorithm was used to prove the binomial theorem. Prove it.

> To calculate $(1 + x)^n$, first write it as the product of the n factors $(1+x)(1+x)\ldots(1+x)$. Next, choose either 1 or x from each of the n terms and multiply them together. Finally, do this in all possible ways, and add the products. The result is $(1 + x)^n$.

[Hint: Induction]

1.67. In addition to the binomial theorem, the Swiss mathematician Jacob Bernoulli (1654–1705) in his book Ars Conjectandi showed the following remarkable fact. The number of ways to make the point m in a roll of n dice is equal to the coefficient of x^m in the expansion in powers of x of the polynomial

$$P(x) \stackrel{def}{=} (x + x^2 + x^3 + x^4 + x^5 + x^6)^n.$$

Prove it.

[Hint: Take a cue from the proof of the binomial theorem.]

1.7. The Gambler's Ruin

Let us look at a problem which interests both gamblers and mathematicians, though for different reasons. It is the problem of the Gambler's Ruin[13].

---

[12]Stefan Banach was a Polish mathematician. In the years just before World War II, he and his colleagues used to meet in the Scottish Coffee Shop in Lvov, Poland, to discuss mathematics. They kept a book there which they called the Scottish Book, in which they wrote down challenging problems. This was the last problem in the book.

[13]Pascal and Fermat posed and solved the Gambler's Ruin some two years after their 1654 correspondence. Both the Gambler's Ruin and the problem of points have the same subtle difficulty: they concern a game whose duration is random, not fixed.

1.7. The Gambler's Ruin

A gambler plays a game of chance repeatedly, against an opponent. The different plays are independent, and the game is fair: both gamblers have an equal chance to win each bet. Each bet is for one dollar. The gambler starts with an initial stake of n dollars, and the opponent starts with m. The game continues until one of the two loses that initial stake, and goes broke. Then it stops. Question: What is the probability that the first gambler goes broke?

The subtlety of this problem is that the game lasts for a random number of plays. For any given integer M, there is a strictly positive probability that the game lasts longer than M plays. This will happen, for instance, if the gambler alternately wins and loses the first M bets. However, the game will end...eventually. (See the problems.)

Let $N = m + n$ be the total amount of money in the game. Suppose that it is fixed. Then the probability that the gambler wins depends only on his initial stake. So let

$$f(k) = P\{\text{Gambler goes broke if initial stake is } k\}\,.$$

Note that $f(0) = 1$, since if his stake is zero, the gambler is ruined before the game starts. Similarly, $f(N) = 0$, for in this case it is the opponent who is ruined. If k is neither 0 nor N, the game lasts at least one turn.

At the first play, the gambler either wins or loses the bet. A win increases his fortune by one and decreases the opponent's fortune by the same amount. A loss does the opposite. The total amount of money in the game remains the same.

The idea is this: reconsider the situation immediately after the first play. The gambler's fortune k has either increased to $k+1$ or decreased to $k-1$, depending on whether the bet was won or lost. This means that the game effectively restarts with a new initial stake, either $k+1$ or $k-1$, depending on the result of the first play. That is, the new situation is equivalent to restarting the game with the same total amount of money in the game and the new initial stakes.

Starting with initial stake k, $0 < k < N$, we have:

(1.7) $P\{\text{ruin}\} = P\{\text{win } 1^{\text{st}} \text{ bet}\}P\{\text{ruin} \mid \text{win } 1^{\text{st}} \text{ bet}\}$
$\qquad\qquad\qquad + P\{\text{lose } 1^{\text{st}} \text{ bet}\}P\{\text{ruin} \mid \text{lose } 1^{\text{st}} \text{ bet}\}.$

From this point on, the gambler's fate depends only on the succeeding bets, so it is independent of the outcome of the first bet. If the gambler wins the first bet, the game effectively restarts with a new initial fortune $k+1$. Thus

$$P\{\text{ruin} \mid \text{win } 1^{\text{st}} \text{ bet}\} = f(k+1).$$

Similarly,
$$P\{\text{ruin} \mid \text{lose 1}^{\text{st}} \text{ bet}\} = f(k-1).$$

The probabilities of winning or losing are one-half each, and the initial probability of ruin is $f(k)$, so we have an equation involving the unknown function f: $f(k) = (1/2)f(k+1) + (1/2)f(k-1)$. Put this together with the boundary conditions to get a system of equations for f:

(1.8)
$$\begin{cases} f(k) = \dfrac{1}{2}\bigl(f(k+1) + f(k-1)\bigr), & k = 1, 2, \ldots, N-1, \\ f(0) = 1, \\ f(N) = 0. \end{cases}$$

Now we are on familiar ground: this is an algebra problem. Specifically, (1.8) is a difference equation with boundary conditions. To solve it, write $f(k+1) = f(k) + (f(k) - f(k-1))$. If we let $q = f(0) - f(1)$, then $f(2) = 1 - q - q = 1 - 2q$, and by induction, $f(k) = 1 - kq$. Use the boundary condition at $k = N$ to see that $0 = f(N) = 1 - Nq$. Thus $q = 1/N$, and $f(k) = \frac{N-k}{N}$ for all k. The gambler's original stake is n, so

(1.9)
$$P\{\text{ruin}\} = f(n) = \frac{N-n}{N}.$$

Notice that the probability of ruin is a linear function of the initial stake.

Example 1.23.7. A friend tells us that he goes to Las Vegas for business each year, and gambles at a casino until he wins enough money to pay for his hotel room. He claims to be be such a good gambler that he has done this the last five years hand-running, and never lost.

Is this believable? Let us see. Suppose that the hotel cost, say, $100 per night, and that he had a bankroll of $1000. Suppose also that he managed to find a fair game. He will be successful if he wins $100 before he loses his stake of $1000. This is exactly a gamblers ruin with $n = 1000$ and $N = 1100$, so the probability that he fails—and loses his entire stake—is, by (1.9), 100/1100 or, roughly, .091. Thus he successfully pays for his hotel about 91% of the time. The probability that he does this five times in a row is about .62, well over half. So it is not at all unlikely, especially since, had he ever gone broke, he would probably not have told us the story.

The assumption that the friend managed to find a fair game in Las Vegas is generous: most gambling games give the house an edge, and are therefore unfavorable to the player. Let us suppose that the gambler played a game in which the probability of winning was $p < 1/2$. Once again, let $f(k)$ be the probability of ruin if the gambler's initial stake is k. In this case, exactly the same analysis holds, with the exception that in (1.7), $P\{\text{win 1}^{\text{st}} \text{ play}\} = p$,

1.7. The Gambler's Ruin

and $P\{\text{lose } 1^{\text{st}} \text{ play}\} = 1 - p$. This leads to the new system

(1.10)
$$\begin{cases} f(k) = pf(k+1) + (1-p)f(k-1), & \text{if } 1 \le k \le N-1, \\ f(0) = 1, \\ f(N) = 0. \end{cases}$$

Once again, this is a linear difference equation with boundary conditions. To solve it, first find all possible solutions of the difference equation *without the boundary conditions*. Then use the boundary conditions to see which of the possible solutions is the true one.

By the basic theory, the difference equation has two linearly independent solutions, and the general solution is a linear combination of the two. To find them, we make an educated guess, and try a solution of the form $f(k) = a^k$ for some $a > 0$. Substitute this in (1.10):
$$a^k = pa^{k+1} + (1-p)a^{k-1}.$$

Now a is strictly positive, so we can divide by a^{k-1}, which leads to $a = (1-q) + qa^2$, or $qa^2 - a + 1 - q = 0$. This is a quadratic equation for a. It has two solutions: $a = 1$ (which leads to $f(k) \equiv 1$, a solution we might have seen immediately) and $a = (1-p)/p$. Thus the general solution f of the difference equation in (1.10) is of the form
$$f(k) = A + B \left(\frac{1-p}{p} \right)^k,$$

where A and B are constants. Use the boundary conditions at $k = 0$ and $k = N$ to find A and B. After some algebra, we see that

(1.11)
$$f(n) = \frac{a^n - a^N}{1 - a^N},$$

where $a = \frac{1-p}{p}$.

Example 1.23.8. Suppose that in the previous example our friend plays a very slightly unfavorable game, where the probability of winning is $p = 49/100$. Then $a = 51/49$ and the probability of losing \$1000 before winning \$100 is
$$\frac{\left(\frac{51}{49}\right)^{1000} - \left(\frac{51}{49}\right)^{1100}}{1 - \left(\frac{51}{49}\right)^{1100}} = .9817.$$

This time, our friend is almost certain to lose. The probability that he pays for the hotel room out of his winnings that night is less than 2%.

For comparison, if the game is fair, the gambler succeeds in paying for the hotel room with probability 91%, and has better than a 50-50 chance of doing it five times in a row. If the game is only very slightly unfair—as it will be in a casino—the probability of success drops drastically, to less than 2%, and the probability of succeeding five times in a row is negligible.

This may seem surprising, but when the bets are only a single dollar, it takes an enormous number of them to either win or lose. The casino only has a small advantage over the gambler in each bet, but over the course of so many bets, the small advantages add up.

One way of decreasing the casino's advantage is to increase the size of the bets. This shortens the game, and increases the gambler's chances.

Suppose the bet is $20 per play. As far as the gamblers ruin is concerned, this is equivalent to dividing the original stakes by 20. This means that the gambler's initial stake becomes 50 and $N = 55$. The probability of ruin becomes
$$\frac{\left(\frac{51}{49}\right)^{50} - \left(\frac{51}{49}\right)^{55}}{1 - \left(\frac{51}{49}\right)^{55}} = .2039.$$

The probability of success is now almost 80%, which is a great improvement. If the bets are $100 each play, the probability of ruin is .1102, almost as good as with a fair game. (Notice that changing the value of the bets does *not* change the probabilities of the gamblers ruin in a fair game. It only changes them if the game is unfair.)

There is a moral to this. If you must play in an unfair game, bet big to minimize the house's take!

Problems 1.7

1.68. (The Birthday Problem.) A class has n students. What is the probability that at least two have the same birthday? (Ignore leap years and twins.) Compute the probability for $n = 10, 15, 22, 30$ and 40. Do you find anything curious about your answers?

1.69. (G. Slade.) A gambler is in desperate need of $1000, but only has $$n$. The gambler decides to play roulette, betting $1 on either red or black each time, until either reaching $1000 (and then stopping) or going broke. The probability of winning a bet on red or black at a US casino is 18/38.

(a) How large must n be for the gambler to have a probability 1/2 of success?

(b) Suppose the gambler starts with this stake. How much can the casino possibly lose?

[Hint: It is instructive to plot the probability of success as a function of the initial stake n.]

1.70. (Continuation.) Suppose the gambler in the preceding problem starts with $950.

(a) What is the probability of succeeding—i.e., reaching $1000 before going broke?

1.7. The Gambler's Ruin

(b) Suppose the gambler is not restricted to bets of one dollar, but can bet all the money he or she has at any play. (Bets are still on either red or black.) Show that the gambler can succeed more than 92% of the time.

1.71. Show that the gambler's ruin eventually ends, i.e. that the gambler's fortune eventually reaches either 0 or N.

1.72. Suppose the gambler's opponent can borrow money at any time, and therefore need never go broke. Suppose that the the probability that the gambler wins a given game is p.

(a) If $p \leq 1/2$, show that the gambler will eventually go broke.

(b) Suppose $p > 1/2$. Find the probability that the gambler eventually goes broke. What do you suppose happens if he does not go broke? Can you prove it?

1.73. There are two urns, one with m balls, the other with n balls, where m and n are strictly positive integers. An urn is chosen at random, and one ball is transferred from it to the other. This is continued until one of the urns is empty. Find the probability that the urn with m balls empties first.

1.74. Suppose the gambler starts with m dollars and the opponent starts with $N - m$ dollars. If the gambler plays the gamblers ruin with $p = 1/2$, find the probability that the maximum value of the gamblers fortune, before the game ends, is equal to n, for $n = m, m+1, \ldots, N$.

[Hint: First find the probability that it is greater than n.]

1.75. A check-out counter at a supermarket will service one customer per unit time if there is anyone in line. Customers arrive at the line: in each unit of time, the probability that a single new customer arrives is 1/3, the probability that two arrive is 1/3, and and the probability that no new customer arrives is also 1/3. There are initially three customers in line. Find the probability that the line empties before it has ten persons in it.

1.76. NASA is developing two top-secret space shuttles. One has two engines, the other has four. All the engines are identical, and have the same probability of failure. Each is designed to fly if at least half of its engines work. A visiting scientist says, "The four-engine shuttle is more reliable, isn't it?" The NASA technician replies that the probability of failure is top secret, but that in fact both shuttles have the same probability of flying. The visitor then says, "Aha! Never mind, now I know both the probability an engine will fail and the probability that the shuttle will fly." How did he figure this out, and what are the two probabilities?

1.77. An experiment is repeated independently. Let A and B be disjoint events of strictly positive probability. Show that the probability that A occurs before B is $\frac{P\{A\}}{P\{A\}+P\{B\}}$.

[Hint: Do the experiment once, then reassess the situation.]

1.78. Find the probability of rolling a six before a seven in a sequence of rolls of a pair of fair dice.

1.79. In the Damon Runyon story "Blood Pressure" a guy named Rusty Charley tries to make the point ten in a craps game. That is, he rolls the dice until he gets either a ten or a seven. He makes the point if he rolls the ten first.

(a) Find the probability that he makes the point ten.

(b) He does this successfully, and a guy who does not see the roll asks him if he makes it the hard way, meaning, does he roll two fives? What is the probability he makes the point ten the hard way?

(c) Given that he makes the point ten, what is the probability that Rusty Charley makes it the hard way?

1.80. Suppose we have a biased coin with $P\{\text{head}\} = p \neq 1/2$. John von Neumann suggested the following method of synthesizing a fair coin. Toss the biased coin twice. If the two tosses are different, take the first of the two as the result. (That is, $HT \to$ heads, $TH \to$ tails.) If both tosses are the same, toss the coin two more times. Continue until one pair of tosses is either HT or TH. That completes the "toss" of the fair coin.

(a) Show that $P\{\text{heads}\} = 1/2$.

(b) Let $n \geq 1$ be an integer. Find the probability that it takes exactly $2n$ tosses of the biased coin to synthesize one toss of a fair coin. (For a deeper look at this, see Problem 9.6.)

1.81. Choose a four-digit number at random. What is the probability that the four digits are in strictly increasing order? (Note that the first place cannot be a zero.)

1.82. Roll a die three times in succession. What is the probability of getting a strictly larger number each time?

1.83. How many ways are there to put n distinguishable balls in n boxes in such a way that exactly one of the boxes is empty? What if the balls are *not* distinguishable?

1.84. Throw a pair of dice until either they are equal or differ by one. (So (2,3), (3,3), and (4,3) all stop the game.) What is the probability that the two dice are equal when the game stops?

1.85. Tom, Dick, and Harry toss a fair coin in turn: Tom tosses first, then Dick, then Harry. The first to toss a head wins. What is the probability that the winner is (a) Tom, (b) Dick, and (c) Harry?

1.86. A student takes a multiple choice text with n possible answers to each question. If the student knows the answer, he or she answers it. Otherwise, the student guesses at random. The probability of knowing the answer to any given question is p. Given that the answer to a given question is correct, what is the probability the student knew the answer to that question?

1.87. An experiment has five equally-likely outcomes, $\{a, b, c, d, e\}$. Can there be independent events A and B which do not have probability either zero or one?

1.88. Bobby's Bibbet Bobbin Business builds bibbet bobbins. Experience shows that four percent of its bibbet bobbins are bobbled. A be-bobbler is supposed to unbobble them, but it fails to unbobble ten percent of the bobbled bibbet bobbins. You buy one. What is the probability that the bibbet bobbin you bought is bobbled?

1.7. The Gambler's Ruin

1.89. (J. L. Doob) Johnny has not studied for tomorrow's test, so his probability of passing it is only 1/4. However, if he cheats, he will raise his probability of passing to 3/4. He hasn't yet decided whether or not to cheat, but he is leaning that way: the probability that he will cheat is 3/4. Suppose he takes the test and passes it. What is the probability that he cheated?

[Doob's question, to answer *after* doing the problem: if he passes, should he fail?]

1.90. A gambler has two coins in his pocket. One is a two-headed coin, and the other is fair. He chooses one at random and flips it.

(a) Suppose it is heads. What is the probability it was the fair coin?

(b) Suppose he flips the same coin a second time and it falls heads again. Now what is the probability it was the fair coin?

1.91. (Tie breaks in tennis.) A seven-point tie-break is played to decide a tied set in tennis. The players play points, and the first to reach seven points wins, provided he or she leads by two points. If the tie-break is tied 6-6, it continues until one player leads by two points. Suppose the players are even, so that each point is equally likely to be won by either player, and that the points are independent.

(a) What is the probability of reaching 6-6?

(b) Find the probability that the tie-break ends with a score $n+8$ to $n+6$, $n = 0, 1, 2, \ldots$.

(c) The longest tie-break on record in professional men's tennis is 20-18. What is the probability that a tie-break goes on this long or longer?

1.92. A box contains two red and two white balls. Draw two at random without replacement.

(a) Which, if either, is more probable: the colors of the two balls are the same, or that they are different?

(b) Find all possible numbers of red and white balls for which the events "both have the same color" and "different colors" are the same.

1.93. Three kinds of shirts are on sale. Abercrombie and Fitch each buy one shirt. How many ways can this be done?

1.94. (Continuation.) Make up your own problem involving Abercrombie, Fitch, and three shirts, whose answer is 2^3, and make up another whose answer is $\binom{4}{3}$.

Chapter 2

Random Variables

2.1. Random Variables and Distributions

Random experiments often have numerical values associated with them, even if the outcome itself is not numeric. These are called **random variables.** A series of coin flips results in a sequence of heads and tails, which is not numeric, but the number of heads in the sequence is. Quantities like the class-average score on an exam, the percentage of the popular vote amassed by the winning party, or the exact amount of money won by the man who broke the bank at Monte Carlo are all random variables.

There is a popular dice game called craps, in which the players throw a pair of dice, and the game proceeds according to the sum on the two dice. Call that X. We constructed a probability space for a pair of dice in Example 1.14.1. The sample space is $\{(m,n),\ m,n = 1,\ldots,6\}$, the σ-field $\mathcal{F}$ is all subsets of Ω, and the outcomes are equally likely. The outcome is is an ordered pair, say (m,n), not a number, and X is their sum. So X is a function on Ω: $X(m,n) = m + n$.

We repeat: the random variable X is a function on Ω.

A moments thought tells us that this is a general fact, not a coincidence. In any experiment, if Ω is the set of all outcomes, then any numerical quantity determined by the outcomes must be a *function* of the outcomes, that is, a function on Ω. This brings us to the definition.

Definition 2.1. Let $(\Omega, \mathcal{F}, P)$ be a probability space. A **random variable** is a real-valued function X on Ω such that for all $x \in \mathbb{R}$, $\{\omega : X(\omega) \leq x\} \in \mathcal{F}$.

This may appear confusing: a variable is a function? But this is just terminology: the words "random variable" were in use long before people discovered the connection between probability and analysis, and it is too late to change them[1].

Basically, a random variable is a function with the property that all interesting things about it involve events—namely elements of $\mathcal{F}$. At the moment, this is a purely technical requirement and, in fact, it will remain purely technical until we start using σ-fields to encapsulate information in Chapter 5. For the moment, we will just mention a reassuring fact: as long as we do reasonable operations on random variables, we end up with other random variables. For instance, sums, products and quotients of random variables are random variables, as the next exercise shows.

Exercise 2.1. Let X and Y be random variables and let a be a real number. Show that the following are also random variables.

(a) aX;

(b) $X+Y$;

(c) XY;

(d) $Z(\omega) \stackrel{\text{def}}{=} \begin{cases} Y(\omega)/X(\omega) & \text{if } X(\omega) \neq 0, \\ 0 & \text{if } X(\omega) = 0. \end{cases}$

[Hint: To show that a function on Ω is a random variable, verify the definition directly. In (b), for example, first prove that $\{\omega : X(\omega) + Y(\omega) > x\} = \bigcup_{r \in \mathbb{Q}} \{\omega : X(\omega) > r, Y(\omega) > x - r\}$. Then verify that this is in $\mathcal{F}$, and conclude that $\{\omega : X(\omega) + Y(\omega) \leq x\} \in \mathcal{F}$ for all x, and $X + Y$ is therefore a random variable.]

One important thing associated with a random variable is its distribution function.

Definition 2.2. The **distribution function** of a random variable X is the function F defined by

$$F(x) = P\{\omega \in \Omega : X(\omega) \leq x\}, \; x \in \mathbb{R}.$$

For example, if X is the number of spots showing on a single roll of a fair die, then its distribution function F is:

$$F(x) = \begin{cases} 0 & \text{if } x < 1, \\ \frac{n}{6} & \text{if } n \leq x < n+1, \; n = 1, 2, \ldots, 5, \\ 1 & \text{if } x \geq 6. \end{cases}$$

Remark 2.3. We can now see why we needed $\{\omega : X(\omega) \leq x\} \in \mathcal{F}$. That is so that its probability—needed to define the distribution function—is defined.

Remark 2.4. A random variable is a function on Ω with values in $\mathbb{R}$, so it can be regarded as a mapping $X : \Omega \mapsto \mathbb{R}$ from Ω to $\mathbb{R}$. It induces an

---

[1] Besides, probabilists take a certain secret enjoyment from these anachronisms.

2.1. Random Variables and Distributions

inverse mapping, X^{-1} from $\mathbb{R}$ to Ω. Since X is not necessarily one-to-one, X^{-1} is best regarded as a map from subsets of $\mathbb{R}$ to subsets of Ω: if $A \subset \mathbb{R}$, $X^{-1}(A)$ is the subset of Ω given by $X^{-1}(A) = \{\omega : X(\omega) \in A\}$.

In terms of X^{-1}, the distribution function F can be written

$$F(x) = P\{X^{-1}(-\infty, x]\}.$$

As a set mapping, X^{-1} is quite nice: it preserves set operations. The reader can verify that, for instance, $X^{-1}(A \cup B) = X^{-1}(A) \cup X^{-1}(B)$ and $X^{-1}(A \cap B) = X^{-1}(A) \cap X^{-1}(B)$. In particular, if A and B are disjoint, so are $X^{-1}(A)$ and $X^{-1}(B)$.

2.1.1. Elementary Properties of Distribution Functions.

Proposition 2.5. *Let X be a random variable and let F be its distribution function. Then for each $x, y \in \mathbb{R}$,*

(i) $0 \leq F(x) \leq 1$;

(ii) $x \leq y \implies F(x) \leq F(y)$;

(iii) $\lim\limits_{x \to -\infty} F(x) = 0$ and $\lim\limits_{x \to \infty} F(x) = 1$.

(iv) $\lim\limits_{\substack{y \to x \\ y > x}} F(y) = F(x)$.

(v) $\lim\limits_{\substack{y \to x \\ y < x}} F(y) \stackrel{\text{def}}{=} F(x-)$ *exists.*

(vi) *F has at most a countable number of discontinuities.*

Proof. (i) is clear, since $F(x)$ is a probability.

(ii) Let $B_x = \{\omega : X(\omega) \leq x\}$. Then $x \leq y \implies B_x \subset B_y$. Therefore, $F(x) = P\{B_x\} \leq P\{B_y\} = F(y)$. Thus F is monotone. Since it is also bounded, the limits in (iii), (iv) and (v) exist. This proves (v), and, for (iii) and (iv), we need only identify the limits by taking convenient sequences.

(iii) Note that $\bigcap_n B_{-n} = \emptyset$, so by Proposition 1.10, $F(-n) = P\{B_{-n}\} \to P\{\emptyset\} = 0$ as $n \to \infty$. Similarly, $\bigcup_n B_n = \Omega$, so by Proposition 1.10 again, $F(n) = P\{B_n\} \longrightarrow P\{\Omega\} = 1$.

(iv) If $y_n \downarrow x$, then $\bigcap_n B_{y_n} = B_x$, so by Proposition 1.10, $F(y_n) = P\{B_{y_n}\} \downarrow P\{B_x\} = F(x)$.

(v) F is an increasing function which is bounded above by 1, so it has a left limit at all points.

(vi) F is monotone, so its only discontinuities are jumps: x is a point of discontinuity iff $F(x) > F(x-)$. Let $\Lambda_k = \{x : F(x) - F(x-) > 1/k\}$. Note that if $x_0 < x_1 < \cdots < x_n$, certainly $\sum_{i=0}^n (F(x_i) - F(x_i-)) \leq F(x_n) - F(x_0-) \leq 1$. Thus there are at most k discontinuities at which $F(x) -$

$F(x-) > 1/k$, $k = 1, 2, \ldots$, so there are at most k points in Λ_k. The set of discontinuities is $\bigcup_k \Lambda_k$, which is a countable union of finite sets, and is therefore countable. □

Notation: A random variable X is a function, $X(\omega)$. The usual convention is to omit the variable ω except when clarity requires it. Since the variable is always there, it is understood, and we do not need to write it explicitly. So, for example, we will write $\{X \leq x\}$ instead of $\{\omega : X(\omega) \leq x\}$. It is easier to write, easier to read, and quicker to understand. From now on, we will omit the variable except when it helps our understanding.

Here are a few probabilities easily computed from the distribution function.

Proposition 2.6. *Let X be a random variable with distribution function F. Then:*

(i) $P\{X < x\} = F(x-)$.

(ii) $P\{X = x\} = F(x) - F(x-)$.

(iii) *If* $a < b$, $P\{a < X \leq b\} = F(b) - F(a)$.

(iv) $P\{X > x\} = 1 - F(x)$.

Proof. (i) Let $B_x = \{X \leq x\}$ and note that $\{X < x\} = \bigcup_n \{X \leq x - 1/n\} = \bigcup_n B_{x-1/n}$. The sets B_x increase with x, so by Proposition 1.10, $P\{X < x\} = \lim_n P\{B_{x-1/n}\} = \lim_n F(x - 1/n) = F(x-)$.

(ii) $P\{X = x\} = P\{X \leq x\} - P\{X < x\} = F(x) - F(x-)$.

(iii) $P\{a < X \leq b\} = P\{X \leq b\} - P\{X \leq a\} = F(b) - F(a)$.

(iv) This follows from the fact that $\{X > x\}$ is the complement of $\{X \leq x\}$. □

Remark 2.7. It follows that if $(a_n, b_n]$, $n = 1, 2, \ldots$ is a disjoint set of intervals and if $A = \bigcup_n (a_n, b_n]$, then

(2.1) $$P\{A\} = \sum_n \bigl(F(b_n) - F(a_n)\bigr).$$

Evidently, the distribution function allows us to calculate $P\{X \in A\}$ for many sets. In fact, we can get the probability for any set we are likely to encounter as limits of probabilities of the form (2.1). This includes the Borel sets $\mathcal{B}$, which we introduced in Definition 1.3.

Proposition 2.8. *Let X be a random variable, and let $A \in \mathcal{B}$. Then $\{X \in A\}$ is an event, i.e., it is in $\mathcal{F}$.*

2.1. Random Variables and Distributions

Proof. We will use a trick here: we define the class $\mathcal{G}$ of all $A \subset \mathbb{R}$ for which $\{X \in A\} \in \mathcal{F}$. Then we show that $\mathcal{G}$ is a σ-field which contains the open intervals. This means that it contains the Borel sets since $\mathcal{B}$ is the minimal such σ-field; this implies the result. The tricky part is that we deal simultaneously with $\mathcal{F}$, a class of subsets of Ω, and $\mathcal{G}$, a class of subsets of $\mathbb{R}$. We know that $\mathcal{F}$ is a σ-field. The problem is to show that $\mathcal{G}$ is too. Let's get to it.

Since $\{a < X < b\}$ is an event, $\mathcal{G}$ contains open intervals. It clearly contains the empty set. If $A \in \mathcal{G}$, then $\{X \in A\} \in \mathcal{F}$ so $\{X \in A^c\} = \{X \in A\}^c \in \mathcal{F}$, since $\mathcal{F}$ is closed under complementation. Therefore, $A^c \in \mathcal{G}$. Finally, if $A_1, A_2, \dots \in \mathcal{G}$, then $\{X \in \bigcup_n A_n\} = \bigcup_n \{X \in A_n\} \in \mathcal{F}$ since $\mathcal{F}$ is also closed under countable unions. Therefore, $\bigcup_n A_n \in \mathcal{G}$. This verifies that $\mathcal{G}$ is a σ-field containing the open intervals, as claimed. $\square$

We can now introduce the idea of a distribution, the companion of the distribution function. The distribution of a random variable gives the probabilities directly.

Definition 2.9. The **distribution** of a random variable X is the set function μ defined by

$$\mu(A) = P\{X \in A\}, \quad A \in \mathcal{B}.$$

The distribution function can be expressed in terms of the distribution: $F(x) = \mu\big((-\infty, x]\big)$, and conversely, the distribution function determines the distribution.

Indeed, if μ and ν are distributions with the same distribution function F, then for each $a < b$, $\mu\big((a,b]\big) = F(b) - F(a) = \nu\big((a,b]\big)$. By additivity, the two agree on finite unions of right semi-closed intervals, and on $(-\infty, a]$ and $(a\infty]$. But this is a field, and it generates the Borel sets by Exercise 1.2. Therefore by Corollary 1.8, $\mu(A) = \nu(A)$ for all Borel sets A.

Proposition 2.10. *If μ is the distribution of a random variable, then $(\mathbb{R}, \mathcal{B}, \mu)$ is a probability space. In particular, μ is a probability measure.*

Proof. The Borel sets are a σ-field of subsets of $\mathbb{R}$, so we only have to verify that μ is a probability measure. Indeed, μ is non-negative, $\mu(\mathbb{R}) = P\{X \in \mathbb{R}\} = 1$, and its countable additivity follows from that of X: if $A_1, A_2, \dots$ are disjoint Borel sets, $\mu(\bigcup_n A_n) = P\{X \in \bigcup_n A_n\} = P\{\bigcup_n \{X \in A_n\}\}$. But, as the A_n are disjoint, so are the events $\{X \in A_n\}$. Thus this equals $\sum_n P\{X \in A_n\} = \sum_n \mu(A_n)$. So μ is indeed a probability measure. $\square$

Definition 2.11. Let $A \subset \Omega$. The **indicator function of A** is the function

$$I_A(\omega) = \begin{cases} 1 & \omega \in A, \\ 0 & \omega \notin A. \end{cases}$$

Example 2.11.1. (i) $X \equiv c$ (c a constant) is a random variable. Its distribution function is
$$F(x) = \begin{cases} 0 & x < c, \\ 1 & x \geq c. \end{cases}$$

(ii) Let $Y(\omega) = I_A(\omega)$. Then
$$P\{Y = 1\} = P\{A\}, \qquad P\{Y = 0\} = 1 - P\{A\}.$$
A random variable which takes on just the values 0 and 1 is called a **Bernoulli** random variable.

(iii) Let Ω be the unit interval $[0, 1]$ and let the probability of a set A be its length. (That is, P is Lebesgue measure.) Consider the probability space $([0, 1], \mathcal{B}(0, 1), P)$ introduced in Example 1.14.2. Let $U(\omega) = \omega$, $0 \leq \omega \leq 1$. Then U is a random variable and $P\{U \leq x\} = P\{[0, x]\} = x$. We say U has a **uniform** distribution on $[0, 1]$, and its distribution function $F(x)$ satisfies $F(x) = 0$ if $x \leq 0$, $F(x) = x$ if $0 \leq x \leq 1$ and $F(x) = 1$ for $x \geq 1$.

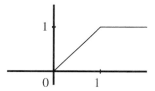

Figure 1. The $U(0, 1)$ distribution function.

(iv) Let us extend the previous example, and consider the probability space $([0, 1], \mathcal{B}, P)$, where P is Lebesgue measure. A random variable on this space is just a real-valued function f on $[0, 1]$. To be a true random variable, it must satisfy $\{x : f(x) \leq \lambda\} \in \mathcal{B}$ for all $\lambda \in \mathbb{R}$. This turns out to be a very useful class of functions. It includes, for example, the continuous functions, for if f is continuous, this set would be closed and therefore in $\mathcal{B}$. Since it is so useful, we give it a name.

Definition 2.12. A real-valued function f on $\mathbb{R}$ is a **Borel function** and is said to be **Borel measurable** if $\{x : f(x) \leq \lambda\} \in \mathcal{B}$ for each $\lambda \in \mathbb{R}$.

Note. This definition extends to functions defined on any sub-interval of $\mathbb{R}$, or even, if required, on a Borel subset of $\mathbb{R}$. In particular, the Borel functions on $[0, 1]$ are random variables on $([0, 1], \mathcal{B}, P)$, and vice-versa.

Exercise 2.2. Show that if f is a Borel function on $\mathbb{R}$ and $A \in \mathcal{B}$, that $\{x : f(x) \in A\} \in \mathcal{B}$. In other words, $f^{-1}(\mathcal{B}) \subset \mathcal{B}$. Conclude that if X is a random variable and f is a Borel function, that $f(X)$ is also a random variable.

2.2. Existence of Random Variables

A random variable is a function. As we remarked, that might look strange[2], but the question is not whether it looks strange, it is whether it works. That is, does this definition encompass all possible random variables? Random variables have distributions. Does that include all possible distributions? All possible distribution functions?

Let us define the most general possible distribution function.

Definition 2.13. A **distribution function** is a function F defined on $\mathbb{R}$ such that

(i) F is right-continuous, monotonically increasing and

(iii) $\lim_{x \to -\infty} F(x) = 0$ and $\lim_{x \to \infty} F(x) = 1$.

We saw in Proposition 2.5 that the distribution function of a random variable is a distribution function in this sense. But is each distribution function the distribution function of a random variable? If the answer is "Yes", then there is reason to believe that our definition is not strange, just efficient. So let us prove it.

Theorem 2.14. *Let F be a distribution function. Then there exists a random variable with distribution function F.*

We will prove this by explicitly producing the random variable. First, let us define the "inverse" of F.

Let F be a distribution function. Define its **left-continuous inverse** $\boldsymbol{G_F}$ by
$$G_F(u) = \inf\{x \in \mathbb{R} : F(x) \geq u\}, \quad 0 < u < 1.$$

Remark 2.15. If F is continuous and strictly increasing, then G_F is the true inverse of F, that is, $F(G_F(x)) = G_F(F(x)) = x$. In general, F is only right-continuous, and G_F is its inverse in a weaker sense. This is explored in Exercise 2.12 below.

Lemma 2.16. *Let $u \in (0, 1)$, $x \in \mathbb{R}$. Then $u \mapsto G_F(u)$ is monotonically increasing and*
$$u \leq F(x) \iff G_F(u) \leq x.$$

Proof. G_F is clearly increasing. Let $\varepsilon > 0$ and note that

---

[2]If it doesn't, either you have seen it before, or you are missing something.

$G_F(u) \leq x \implies \exists t \leq x + \varepsilon \ni F(t) \geq u \implies F(x+\varepsilon) \geq u$, since F is increasing. Let $\varepsilon \downarrow 0$ to see that $F(x) = \lim_{\varepsilon \downarrow 0} F(x+\varepsilon) \geq u$. Conversely, $u \leq F(x) \implies x \geq \inf\{t : F(t) \geq u\} = G_F(u)$. □

Proof. (of Theorem 2.14). Consider the probability space $([0,1], \mathcal{B}(0,1), P)$ of Example 1.14.2, where $P(dx) = dx$ is Lebesgue measure.

Let $U(\omega) = \omega$, $0 \leq \omega \leq 1$. Then U is uniform on $(0,1)$. Define $X(\omega) = G_F(U(\omega))$, $0 \leq \omega \leq 1$. Then X is a function on our probability space, and therefore a random variable. We claim that its distribution function is F.

Fix x for which $0 < F(x) < 1$. From Lemma 2.16 we see that
$$\{G_F(U) \leq x\} = \{U \leq F(x)\}.$$
Thus
$$P\{X \leq x\} \equiv P\{G_F(U) \leq x\} = P\{U \leq F(x)\} = F(x)$$
because U is uniform. □

This not only shows the existence of a random variable with distribution function X, it identifies the random variable. For future reference, let us state this as a corollary.

Corollary 2.17. *Let F be a distribution function and U a $U(0,1)$ random variable. Then $G_F(U)$ is a random variable with distribution function F.*

Remark 2.18. If U is $U(0,1)$, then $G_F(U)$ has distribution function F. The converse of this result is called the *probability integral transformation*: if X is a random variable with a continuous distribution function F, then $F(X)$ is $U(0,1)$.

To see this, note that if $y \in (0,1)$, then the set $\{x : F(x) = y\}$ is a non-empty (by the intermediate value theorem) closed (F is continuous) interval (F is monotone), $[x_0, x_1]$, possibly degenerate. Thus $F(x) \leq y \iff x \leq x_1$. Thus $P\{F(X) \leq y\} = P\{X \leq x_1\} = F(x_1) = y$. This is true for all $y \in (0,1)$, so $F(X)$ is indeed uniform. □

Problems 2.2

2.3. Let X be a constant random variable: $X \equiv 3/2$. Find its probability mass function and distribution function.

2.4. Let U be a uniform random variable on $[0,1]$. Let $X = U^2$. Find the distribution function of X.

2.5. Let X be a random variable with distribution function
$$F(x) = \begin{cases} 0 & \text{if } x < 0, \\ x^2 & \text{if } 0 \leq x \leq 1, \\ 1 & \text{if } x > 1. \end{cases}$$
Find $P\{1/4 < X < 5\}$, $P\{.2 < X < .8\}$, and $P\{X = 1/2\}$.

2.6. Toss a coin three times. Let $X = \#\,\text{Heads} - \#\,\text{Tails}$. Find the distribution function of X.

2.7. Let X be a random variable with distribution function
$$F(x) = \begin{cases} 0 & \text{if } x < 0, \\ x & \text{if } 0 \leq x < \tfrac{1}{2}, \\ 1 & \text{if } x \geq \tfrac{1}{2}. \end{cases}$$
Find $P\{X < \tfrac{1}{2}\}$, $P\{.2 \leq X \leq .7)\}$, and $P\{X = \tfrac{1}{2}\}$.

2.8. Which of the following three functions on $\mathbb{R}$ are distribution functions?
  (a) $F(x) = I_{(2,\infty)}(x)$.
  (b) $F(x) = \frac{x^2}{1+x^2}$.
  (c) $F(x) = \frac{1}{\pi}\left(\arctan(x) + \frac{\pi}{2}\right)$.

2.9. Let $P\{X = n\} = \frac{1}{n(n+1)}$, $n \geq 1$. Is this a probability distribution?

2.10. Five boys and five girls arrive at random for a rock concert and form a line at the ticket window. Number them from 1 to 10, starting at the front of the line. Let X be the number of the first girl in line. (If a girl is at the head of the line, $X = 1$. If a boy is at the head and a girl is next, $X = 2$.) Find the distribution of X.

2.11. Let A and B be sets and let f be a function taking A into B. Show that the inverse mapping f^{-1} preserves set operations, in the sense that $f^{-1}(A^c) = f^{-1}(A)^c$, $f^{-1}(A \cup B) = f^{-1}(A) \cup f^{-1}(B)$ and $f^{-1}(A \cap B) = f^{-1}(A) \cap f^{-1}(B)$.

2.12. Show that G_F is monotone increasing, left-continuous, and for each $x \in \mathbb{R}$ for which $0 < F(x) < 1$, and each $u \in (0,1)$,
$$G_F\bigl(F(x)-\bigr) \leq x \leq G_F\bigl(F(x)\bigr),$$
$$F\bigl(G_F(u)-\bigr) \leq u \leq F\bigl(G_F(u)\bigr).$$

2.13. Let F be the distribution function of a random variable X. Prove the following.
  (a) $P\{X = a\} = \lim_{\varepsilon \downarrow 0}\bigl(F(a+\varepsilon) - F(a-\varepsilon)\bigr)$.
  (b) $P\{a < X < b\} = \lim_{\varepsilon \downarrow 0}\bigl(F(b-\varepsilon) - F(a+\varepsilon)\bigr)$.

2.14. A distribution function F is said to be **purely discontinuous** if it equals the sum of its jumps, i.e., if $F(x) = \sum_{y \leq x}\bigl(F(y) - F(y-)\bigr)$ for all x.

Show that any distribution function F can be written as $F = aF_c + (1-a)F_d$, where F_c and F_d are continuous and purely discontinuous distribution functions respectively, and $0 \leq a \leq 1$.

[Hint: Remove the sum of the jumps and show what remains is continuous and increasing.]

2.15. Prove the following.
  (a) A distribution function F is purely discontinuous iff $\sum_y (F(y) - F(y-)) = 1$.
  (b) The distribution function of a discrete random variable is purely discontinuous.

2.16. It is not enough just to handle finite families of random variables. We must also handle sequences. Let $X_1, X_2, \ldots$ be an infinite sequence of random variables defined on some probability space $(\Omega, \mathcal{F}, P)$. Show that the following are also random variables.

(a) $\sup_n X_n$ and $\inf_n X_n$.

(b) $\limsup_{n \to \infty} X_n$ and $\liminf_{n \to \infty} X_n$.

(c) Show that $\{\omega : \lim_{n \to \infty} X_n \text{ exists}\} \in \mathcal{F}$.

2.17. Prove that a continuous function on $\mathbb{R}$ is Borel measurable.

2.18. (Continuation.) (a) Prove that a right-continuous function on the line is Borel measurable.

(b) A function is said to be **lower semi-continuous** if it is the increasing limit of a sequence of continuous functions. Show that a lower semi-continuous function on the line is Borel measurable.

2.19. (a) Show that for every distribution function F there exists a probability measure μ on $(\mathbb{R}, \mathcal{B})$ such that $F(x) = \mu\{(-\infty, x]\}$.

[Hint: Proposition 2.10 might help.]

(b) Show that for any probability measure μ on $(\mathbb{R}, \mathcal{B})$, the random variable $X(x) = x$ on the probability space $(\mathbb{R}, \mathcal{B}, P)$ has distribution μ. (This is called the **canonical representation** of the random variable.)

2.3. Independence of Random Variables

Definition 2.19. (*i*) Two random variables X and Y are **independent** if for all x and $y \in \mathbb{R}$, $P\{X \leq x, Y \leq y\} = P\{X \leq x\} P\{Y \leq y\}$.

(*ii*) A finite family $X_1, \ldots, X_n$ of random variables is **independent** if, for each subsequence $i_1, \ldots i_k$ of $1, \ldots, n$ and each $x_{i_1}, \ldots, x_{i_k}$, $P\{X_{i_j} \leq x_{i_j}, j = 1, \ldots, k\} = \prod_{j=1}^{k} P\{X_{i_j} \leq x_{i_j}\}$.

(*iii*) An arbitrary family $\{X_\alpha, \alpha \in I\}$ is **independent** if each finite sub-family is.

The definition only involves the events $\{X \leq x\}$ and $\{Y \leq y\}$, but it extends to all events associated with X and Y.

Theorem 2.20. *Let X and Y be random variables. Then X and Y are independent if and only if for all Borel sets A and B,*

(2.2) $$P\{X \in A, Y \in B\} = P\{X \in A\} P\{Y \in B\}.$$

Proof. One direction is easy: (2.2) implies independence, since $A = (-\infty, x]$ and $B = (-\infty, y]$ are Borel sets.

The converse takes some work. We will apply a trick we used in the proof of Proposition 2.8 not once, but twice.

Let $A = (-\infty, x]$ and let $\mathcal{G}$ be the class of all $B \subset \mathbb{R}$ for which (2.2) holds.

(i) $\mathcal{G}$ contains $\emptyset$, $\mathbb{R}$, and all sets of the form $(-\infty, y]$ by the definition of independence.

(ii) $\mathcal{G}$ contains the intervals $(a, b]$. Indeed, letting $I_x = (-\infty, x]$, we have
$$P\{X \in A, Y \in (a,b]\} = P\{X \in A, Y \in I_b\} - P\{X \in A, Y \in I_a\}$$
$$= P\{X \in A\}P\{Y \in I_b\} - P\{X \in A\}P\{Y \in I_a\} = P\{X \in A\}P\{Y \in (a,b]\}.$$

(iii) If B_1, B_2 are disjoint and in $\mathcal{G}$, then $B_1 \cup B_2 \in \mathcal{G}$, for
$$P\{X \in A, Y \in B_1 \cup B_2\} = P\{\{X \in A, Y \in B_1\}\} + P\{X \in A, Y \in B_2\}$$
$$= P\{X \in A\}P\{Y \in B_1\} + P\{X \in A\}P\{Y \in B_1\}$$
$$= P\{X \in A\}P\{Y \in B_1 \cup B_2\}.$$

So $B_1 \cup B_2 \in \mathcal{G}$. In particular, $\mathcal{G}$ contains all finite unions of sets of the form $(-\infty, a]$, $(a, b]$, and (b, ∞). These sets form a field. Call it $\mathcal{F}_0$.

But if B_n is a sequence of sets in $\mathcal{G}$ such that either $B_1 \subset B_2 \subset \ldots$, or $B_1 \supset B_2 \supset \ldots$, then $P\{X \in A, Y \in B_n\} = P\{X \in A\}P\{Y \in B_n\}$. Let $n \to \infty$. The left- and right-hand sides both converge, and we get $P\{X \in A, Y \in \lim_n B_n\} = P\{X \in A\}P\{Y \in \lim_n B_n\}$, so that $\lim_n B_n \in \mathcal{G}$.

This implies that $\mathcal{G}$ is a monotone class containing the field $\mathcal{F}_0$. Therefore, by the monotone class theorem (Theorem 1.5) $\mathcal{G}$ contains $\sigma(\mathcal{F}_0) = \mathcal{B}$. Thus (2.2) holds for all Borel B.

Repeat this argument: fix a Borel set B, and redefine $\mathcal{G}$ to be the class of all sets A for which (2.2) holds. We have just shown that $\mathcal{G}$ contains the intervals $(-\infty, x]$. Now exactly the same calculation, with A and B interchanged shows that $\mathcal{G}$ is a monotone class which contains the field $\mathcal{F}_0$, and therefore contains $\sigma(\mathcal{F}_0) = \mathcal{B}$. So, finally, (2.2) holds for all Borel sets A and B, as claimed. $\square$

Exercise 2.20. Let X and Y be independent random variables and f and g Borel functions. Show that $f(X)$ and $g(Y)$ are independent.

2.4. Types of Distributions

Definition 2.21. (i) A random variable X has a **discrete distribution** if there exists a countable subset $\{x_n\}$ of $\mathbb{R}$ such that
$$\sum_{n=1}^{\infty} P\{X = x_n\} = 1.$$

(ii) A random variable has a **continuous distribution** if its distribution function is continuous.

(*iii*) A random variable X has an **absolutely continuous distribution** if its distribution function is absolutely continuous, that is, if there is a function $f(x)$ on $\mathbb{R}$ such that the distribution function $F(x)$ of X is

$$F(x) = \int_{-\infty}^{x} f(y)\,dy, \quad -\infty < x < \infty.$$

Then f is called the **density** of X.

Remark 2.22. (*i*) We also say that a random variable or its distribution function is discrete, continuous, or absolutely continuous if the corresponding distribution is.

(*ii*) An absolutely continuous distribution is continuous, but there are continuous distributions which are not absolutely continuous. See Exercise 3.45.

(*iii*) If X has a continuous distribution, then for any x, $P\{X = x\} = 0$. Indeed, its distribution function F is continuous, so $P\{X = x\} = F(x) - F(x-) = 0$.

(*iv*) Not every distribution function is of one of the above types, but any distribution function F can be written as a sum $F = aF_d + (1-a)F_c$, where F_d is a discrete distribution function, F_c is a continuous distribution function, and $0 \le a \le 1$. F_c can be further broken down into the sum of an absolutely continuous distribution function, and another, called a *purely singular* distribution function.

Definition 2.23. If X has a discrete distribution, define its **probability mass function** $p(x)$ by

$$p(x) = P\{X = x\}, \quad x \in \mathbb{R}.$$

Remark 2.24. (*i*) The probability mass function vanishes except for countably many points, and $\sum_y p(y) = 1$. (The sum is understood to be over non-zero values only.)

(*ii*) If F is an absolutely continuous distribution function with density f and if f is continuous at x, then

$$f(x) = \frac{dF}{dx}(x).$$

(*iii*) If x is a continuity point of the density f, then $F(x+\Delta x) - F(x) = \int_x^{x+\Delta x} f(y)\,dy \sim f(x)\,\Delta x$. Intuitively, if X has a density f, we can think of the probability that X is in an infinitesimal interval $(x, x+dx)$ as $f(x)\,dx$.

Here are several examples of absolutely continuous distributions.

Example 2.24.1. A continuous uniform distribution can be defined on any interval. The **uniform distribution** on $[a, b]$ is denoted $U(a, b)$. It has a density $f(x)$ which is constant on (a, b), and zero outside. Since the density integrates to 1, we must have $f(x) = I_{[a,b]}(x)/(b - a)$. It follows that the distribution function $F(x)$ is zero for $x < a$, $F(x) = (x - a)/(b - a)$ for $a \leq x \leq b$, and $F(x) = 1$ for $x \geq b$.

The uniform distribution can be recognized by its symmetry. It has no preference: any value in (a, b) is as likely as any other, and the probability that U is in some sub-interval is proportional to its length.

We defined the uniform distribution on a closed interval, but it could just as well have been on the open interval, since if X is uniform, the probability that it equals either end point is zero. Similarly, the exact value of the density at the endpoints, where it is discontinuous, is unimportant.

Example 2.24.2. *(i)* Toss a needle on a blanket, and let θ be the angle, measured in radians, that it makes with North. Then θ is a random variable whose values are in $[0, 2\pi)$. If we toss the needle a reasonable distance with some spin, there is no real preference for direction in the way it falls—any angle is as likely as any other. So the distribution should be uniform on $[0, 2\pi)$. θ has density $I_{[0,2\pi]}/2\pi$, and distribution function $F_\theta(x) = x/2\pi$ for $0 \leq x \leq 2\pi$, $F_\theta(x) = 0$ for $x < 0$ and $F_\theta(x) = 1$ for $x \geq 2\pi$.

(ii) Let $Y = \theta^2$. To find its density, first find its distribution function. If $0 \leq y \leq 4\pi^2$, $F_Y(y) = P\{Y \leq y\} = P\{\theta^2 \leq y\} = P\{\theta \leq \sqrt{y}\} = \sqrt{y}/2\pi$. Differentiate to get the density of Y:

$$f_Y(y) = \frac{dF_Y}{dy} = \frac{1}{4\pi\sqrt{y}} I_{[0,4\pi^2]}(y).$$

Example 2.24.3. Buffon's Needle Problem[3]. Toss a needle of length ℓ on an American flag. The width of the stripes is D. Assume $\ell < D$. What is the probability that the needle intersects the boundary of a stripe? (Since $\ell < D$, the needle can intersect at most one boundary.)

Answer. Describe the fallen needle's position by two things:

1° the distance X from the center of the needle to the nearest stripe-boundary, and

2° the angle θ that the needle makes with the boundary of the stripe.

If the stripes are fairly narrow compared to the length of the toss, the distribution of the landing-spot of the needle's center will be smeared out over a number of stripes. By symmetry, we would expect that its distance

---

[3] It was proposed by G-L LeClerc, Compte de Buffon, in 1733. Without an American flag, of course.

from the nearest stripe boundary should be (nearly) uniform. So let us assume that X is, in fact, uniform on $[0, D/2]$. As before, θ is uniform, this time on $[0, 2\pi]$. There is no evident connection between θ and X, so we assume they are independent.

A good choice of sample space helps. Let Ω be the rectangle $[0, D/2] \times [0, 2\pi]$. Define X and θ on Ω: $X(x, y) = x$ and $\theta(x, y) = y$. The outcome of the toss is a point (x, y) in Ω.

Let us find the probability measure P on Ω. Consider a rectangle $[a, b] \times [c, d] \subset \Omega$: $P\{a \leq x \leq b,\ c \leq \theta \leq d\} = P\{(x, \theta) \in \Omega : a \leq X \leq b,\ c \leq \theta \leq d\} = P\{a \leq X \leq b\}P\{c \leq \theta \leq d\}$ since the events are independent. But X and θ are uniform, so for any rectangle $[a, b] \times [c, d] \subset \Omega$,

$$P\{[a, b] \times [c, d]\} = \frac{2(b-a)}{D} \frac{d-c}{2\pi}.$$

Notice that this is just $1/\pi D$ times the area of the rectangle. If A is a sub-rectangle of Ω,

(2.3) $$P\{A\} = \frac{\text{area of } A}{\pi D}.$$

By additivity, (2.3) holds for disjoint unions of rectangles, and, in fact, one can show that it holds for all events.

The needle will intersect a boundary between two stripes if $X < \frac{\ell}{2}|\sin\theta|$. Therefore,

$$\begin{aligned} P\{\text{ intersection }\} &= \frac{1}{\pi D} \text{Area}\left\{(x, y) : x < \frac{\ell}{2}|\sin y|\right\} \\ &= \frac{\ell}{\pi D} \int_0^\pi \sin y\, dy \\ &= \frac{2\ell}{\pi D}. \end{aligned}$$

This formula suggests yet another way to calculate π, this time by Monte Carlo methods. Toss the needle a large number of times. Count the number of intersections in the first n tosses, and divide by n. This is an estimate of the true probability $2\ell/\pi D$. Thus

$$\pi \sim \frac{2n\ell}{\#\{\text{intersections in first } n \text{ tosses}\} \cdot D}.$$

The Italian mathematician, M. Lazzerini, reportedly did this in 1901. With $\ell = 5/6\, D$, he counted 1808 intersections in 3408 tosses, and estimated π to seven decimal places by 3.1415929. The true value is 3.1415927, so his error is less than 3 in the seventh decimal... which is suspiciously good, to say the least.

2.4. Types of Distributions

Problems 2.4

2.21. Let X and Y be independent Bernoulli $(1/2)$ random variables. Let $Z = (Y - X)^2$. What is the distribution of Z? Show that Z and X are independent. Show that Z and Y are also independent, but that X, Y and Z are not independent.

2.22. (a) Choose one of the points $(0,0)$, $(0,1)$, $(1,0)$ and $(1,1)$ at random. Let X be the first coordinate and Y the second. Are X and Y independent?

(b) Do the same for the points $(0,1)$, $(1,0)$, $(-1,0)$ and $(0,-1)$. Are X and Y independent?

2.23. Let X be a random variable and f a Borel function. Suppose that $f(X)$ is not a.s. constant. Is it possible for X and $f(X)$ to be independent?

2.24. Let $\Omega = \{\omega_1, \omega_2, \omega_3\}$ with $P\{\omega_1\} = P\{\omega_2\} = P\{\omega_3\} = 1/3$. Define random variables X, Y, and Z by

$$X(\omega_1) = 1, \quad X(\omega_2) = 2, \quad X(\omega_3) = 3,$$
$$Y(\omega_1) = 2, \quad Y(\omega_2) = 3, \quad Y(\omega_3) = 1,$$
$$Z(\omega_1) = 3, \quad Z(\omega_2) = 1, \quad Z(\omega_3) = 2.$$

Find the distributions of $X + Y$, $Y + Z$, $Z + X$ and $X + Y + Z$.

2.25. Roll two fair dice. Let X be the smaller of the pips showing on the two, and Y the larger. Find the probability mass function of $Z \stackrel{\text{def}}{=} Y - X$.

2.26. The following are probability mass functions.

(a) $p_X(x) = kx$ for $x = 1, 2, 3, 4, 5$, $p_X(x) = 0$ otherwise. Find k.

(b) $p_Y(y) = cy(1 - y)$ for $y = 1/4, \ 1/2$ and $3/4$, and $p_Y(y) = 0$ otherwise. Find c.

2.27. A random variable X has an absolutely continuous distribution whose density $f(x)$ is proportional to x on $0 \leq x \leq 1$, and $f(x) = 0$ otherwise. What is $P\{X \in (1/3, 7/9)\}$?

2.28. Let X be absolutely continuous with density $f(x) = cx(1 - x)$ if $0 \leq x \leq 1$, $f(x) = 0$ otherwise. Find c. Let $Y = X^4$. Find the density f_Y of Y.

2.29. Four fair dice are rolled. Let X be the product of the four rolls. If p_X is its probability mass function, find $p_X(1)$, $p_X(2)$, and $p_X(4)$.

2.30. Toss a fair coin four times independently. Let X be the number of heads minus the number of tails. Find the distribution function of X.

2.31. Suppose that X has the density $f(x) = 2x$ on the interval $[0, 1]$, zero elsewhere. Let $Y = |X - 1/2|$. Find the density of Y.

2.32. Let X and Y be independent, uniform random variables on $[0, 1]$. Explain why the pair (X, Y) fits the description of "a point randomly chosen in the unit square".

2.33. Let X and Y be independent random variables, uniform on $[0, 1]$. Find $P\{|Y - X| \leq 1/2\}$.
[Hint: Use the previous problem.]

2.34. Let X and Y be independent random variables, uniform on $[0,1]$. If the smaller is less than one-fourth, find the conditional probability that the larger is greater than $3/4$.

2.35. A grid of lines a distance a apart and another grid of lines b apart are perpendicular, dividing the plane into rectangles. A needle of length $\ell < a \wedge b$ is thrown at random onto the grid. What is the probability it intersects lines in both grids simultaneously?

2.36. Let X and Y be i.i.d. integer-valued random variables and let $p_n = P\{X = n\}$. Find $P\{X = Y\}$ and $P\{X \leq Y\}$.

2.5. Expectations I: Discrete Random Variables

The expectation of a random variable is its average value. We will begin with discrete random variables. A discrete random variable can take on infinitely many values, though, so we must take care. In particular, we need to use series.

Let us recall some facts about series. A sequence $x_1, x_2, \ldots$ converges if there is $L \in \mathbb{R}$ such that $\lim |x_n - L| = 0$. In particular, "converges" means "converges to a finite limit." (This warning is necessary because we will encounter situations in which the limit can be infinite. But we will not use the word "converge" then. "Converge" will always mean "Converge to a finite limit.")

- $\sum_{n=1}^{\infty} a_n$ **converges** if $\lim_{N \to \infty} \sum_{n=1}^{N} a_n$ exists and is finite; it **diverges** otherwise. If it converges, we define $\sum_{n=1}^{\infty} a_n \stackrel{\text{def}}{=} \lim_{N \to \infty} \sum_{n=1}^{N} a_n$.

- $\sum_{n=1}^{\infty} a_n$ **converges absolutely** if $\lim_{N \to \infty} \sum_{n=1}^{N} |a_n| < \infty$. Otherwise, we say that $\sum_{n=1}^{\infty} |a_n|$ **diverges to infinity,** and write $\sum_{n=1}^{\infty} |a_n| = \infty$.

- Absolute convergence implies convergence. A convergent series which does not converge absolutely is said to be **conditionally convergent**.

- A series of positive terms, convergent or divergent, can be rearranged without changing its sum; so can an absolutely convergent series.

Let $(\Omega, \mathcal{F}, P)$ be a probability space, and X a random variable. If X is discrete, there is a countable set $\{x_n\}$ such that $P\{X \notin \bigcup_{n=1}^{\infty}\{x_n\}\} = 0$. Ignoring this set of probability zero, we have

2.5. Expectations I: Discrete Random Variables

- X takes on only countably many values.
- There is a partition of Ω, $\Omega = \Lambda_1 \cup \Lambda_2 \cup \ldots$ where the Λ_i are disjoint, such that X is constant on each Λ_i.

We could take $\Lambda_i = \{X = x_i\}$, for instance. We can get other partitions by dividing $\{X = x_i\}$ into smaller sets. In any case, we can suppose that X is of the form

$$(2.4) \qquad X(\omega) = \sum_{i=0}^{\infty} x_i I_{\Lambda_i}(\omega),$$

where the x_i are real and the Λ_i are disjoint.

Definition 2.25. A random variable X of the form (2.4) is **integrable** if

$$\sum_{i=1}^{\infty} |x_i| P\{\Lambda_i\} < \infty.$$

If X is integrable, then the **expectation** of X is

$$(2.5) \qquad E\{X\} = \sum_{i=1}^{\infty} x_i P\{\Lambda_i\}.$$

Remark 2.26. (i) A random variable is integrable if and only if the series (2.5) is *absolutely* convergent. Conditional convergence is not enough.

(ii) The sets Λ_i in the representation are not unique, so it is not obvious that the expectation is well-defined. We must show that it does not depend on the particular sets Λ_i.

To see this, note that as X is constant on Λ_i, Λ_i must be a subset of $\{X = x_j\}$ for some j. Thus, for each j there is a subsequence (j_k) such that $\{X = x_j\} = \bigcup_k \Lambda_{j_k}$. Notice that $x_{j_k} = x_j$ for each k. The series (2.5) is absolutely convergent and can therefore be rearranged without changing its sum, so that we can write

$$\sum_i x_i P\{\Lambda_i\} = \sum_j \sum_k x_{j_k} P\{\Lambda_{j_k}\}.$$

Now $x_{j_k} = x_j$ for each k, so we can factor it out:

$$= \sum_j x_j \sum_k P\{\Lambda_{j_k}\}$$
$$= \sum_j x_j P\{X = x_j\},$$

since $P\{X = x_j\} = \sum_k P\{\Lambda_{j_k}\}$ by the countable additivity of P. This is independent of the initial choice of the Λ_i. $\square$

In terms of the probability mass function $p(x) = P\{X = x\}$,
$$E\{X\} = \sum_x x\, p(x)\,.$$

This[4] can be extended into what is called the *law of the unconscious statistician*.

Proposition 2.27. *Let f be a real-valued function, and let X be a discrete random variable with probability mass function p. Then the random variable $f(X)$ is integrable iff $\sum_x |f(x)|\, p(x) < \infty$, in which case*
$$E\{f(X)\} = \sum_x f(x)\, p(x)\,.$$

Proof. This is called the law of the unconscious statistician because people use it unconsciously, without realizing it needs proof. The proof here is quick, but that is because we already did the real work when we verified that the expectation was well-defined.

Let $\Lambda_i = \{X = x_i\}$, where (x_i) are the possible values of X. Notice that the Λ_i are disjoint, and—ignoring a set of probability zero—form a partition of Ω. Moreover, $f(X)$ is a random variable which is constant on each Λ_i. Therefore, by definition, $f(X)$ is integrable if and only if $\sum_i |f(x_i)|\, P\{\Lambda_i\} < \infty$. If it is integrable,
$$E\{f(X)\} = \sum_i f(x_i)\, P\{\Lambda_i\} = \sum_i f(x_i)\, p(x_i)\,.$$
$\square$

Exercise 2.37. Let X be a non-negative random variable with integer values. Show that
$$E\{X\} = \sum_{n=1}^{\infty} P\{X \geq n\}\,.$$

Here are some of the basic properties of expectations.

Proposition 2.28. *Let X and Y be discrete random variables and $a, b \in \mathbb{R}$. Then*

(i) X *is integrable* $\iff |X|$ *is integrable.*

(ii) *If X and Y are integrable, so is $aX + bY$ and*
$$E\{aX + bY\} = aE\{X\} + bE\{Y\}\,.$$

(iii) *If $|X| \leq |Y|$ and Y is integrable, then X is integrable.*

(iv) *If X and Y are integrable and $X \leq Y$, then $E\{X\} \leq E\{Y\}$.*

(v) *If X is integrable, $E\{X\} \leq E\{|X|\}$.*

---

[4]By convention, $\sum_x g(x)$ means the sum over values of x for which $g(x) \neq 0$.

2.5. Expectations I: Discrete Random Variables

Proof. Partition Ω into events Λ_i such that *both* X and Y are constant on each Λ_i, say $X = x_i$ and $Y = y_i$ on Λ_i. (This can be done. Why?)

(i) This is part of the definition. We just state it for emphasis.

(ii) First, let us verify integrability. Note that $aX + bY = ax_i + by_i$ on Λ_i, so

$$E\{|aX+bY|\} = \sum_i |ax_i + by_i| P\{\Lambda_i\} \leq |a| \sum_i |x_i| P\{\Lambda_i\} + |b| \sum_i |y_i| P\{\Lambda_i\}$$
$$= |a|E\{|X|\} + |b|E\{|Y|\}.$$

But X and Y are integrable, so the last two expectations are finite, and $aX + bY$ is integrable. Now redo the above calculation without the absolute value signs to see that (ii) holds.

(iii) In this case, on each Λ_i, $|x_i| \leq |y_i|$, so that $\sum_i |x_i| P\{\Lambda_i\} \leq \sum_i |y_i| P\{\Lambda_i\} < \infty$.

(iv) $x_i \leq y_i$ on each Λ_i, so this follows from the calculation in (iii) with the absolute values signs removed.

Finally, (v) follows from (iv) since $X \leq |X|$. □

Remark 2.29. The second conclusion can be restated: "The expectation is linear." Notice that there is no assumption about the relations between X and Y. The conclusion holds whether or not they are independent.

Special cases: Let a, b, and c be constants.

1. $X \equiv c \implies E\{X\} = c$.
2. $a \leq X \leq b \implies a \leq E\{X\} \leq b$.
3. Roll one fair die. Then $E\{$ number of pips showing $\} = 7/2$. Roll two dice. Then $E\{$ total number of pips showing $\} = 7$.

Indeed, for one die, the sides all have probability $1/6$, so the expectation is $(1+2+3+4+5+6)/6 = 7/2$. The sum of the two dice is twice that. (Notice that the two rolls do not have to be independent for this to be true.)

4. The linearity of the expectation extends by induction to any number of random variables. If $X_1, \ldots, X_n$ are integrable random variables, so is their sum and

$$E\left\{\sum_{i=1}^n X_i\right\} = \sum_{i=1}^n E\{X_i\}.$$

This is the basis of a number of clever tricks, based on writing random variables in terms of indicator functions. Consider the following two problems.

Example 2.29.1. Scientists use what is called a capture-recapture scheme to estimate wildlife populations. They catch a number of animals, tag them, and return them to the wild. They then catch a second batch of animals and count the number of tagged animals among them. The more tagged animals they find, the smaller the population is likely to be. Suppose that there are actually fifty animals in the population and that the scientists tag and release ten of them. They then recapture a second sample of ten animals, at random. What is the expected number of tagged animals recovered?

Answer: Let $X_1 = 1$ if the first animal recaptured is tagged, zero otherwise. Similarly, let X_k be the indicator function of the event "The kth animal recaptured is tagged." Then the total number of tagged animals recaptured is $X_1 + \cdots + X_n$. There is no need to keep track of the order of recapture, for it is irrelevant: the kth animal recaptured is exactly as likely to be tagged as the first. That means that the X_i all have the same distribution. But we know the distribution of the first, since, at that point, ten of the fifty animals are tagged, so the probability that $X_1 = 1$ is $1/5$. But the expected number of tagged animals is $E\{X_1 + \cdots + X_{10}\} = E\{X_1\} + \cdots + E\{X_{10}\} = 10E\{X_1\} = 2$.

Example 2.29.2. Karen has ten pair of socks in her sock drawer. Once, when the room was dark and she could not see anything, she took six individual socks out of the drawer, totally at random. What is the expected number of pairs among the six?

Answer: Let us look at the situation immediately after Karen takes the first sock. What is the probability that it is eventually paired?

Once she has taken the first sock, there are nineteen left, and only one of them will pair the first. She draws five more socks, and the probability that one of them will pair the first sock is $5/19$.

Let A_i be the event "The ith sock drawn is paired," and let $X_i = I_{A_i}$. If X is the total number of paired socks, $X = X_1 + \cdots + X_6$. But this is *twice* the number of pairs, so that the number of pairs is $(X_1 + \cdots + X_6)/2$. Thus

$$E\{\# \text{ pairs drawn}\} = \frac{1}{2} E\{X_1 + \cdots + X_6\} = \frac{1}{2}\Big(E\{X_1\} + \cdots + E\{X_6\}\Big).$$

But $E\{X_1\} = \cdots = E\{X_6\}$ and $E\{X_1\} = P\{A_1\} = 5/19$, so the expected number of pairs is $15/19$.

Example 2.29.3. (The Inclusion-Exclusion Principle.) Sometimes we can let algebra do the heavy counting. The inclusion-exclusion principle generalizes the familiar formula $P\{A \cup B\} = P\{A\} + P\{B\} - P\{A \cap B\}$ to

2.5. Expectations I: Discrete Random Variables

n events, $A_1, \ldots, A_n$:

$$(2.6) \quad P\{A_1 \cup \cdots \cup A_n\} = \sum_{j=1}^{n} P\{A_j\} - \sum_{j,k=1}^{n} P\{A_j \cap A_k\}$$
$$+ \sum_{i,j,k=1}^{n} P\{A_i \cap A_j \cap A_k\} - \cdots \pm P\{A_1 \cap \cdots \cap A_n\}.$$

If $n = 2$, this is the familiar formula, and we can see it directly by drawing pictures of the sets. It is also possible, but harder, to see in the case $n = 3$. Beyond that, it is best to put our faith in algebra, as follows. (The reader should go through the following calculation in the case $n = 3$ to see how it works, and to see what is being included and excluded.) The key idea is to notice that, as $\bigcup_j A_j = (\bigcap_j A_j^c)^c$, that

$$I_{\cup_j A_j} = 1 - \prod_{j=1}^{n}(1 - I_{A_j}).$$

Multiply this out, as was done in the Binomial Theorem, to see that it is

$$= 1 - \left(1 - \sum_{j=1}^{n} I_{A_j} + \sum_{j,k=1}^{n} I_{A_j} I_{A_k} - \sum_{i,j,k=1}^{n} I_{A_i} I_{A_j} I_{A_k} + \cdots \pm \prod_{j=1}^{n} I_{A_j}\right).$$

Notice that the product of indicator functions equals the indicator function of the intersection, and take expectations of both sides to get (2.6).

Example 2.29.4. (Probability Generating Function) Generating functions are essentially tricks, but extremely useful tricks. They allow us to code a lot of information in a small package. We will encounter several different types. This one encodes the distribution of an integer-valued random variable in a single function, called the probability generating function.

Let X be a positive integer-valued random variable. Define the **probability generating function** $G(t)$ of X by

$$G(t) = E\{t^X\}.$$

Then $G(t) = \sum_{n=0}^{\infty} p(n) t^n$ where $p(n)$ is the probability mass function of X. This is a power series, so it has a radius of convergence $0 \leq R \leq \infty$, say, and it converges for $|t| < R$ and diverges for $|t| > R$. It is differentiable term-by-term inside its radius of convergence, and the derivative series has the same radius of convergence. This tells us that:

(i) $G(t)$ is a power series whose radius of convergence is at least one.
(ii) $G(0) = P\{X = 0\}$, $G(1) = 1$.
(iii) $p(n) = \frac{1}{n!} G^{(n)}(0)$, $n = 1, 2, \ldots$.

(iv) If the radius of convergence is strictly greater than 1, $E\{X\} = G'(1)$ and the higher derivatives generate what are called factorial moments: $E\{X(X-1)\ldots(X-k+1)\} = G^{(k)}(1)$.

(v) In case the radius of convergence is 1, (iv) still holds if we interpret $G^{(k)}(1)$ as $\lim_{t\uparrow 1} G^{(k)}(t)$.

Since $\sum_n p(n) = 1$, G converges for $t = 1$, so (i) and (ii) are clear. Then (iii) is the usual formula for the coefficients of a power series. To see (iv), differentiate the series term-by-term:

$$G'(t) = \sum_{n=1}^{\infty} np(n)t^{n-1},$$

and set $t = 1$ to get $\sum_n np(n) = E\{X\}$. If we differentiate k times we get $\sum_{n=k}^{\infty} n(n-1)\ldots(n-k+1)p(n)t^{(n-k)}$. Set $t = 1$ to get (iv). If the radius of convergence is exactly one, G is not differentiable at $t = 1$ since it isn't defined for $t > 1$. In that case we call on Abel's theorem, which tells us that since $f(t) = \sum_{n=0}^{\infty} b_n t^n$ is a power series with radius of convergence at least one, and since all $b_n \geq 0$, then $f(1) = \lim_{t\uparrow 1} G(t)$, whether the limit is finite or infinite. That verifies (v).

2.6. Moments, Means and Variances

The mean is the average value of a random variable, in some sense the center of the distribution. There is a companion quantity, called the variance, which measures how much the distribution is "spread out" from its center.

Definition 2.30. Let X be a random variable with a finite mean. The **variance** of X is $\text{Var}(X) = E\{(X - E\{X\})^2\}$. The **standard deviation** of X is $\sqrt{\text{Var}(X)}$.

Notation: The mean and variance are often denoted $\mu = E\{X\}$ and $\sigma^2 = \text{Var}(X)$, respectively. If we need to specify the random variable, we will write μ_X and σ_X^2 for the mean and variance of X.

Here is a useful formula for the variance.

Proposition 2.31. $\text{Var}(X) = E\{X^2\} - E\{X\}^2$.

Proof. Expand the square: $E\{(X - E\{X\})^2\} = E\{X^2\} - E\{2XE\{X\}\} + E\{X\}^2$. But $E\{2XE\{X\}\} = 2E\{X\}E\{X\} = 2E\{X\}^2$ so the above equals $E\{X^2\} - E\{X\}^2$, as claimed. □

Proposition 2.32. *Let X be a random variable with mean μ and variance σ^2 Let a and b be real numbers. Then*

$$\text{Var}(aX + b) = a^2 \text{Var}(X).$$

2.6. Moments, Means and Variances

Proof. $E\{(aX + b - E\{aX + b\})^2\} = E\{(aX - aE\{X\})^2\} = E\{a^2(X - E\{X\})^2\} = a^2 \text{Var}(X)$. □

Definition 2.33. Let X be a random variable and $k \geq 1$ an integer. Then $E\{|X|^k\}$ is the kth **absolute moment** of X. If the kth absolute moment of X is finite, $E\{X^k\}$ is the kth **moment**, and $E\{(X - E\{X\})^k\}$ is the kth **central moment** of X.

The first moment ($k = 1$) is the expectation, or, as it is often called, the **mean** of the random variable. The second central moment is called the variance.

Example 2.33.1. The existence of moments is equivalent to the existence of absolute moments, but it is by no means automatic that either exists. Let k be a non-negative integer and set $p_k(n) = C_k/n^{k+2}$, $n = 1, 2, \ldots$ where C_k is chosen to make the p_k sum to one. Let X_k be a random variable with probability mass function (p_k). Then:

- X_0 has no mean.
- X_1 has a mean but no variance.
- X_2 has a mean and variance, but no higher moments.
- In general, X_k has moments of order k or less, but the moments of order $k + 1$ or more do not exist.

Indeed, if $m \geq 0$ is an integer, $E\{|X_k|^m\} = \sum_{n=1}^{\infty} C_k/n^{k-m+2}$. This converges if $k - m + 2 > 1$ and diverges otherwise, so that the mth moment exists if $m \leq k$, and fails to exist otherwise.

Proposition 2.34. *Let $k \geq 0$ be an integer. If random variables X and Y both have k^{th} moments, so does $X + Y$.*

Proof. It is enough to show the absolute moments exist. Note that for any reals x and y, $x \leq \max(x, y)$ and $y \leq \max(x, y)$, so that $|x + y|^k \leq (2\max(|x|, |y|))^k \leq 2^k|x|^k + 2^k|y|^k$. Thus $E\{|X + Y|^k\} \leq 2^k E\{|X|^k\} + 2^k E\{|Y|^k\} < \infty$. □

Proposition 2.35. *Let X be a random variable and let $k \geq 1$ be an integer. Suppose that the kth absolute moment of X exists. Then the jth moment, the jth absolute moment, and the jth central moment all exist for each integer $0 \leq j \leq k$.*

Proof. If $j \leq k$ and $x \geq 1$, then $x^j \leq x^k$. If $0 \leq x \leq 1$, then $x^j \leq 1$. Thus $|X|^j \leq |X|^k + 1$, so $E\{|X|^j\} \leq E\{|X|^k\} + 1 < \infty$.

Since a constant random variable automatically has all moments, Proposition 2.34 implies that if X has a jth moment, so does $X - E\{X\}$, which means that X has a jth central moment as well. □

Notice that "X has a kth moment," "X has a kth absolute moment" and "X has a kth central moment" are all synonyms for "X^k is integrable."

Definition 2.36. The **moment generating function** $M(\theta)$ of a random variable X is

$$M(\theta) = E\{e^{\theta X}\}.$$

Remark 2.37. (i) The moment generating function is closely related to the probability generating function $G(t)$: for positive integer-valued random variables, $M(\theta) = G(e^\theta)$.

(ii) The moment generating function is defined for all θ for which $e^{\theta X}$ is integrable.

(iii) $M(0) = 1$ always exists.

(iv) Let X_k be the random variable defined in Example 2.33.1. Then the moment generating function for X does not exist for $\theta > 0$, since, as $e^{n\theta}/n^k$ goes to infinity as $n \to \infty$, the series $M(\theta) = C_k \sum_n e^{n\theta}/n^{k+2}$ diverges.

(v) If $M(\theta_0)$ exists for some $\theta_0 > 0$, then $M(\theta)$ exists for all $0 \leq \theta \leq \theta_0$.

(vi) If $M(\theta_1)$ exists for some $\theta_1 < 0$, then $M(\theta)$ exists for all $\theta_1 \leq \theta \leq 0$.

Indeed, (iii) is clear since $e^0 = 1$, and (iv) is proved above. For (v), note that if $0 \leq \theta \leq \theta_0$, $e^{\theta X} \leq 1$ on $\{X < 0\}$, and $e^{\theta X} \leq e^{\theta_0 X}$ on $\{X \geq 0\}$. Thus $0 \leq e^{\theta X} \leq 1 + e^{\theta_0 X}$, and $E\{e^{\theta X}\} \leq 1 + E\{e^{\theta_1 X}\} = 1 + M(\theta_1) < \infty$.

(vi) holds for the same reason. (Just replace X by $-X$.)

The key result about moment-generating functions is this: they generate the moments.

Theorem 2.38. *Suppose $M(\theta)$ exists for θ in a neighborhood of the origin. Then:*

(i) X^n *is integrable for all* $n = 0, 1, 2, \ldots$.

(ii) $E\{X^n\} = \dfrac{d^n M(\theta)}{d\theta^n}\bigg|_{\theta=0}$.

(iii) $M(\theta) = \sum_{n=0}^{\infty} \dfrac{\theta^n}{n!} E\{X^n\}$.

Proof. (i) Choose $\theta_0 > 0$ so that both $M(\theta_0)$ and $M(-\theta_0)$ exist. If $x > 0$, $\frac{x^n \theta_0^n}{n!} \leq e^{\theta_0 x}$ so, in general,

$$(2.7) \qquad E\{|X|^n\} \leq \frac{n!}{\theta_0^n} E\{e^{\theta_0 X} + e^{-\theta_0 X}\}$$

$$= \frac{n!}{\theta_0^n} \big(M(\theta_0) + M(-\theta_0)\big) < \infty.$$

(ii) Suppose $|\theta| < \theta_0$. Then $\frac{dM(\theta)}{dx} = \frac{d}{dx} E\{e^{\theta X}\}$. Now *if* we can interchange expectation and differentiation, this is $= E\{\frac{d}{dx} e^{\theta X}\} = E\{Xe^{\theta X}\}$. Therefore, $M'(0) = E\{X\}$. Differentiating further, we get $M^{(k)}(\theta) = E\{X^k e^{\theta X}\}$, and (ii) follows by setting $\theta = 0$.

(iii) If $|\theta| < \theta_0$, by (2.7), $\sum_{n=0}^{\infty} \frac{\theta^n}{n!} E\{|X|^n\} \leq \sum_{n=0}^{\infty} \big(\frac{\theta}{\theta_0}\big)^n < \infty$. Thus the series in (iii) converges absolutely for $|\theta| < \theta_0$.

But $M(\theta) = E\{e^{\theta X}\} = E\{\lim_{N \to \infty} \sum_{n=1}^{N} \frac{\theta^n}{n!} X^n\}$. Once again *if* we can interchange limit and expectation, this equals $\lim_{N \to \infty} \sum_{n=0}^{N} \frac{\theta^n}{n!} E\{X^n\}$, which verifies (iii). (The interchanges in (ii) and (iii) are not obvious. We will prove some limit theorems later which justify them. See Proposition 4.21 for (ii) and and Exercise 4.30 for (iii).) □

2.7. Mean, Median, and Mode

What is the average value of a distribution? Or a typical value? Or the most likely value? Questions like these seem simple and straightforward, but they can generate an amazing amount of controversy. For they have several possible answers. The usual choices are the mean, the median, and the mode, respectively. In nice cases, they are all nearly equal, but there are distributions for which they are strikingly different.

The **mean**, or average value, of the random variable X is $E\{X\}$. The median is the mid-point of the distribution: for a continuous distribution, X will have probability $1/2$ of being greater than the median, and probability $1/2$ of being less. The mode is the value which has the greatest probability. More exactly,

Definition 2.39. The value x is a **median** of the distribution of a random variable X if $P\{X \leq x\} \geq 1/2$ and $P\{X \geq x\} \geq 1/2$.

Definition 2.40. If X has a discrete distribution, then x is a **mode** of the distribution if $P\{X = x\} \geq P\{Y = y\}$ for all other values y.

Neither the median nor the mode is necessarily unique. For instance, if X equals ± 1 with probability $1/2$ each, then the mean is zero, but both -1 and 1 are modes, and *every* point in the closed interval $[-1, 1]$ is a median. But often, particularly with nice, symmetric distributions, the mean, median, and mode are all close, and indicate a "typical value" of the distribution.

Still...take care. The three can be quite different. The mean is sensitive to exceptionally large values; the median is much less so. This is particularly true for things like the distribution of income. For instance, if Bill Gates, the CEO of Microsoft, were to board a bus in Redmond, Washington, the average annual income on the bus would go up sharply up when he boarded, and go down just as sharply when he left, but the median annual income would probably not change much at all.

Problems 2.7

2.38. Toss a coin n times. Let $X_n = $ #heads $-$ #tails. If the probability of heads is p, find the distribution and expectation of X_n.

2.39. Consider the Bernoulli (p) distribution. Show that if $p > 1/2$, the median is 1; if $p < 1/2$, the median is zero, and if $p = 1/2$, every $0 \leq x \leq 1$ is a median.

2.40. A fair coin is tossed independently until the nth head appears. Find the moment generating function for the total number of tosses, and find the variance of the total number of tosses.
[Hint: It might help to treat the case $n = 1$ first.]

2.41. Pick n independent points at random in $[0, 1]$. Find the expected value of their maximum, their minimum, and their range (= maximum $-$ minimum).

2.42. Let $X_1, X_2, \ldots, X_n$ be i.i.d. random variables with (unknown) mean μ and variance σ^2. We want to estimate μ and σ^2. Show that $\bar{S}_n \stackrel{\text{def}}{=} (X_1 + \cdots + X_n)/n$ has expectation μ and $\frac{1}{n-1}\sum_{j=1}^n (X_j - \bar{S}_n)^2$ has expectation σ^2. (Note: $n-1$, not n.) These are called **unbiased estimates** of μ and σ^2, respectively.

2.43. $A_1, \ldots, A_n$ be independent events in an experiment, with $P\{A_i\} = p_i$, $i = 1, \ldots, n$. Do the experiment and let N be the number of events A_i that occur. Find the moment generating function of N, and use it to compute $E\{N\}$. Use the indicator functions of the A_i to give a second, simple derivation of the value of $E\{N\}$.

2.44. Find the expected number of spades in a bridge hand of 13 cards. Find the expected number of aces in the same hand.

2.45. n players each roll a single die. Any pair of them who roll the same number score one point. Let A_{ij} be the event on which player i and player j roll the same number.

(a) Show that A_{12} and A_{23} are independent.

(b) Find the mean and variance of the total score of the group.

2.46. A city bus in Redmond, Washington contained 29 passengers. A survey revealed that their average yearly income was $40,000, they had a unique median income of $38,000. (a) How large could the maximum income be?

(b) The mode of their yearly income was also $38,000, for two passengers earned exactly that, and one passenger earned just seventeen dollars more, $38,017. Then

2.8. Special Discrete Distributions

Bill Gates boarded the bus. His income that year was $30,000,000. What changes in the mean, median, and mode did his boarding cause?

2.47. The Fahrenheit-Celsius conversion formula is $F = \frac{9}{5}C + 32$. Suppose the temperature measured in Celsius has mean μ and variance σ^2. What are the mean and variance of the temperature in Fahrenheit degrees?

2.48. Show that the set of medians is either a single point or a finite non-empty closed interval. In particular, show that there always exists at least one median. [Hint: First show x is a median iff $F(x) \geq 1/2$ and $1 - F(x-) \geq 1/2$.]

2.49. Let $x_1, \ldots, x_n$ be a sequence of numbers, where n is odd. Let m be a median. (a) Show that there must be at least one j such that $x_j = m$, and the median is unique. (b) Suppose one more number, x_{n+1}, is added to the sequence. Show that m is still a median of the new sequence. Give an example in which the new median is not unique.

2.50. Let $X_1, X_2, X_3, \ldots$ be i.i.d. positive-integer-valued random variables. Show that
$$E\{\min\{X_1, \ldots, X_m\}\} = \sum_{n=1}^{\infty} P\{X_1 \geq n\}^m.$$

2.8. Special Discrete Distributions

Certain distributions occur often enough in nature to have their own names. They are often easily recognized.

a. A random variable X has a **Bernoulli** distribution if
$$P\{X = 1\} = p \quad \text{and} \quad P\{X = 0\} = 1 - p \stackrel{\text{def}}{=} q$$
for some $p \geq 0$.

These are sometimes called *coin tossing* random variables, since they only take on two values, 0 and 1. For example, if A is an event, then I_A is a Bernoulli random variable.

For a Bernoulli (p) random variable, $E\{X\} = E\{X^2\} = p$. Therefore $\text{Var}(X) = pq$. The moment generating function is $M(\theta) = q + pe^\theta$. For a quick check, note that $M'(0) = p$ and $M''(0) = p$ are the first two moments.

If $p > 1/2$, the mode and median both equal one; if $p < 1/2$, the mode and median both equal zero. If $p = 1/2$, both zero and one are modes, and all numbers between zero and one, inclusive, are medians.

b. X has a **binomial** distribution with parameters n and p if
$$P\{X = k\} = \binom{n}{k} p^k q^{n-k}, \qquad k = 0, 1, 2, \ldots$$
where $q = 1 - p$. This distribution is denoted $B(n, p)$.

Let us first calculate the moment generating function.

$$M(\theta) = \sum_{k=0}^{n} \binom{n}{k} p^k q^{n-k} e^{k\theta}$$

$$= \sum_{k=0}^{n} \binom{n}{k} (pe^\theta)^k q^{n-k}$$

$$= (pe^\theta + q)^n$$

from the Binomial Theorem (Example 1.23.5). Differentiate and set $\theta = 0$ to get the first two moments. $M'(0) = n(pe^\theta + q)^{n-1} pe^\theta \big|_{\theta=0} = np$, and $M''(0) = n(n-1)p^2$. Thus $E\{X\} = np$, $E\{X^2\} = n(n-1)p^2$ and Var$(X) = n(n-1)p^2 - np^2 = npq$.

The binomial distribution arises from counting: consider an experiment with probability p of success. Repeat the experiment n times independently, and let X be the number of successes. To find the probability that $X = k$, note that the probability of having k successes in a row, followed by $n - k$ failures in a row is $p^k q^{n-k}$. In fact, the probability of any specific sequence which contains k successes and $n - k$ failures is the same. But there are exactly $\binom{n}{k}$ such sequences. (The sequence is determined by the k spots where the successes are placed, and there are $\binom{n}{k}$ ways to place k among the n spots.) Thus $P\{X = k\} = \binom{n}{k} p^k q^{n-k}$, so X has a $B(n,p)$ distribution.

Note that we can let $X_j = 1$ if the kth trial is a success, and $X_j = 0$ if not. Then $X_1, \ldots, X_n$ are independent Bernoulli (p) random variables, and $X_1 + X_2 + \cdots + X_n$ is the total number of successes. Thus $X = X_1 + \cdots + X_n$. In other words...

Remark 2.41. A $B(n,p)$ random variable has the same distribution as a sum of n independent Bernoulli (p) random variables.

c. The **Poisson distribution with parameter** λ is

$$P\{X = k\} = \frac{\lambda^k}{k!} e^{-\lambda}, \quad k = 0, 1, 2 \ldots.$$

Its moment generating function is:

$$M(\theta) = \sum_{k=0}^{\infty} \frac{\lambda^k}{k!} e^{-\lambda} e^{k\theta}$$

$$= e^{-\lambda} \sum_{k=0}^{\infty} \frac{(\lambda e^\theta)^k}{k!}$$

$$= e^{-\lambda} e^{\lambda e^\theta}$$

$$= e^{\lambda(e^\theta - 1)}.$$

2.8. Special Discrete Distributions

Then $E\{X\} = M'(0) = \lambda e^\theta e^{\lambda(e^\theta-1)}\big|_{\theta=0} = \lambda$. Similarly, $E\{X^2\} = M''(0) = \lambda + \lambda^2$. Thus $\text{Var}(X) = \lambda$. The mean and variance of a Poisson random variable are equal.

It can be shown that both the Poisson and the Binomial distributions have medians and modes which are close to their means.

There is an important connection between the Poisson and binomial distributions. The Poisson distribution is a limiting case of the binomial $B(n,p)$, when $n \to \infty$ and $np \to \lambda$. This means that the Poisson is a good approximation to the binomial $B(n,p)$ when n is large and p is small. (See Theorem 2.42.) So the Poisson distribution comes up quite often, for example, in things involving large populations and fairly unlikely events. For instance, the following have Poisson—or very-nearly Poisson—distributions.

- The number of clicks in a ten second period of a Geiger counter held near a weak radiation source.
- The number of telephone calls passing through a given trunk line in a given period.
- The number of times a given internet server is accessed in a fixed time interval.
- The number of left-handed tuba players in Chicago.
- The number of typos in the first draft of a manuscript.
- The number of Prussian cavalrymen killed by horse kicks in 1867.
- The number of V1 rockets landing in a given block of London in one week during the blitz in WW II.

Indeed, the Geiger counter clicks come from decaying atoms. While each individual atom is extremely unlikely to decay in that period, there are so many of them—on the order of 10^{23}—that a few will. The telephone and the internet both have many users, and each individual is fairly unlikely to use a given portion of the network in a small time period, but there are enough users that a few will. There are a lot of citizens in Chicago, and each one is unlikely to be both left-handed and play the tuba—but it is likely that there will be a few. There are many words in a manuscript, and each is likely to be spelled correctly, but mistakes creep in, and it is not surprising that the number should be Poisson. The Poisson nature of typos was reportedly verified by some grad students who checked one of their professor's manuscripts. The deaths of Prussian cavalrymen was famously found to follow a Poisson distribution by Ladislaus Bortkiewicz. And the Poisson distribution of rocket impacts forms a key part of the plot of Gravity's Rainbow, by Thomas Pynchon.

The Poisson limit theorem suggests why the Poisson distribution arises so often in practice: it is the large-sample limit of a binomial distribution.

Theorem 2.42 (Poisson Limits). *Let $n \to \infty$ and $np \to \lambda > 0$. Then, for each k,*
$$\lim_{\substack{n \to \infty \\ np \to \lambda}} \binom{n}{k} p^k (1-p)^{n-k} = \frac{\lambda^k}{k!} e^{-\lambda}.$$

We will leave the proof of this as an exercise, but let us see what happens to the moment-generating function.

The moment generating function of the $B(n,p)$ is $M(\theta) = (pe^\theta + 1 - p)^k$. Set $p = \lambda/n$, let $n \to \infty$, and recall the famous limit: $\lim_{n\to\infty}(1 + \frac{x}{n})^n = e^x$. Then, with $x = \lambda(e^\theta - 1)$,
$$\lim_{n \to \infty} \left(1 + \frac{\lambda(e^\theta - 1)}{n}\right)^n = e^{\lambda(e^\theta - 1)},$$
which is the moment generating function of a Poisson (λ).

d. The geometric distribution.

Do an experiment repeatedly and independently with probability p of success. Let X be the number of times it takes to get the first success. Then X has a geometric distribution with parameter p.

In order for X to equal $k \geq 1$, the experiment must have failed the first $k-1$ times—probability $(1-p)^{k-1}$—and succeeded on the kth—probability p—leading to the probability mass function
$$p(k) \stackrel{\text{def}}{=} P\{X = k\} = (1-p)^{k-1} p, \text{ for } k = 1, 2, 3, \ldots.$$

Remark 2.43. There is a possible confusion in terminology here, but if we are aware of it, it should cause no problems. Let Y be the number of failures *before* the first success. Then Y takes values in $0, 1, 2, \ldots$. Y is also called a geometric random variable. So, when confronted with a geometric random variable, check the set of possible values to see which type is meant. If the possible values are $1, 2, 3, \ldots$, then the variable is X. If they are $0, 1, 2, \ldots$, the variable is Y. The relation between X and Y is so simple: $Y = X - 1$, that it is not worthwhile to give them different names.

The moment generating function is
$$M(\theta) = E\{e^{\theta X}\} = \sum_{k=1}^{\infty} pe^{k\theta}(1-p)^{k-1}.$$

We can sum this geometric series to see that
$$M(\theta) = \frac{pe^\theta}{1 - (1-p)e^\theta}, \text{ if } \theta < \log \frac{1}{1-p}.$$

2.8. Special Discrete Distributions

We get the moments by differentiation: $M'(0) = \frac{1}{p}$, $M''(0) = \frac{2-p}{p^2}$, leading to

$$E\{X\} = \frac{1}{p}, \qquad \text{Var}(X) = \frac{1-p}{p^2}.$$

The mode of the geometric distribution is 1.

The median can be determined directly, noting that the value of the distribution function F at an integer n is $F(k) = p\sum_{k=1}^{n}(1-p)^{k-1} = 1-(1-p)^n$. This leads to the conclusion that n is a median if $(1-p)^{n-1} \geq 1/2 \geq (1-p)^n$. Take logs to see that the median is the first value of n which is greater than $\log(1/2)/\log(1-p)$. (If this ratio happens to be an integer, the median is not unique.)

In this case, the mean, median and mode are not necessarily close together. If $p = 1/7$, for example, the mode is 1, the median is 5, and the mean is 7.

e. The negative binomial distribution. Repeat an experiment independently, over and over. Let X_k be the time of the kth success. Then X_k has a negative binomial distribution. In order that $X_k = n$, there must have been exactly $k-1$ successes in the first $n-1$ tries, and the nth try is a success. This has probability

$$P\{X_k = n\} = \binom{n-1}{k-1} p^k (1-p)^{n-k}, \quad n = k, k+1, \ldots.$$

There are two parameters: k, the number of successes, and p, the probability of success. If $k = 1$, this is a geometric distribution.

f. The hypergeometric distribution. The hypergeometric distribution comes up in choosing a sample from a population.

Consider a population of N objects. Suppose that m of these objects are red and $N-m$ are green, but that they are otherwise indistinguishable. Draw a sample of n objects at random without replacement from these N objects. Let X be the number of red objects in the sample. Note that $X \leq m \wedge n$, and, as the sample can include at most $N-m$ green objects, the number of reds has to be at least $n-(N-m)$. Then X has a hypergeometric distribution with parameters N, n, m.

To find $P\{X = k\}$, note that there are $\binom{N}{n}$ possible samples of n objects chosen from N. Of these, there are $\binom{m}{k}$ ways of choosing k red objects from m, and $\binom{N-m}{n-k}$ ways to choose the $n-k$ green objects from the original $N-m$ green objects. Thus

$$P\{X = k\} = \frac{\binom{m}{k}\binom{N-m}{n-k}}{\binom{N}{n}}, \quad (n+m-N)^+ \leq k \leq m \wedge n.$$

For $n = 1$, the hypergeometric distribution is Bernoulli (m/N). For larger n, it comes up in various ways: the number of black marbles chosen in ten draws from an urn with fifty black and fifty white marbles, the number of defective items found by an inspector sampling from a shipment of a gross with 20 defectives, the number of spades in a bridge hand, the number of aces in a poker hand. The important thing is that the sampling is done without replacement.

Example 2.43.1. A pollster chooses n different individuals at random and asks them which type of cola they prefer. Suppose that the total population is N, and m of them prefer Pepsi to Coke. As the respondents are chosen without replacement—the pollster never asks the same person twice—the total number preferring Pepsi is hypergeometric with parameters N, n, m.

Notice also that, if N is far, far larger than n, and if we sample with replacement, the probability of choosing the same person twice is small, so that there is little difference in sampling with and without replacement. But the reader can verify that sampling with replacement leads to a binomial distribution. Thus for large N the hypergeometric is well-approximated by a binomial $B(n,p)$, with $p = m/N$. When major polls sample from the total population of a city or a country, they always make this approximation. However, if the total population is small—if the pollster were to sample from a typical college math class, for instance—the binomial would be a poor approximation, and the pollster would use the hypergeometric distribution.

Exchangeability★

Before deriving the hypergeometric mean and variance, let us note something about choosing objects at random without replacement and its connection with exchangeability.

Consider a collection of N objects numbered 1 to N, but otherwise identical. Draw them randomly in order and keep track of the numbers as drawn. Let $Y_1, \ldots, Y_N$ be the successive numbers drawn. Because the objects are identical, there should be no preference of one over the other in any draw. The Y_i should be identically distributed. In fact, they are something more. They are exchangeable. Exchangeability is basically a symmetry property for random variables.

Definition 2.44. Random variables $X_1, \ldots, X_N$ are **exchangeable** if, for every permutation σ of $1, 2, \ldots, N$, $(X_{\sigma 1}, \ldots, X_{\sigma N})$, has the same joint distribution as $(X_1, \ldots, X_N)$.

They are called "exchangeable" since they can be exchanged without changing the joint distribution. So for $N = 2$, (X_1, X_2) and (X_2, X_1) have the same distribution. For $N = 3$, (X_1, X_2, X_3) taken in each of the six orders has the same distribution.

2.8. Special Discrete Distributions

The following are easily verified:

(i) if $X_1, \ldots, X_N$ are exchangeable, then the X_i are identically distributed.

(ii) So are the pairs (X_i, X_j) for $i \neq j$. This extends to k-tuples.

(iii) I.i.d. random variables are exchangeable, but the converse is false. Exchangeability has more to do with symmetry than independence. There are exchangeable random variables which are not independent. The Y_i above are examples of this, as we shall see.

Let us return to the experiment of drawing N numbers in order without replacement, where $Y_1, \ldots, Y_N$ are the numbers drawn in order. We claim that the Y_i are exchangeable. We can describe this experiment with the probability space $(\Omega, \mathcal{F}, P)$, where Ω is the set of permutations of the integers from 1 to N, representing the successive numbers drawn, and $\mathcal{F}$ is all subsets of Ω:

$$\Omega = \{\omega : \omega = (\omega_1, \ldots, \omega_N)\}.$$

Each ω_i is an integer from 1 to N, and $Y_k(\omega)$, being the number of the kth chosen, is

$$Y_k(\omega) \stackrel{\text{def}}{=} \omega_k.$$

There are $N!$ points in Ω, one for each permutation of $1, \ldots, N$. To see that the probability measure P is uniform on Ω, note that the probability of choosing a given integer first, say i_1, is $1/N$. Having chosen it, there are $N-1$ numbers left to choose second, each equally likely, so $P\{Y_2 = i_2\}$ is $1/(N-1)$, and so on, so that for any permutation $(i_1, \ldots, i_n)$ of $1, \ldots, N$, we have: $P\{Y_j = i_j, j = 1, \ldots, N\} = P\{\omega : \omega_j = i_j, j = 1 \ldots, N\} = 1/N!$. Equivalently, for each $\omega \in \Omega$, $P\{\omega\} = 1/N!$. $Y_k(\omega) = \omega_k$. Then we have:

Proposition 2.45. *The random variables $Y_1, \ldots, Y_N$ are exchangeable.*

Intuitively, this has to be true. The objects are identical, and so no one order of choice is preferable to another. All orders *have* to be equally likely, and the Y_i just keep track of the order, so we should be able to interchange them: any order should be as probable as any other. But we should check this formally.

Proof. Careful: the following turns in very small logical circles. We have to relate permutations of the Y_k to permutations of the ω_k.

Let σ be a permutation of $1, 2, \ldots, N$. If $Y = (Y_1, \ldots, Y_n)$ and $\omega = (\omega_1, \ldots, \omega_n)$, let $\sigma Y \stackrel{\text{def}}{=} (Y_{\sigma 1}, Y_{\sigma 2}, \ldots, Y_{\sigma N})$. Thus $(\sigma Y)_k = Y_{\sigma k}$. Similarly, let $\sigma\omega = (\omega_{\sigma 1}, \ldots, \omega_{\sigma N})$. Thus $(\sigma Y)_k(\omega) = \omega_{\sigma k}$, which means, in terms of Y, that $\sigma Y(\omega) = \sigma \omega$.

We must show that σY and Y have the same distribution.

Choose $\hat{\omega} \in \Omega$, and calculate the probability that σY equals $\hat{\omega}$.
$$P\{\omega : \underbrace{\sigma Y(\omega)}_{\sigma\omega} = \hat{\omega}\} = P\{\omega : \sigma\omega = \hat{\omega}\}.$$

But σ is invertible, so $\sigma\omega = \hat{\omega} \Rightarrow \omega = \sigma^{-1}\hat{\omega}$, so this is
$$= P\{\sigma^{-1}\hat{\omega}\} = \frac{1}{N!} = P\{Y = \hat{\omega}\}$$
since all outcomes are equally likely. This is true for all σ and all $\hat{\omega}$, so that σY has the same distribution as Y. □

Let us now return to the hypergeometric distribution with parameters N, m, and n, and find its mean and variance. Let $p = m/N$. The mean and variance of the hypergeometric distribution are
$$E\{X\} = np,$$
$$\text{Var}(X) = np(1-p)\frac{N-n}{N-1}.$$

To see why, order the objects so that the first m are red, the rest green. Define $Y_1, Y_2, \ldots$ as above. They are exchangeable. Now the ith draw is a red object if $Y_i \leq m$, so $X = \sum_{i=1}^n I_{\{Y_i \leq m\}}$. Then $E\{X\} = \sum_{i=1}^n E\{I_{\{Y_i \leq m\}}\} = \sum_{i=1}^n P\{Y_i \leq m\}$. But the Y_i are identically distributed, so this equals $nP\{Y_1 \leq m\}$, that is, n times the probability that the first draw is red. But when we start, m out of the N objects are red, so the probability that the first one picked is red is m/N. Thus $E\{X\} = nm/N = np$.

To find the variance, first calculate $E\{X^2\}$.
$$E\{X^2\} = E\left\{\left(\sum_{i=1}^n I_{\{Y_i \leq m\}}\right)^2\right\}$$
$$= \sum_{i=1}^n \sum_{j=1}^n E\{I_{\{Y_i \leq m\}} I_{\{Y_j \leq m\}}\}$$
$$= \sum_{i=1}^n \sum_{j=1}^n P\{Y_i \leq m, Y_j \leq m\}$$
$$= \sum_{i=1}^n P\{Y_i \leq m\} + \sum_{i \neq j}^n P\{Y_i \leq m, Y_j \leq m\}.$$

We just saw that the first sum equals np. There are $n(n-1)$ terms in the second sum. Since the Y_i are exchangeable, (Y_i, Y_j) has the same distribution as (Y_1, Y_2). But we can calculate this: the first object is red with probability m/N. If the first is red, there remain $N-1$ objects of which $m-1$ are red, so the probability that both are red is $m/N \times (m-1)/(N-1)$. Thus

2.8. Special Discrete Distributions

$E\{X^2\} = np + np(m-1)/(N-1)$. Thus

$$\mathrm{Var}(X) = E\{X^2\} - E\{X\}^2 = np + n(n-1)p\frac{m-1}{N-1} - n^2p^2\,;$$

but $m = Np$, so this is

$$= np(1-p)\frac{N-n}{N-1}\,.$$

Problems 2.8

2.51. It is said that many are called and few are chosen. Suppose that the number called is Poisson (λ). Each person called is chosen, independently, with probability p. Show that the distribution of the number chosen is Poisson $(p\lambda)$.

2.52. Let Let X and N be random variables. Suppose that N is binomial $B(M,q)$ and, given N, X is $B(N,p)$. Show X is $B(M,pq)$. Do this both analytically and probabilistically.

2.53. Prove Theorem 2.42 by direct calculation.

2.54. Toss a fair coin 10 times. Let N be the number of times the pattern $HHHH$ occurs. (For example, in the sequence $HHHHHHT$ it appears three times, once each for positions 1, 2, and 3.) Find $E\{N\}$.

2.55. The probability that a typist makes an error on a given character is $1/1000$. Suppose a page contains 2000 characters, and let N be the number of errors.

(a) What is the exact distribution of N? What assumption are you making?

(b) Find an approximate expression for $P\{N \geq 2\}$.

(c) Suppose the page must be retyped if $N \geq 2$. Let X be the number of sheets of paper used to make one good page. Find the distribution of X. What is $E\{X\}$?

2.56. N people live in a neighborhood. Their mean income is m and the mean of the *squares* of their income is S^2. Of these, n are chosen at random. Find the mean and variance of the total income of these n people.

[Hint: The individual incomes are fixed. The only randomness is in the choice of the n people.]

2.57. There are N chips, numbered $1, \ldots, N$. Choose n of them at random without replacement. Let X be the largest number among them. Find the distribution and expectation of X.

2.58. Let $(\Omega, \mathcal{F}, P)$ be a probability space. Suppose that each $\Lambda \in \mathcal{F}$ has probability zero or one. Show that if X is a random variable, it is constant a.s.

2.59. In n independent trials, the probability of success is p, the probability of failure is $q = 1 - p$. Find the probability that there are an even number of successes. [Hint: First expand $(q+p)^n$ and $(q-p)^n$ by the binomial theorem.]

2.60. Toss a coin repeatedly and independently. Show that heads eventually appears. Show that any fixed finite sequence such as HTHHHT eventually appears. Use this to show that the Gambler's Ruin eventually ends.

2.61. (Continuation.) Show that if the series of tosses is continued forever, that heads will appear not just once, but infinitely often.

2.62. Find the probability of k heads in n tosses of a fair coin. Use this to show that $\sum_{k=0}^{n} \binom{n}{k} = 2^n$.

2.63. Show that if f is a continuous function on the real line such that for all x and y, $f\big((x+y)/2\big) = \big(f(x) + f(y)\big)/2$, then $f(x) = ax + b$ for some real a and b.
Note: This actually holds for all Borel f, but not necessarily for non-measurable f, where very strange things can happen.
[Hint: First prove it for rationals.]

2.64. If X is a positive random variable with the "memoryless" property, (3.14), then X has an exponential distribution.
[Hint: Reduce it to the previous problem.]

2.65. A jar contains n chips, numbered from 1 to n. A chip is drawn at random, its number is noted, and it is replaced in the jar. This continues until a chip is drawn for the second time. Let X be the number of the draw on which this happens. Find the distribution of X.

2.66. Fred matches coins five times with fair coins. What is the probability that he ends up with exactly one more coin than he started with? (If both coins show the same face, he gets both. If they show different faces, he loses his.)

2.67. n couples attend a dance. Each wife has a dance partner chosen at random from the n husbands. Let E_i be the event that wife i dances with her own husband, and let p_n be the probability that none of the n couples dance together.

(a) Find p_2 and p_3 by expressing them in terms of the E_i.

(b) Find $\lim_{n \to \infty} p_n$.

[Hint: Calculate the probability that the jth wife dances with her husband, that both the jth and the kth wives do, and so on. The inclusion-exclusion formula gives an answer which links this to the Taylor's expansion of a certain well-known function.]

Chapter 3

Expectations II: The General Case

3.1. From Discrete to Continuous

We have defined the expectation of discrete random variables. We will extend it to the general case by approximating an arbitrary random variable above and below by discrete random variables[1].

Definition 3.1. Let X be a random variable. Its **upper and lower dyadic approximations** $\bar{X}_n$ and $\underline{X}_n$, $n = 0, 1, \ldots$ are

$$\bar{X}_n(\omega) = \frac{k+1}{2^n} \quad \text{if } k2^{-n} < X(\omega) \leq (k+1)2^{-n},$$

$$\underline{X}_n(\omega) = \frac{k}{2^n} \quad \text{if } k2^{-n} < X(\omega) \leq (k+1)2^{-n}.$$

If ω is fixed, then $(\bar{X}_n(\omega))$ is just a sequence of real numbers.

The following three statements are clear from the definition.

(1°) Both $\underline{X}_n$ and $\bar{X}_n$ are discrete random variables.
(2°) $\underline{X}_n(\omega) < X(\omega) \leq \bar{X}_n(\omega)$.
(3°) $\underline{X}_n(\omega) = \bar{X}_n(\omega) - 2^{-n}$.

The next four take a little more thought:

(4°) $\underline{X}_n(\omega) \leq \underline{X}_{n+1}(\omega) \leq \bar{X}_{n+1}(\omega) \leq \bar{X}_n(\omega)$.

---

[1]Our construction of the expectation follows that of the Lebesgue integral, albeit in different language. See Royden [**31**] Chapters 4 and 11 for an account of the classical and general Lebesgue integral respectively. The principle difficulty is to show that Lebesgue measure is countably additive, something we assumed. See [**31**] Chapter 3 for a proof.

(5°) $\lim_{n\to\infty} \underline{X}_n(\omega) = \lim_{n\to\infty} \bar{X}_n(\omega) = X(\omega)$, and the limits are monotone.

(6°) The $\bar{X}_n$ and $\underline{X}_n$ are either all integrable, or all non-integrable.

(7°) If one (and therefore all) of the discrete approximations are integrable,

$$E\{\underline{X}_0\} \leq E\{\underline{X}_1\} \leq E\{\underline{X}_2\} \leq \cdots \leq E\{\bar{X}_2\} \leq E\{\bar{X}_1\} \leq E\{\bar{X}_0\},$$
$$E\{\bar{X}_n\} - E\{\underline{X}_n\} = 2^{-n}.$$

To see 4°, note that if, say, $X(\omega) \in (k2^{-n}, (k+1)2^{-n}]$ that $\bar{X}_n(\omega) = (k+1)2^{-n}$, and the only two possibilities for $\bar{X}_{n+1}(\omega)$ are $(k+1)2^{-n}$ and $(k+1/2)2^{-n}$, depending on whether $X(\omega) \in ((k+1/2)2^{-n}, (k+1)2^{-n}]$ or $X(\omega) \in (k2^{-n}, (k+1/2)2^{-n}]$, respectively. In either case, $\bar{X}_{n+1} \leq \bar{X}_n$. The same reasoning shows that $\underline{X}_n(\omega) \leq \underline{X}_{n+1}(\omega)$.

For each ω the sequences $\bar{X}_n(\omega)$ and $\underline{X}_n(\omega)$ are monotone (by 4°) and bounded, so that they have limits. The limits are equal by 3°. Finally, 6° follows since all the $\bar{X}_n$ and $\underline{X}_n$ differ by at most one. The last conclusion is now immediate from 4° and 3°.

We have defined integrability for discrete random variables. Let us extend it to general random variables.

Definition 3.2. X is **integrable** if $\bar{X}_0$ is integrable.

The following definition is now inevitable.

Definition 3.3. If X is integrable, then

$$E\{X\} = \lim_{n\to\infty} E\{\bar{X}_n\} = \lim_{n\to\infty} E\{\underline{X}_n\}.$$

If X is discrete, its expectation has already been defined by (2.5). To see that the new definition gives the same value, note that the $\bar{X}_n$ and $\underline{X}_n$ are all between $X-1$ and $X+1$, so they will be integrable if and only if X is. If X is integrable with expectation L, say, then $E\{\underline{X}_n\} \leq L \leq E\{\bar{X}_n\} = E\{\underline{X}_n\} + 2^{-n}$. It follows that $\lim_n E\{\bar{X}_n\} = L$, so the two definitions of expectation are indeed equal.

Proposition 2.28 gave the elementary properties of the expectation of discrete random variables. But the expectation of any random variable is the limit of expectations of discrete random variables, so we can essentially take limits in Proposition 2.28. However, the proof is not quite immediate, since we must deal directly with the dyadic approximations.

Before stating the theorem, notice that, as $|\bar{X}_n - X| \leq 2^{-n}$,

(3.1) $$|\bar{X}_n| - 2^{-n} \leq |X| \leq |\bar{X}_n| + 2^{-n}.$$

Theorem 3.4. *Let X and Y be random variables and let $a \in \mathbb{R}$. Then*

(i) If $X = Y$ a.s., then Y is integrable iff X is, and, if so, $E\{X\} = E\{Y\}$.

(ii) If $|X| \leq |Y|$ a.s. and Y is integrable, then X is integrable. In particular, X is integrable if and only if $|X|$ is.

(iii) If X is integrable, so is aX, and $E\{aX\} = aE\{X\}$.

(iv) If X and Y are integrable, so is $X + Y$, and $E\{X + Y\} = E\{X\} + E\{Y\}$.

(v) If $X \geq 0$ a.s. and X is integrable, then $E\{X\} \geq 0$.

(vi) If X and Y are integrable and $X \leq Y$ a.s., then $E\{X\} \leq E\{Y\}$.

Proof. These conclusions are all known in the discrete case. We can *almost* just say that they all follow by taking limits. There is a slight technical difficulty: integrability is defined in terms of dyadic approximations of X, so we have to work with them. Notice that once we have shown (i), we can ignore the "a.s."s in the succeeding parts.

(i) If $X = Y$ a.s., then $\bar{X}_n = \bar{Y}_n$ a.s., and $E\{|\bar{X}_n|\} = \sum_j |j| 2^{-n} P\{\bar{X}_n = j2^{-n}\} = \sum_j |j| 2^{-n} P\{\bar{Y}_n = j2^{-n}\} = E\{|\bar{Y}_n|\}$. Thus $\bar{X}_n$, and hence X, is integrable iff $\bar{Y}_n$, and hence Y, is. If so, the same calculation without absolute value signs shows that $E\{\bar{X}_n\} = E\{\bar{Y}_n\}$. The result follows by letting $n \to \infty$.

(ii) Apply (3.1) with $n = 0$ to X and Y: $|\bar{X}_0| - 1 \leq |X| \leq |Y| \leq |\bar{Y}_0| + 1$. The right side is integrable by hypothesis. By Proposition 2.28, the discrete random variable $\bar{X}_0$ is also integrable. Then X is integrable by definition.

(iii) $|aX| = |a||X| \leq |a|(|\bar{X}_0| + 1)$. Since the right-hand side is integrable, so is $|aX|$. To see the integrals are equal, let $Z = aX$. Then

$$|\bar{Z}_n - a\bar{X}_n| \leq |\bar{Z}_n - Z| + |Z - aX| + |aX - a\bar{X}_n| \leq 2^{-n} + 0 + |a|2^{-n}.$$

Thus $|E\{\bar{Z}_n\} - aE\{\bar{X}_n\}| \leq (|a| + 1)2^{-n} \to 0$. It follows that $E\{aX\} = \lim_n E\{\bar{Z}_n\} = a \lim_n E\{\bar{X}_n\} = aE\{X\}$.

(iv) Let $Z = X + Y$, and apply (3.1) with $n = 0$:

$$|\bar{Z}_0| \leq |Z| + 1 \leq |X| + |Y| + 1 \leq |\bar{X}_0| + |\bar{Y}_0| + 3.$$

The right-hand side is integrable since X and Y are, so $|\bar{Z}_0|$, and hence Z, is integrable.

To check equality of the integrals, write the difference $|\bar{Z}_n - (\bar{X}_n + \bar{Y}_n)| \leq |Z_n - Z| + |Z - (X+Y)| + |X - \bar{X}_n| + |Y - \bar{Y}_n| \leq 3/2^n$. This goes to zero, so that the expectation of Z_n has the same limit as the sum of the expectations of $\bar{X}_n$ and $\bar{Y}_n$, which gives the desired equality.

(v) This is clear since $X \geq 0 \implies \bar{X}_n \geq 0$.

(vi) This follows from (iv), (iii), and the fact that $Y - X \geq 0$. □

If X and Y are integrable, it does not follow that their product XY is. (If, for instance, $Y \equiv X$, $XY = X^2$, and the integrability of X does not imply integrability of X^2.) However, if they are independent, it is true. Moreover, we have:

Theorem 3.5. *Let X and Y be independent random variables. If both X and Y are integrable, so is XY, and*

(3.2) $$E\{XY\} = E\{X\}\,E\{Y\}.$$

Proof. Suppose first that X and Y are discrete with possible values (x_i) and (y_j), respectively. Then

$$E\{|XY|\} = \sum_{i,j} |x_i|\,|y_j|P\{X = x_i,\ Y = y_j\}$$

$$= \sum_{i,j} |x_i|\,|y_j|P\{X = x_i\}P\{Y = y_j\}$$

$$= \sum_i |x_i|P\{X = x_i\} \sum_j |y_j|P\{Y = y_j\}$$

$$= E\{|X|\}\,E\{|Y|\},$$

where the change of order in the summation is justified since the summands are positive. Both X and Y are integrable, so this is finite. Thus XY is integrable. Now remove the absolute value signs. The series converges absolutely, so that the terms can be rearranged, and the same calculation (sans absolute value signs) gives (3.2).

Next, if X and Y are integrable, so are the dyadic approximations $\bar{X}_n$ and $\bar{Y}_n$, and, as $\bar{X}_n - X$ and $\bar{Y}_n - Y$ are both positive and less than 2^{-n},

$$E\{|\bar{X}_n \bar{Y}_n - XY|\} \leq E\{|\bar{X}_n|\,|\bar{Y}_n - Y|\} + E\{|Y|\,|\bar{X}_n - X)|\}$$
$$\leq 2^{-n}\bigl(E\{|\bar{X}_n|\} + E\{|Y|\}\bigr)$$
$$\leq 2^{-n}\bigl(2 + E\{|\bar{X}_0|\} + E\{|\bar{Y}_0|\}\bigr) \to 0$$

where we have used the facts that $|\bar{X}_n| \leq 1 + |\bar{X}_0|$ and $|Y| \leq 1 + |\bar{Y}_0|$. In particular, since $\bar{X}_n\bar{Y}_n$ is integrable, so is XY. Therefore

$$E\{XY\} = \lim_n E\{\bar{X}_n \bar{Y}_n\} = \lim_n E\{\bar{X}_n\}\,E\{\bar{Y}_n\} = E\{X\}\,E\{Y\}.$$

□

Corollary 3.6. *If X and Y are independent random variables with finite variances, then*

$$\mathrm{Var}\,(X + Y) = \mathrm{Var}\,(X) + \mathrm{Var}\,(Y).$$

Proof. $E\{(X+Y)^2\} - E\{X+Y\}^2 = E\{X^2\} + 2E\{XY\} + E\{Y\}^2 - (E\{X\} + E\{Y\})^2 = E\{X^2\} - E\{X\}^2 + E\{Y^2\} - E\{Y\}^2 = \text{Var}(X) + \text{Var}(Y)$. □

Example 3.6.1. Generating Functions. We have seen probability generating functions and moment generating functions, and we shall soon see a third type, characteristic functions. Let X be a random variable. The general form of these generating functions is this:

$$(3.3) \qquad G(t) = E\{e^{g(t)X}\}.$$

For the moment generating function, $g(t) = t$, for the probability generating function, $g(t) = \log t$, and for the characteristic function, $g(t) = it$. They have several things in common: they are power series in t whose coefficients have important connections with the distribution of X. Perhaps more important is that, leaving existence questions aside, they actually determine the distribution of X. The third common factor, which we will establish here, is that they work well with sums of independent random variables.

Proposition 3.7. *Let $X_1, X_2, \ldots, X_n$ be independent random variables with generating functions $G_{X_1}(t), G_{X_2}(t), \ldots, G_{X_n}(t)$. Suppose these exist at t. Then the generating function for the sum $X_1 + \cdots + X_n$ exists at t, and*

$$G_{X_1 + \cdots + X_n}(t) = \prod_{i=1}^{n} G_{X_i}(t).$$

Proof. $E\{e^{g(t)(X_1 + \cdots + X_n)}\} = E\{\prod_{i=1}^{n} e^{g(t)X_i}\}$. The random variables $e^{g(t)X_i}$ are independent, since the X_i are, and, by hypothesis, they are integrable. By Theorem 3.5 their product is also integrable, and its expectation is the product of the expectations, so $G_{X_1 + \cdots + X_n}(t) = \prod_{i=1}^{n} E\{e^{g(t)X_i}\} = \prod_{i=1}^{n} G_{X_i}(t)$, as claimed. □

Example 3.7.1. Application to Analysis. Consider the probability space

$$([0,1], \mathcal{B}, P),$$

where $\mathcal{B}$ is the Borel sets and P is Lebesgue measure: $P(dx) = dx$.

Let g be a continuous function on $[0,1]$. Then g is a random variable! (In fact, as we saw in Exercise 2.2, any Borel function of a random variable is a random variable, and continuous functions are Borel.) It is bounded, so it is integrable and $E\{g\}$ exists. Let us compare the expectation with something familiar, the Riemann integral $\int_0^1 g(x)\,dx$. We claim that $\int_0^1 g(x)\,dx = E\{g\}$. To see why, write the integral as a limit of Riemann sums.

$$(3.4) \qquad \int_0^1 g(x)\,dx = \lim_{n \to \infty} \sum_{k=1}^{n} g(k/n) \frac{1}{n}.$$

Let $A_0 = [0, 1/n]$ and set $A_k = (k/n, (k+1)/n]$ for $k \geq 1$. Define $g_n(x) = g(k/n)$ if $x \in A_k$. In other words,

$$g_n(x) = \sum_{k=0}^{n-1} g(k/n) I_{A_k}(x).$$

Thus g_n is not only a step-function, it is a discrete random variable as well. Moreover, g_n converges uniformly to g, and its expectation is

$$E\{g_n\} = \sum_{n=0}^{n-1} g(k/n) P\{A_n\} = \sum_{n=0}^{n-1} g(k/n) \frac{1}{n}.$$

But this is the Riemann sum in (3.4)! The convergence is uniform so $E\{g\} = \lim_n E\{g_n\}$. Moreover, the Riemann integral is the limit of its Riemann sums, which proves the following proposition.

Proposition 3.8. *Let g be continuous on $[0, 1]$. Then $\int_0^1 g(x)\,dx = E\{g\}$.*

Note. This works for highly discontinuous g, too, such as the indicator function of the rationals, which, itself, is not even Riemann integrable. Still, we can compute its expectation. Indeed, let (r_n) be an enumeration of the rationals, and put $g(x) = I_\mathbb{Q} = \sum_n I_{\{r_n\}}(x)$. Then $E\{g\} = \sum_n P\{r_n\} = 0$, since, as P is Lebesgue measure, the probability of any singleton is zero. So this is a genuine extension of the Riemann integral. In fact, it **is** the Lebesgue integral.

To make this formal, if g is a Borel function on $[0, 1]$ (and therefore a random variable) we say it is *Lebesgue integrable* if $E\{|g|\} < \infty$, and define $\int_0^1 g(x)\,dx \stackrel{\text{def}}{=} E\{g\}$.

This is almost, but not quite the usual Lebesgue integral, which integrates Lebesgue measurable functions, a larger class than the Borel functions. But we can simply replace the Borel sets $\mathcal{B}$ by the Lebesgue sets $\mathcal{L}$ above. Then Lebesgue measurable functions are random variables on the probability space $([0, 1], \mathcal{L}, P)$, and the expectation gives the classical Lebesgue integral.

The restriction to $[0, 1]$ was purely for ease of exposition. The same idea works for any finite interval $[a, b]$. Indeed, normalized Lebesgue measure $P(dx) = dx/(b-a)$ is a probability measure on $([a, b], \mathcal{B})$, a Borel function g on $[a, b]$ is a random variable on $([a, b], \mathcal{B}, P)$ and, taking the normalization into account, $\int_a^b g(x)\,dx = (b-a)E\{g\}$.

Extended Real Values★. Let X be a positive random variable. We have defined $E\{X\}$ if X is integrable. If X is not integrable, we say $E\{X\} = \infty$.

So far, our random variables have had finite values. When we deal with limits, we will meet random variables with infinite values, and we have to be able to handle them.

Let X be a random variable with values in the extended interval $[0, \infty]$. If $P\{X = \infty\} > 0$, we say $E\{X\} = \infty$. If $P\{X = \infty\} = 0$, define $\hat{X}$ by

$$\hat{X}(\omega) = \begin{cases} X(\omega) & \text{if } X(\omega) < \infty, \\ 0 & \text{if } X(\omega) = \infty. \end{cases}$$

$\hat{X}$ equals X a.e., and it has finite values, so we define

(3.5) $$E\{X\} \stackrel{\text{def}}{=} E\{\hat{X}\}.$$

Intuitively, events of probability zero do not happen, and the values of a random variable on a set of probability zero are irrelevant. In the extreme case where $P\{\Lambda\} = 0$ and $X = \infty \cdot I_\Lambda$, the usual formula gives us $E\{X\} = \infty \cdot P\{\Lambda\} = \infty \cdot 0$. This is indeterminate, but our definition (3.5) says that it equals zero. In effect, we have decreed that infinity times zero is zero! It turns out that this is needed to make the theory work. Consider, for example, the sequence of random variables $X_n = nI_\Lambda$, $n = 1, 2, \ldots$, where $P\{\Lambda\} = 0$. These are finite-valued, so $E\{X_n\} = 0$, and $X_n \uparrow X$. If we wish—and we do[2]—to have $E\{X_n\} \uparrow E\{X\}$, we must have $E\{X\} = 0$.

For a general random variable X, let $X^+ = \max(X, 0)$ and $X^- = \max(-X, 0)$ be the positive and negative parts of X. Both are positive, and $X = X^+ - X^-$. We then define

$$E\{X\} = E\{X^+\} - E\{X^-\},$$

whenever we can make sense of the right-hand side. That is, $\infty - \text{finite} = \infty$, $\text{finite} - \infty = -\infty$, and $\infty - \infty$ is indeterminate. In the last case, we simply say that the integral does not exist, even in the extended sense. In all other cases, we can give $E\{X\}$ a (possibly infinite) value.

3.2. The Expectation as an Integral

The use of the word "integrable" suggests that the expectation is directly connected with an integral in some way. Example 3.7.1 shows that this is true on the unit interval, where the expectation actually gives the Lebesgue integral. But the truth is even more general: the expectation *is* an integral. Indeed, we can write

$$E\{X\} = \int_\Omega X(\omega)\, P(d\omega), \quad \text{or simply} \quad E\{X\} = \int X\, dP,$$

---

[2] See the Monotone Convergence Theorem, Theorem 4.12.

where the right-hand side is a classical Lebesgue integral with respect to the measure P.

This is not the only connection between expectations and integrals. There is a second one, involving the distribution of X, which leads to the general form of the law of the unconscious statistician. Recall that (see Exercise 2.2) if X is a random variable and f is a Borel function, then $f(X)$ is also a random variable. The class of Borel functions includes most functions we can think of. For instance, a continuous function is Borel: if f is continuous, then $\{x : f(x) < \lambda\}$ is open, and therefore a Borel set.

Let X be a random variable on a probability space $(\Omega, \mathcal{F}, P)$, with distribution function F and distribution μ. As we saw in Proposition 2.10, μ is a probability measure on $\mathbb{R}$ and $(\mathbb{R}, \mathcal{B}, \mu)$ is also a probability space, where $\mathcal{B}$ is the σ-field of Borel sets. A random variable on this space is a real-valued function f on $\mathbb{R}$. Technically, to be a random variable, the function f must satisfy one mild condition: $\{x : f(x) < \lambda\} \in \mathcal{B}$ for all x. This means that f must be a Borel function. So, on this probability space, random variables are Borel functions.

This is of more than passing interest, since it means that we have a ready-made integration theory. The integral of f is just its expectation. Let us use E^μ for the expectation on $(\mathbb{R}, \mathcal{B}, \mu)$. Then we define an integral with respect to the distribution function:

$$(3.6) \qquad \int_{-\infty}^{\infty} f(x)\, dF(x) \stackrel{\text{def}}{=} E^\mu\{f\}.$$

This is a Lebesgue-Stieltjes integral. There are other ways of writing it[3]. We use the distribution function F instead of the distribution μ since it leads naturally to the probability density. The point is that we do not have to reconstruct the integral: it is already defined.

To see why the notation $\int dF$ makes sense, consider what happens when we integrate over an interval, say $(a, b]$. Certainly the integral of dF over this should be $F(b) - F(a)$, and it is: to see that, just choose $f = I_{(a,b]}$ and note that $E^\mu\{f\} = \mu((a,b])$; but $\mu((a,b]) = F(b) - F(a)$. This in turn is equal to $\int_a^b dF(x) = \int f(x)\, dx$.

This brings us to the law of the unconscious statistician.

---

[3]T. J. Stieltjes defined this to generalize the Riemann integral in 1894. It was later extended to Borel (even Lebesgue) functions using the ideas of Lebesgue. Other notations for this are $\int_{\mathbb{R}} f(x)\, \mu(dx)$ or simply $\int f\, d\mu$. In this notation, it would be called a Lebesgue integral. While the ideas behind the two forms of integral are different, the integrals themselves are, in the end, the same.

3.2. The Expectation as an Integral

Theorem 3.9. *Let X be a random variable with distribution function F, and let g be a positive Borel function. Then*

$$(3.7) \qquad E\{g(X)\} = \int_{-\infty}^{\infty} g(x)\, dF(x)\,.$$

In particular, $E\{X\} = \int_{-\infty}^{\infty} x\, dF(x)$.

Proof. Consider g as a random variable on $(\mathbb{R}, \mathcal{B}, \mu)$. We must show that the expectation of g—that is, $E^\mu\{g\}$—equals the expectation $E\{g(X)\}$ of $g(X)$ on $(\Omega, \mathcal{F}, P)$. Recall the dyadic approximation of g: $\bar{g}_n(x) = (k+1)2^{-n}$ if $k2^{-n} < g(x) \le (k+1)2^{-n}$. Let $\Lambda_k = \{x : \bar{g}_n(x) = k2^{-n}\} \subset \mathbb{R}$ and $A_k = \{X \in \Lambda_k\} \subset \Omega$. Thus $A_k = \{\bar{g}_n(X) = k2^{-n}\}$, and $P\{A_k\} = \mu(\Lambda_k)$. Then

$$E\{\bar{g}_n(X)\} = E\Big\{\sum_{k=1}^\infty k2^{-n} I_{A_k}\Big\}$$
$$= \sum_{k=1}^\infty k2^{-n} P\{A_k\}$$
$$= \sum_{k=1}^\infty k2^{-n} \mu(\Lambda_k)$$
$$= E^\mu\{\bar{g}_n\}\,.$$

If we let $Z = g(X)$, then $\bar{Z}_n = \bar{g}_n(X)$. Let $n \to \infty$ on both sides; $E\{\bar{g}_n(X)\} \to E\{g(X)\}$ and $E^\mu\{\bar{g}_n\} \to E^\mu\{g\} \equiv \int g(x)\,dF(x)$. $\square$

Remark 3.10. We have assumed g is positive, so that the integrals always have a value, though it might be infinite.

Corollary 3.11. *If X has a density f and if $g \ge 0$ is Borel measurable, then*

$$(3.8) \qquad E\{g(X)\} = \int_{-\infty}^{\infty} g(x) f(x)\, dx\,.$$

In particular, $E\{X\} = \int_{-\infty}^{\infty} x f(x)\, dx$.

Remark 3.12. If f and g are bounded and continuous, the distribution function F is differentiable, and integrating (3.7) by parts gives (3.8).

The general proof of Corollary 3.11 follows from Exercise 3.1 below.

Remark 3.13. If a random variable X has a density f, then by (3.8), its moment generating function is

$$M(\theta) = \int_{-\infty}^{\infty} e^{\theta x} f(x)\, dx\,.$$

Modulo a change of sign of θ, this is the bi-lateral (i.e., 2-sided) Laplace transform of f. More generally, even if there is no density, we can write

$$M(\theta) = \int_{-\infty}^{\infty} e^{\theta x} \, dF(x),$$

so that the moment generating function is the bilateral Laplace transform of the distribution of X.

The importance of these remarks is that there is an inversion formula for the Laplace transform, so that two functions having the same Laplace transform must be equal a.e. Thus, modulo questions of existence[4], the moment generating function determines the distribution.

Problems 3.2

3.1. Let μ be an absolutely continuous probability distribution with distribution function F and density f. For any Borel set Λ and $a < b$, let $\nu(\Lambda \cap [a,b]) = \int_a^b I_\Lambda(x) f(x) \, dx$. Show that this defines a probability measure ν on the Borel sets of $\mathbb{R}$, and that $\nu = \mu$. Conclude that for positive Borel functions g, $\int_a^b g(x) \, dF(x) = \int_a^b g(x) f(x) \, dx$.
[Hint: Let $\mathcal{G}$ be the class of sets on which $\nu = \mu$. Then intervals of the form $(a, b]$ are in $\mathcal{G}$.]

3.2. Let X be a random variable and let $\bar{X}_n$ and $\underline{X}_n$ be the upper and lower dyadic approximations to X. Show that $\bar{X}_n$ is the smallest random variable with values in $2^{-n}\mathbb{Z}$ which is greater than or equal to X, and that $\underline{X}_n$ is the greatest random variable with values in $2^{-n}\mathbb{Z}$ which is strictly smaller than X.

3.3. Let $X \geq 0$ be an integrable random variable. Show that if $E\{X\} = 0$, then $X = 0$ a.e.

3.4. Let X be a positive random variable with distribution function F. Show that

(3.9) $$E\{X\} = \int_0^\infty (1 - F(x)) \, dx.$$

Conclude that if $c > 0$, then $E\{X \wedge c\} = \int_0^c (1 - F(x)) \, dx$.

3.5. (Continuation.) Show that if X is an integrable random variable, then $\lim_{x \to \infty} x P\{|X| > x\} = 0$.
[Hint: Try integration by parts.]

3.6. Let $X \geq 0$ be a random variable. Show that for any $M > 0$ that

$$M \sum_{k=1}^{\infty} P\{X > kM\} \leq E\{X\} \leq M \sum_{k=0}^{\infty} P\{X > kM\}.$$

---

[4]These questions limit its utility, however. In practice, one replaces the Laplace transform by the Fourier transform, where existence is not a problem. This gives what it is called the characteristic function. We will meet it in Section 6.1.

3.3. Some Moment Inequalities

Chebyshev's Inequality.

Let us give several well-known inequalities involving moments. The first one bounds the probability that a random variable takes on large values. This is a very easy, but also very useful result, Chebyshev's inequality[5].

Theorem 3.14 (Chebyshev's Inequality). *Let $p > 0$, $\lambda > 0$, and let X be a random variable. Then*

$$(3.10) \qquad P\{|X| \geq \lambda\} \leq \frac{1}{\lambda^p} E\{|X|^p\}.$$

Proof. $P\{|X| \geq \lambda\} = P\{|X|^p \geq \lambda^p\} = \int_{\{|X|^p \geq \lambda^p\}} 1\, dP$. But $1 \leq |X|^p/\lambda^p$ on the set $\{|X|^p \geq \lambda^p\}$, so this is

$$\leq \int_{\{|X|^p \geq \lambda^p\}} \frac{|X|^p}{\lambda^p}\, dP \leq \int_\Omega \frac{|X|^p}{\lambda^p}\, dP = \frac{1}{\lambda^p} E\{|X|\}.$$

□

If we replace X by $X - E\{X\}$, and take $p = 2$, we get

Corollary 3.15. $P\{|X - E\{X\}| \geq \lambda\} \leq \dfrac{1}{\lambda^2}\text{Var}(X)$.

Let X and Y be square-integrable random variables, which is another way of saying that X^2 and Y^2 are integrable. Here is a useful and venerable inequality, called—depending on the country—the Schwarz inequality, the Cauchy-Schwarz inequality, or the Cauchy-Schwarz-Buniakovski inequality[6].

Theorem 3.16 (Schwarz Inequality). *Suppose X and Y are square integrable. Then*

$$(3.11) \qquad E\{XY\} \leq \sqrt{E\{X^2\}}\sqrt{E\{Y^2\}}.$$

There is equality if and only if either X or Y vanishes a.s., or if $X = \lambda Y$ for some constant λ.

Proof. In spite of Doob's opinion, the proof is rather nice. If either of X or Y vanishes almost surely, (3.11) holds with equality. Suppose neither does and let $\lambda \geq 0$. Note that

$$0 \leq E\{(X - \lambda Y)^2\} = E\{X^2\} - 2\lambda E\{XY\} + E\{Y^2\},$$

---

[5] There are a number of similar inequalities which follow from (3.10) by simple substitutions for X. We call them all Chebyshev's inequality. However, many authors call (3.10) with $p = 1$ Markov's inequality, and reserve "Chebyshev's Inequality" for its corollary.

[6] When presenting this in his classes, J.L. Doob would remark that he didn't know why they were fighting about it, it was trivial anyway.

so
$$2\lambda E\{XY\} \leq E\{X^2\} + \lambda^2 E\{Y^2\}.$$

Divide by 2λ:

(3.12) $$E\{XY\} \leq \frac{1}{2\lambda}E\{X^2\} + \frac{\lambda}{2}E\{Y^2\}.$$

Set $\lambda = \frac{\sqrt{E\{X^2\}}}{\sqrt{E\{Y^2\}}}$ to get (3.11).

If neither X nor Y vanishes a.e., then there is equality iff there is equality in (3.12), which happens iff $E\{(X - \lambda Y)^2\} = 0$, i.e., iff $X = \lambda Y$ a.e. □

3.4. Convex Functions and Jensen's Inequality

Definition 3.17. A function ϕ defined on an open interval (a, b) is **convex** if for each $a < x < y < b$ and each $\lambda \in [0, 1]$,
$$\phi(\lambda x + (1 - \lambda)y) \leq \lambda \phi(x) + (1 - \lambda)\phi(y).$$

The geometrical interpretation of this is that the chord of the graph between x and y lies above the graph. It implies that the graph of ϕ is roughly cup-shaped. Convex functions are sometimes called "concave up", which is a good way to think about them. The negative of a convex function is sometimes called "concave down" or just concave.

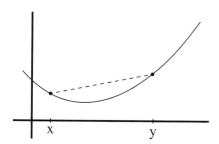

Figure 1. The graph of ϕ lies under the chord.

Functions like $|x|$, x^2, e^x, and e^{-x} are convex on the line, $1/x$ is convex on $(0, \infty)$ and $1/x^2$ is convex on both $(0, \infty)$ and $(-\infty, 0)$ (but not on $(-\infty, \infty)$!) Linear functions are both convex and concave.

Roughly speaking, convex functions generalize functions with positive second derivatives. However, they need not be differentiable: $|x|$ is convex, for instance. But they are *almost* differentiable, as the following shows.

3.4. Convex Functions and Jensen's Inequality

Proposition 3.18. *Let ϕ be convex on (a,b). Then ϕ is continuous and has both left and right-hand derivatives $D_-\phi$ and $D_+\phi$ at each point of (a,b). For each $x \in (a,b)$,*
$$D_-\phi(x) \leq D_+\phi(x).$$
Moreover, both $D_-\phi$ and $D_+\phi$ are increasing functions.

In order to prove this, we need a lemma involving the slopes of chords to the graph.

Lemma 3.19. *Let ϕ be convex on (a,b) and let x, y, x', y' be points of (a,b) with $x \leq x' < y'$ and $x < y \leq y'$. Then the chord over (x', y') has larger slope than the chord over (x, y). That is,*
$$\frac{\phi(y) - \phi(x)}{y - x} \leq \frac{\phi(y') - \phi(x')}{y' - x'}.$$

Proof. Consider three chords:

C_1, the chord from $A = (x, \phi(x))$ to $(y, \phi(y))$;

C_2, the chord from $C = (x', \phi(x'))$ to $D = (y', \phi(y'))$;

C_3, the chord from A to D.

The lemma is proved if we can show that
$$\text{slope}(C_1) \leq \text{slope}(C_3) \leq \text{slope}(C_2).$$

By convexity, the graph of ϕ lies below C_3 in the interval (x, y'), so both chords C_1 and C_2 lie below C_3. Moreover, the left-hand ends of C_1 and C_3 coincide, while the right-hand end of C_1 lies below C_2. This implies (see Figure 2) that the slope of C_1 is less than or equal to that of C_2. Similarly, the right-hand ends of C_2 and C_3 coincide, while the left-hand end of C_1 lies below C_3, implying that the slope of C_2 is greater than that of C_3. □

Now we can prove the proposition.

Proof. Fix $x \in (a,b)$ and choose $h > 0$ small enough that $[x-h, x+h] \subset (a,b)$. By the lemma
$$\frac{\phi(x) - \phi(x-h)}{h} \leq \frac{\phi(x+h) - \phi(x)}{h}.$$
Moreover, the lemma also implies that the left side of this inequality increases as h decreases, while the right side decreases. Clearly the left side is bounded above (by the right side) and the right side is bounded below by the left. Therefore, both sides have limits as h decreases to 0. The limits are the one-sided derivatives, and we have
$$D_-\phi(x) = \lim_{h \downarrow 0} \frac{\phi(x) - \phi(x-h)}{h} \leq \lim_{h \downarrow 0} \frac{\phi(x+h) - \phi(x)}{h} = D_+\phi(x).$$

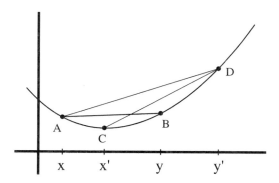

Figure 2. Graph of ϕ with chords $C_1 = \overline{AB}$, $C_2 = \overline{CD}$ and $C_3 = \overline{AD}$.

Thus both left and right-hand derivatives exist. It follows that ϕ is continuous from both the left and right, and therefore continuous. The reader can readily verify from the lemma that both $D_-\phi$ and $D_+\phi$ are increasing functions. □

Notice that the left and right-hand derivatives need not be equal, so that ϕ is not necessarily differentiable everywhere. For example, $|x|$ is convex but not differentiable at zero. However, something almost as good is true. The graph of ϕ has a tangent—sometimes several of them—at each point, and the graph lies above the tangent. Even better, these "tangents" lie entirely below the graph, and are called **supporting lines**.

Proposition 3.20. *Let ϕ be convex on (a,b). Then ϕ has a supporting line at each point $x_0 \in (a,b)$. That is, for each $x_0 \in (a,b)$, there exists m such that for all $x \in (a,b)$, $\phi(x) \geq \phi(x_0) + m(x - x_0)$.*

Proof. Choose m such that $D_-\phi(x_0) \leq m \leq D_+\phi(x_0)$, which we can do by Proposition 3.18. Then if $y \geq x$, $\frac{\phi(y)-\phi(x_0)}{x-x_0} \geq D_+\phi(x_0) \geq m$, so

$$\phi(x) \geq \phi(x_0) + m(x - x_0)$$

Similarly, if $x < x_0$, then $\frac{\phi(x_0)-\phi(x)}{x_0-x} \leq D_-\phi(x_0) \leq m$, so that, again,

$$\phi(x) \geq \phi(x_0) + m(x - x_0)$$

as claimed. □

Theorem 3.21 (Jensen's Inequality). *Let ϕ be a convex function on an open interval (a,b). Suppose X is a random variable whose values are a.s. in (a,b). Suppose that both X and $\phi(X)$ are integrable. Then*

(3.13) $$\phi(E\{X\}) \leq E\{\phi(X)\}.$$

Proof. Let $x_0 = E\{X\}$. This is in (a, b). Let $y = \phi(x_0) + m(x - x_0)$ be a supporting line to the graph of ϕ at x_0. Then $\phi(x_0) + m(X - x_0) \leq \phi(X)$. Take the expectation of both sides to get (3.13). □

Remark 3.22. Jensen's inequality explains some familiar facts: $E\{X\}^2 \leq E\{X^2\}$, for instance, and some less familiar: $1/\{E\{|X|\} \leq E\{1/|X|\}$, $e^{E\{X\}} \leq E\{e^X\}$, and $e^{-E\{X\}} \leq E\{e^{-X}\}$. Providing, of course, the relevant random variables are integrable.

We will use it to prove another set of inequalities which are important in the theory of L^p spaces, called **Lyapounov's Inequalities.**

Corollary 3.23 (Lyapounov's Inequalities). *Let $1 \leq p < q$ and let X be a random variable with finite q^{th} moment. Then*

(i) $E\{|X|\}^q \leq E\{|X|^q\}$;

(ii) $E\{|X|^p\}^{\frac{1}{p}} \leq E\{|X|^q\}^{\frac{1}{q}}$.

Proof. Let $\phi(x) = |x|^q$. Since $q \geq 1$, ϕ is convex, and (i) is Jensen's inequality with $\phi(x) = |x|^q$.

For (ii), apply Jensen's inequality with x replaced by $|x|^p$ and $\phi(x) = |x|^{q/p}$. Since $q/p > 1$, ϕ is convex so

$$E\{|X|^p\}^{\frac{q}{p}} \leq E\{|X|^{p\frac{q}{p}}\} = E\{|X|^q\}\,.$$

Raise both sides to the $1/q^{th}$ power to get (ii). □

3.5. Special Continuous Distributions

a. $U(a, b)$, the **uniform distribution on $[a, b]$**. This has density

$$f(x) = \begin{cases} \frac{1}{b-a} & \text{if } a \leq x \leq b, \\ 0 & \text{otherwise.} \end{cases}$$

The median and mean are equal, and

$$E\{X\} = \frac{1}{b-a} \int_a^b x\,dx = \frac{a+b}{2},$$

$$E\{X^2\} = \frac{1}{b-a} \int_a^b x^2\,dx = \frac{1}{3}(b^2 + ab + a^2),$$

leading to

$$\operatorname{Var}(X) = \frac{(b-a)^2}{12}.$$

b. Exponential (λ). This is defined for $\lambda > 0$ by $P\{X > t\} = e^{-\lambda t}$, $t \geq 0$. It is often used for the first time some event happens. It has the "memoryless property":

(3.14) $$P\{X > s+t \mid X > s\} = P\{X > t\}.$$

This property is called "memoryless" because, if we expect to wait a certain time for an event to happen, and if the event hasn't happened by time s, then we forget the value of s: the additional time we must wait has exactly the same distribution as it did at the beginning.

To see this, write:

$$P\{X > s+t \mid X > s\} = \frac{P\{X > s+t\}}{P\{X > s\}} = \frac{e^{-\lambda(s+t)}}{e^{-\lambda s}} = P\{X > t\}.$$

In fact, (3.14) characterizes the exponential distribution. (See the problems.)

The distribution function is $F(t) = 1 - e^{-\lambda t}$, and the density is $f(t) = dF/dt = \lambda e^{-\lambda t}$.

Its expectation and second moment are

$$E\{X\} = \lambda \int_{-\infty}^{\infty} t e^{-\lambda t}\, dt = \frac{1}{\lambda},$$

$$E\{X^2\} = \lambda \int_{-\infty}^{\infty} t^2 e^{-\lambda t}\, dt = \frac{2}{\lambda^2},$$

so that

$$\text{Var}(X) = \frac{2}{\lambda^2} - \left(\frac{1}{\lambda}\right)^2 = \frac{1}{\lambda^2}.$$

To find the median, set $e^{-\lambda t} = 1/2$. Take logs and solve for t to see that the median is $(\ln 2)/\lambda$, which is less than the expectation.

c. $N(\mu, \sigma^2)$, the **normal or Gaussian distribution** with mean μ and variance σ^2. This often comes up in the following way. Consider a *large* number $Y_1, \ldots, Y_n$ of *small* independent random variables. Then the sum $Y_1 + \cdots + Y_n$ has a nearly normal distribution. This is the content of the Central Limit Theorem (see Section 6.4.)

The normal distribution has two parameters, μ, the mean, and σ^2, the variance. The $N(\mu, \sigma^2)$ density is

(3.15) $$f(x) = \frac{1}{\sqrt{2\pi\sigma^2}} e^{-\frac{(x-\mu)^2}{2\sigma^2}}.$$

The standard normal distribution, $N(0,1)$, has the density

$$f(x) = \frac{1}{\sqrt{2\pi}} e^{-\frac{1}{2}x^2}.$$

3.5. Special Continuous Distributions

Here are some facts about the normal distribution.

1° *f is indeed a probability density, that is, it is positive and its integral is one.*

This proof makes the highlight reel. It is short, so sit back and enjoy. Let us calculate
$$I \stackrel{\text{def}}{=} \int_{-\infty}^{\infty} e^{-\frac{x^2}{2}}\, dx\,.$$

There is a problem. We cannot do the integral. Indeed, the function $e^{-x^2/2}$ cannot be integrated in terms of elementary functions. However, let us consider, not I, but I^2, and write it as an iterated integral:
$$I^2 = \int_{-\infty}^{\infty}\int_{-\infty}^{\infty} e^{-\frac{x^2+y^2}{2}}\, dx\, dy\,.$$

This is an integral over all of $\mathbb{R}^2$. Let $r^2 = x^2 + y^2$ and change to polar coordinates:
$$= \int_0^{2\pi}\int_0^{\infty} e^{-\frac{r^2}{2}} r\, dr\, d\theta\,.$$

Aha! The integral over θ gives 2π and the integral over r can be done by substituting $u = r^2$, giving $I^2 = 2\pi$, or $I = \sqrt{2\pi}$. Thus $\int_{-\infty}^{\infty} f(x)\, dx = 1$. Done!

2° *If X is $N(\mu, \sigma^2)$ and $Y = (X - \mu)/\sigma$, then Y is $N(0,1)$. Conversely, if Y is $N(0,1)$, then $\sigma Y + \mu$ is $N(\mu, \sigma)$.*

Indeed, to see 2°,
$$P\{Y \leq x\} = P\left\{\frac{X-\mu}{\sigma} \leq x\right\}$$
$$= P\{X \leq \sigma x + \mu\}$$
$$= \int_{-\infty}^{\sigma x + \mu} \frac{1}{\sqrt{2\pi\sigma^2}} e^{-\frac{(y-\mu)^2}{2\sigma^2}}\, dy\,.$$

Let $u = (y-\mu)/\sigma$, to see this is
$$= \frac{1}{\sqrt{2\pi}} \int_{-\infty}^{x} e^{-\frac{u^2}{2}}\, du\,,$$

so that Y is $N(0,1)$. The second statement follows by inverting the transformation. It follows easily that if X is $N(\mu, \sigma^2)$ and a and b are real, then $aX + b$ is $N(a\mu + b, a^2\sigma^2)$. This gives us a vital fact about Gaussians:

3° *Any linear function of a Gaussian random variable is Gaussian.*

4° *Let X be $N(0,1)$. Its moment generating function is*
$$M(\theta) = e^{\frac{1}{2}\theta^2}\,.$$

Its moments are: $E\{X^k\} = 0$ if k is odd, $E\{X^2\} = 1$, $E\{X^4\} = 3$, and, in general, $E\{X^{2k}\} = 1 \cdot 3 \cdot 5 \cdots (2k-1)$.

Indeed, if we complete the square in the exponent, we see that

$$M(\theta) = \frac{1}{\sqrt{2\pi}} \int_{-\infty}^{\infty} e^{-\frac{x^2}{2}} e^{\theta x} \, dx$$
$$= \left(\frac{1}{\sqrt{2\pi}} \int_{-\infty}^{\infty} e^{-\frac{1}{2}(x-\theta)^2} \, dx\right) e^{\frac{\theta^2}{2}}$$
$$= e^{\frac{\theta^2}{2}}.$$

To verify the claim about the limits, expand $e^{\frac{\theta^2}{2}}$ in a Taylor series and compare it with the expansion of $M(\theta)$ from Theorem 2.38 (iii):

$$M(\theta) = \sum_{k=0}^{\infty} \frac{\theta^{2k}}{2^k k!} = \sum_{k=0}^{\infty} \frac{\theta^k}{k!} E\{X^k\}.$$

Since the coefficients of the power series must be equal, the claim follows.

Finally, the $N(\mu, \sigma^2)$ density is symmetric about μ and has its maximum there, so

5° The mean, median and mode of a $N(\mu, \sigma^2)$ all equal μ.

This allows us to reduce problems about general Gaussian distributions to questions about the $N(0,1)$ distribution. The **standard normal distribution function** $\Phi(x)$ is given by

$$\Phi(x) = \frac{1}{\sqrt{2\pi}} \int_{-\infty}^{x} e^{-\frac{y^2}{2}} \, dy.$$

There is no formula for Φ in terms of elementary functions, but there are tables of its values. When the values of Φ are needed, one can look them up in the table.

Example 3.23.1. Let Y be $N(\mu, \sigma^2)$. Then $Z = (Y - \mu)/\sigma$ is $N(0,1)$, so its distribution function is Φ. Moreover, $Y \leq x \iff Z \leq (x-\mu)/\sigma$.

Then the distribution function F_Y of Y is

$$F_Y(x) = \Phi\left(\frac{x-\mu}{\sigma}\right).$$

So, for instance, if Y is $N(2, 4)$, $P\{Y \leq 1\} = \Phi\left(\frac{1-2}{\sqrt{4}}\right) = \Phi(-1/2)$. We look up $\Phi(-.5)$ in the tables of Φ, to see that $P\{Y \leq 1\} = .4025$.

d. The Cauchy distribution. Let $f(x) = \frac{1}{\pi(1+x^2)}$, $-\infty < x < \infty$. As the reader can verify, this is a probability density. It is symmetric about zero, so its median is zero, and, if it has a mean, by symmetry it would have to be zero. But... it doesn't. Indeed,

$$E\{|X|\} = \frac{1}{\pi} \int_{-\infty}^{\infty} \frac{|x|}{1+x^2} \, dx = \infty.$$

So the Cauchy distribution is not integrable. It has a median, but no mean. The expectation simply does not exist.

More generally, there is a family of symmetric Cauchy distributions with parameter $a > 0$ and density:

$$f_a(x) = \frac{a}{\pi} \frac{1}{a^2 + x^2}.$$

Example 3.23.2. Consider a line source of light—say a long horizontal fluorescent tube—placed a distance a from a wall. The intensity of light on the wall is constant horizontally by symmetry, but vertically, it is Cauchy. Indeed, let θ be the angle with the horizontal. The source emits light uniformly in theta. If the total amount emitted in the direction of the wall is one, the amount emitted in the angles $(\theta, \theta + d\theta)$ is $d\theta/\pi$, $-\pi/2 < \theta < \pi/2$. Write $y = a \tan \theta$ and note that the intensity at y is $(1/\pi)d\theta/dy = 1/(\pi dy/d\theta) = 1/(a \sec^2 \theta) = a/\pi(a^2 + y^2)$. But this is the Cauchy (a) density.

e. The Gamma Distribution. The sum of n independent exponential (λ) random variables has a gamma distribution with parameters (n, λ), denoted **$\Gamma(n, \lambda)$**. It is a positive random variable with density

$$f(x) = \lambda e^{-\lambda x} \frac{(\lambda x)^{n-1}}{(n-1)!}, \quad x \geq 0.$$

The density vanishes for $x < 0$. (See Exercise 3.12.) More generally, a gamma distribution can have non-integer values for its parameters. If $r > 0$, $\lambda > 0$, a density $f(x)$ is *gamma with parameters* (r, λ) if

$$f(x) = \lambda e^{-\lambda x} \frac{(\lambda x)^{r-1}}{\Gamma(r)}, \quad x \geq 0,$$

and $f(x) = 0$ for $x < 0$. Here, Γ is the gamma function, which extends the factorial to non-integers: for integer values of r, $\Gamma(r+1) = r!$.

Let X be a random variable with a gamma distribution, parameters (r, λ). Then

$$E\{X\} = \frac{r}{\lambda},$$
$$\text{Var}(X) = \frac{r}{\lambda^2}.$$

For integer r, this follows from the fact that $X = X_1 + \cdots + X_r$, where the X_i are independent exponential (λ) random variables. For the expectation is the sum of the expectation of the X_i's, namely $r \times 1/\lambda$, and the variance is the sum of the variances, namely $r \times 1/\lambda^2$.

Example 3.23.3. The life of a certain kind of light bulb is exponentially distributed with parameter λ. If the light bulb is replaced with another, independent, light bulb as soon as it burns out, the total lifetime of n light bulbs will have a gamma distribution with parameters (n, λ).

3.5.1. Transformation of Densities.

Example 3.23.4. Suppose X is $U(-1, 1)$. Let $Y = X^2$. What is the density of Y?

Answer. Start with the distribution function, not the density. $P\{Y \leq y\} = P\{X^2 \leq y\} = P\{-\sqrt{y} \leq X \leq \sqrt{y}\} = \sqrt{y}$. Therefore, the density of Y is

$$f_Y(y) = \frac{d}{dy}\sqrt{y} = \frac{1}{2\sqrt{y}}, \quad y > 0.$$

Note that the density blows up as $y \to 0$, and it is not even defined at the origin. That is not a problem, since the defining property of a density is that it integrates to the distribution function. The value at a single point will not change the integral, so densities can be undefined at a few points.

We can do this more generally.

Proposition 3.24. *Let X be a random variable with probability density $f_X(x)$. Let $g(x)$ be a smooth, one-to-one function, i.e., strictly monotone and continuously differentiable. Set $Y = g(X)$. Then the density of Y is*

$$(3.16) \qquad f_Y\bigl(g(x)\bigr) = f_X(x)\frac{1}{|g'(x)|}.$$

Proof. Let us first suppose that g is increasing. Then its inverse function, g^{-1} exists. Making free use of the chain rule, we see that

$$F_Y(y) = P\{Y \leq y\} = P\{g(X) \leq y\} = P\{X \leq g^{-1}(y)\} = F_X(g^{-1}(y)).$$

Now differentiate:

$$f_Y(y) = \frac{d}{dy}F_Y(y) = \frac{d}{dy}F_X(g^{-1}(y)) = F'_X(g^{-1}(y))\frac{d}{dy}g^{-1}(y).$$

But note that if $y = g(x)$, then $x = g^{-1}(y)$, so $F'_X(g^{-1}(y)) = F'_X(x) = f_X(x)$, and, by the usual calculus formula, $\frac{d}{dy}g^{-1}(y) = 1/g'(x)$, giving us

$$f_Y\bigl(g(x)\bigr) = f_X(x)\frac{1}{g'(x)}.$$

This assumes that g is increasing, and hence that g' is positive. In case g is decreasing, we just replace g by $-g$. □

3.5. Special Continuous Distributions

Of course, this proposition gives the wrong answer in Example 3.23.4, because the function x^2 is not monotone. Always check monotonicity before applying this result. In general, non-monotone functions require some thought.

Problems 3.5

3.7. Let X be uniform on $[0, L]$. Find the mean and standard deviation of $|X - \mu|$, where $\mu = E\{X\}$.

3.8. Let $a > 0$ and let X have a Cauchy (a) distribution. Show that aX has a Cauchy distribution. Verify that the density integrates to one.

3.9. Show that the exponential (λ) moment generating function is $\lambda/(\lambda - \theta)$, $0 \le \theta < \lambda$.

3.10. The χ-squared distribution is important in statistics. If $X_1, \ldots, X_n$ are independent $N(0, 1)$ random variables, then the distribution of $X_1^2 + \cdots + X_n^2$ is said to be **chi-squared with n degrees of freedom**, written $\chi^2(n)$.

(a) Find the density of X_1^2.

(b) Show by induction that the $\chi^2(n)$ density is of the form $Cx^{\frac{n}{2}-1}e^{-\frac{x}{2}}$, where C is a normalizing constant.

(c) Show that $\chi^2(n) = \Gamma(\frac{n}{2}, 2)$.

3.11. An airport check-in counter has a single line which feeds into two service counters. Whenever a service counter is empty, the person at the head of the line goes to it. The service time at each counter is exponential with a mean of four minutes, and the two counters are independent.

(a) What is the density of the waiting time until arriving at the service counter for the person at the head of the line? Find the moment generating function of this time.

(b) Beth is the eleventh person in the line. What is the moment generating function for her waiting time?

(c) Find the expectation and variance of the time Beth must wait before arriving at the service counter.

3.12. Show by induction that the sum of n independent exponential (λ) random variables has a $\Gamma(n, \lambda)$ distribution.

3.13. Let $x_1, \ldots, x_n$ be strictly positive numbers. Their *arithmetic average* is $A \stackrel{\text{def}}{=} \frac{1}{n}\sum_{i=1}^{n} x_i$ and their *geometric average* is $G \stackrel{\text{def}}{=} \left(\prod_{i=1}^{n} x_i\right)^{1/n}$. Use Jensen's inequality to show that $G \le A$: i.e., the geometric average is always less than or equal to the arithmentic average. Show, moreover, that there is equality if and only if $x_1 = \cdots = x_n$.
[Hint: For equality, draw the graph-chord picture for the logarithm.]

3.14. A random variable X has density $f_X(x) = \frac{1}{2}e^{-|x|}$, $-\infty < x < \infty$. Find the moment generating function and the variance of X.

3.15. Show that the $\Gamma(n, \lambda)$ moment-generating function is $(\lambda/(\lambda - \theta))^n$, $\theta < \lambda$.

3.16. A random variable X satisfies $|X| \leq 1$ a.s. Show that $\text{Var}(X) \leq 1$, and that the maximum is attained.

3.17. Let X and Y be positive random variables with $E\{X\} \leq 1$ and $E\{Y\} \leq 1$. Prove or disprove: for each $\varepsilon > 0$, $P\{X \leq 2+\varepsilon,\ Y \leq 2+\varepsilon\} > 0$. (Note that X and Y need not be independent.)

3.18. Let U have a $U(0,1)$ distribution. Find the density of U^2 and e^U.

3.19. Let X and Y be square-integrable random variables, and define the **L^2-norm** of X to be $\|X\|_2 \stackrel{\text{def}}{=} E\{X^2\}^{1/2}$. Prove **Minkowski's inequality**:
$$\|X+Y\|_2 \leq \|X\|_2 + \|Y\|_2.$$
[Hint: Compute $E\{(X+Y)^2\}$ and use the Schwarz inequality on the cross term. Then rewrite it in terms of norms.]

3.6. Joint Distributions and Joint Densities

Let X and Y be random variables. We want to look at the distribution of the two together, which is called the **joint distribution.** Let us start with the distribution function of the two, called the joint distribution function.

Definition 3.25. Let X and Y be random variables. Their **joint distribution function** $F(x, y)$ is
$$F(x,y) = P\{X \leq x,\ Y \leq y\}.$$

Notation. $x \uparrow y$ means that $x \to y$ and $x \leq y$, and $x \uparrow\uparrow y$ means that $x \to y$ and $x < y$.

Note. Most of what we say in this section extends to more than two random variables. For instance, the joint distribution function of $X_1, \ldots, X_n$ is $F(x_1, \ldots, x_n) = P\{X_i \leq x_i,\ i=1,2,\ldots,n\}$, and the joint density, if it exists, is $f(x_1, \ldots, x_n) = \frac{\partial^n}{\partial x_1 \ldots \partial x_n} F(x_1, \ldots, x_n)$.

Joint distribution functions have many of the properties of distribution functions, such as monotonicity and right-continuity, but they are multi-parameter functions and there are several possible limits to consider. The reader should be able to verify the following.

Exercise 3.20. Show that the joint distribution function of X, Y satisfies:

(a) $x \mapsto F(x,y)$ and $y \mapsto F(x,y)$ are both right-continuous and monotone increasing functions.

(b) $F(x-, y) \stackrel{\text{def}}{=} \lim_{u \uparrow\uparrow x} F(u,y)$, $F(x, y-) \stackrel{\text{def}}{=} \lim_{v \uparrow\uparrow y} F(x,v)$, and $F(x-, y-) \stackrel{\text{def}}{=} \lim_{u \uparrow\uparrow x,\, v \uparrow\uparrow y} F(u,v)$ all exist.

(c) So do $F(x, \infty) \stackrel{\text{def}}{=} \lim_{y \to \infty} F(x,y)$ and $F(\infty, y) \stackrel{\text{def}}{=} \lim_{x \to \infty} F(x,y)$.

(d) For each x and y, $\lim_{x \to -\infty} F(x,y) = \lim_{y \to -\infty} F(x,y) = 0$.

3.6. Joint Distributions and Joint Densities

(e) $\lim_{x \to \infty, y \to \infty} F(x,y) = 1$.

(f) If $a_1 < b_1$, $a_2 < b_2$, $P\{a_1 < X \le b_1, a_2 < Y \le b_2\} = F(b_1, b_2) - F(b_1, a_2) - F(a_1, b_2) + F(a_1, a_2)$.

(g) $P\{X = x, Y = y\} = F(x,y) - F(x-,y) - F(x,y-) + F(x-,y-)$.

Let us first look at the case of discrete random variables, which is the easiest case to explain, and also the easiest to understand.

Let X and Y be discrete random variables. Then we define the **joint probability mass function** of X and Y by

$$p_{XY}(x,y) = P\{X = x, Y = y\}.$$

Then $p_{XY}(x,y) = 0$ unless x and y are possible values for X and Y, respectively. Let us make the convention that sums like $\sum_{x,y} p_{XY}(x,y)$ are summed only over non-zero values of $p_{XY}(x,y)$. (This saves us from worry about what is meant by the sum of uncountably many zeros.) If $A \subset \mathbb{R}^2$, we will also write $\sum_{(x,y) \in A} p_{X,Y}(x,y)$.

Given the joint probability mass function, we can recover the probability mass functions p_X and p_Y of X and Y—in this context, they are called **marginal probability mass functions**—by noting that $P\{X = x\} = P\{X = x, -\infty < Y < \infty\} = \sum_y P\{X = x, Y = y\}$, or

(3.17) $$p_X(x) = \sum_y p_{XY}(x,y),$$

(3.18) $$p_Y(y) = \sum_x p_{XY}(x,y).$$

The corresponding joint distribution function is given by

(3.19) $$F_{XY}(x,y) = \sum_{u \le x,\, v \le y} p_{XY}(u,v).$$

Now let us consider the absolutely continuous distributions, which have probability densities. Suppose X and Y are continuous variables with joint distribution function $F(x,y)$.

Definition 3.26. If there is a non-negative function $f(x,y)$ such that the joint distribution function can be written $F(x,y) = \int_{-\infty}^{y} \int_{-\infty}^{x} f(u,v) \, du \, dv$, then we say that $f(x,y)$ is the **joint density** of X and Y.

It is easy to see from the definition that $P\{a < X \le b,\, c < Y \le d\} = \int_c^d \int_a^b f(x,y) \, dx \, dy$. More generally, for a set A in the plane,

$$P\{(X,Y) \in A\} = \iint_A f(x,y) \, dx \, dy.$$

Remark 3.27. Let $f(x,y)$ be the joint density of X and Y. Then
(i) $f(x,y) \geq 0$.
(ii) $\int_{-\infty}^{\infty} \int_{-\infty}^{\infty} f(x,y)\, dx\, dy = 1$.

Indeed, (i) is clear, and (ii) just says that $P\{(X,Y) \in \mathbb{R}^2\} = 1$.

Remark 3.28. Note that if f is continuous, then

$$(3.20) \qquad f(x,y) = \frac{\partial^2}{\partial x\, \partial y} F(x,y).$$

As in Corollary 3.11, we can show that if $g(x,y)$ is a function such that $g(X,Y)$ is integrable, then

$$E\{g(X,Y)\} = \int_{-\infty}^{\infty} \int_{-\infty}^{\infty} g(x,y)\, f(x,y)\, dx\, dy.$$

Suppose we know the joint distribution function $F(x,y)$ of X and Y. Can we recover the original distribution functions of the two random variables?

We can. Indeed, notice that $P\{X \leq x\} = P\{X \leq x, -\infty < Y < \infty\}$. But this equals $\lim_{y \to \infty} P\{X \leq x, Y \leq y\}$ which in turn is $\lim_{y \to \infty} F(x,y) \equiv F(x,\infty)$. A distribution function of Y is similar. (In this context, they are called "marginal" distribution functions.) This leads to:

Definition 3.29. The **marginal distribution function** of X is $F_X(x) \stackrel{\text{def}}{=} F(x,\infty)$, and the **marginal distribution function** of Y is $F_Y(y) \stackrel{\text{def}}{=} F(\infty,y)$.

Once we have the marginal distribution functions, we can find the original densities too.

Proposition 3.30. *Suppose X and Y have a joint density $f(x,y)$. Then both X and Y have densities, called the **marginal densities**.*

The marginal density of X is $f_X(x) \stackrel{\text{def}}{=} \int_{-\infty}^{\infty} f(x,y)\, dy$.

The marginal density of Y is $f_Y(y) \stackrel{\text{def}}{=} \int_{-\infty}^{\infty} f(x,y)\, dx$.

Proof. By symmetry, it is enough to prove it for X. To show that $f_X(x)$ is the marginal density, we will show that its integral is the marginal distribution function.

$$\int_{-\infty}^{x} f_X(u)\, du = \int_{-\infty}^{x} \int_{-\infty}^{\infty} f(u,v)\, dv\, du = P\{X \leq x, -\infty < Y < \infty\},$$

which equals $F_X(x)$, so f_X is indeed the density of F_X. □

3.6. Joint Distributions and Joint Densities

Remark 3.31. If X and Y have a joint density, then each has a marginal density. However, it is possible that both X and Y have densities, but that they have no joint density. For example, suppose that X is uniform on $(0,1)$ and that $Y = X$. Then the distribution of the pair (X, Y) is concentrated on the diagonal of the unit square. If they had a joint density $f(x, y)$, it would have to be zero off the main diagonal. But the main diagonal has zero area, so the integral of $f(x, y)$ would be zero, not one. So, they have no joint density.

Proposition 3.32. *Suppose random variables X and Y have a continuous joint density $f(x,y)$. Then they are independent if and only if $f(x,y) = f_X(x) f_Y(y)$ for all x and y.*

Proof. From Definition 2.19, random variables X and Y are independent if and only if $F_{XY}(x, y) = F_X(x) F_Y(y)$ for all x and y. To get this in terms of densities, note that by (3.20), the joint density is the mixed second partial of the joint distribution function, so

$$f_{XY}(x,y) = \frac{\partial^2}{\partial x\, \partial y} F_{XY}(x,y) = \frac{\partial}{\partial x} F_X(x) \frac{\partial}{\partial y} F_Y(y) = f_X(x) f_Y(y)\,.$$

$\square$

Remark 3.33. The proposition is true even if the densities are not continuous, but the statement needs to be changed to allow exceptional xy-null-sets: it should say "almost all (x, y)" instead of "all (x, y)". This is a technicality which we can safely ignore at this point. For now, we are happy to assume all densities are continuous.

3.6.1. Covariances and Correlations.

We will introduce two quantities which measure to some extent the relation between two random variables. They are called the covariance and the correlation. While the covariance is in some sense more fundamental—the correlation is a normalization of the covariance—one hears more about the correlation: "Cigarette smoking and cancer are correlated." "Education and income are correlated." "There is a correlation between the price of housing in Vancouver and the price of tea in China[7]." The correlation simply measures an average amount of agreement between two random variables. If the correlation between random variables X and Y is positive, then if X is large, Y is likely to be large, too. If the correlation is negative, the opposite is true.

Definition 3.34. Let X and Y be square-integrable random variables. Then the **covariance** of X and Y is

$$\operatorname{Cov}(X, Y) = E\{(X - E\{X\})(Y - E\{Y\})\}\,.$$

---

[7] I just made that up, but it is probably true. Prices are rising everywhere, so it is likely that both the price of housing and the price of tea are increasing as you read this.

The **correlation** of X and Y is
$$\operatorname{Cor}(X,Y) = \frac{\operatorname{Cov}(X,Y)}{\sqrt{\operatorname{Var}(X)}\sqrt{\operatorname{Var}(Y)}}.$$

Since X and Y are square-integrable, they are also integrable, so they have expectations. Then the covariance is finite by Schwarz' inequality. Here are some elementary properties.

Proposition 3.35. *If X and Y are square-integrable, then*

(i) $-1 \leq \operatorname{Cor}(X,Y) \leq 1$.

(ii) $\operatorname{Cor}(X,Y) = 1 \iff Y = aX + b$, *a.s., for some constants* $a > 0$ *and* b.

(iii) $\operatorname{Cor}(X,Y) = -1 \iff Y = aX + b$, *a.s., for some constants* $a < 0$ *and* b.

(iv) *The correlation is the covariance of the normalized random variables:* $\operatorname{Cor}(X,Y) = \operatorname{Cov}\left(\frac{X}{\sigma_X}, \frac{Y}{\sigma_Y}\right)$.

(v) *If X and Y are independent,* $\operatorname{Cor}(X,Y) = \operatorname{Cov}(X,Y) = 0$.

(vi) $\operatorname{Cov}(X,Y) = E\{XY\} - E\{X\}E\{Y\}$.

Proof. By the Schwarz inequality, $|\operatorname{Cov}(X,Y)| = |E\{(X - E\{X\})(Y - E\{Y\})\}| \leq \sqrt{\operatorname{Var}(X)}\sqrt{\operatorname{Var}(Y)}$ which implies that $|\operatorname{Cor}(X,Y)| \leq 1$. If $|\operatorname{Cor}(X,Y)| = 1$, there must be equality in the Schwarz inequality. This happens iff $Y - E\{Y\} = \lambda(X - E\{X\})$ a.s., for some λ. Thus $Y = \lambda X + E\{Y\} - \lambda E\{X\}$, i.e., Y is a linear function of X. We leave it to the reader to check that if $Y = aX + b$, then $\operatorname{Cor}(X,Y)$ equals plus or minus one according to the sign of a.

Finally, just expand the formula for the covariance to see that (vi) holds, and note that (v) follows since $E\{XY\} = E\{X\}E\{Y\}$. □

Definition 3.36. *If $\operatorname{Cor}(X,Y) = 0$, we say X and Y are* **orthogonal.**

We have seen that independence implies orthogonality. The converse is not true, though: random variables can be orthogonal without being independent.

Example 3.36.1. Let U be uniform on $(0, 2\pi)$. Set $X = \cos U$, $Y = \sin U$. Then $E\{X\} = (2\pi)^{-1} \int_0^{2\pi} \cos u \, du = 0$, $E\{Y\} = (2\pi)^{-1} \int_0^{2\pi} \sin u \, du = 0$, and $\operatorname{Cov}(X,Y) = (2\pi)^{-1} \int_0^{2\pi} \cos u \sin u \, du = 0$, so X and Y are uncorrelated. However, they are not independent, since $X^2 + Y^2 = 1$. (So, for instance, $P\{X > 1/\sqrt{2}, Y > 1/\sqrt{2}\} = 0$, while $P\{X > 1/\sqrt{2}\}P\{Y > 1/\sqrt{2}\} = 1/16$.)

3.6. Joint Distributions and Joint Densities

In practice, correlations are easy to compute from data, and are widely used—and often misused—to infer connections between different quantities. A positive correlation may indicate a causal relationship. For example, it seems all too well established that the positive correlation between alcohol consumption and the motor vehicle accident rate is more than a coincidence, and that an increase in the former leads to an increase in the latter. However, a correlation may also come about because both quantities are affected by a third. The fact that there is a positive correlation between shoe size and hat size does not indicate—as much as those with size 12 brogues would like it—that people with large feet are exceptionally intelligent; it merely reflects the fact that bigger people tend to have larger heads and larger feet than smaller people.

3.6.2. Transformation of Joint Densities. Let X and Y be random variables. Make the transformation $(X, Y) \mapsto (U, V) = g(X, Y)$, where $g : \mathbb{R}^2 \mapsto \mathbb{R}^2$ is a smooth one-to-one function, $g(x, y) = (g_1(x, y), g_2(x, y))$. The Jacobian matrix $J(x, y)$ of this transformation is

$$J(x, y) = \begin{pmatrix} \frac{\partial g_1}{\partial x} & \frac{\partial g_1}{\partial y} \\ \frac{\partial g_2}{\partial x} & \frac{\partial g_2}{\partial y} \end{pmatrix}.$$

Now $U = g_1(X, Y)$ and $V = g_2(X, Y)$. Let $\phi(x, y)$ be a bounded smooth function. Let's calculate the expectation of $\phi(U, V)$ two ways, first in the uv-coordinates, then in xy-coordinates:

$$E\{\phi(U, V)\} = \iint_{\mathcal{R}^2} \phi(u, v) f_{UV}(u, v) \, du \, dv \, .$$

Change variables: $(u, v) = g(x, y)$:

$$= \iint \phi(g(x, y)) f_{UV}(g(x, y)) \, |J(x, y)| \, dx \, dy \, ,$$

where $|J(x, y)|$ is the Jacobian, i.e., the determinant of $J(x, y)$. On the other hand,

$$E\{\phi(g(X, Y))\} = \int \phi(g(x, y)) f_{XY}(x, y) \, dx \, dy \, .$$

Compare the two equations. Since ϕ is essentially an arbitrary function, the two integrands must be equal. That leads to

Proposition 3.37. *Let X and Y be random variables. Let g be a smooth one-to-one function of $\mathbb{R}^2$ onto $\mathbb{R}^2$. Let $f_{UV}(u, v)$ and $f_{XY}(x, y)$ be the joint densities of U, V and of X, Y, respectively. Then*

(3.21) $$f_{UV}(g(x, y)) = f_{XY}(x, y) \frac{1}{|J(x, y)|} \, .$$

Note. This can also be done for more than two random variables. It just depends on the formula for the change of variables in terms of the Jacobian, which is valid in $\mathbb{R}^n$ for any $n \geq 1$. Then (3.21) is valid with the obvious changes for joint densities of n random variables.

Example 3.37.1. Suppose X and Y are independent $N(0,1)$ random variables. Their joint density is $f_{XY}(x,y) \equiv \frac{1}{\sqrt{2\pi}} e^{-\frac{1}{2}(x^2+y^2)}$. Let (R, Θ) be the point (X, Y) in polar coordinates. Let us find the density $f_{R\Theta}(r, \theta)$. The transformation is $x = r\cos\theta$, $y = r\sin\theta$. It is easier to apply the change-of-variables formula in the opposite direction: $(r, \theta) \mapsto (x, y)$. The Jacobian of the transformation is

$$|J(r,\theta)| = \begin{vmatrix} \cos\theta & -r\sin\theta \\ \sin\theta & r\cos\theta \end{vmatrix} = r.$$

Thus

$$f_{R\Theta}(r,\theta)\frac{1}{r} = f_{XY}(x,y) = \frac{1}{2\pi}e^{-\frac{x^2+y^2}{2}},$$

or, noting that $x^2 + y^2 = r^2$,

$$f_{R\Theta}(r,\theta) = \frac{1}{2\pi}re^{-\frac{r^2}{2}}, \quad 0 \leq \theta \leq 2\pi, \ r \geq 0.$$

Problems 3.6

3.21. Let $p(x,y)$ be the joint probability mass function of X and Y. Show that the marginal probability mass function of X is $p_X(x) = \sum_y p(x,y)$.

3.22. Let X and Y be independent $U(0,1)$ random variables.

(a) Find the joint density of $\min(X,Y)$ and $\max(X,Y)$.

(b) Divide $(0,1)$ into three line segments, where X and Y are the dividing points. What is the probability that the three line segments can form a triangle?

(c) Show that the lengths of all three segments have the same distribution.

3.23. Let μ be a probability measure on $(\mathbb{R}^2, \mathcal{B})$. Define X and Y on the probability space $(\mathbb{R}^2, \mathcal{B}, \mu)$ by $X(x,y) = x$, $Y(x,y) = y$. Show that X and Y have the joint distribution μ. Conclude that there exist random variables whose joint distribution is μ.

3.24. Suppose X and Y have a joint density $f_{XY}(x,y)$. Show that if there exist functions $f(x)$ and $g(y)$ such that $f_{XY}(x,y) = f(x)g(y)$ for all x and y, that X and Y are independent. (It is not necessary that f and g be the marginal densities.)

3.25. Let Y be a random variable with density $f_Y(y) = 2y$ if $0 < y < 1$, and zero otherwise. Let X have a $U(0,1)$ distribution. Find a function g such that $g(X)$ has the same distribution as Y.

3.26. Let $X_1, X_2, \ldots$ be i.i.d. random variables, and let $N \geq 1$ be an integer-valued random variable which is independent of the X_i. Define $S_N = \sum_{i=1}^{N} X_i$, i.e., S_N equals $X_1 + \cdots + X_n$ on the set $\{N = n\}$. Show that $E\{S_N\} = E\{N\}E\{X_1\}$ and $\text{Var}(S_N) = E\{N\}\text{Var}(X_1) + E\{X_1\}^2 \text{Var}(N)$.

3.27. Suppose n darts are thrown at a circular target of radius one. Assume the throws are independent and the landing place of each dart is uniform on the target. Let R_n be the distance from the center to the closest dart. Show that $E\{R_n\} = \int_0^1 (1-x^2)^n \, dx$.

3.28. Let $f_{XY}(x,y) = \frac{3y}{2}$ be the joint density of X and Y on $\{0 \leq y \leq 1, |x| \leq y\}$. Find

(a) the marginal density $f_Y(y)$ of Y;
(b) $P\{X + Y \geq 1\}$;
(c) $E\{X^2 Y\}$;
(d) the conditional expectation $E\{Y \mid X = 0\}$.

3.29. A square has side X, where X has density $f_X(x) = 3x^2$, $0 \leq x \leq 1$. Find its expected area in two ways.

3.30. Let (X, Y) be uniformly distributed on $[0,1]^2$. Let Z be the shortest distance from (X, Y) to the boundary (i.e., the minimum of $X, Y, 1-X$ and $1-Y$.) Find the distribution function, density, and median of Z.

3.7. Conditional Distributions, Densities, and Expectations

Let X and Y be random variables. Suppose we are given that $X = x$. What can we say about the distribution of Y? This involves conditional probabilities: $P\{Y \in A \mid X = x\}$. If X is discrete, this is no problem, and we can immediately get conditional versions of the distribution, distribution function, and expectation. Indeed, if $P\{X = x\} > 0$, then the conditional distribution of Y given X is

$$P\{Y \in A \mid X = x\} = \frac{P\{Y \in A, X = x\}}{P\{X = x\}}.$$

The **conditional distribution function** is

$$F_{Y|X}(y \mid x) = P\{Y \leq y \mid X = x\},$$

and if Y is integrable, we can also define the **conditional expectation** of Y given that $X = x$ by

$$E\{Y \mid X = x\} \overset{\text{def}}{=} \frac{E\{Y I_{\{X=x\}}\}}{P\{X = x\}}.$$

Moreover, if Y is also discrete, it has a **conditional probability mass function** $p_{Y|X}(y \mid x)$:

$$p_{Y|X}(y \mid x) = \begin{cases} \frac{p_{YX}(y,x)}{p_X(x)} & \text{if } p_X(x) > 0, \\ 0 & \text{otherwise}. \end{cases}$$

This does not work when X has a continuous distribution, since then $P\{X = x\} = 0$ for every x, and the above formulas are moot: they are of the form 0/0. This is a serious problem, and its brilliant solution[8] came quite late in the development of the subject. We are not ready to deal with it yet, but we can handle one important case right now: where X and Y have a continuous joint density.

Suppose that X and Y have a joint density $f_{XY}(x,y)$. Let $f_X(x)$ be the marginal density of X. Let us calculate the conditional distribution of Y given that $X = x$. We will sneak up on it by an approximation. For $h > 0$, $x, y \in \mathbb{R}$, we have

$$P\{Y \leq y \mid x \leq X \leq x+h\} = \frac{\int_{-\infty}^{y} \int_{x}^{x+h} f_{XY}(u,v)\,du\,dv}{\int_{x}^{x+h} f_X(u)\,du}.$$

If h is small and f_{XY} and f_X are continuous, this is

$$\sim \frac{h \int_{-\infty}^{y} f_{XY}(x,v)\,dv}{h f_X(x)}$$

$$= \int_{-\infty}^{y} \left(\frac{f_{XY}(x,v)}{f_X(x)}\right) dv.$$

Notice that we have calculated a distribution function—a conditional distribution function, but a distribution function nevertheless—and that this distribution function is an integral. Therefore, the term in parentheses is its density. Let us call the conditional distribution function $F_{Y|X}(y \mid x)$, and its density $f_{Y|X}(y \mid x)$. To recap, if X and Y have a joint density, then

Definition 3.38. The **conditional distribution function** of Y given that $X = x$ is

$$F_{Y|X}(y \mid x) = \begin{cases} \int_{-\infty}^{y} \frac{f_{XY}(x,v)}{f_X(x)}\,dv & \text{if } f_X(x) > 0, \\ 0 & \text{if } f_X(x) = 0. \end{cases}$$

The **conditional density** of Y given that $X = x$ is

$$f_{Y|X}(y \mid x) = \begin{cases} \frac{f_{XY}(x,y)}{f_X(x)} & \text{if } f_X(x) > 0, \\ 0 & \text{if } f_X(x) = 0. \end{cases}$$

We can use this to define a conditional expectation:

$$E\{Y \mid X = x\} \stackrel{\text{def}}{=} \int_{-\infty}^{\infty} y\, f_{Y|X}(y \mid x)\,dy.$$

Notation. We have defined conditional probabilities and expectations given that $X = x$. It is useful to have the "conditional probabilities given X" rather than just the conditional probabilities given $X = x$—i.e., without

---

[8]This was Kolmogorov's introduction of the conditional expectation in [22]. See Chapter 8.

3.7. Conditional Distributions, Densities, and Expectations

specifying the value of x. This is simple: just set $E\{Y \mid X\}$ equal to $E\{Y \mid X = x\}$ on the set $\{X = x\}$. More formally, let us first define
$$\psi(x) \stackrel{\text{def}}{=} E\{Y \mid X = x\}.$$
Then we define

Definition 3.39. Suppose that either X is a discrete random variable or that X and Y have a joint density. If Y is integrable, then the **conditional expectation of Y given X** is

(3.22) $$E\{Y \mid X\} \stackrel{\text{def}}{=} \psi(X).$$

We can define the other conditional quantities, such as the conditional distribution $P\{Y \in A \mid X\}$, and the conditional density $F_{Y|X}(y \mid X)$, similarly.

Remark 3.40. (i) The conditional expectation $E\{Y \mid X\}$ is a function of X, so it is a random variable itself.

Proposition 3.41. *Let X and Y be random variables. Suppose that they have a continuous joint density $f_{XY}(x,y)$ and that X has a strictly positive continuous density $f_X(x)$. If Y is integrable, then $E\{Y|X\}$ is also integrable, and*
$$E\Big\{E\{Y \mid X\}\Big\} = E\{Y\}.$$

Proof. Let us suppose that Y is positive, so that it is integrable iff $E\{Y\}$ is finite, and write
$$E\Big\{E\{Y \mid X\}\Big\} = E\{\psi(X)\}$$
$$= \int_{-\infty}^{\infty} \psi(x) f_X(x) \, dx$$
$$= \int_{-\infty}^{\infty} \left[\int_{-\infty}^{\infty} y f_{Y|X}(y \mid x) \, dy\right] f_X(x) \, dx$$
$$= \int_{-\infty}^{\infty} \int_{-\infty}^{\infty} y \frac{f_{XY}(x,y)}{f_X(x)} \, dy \, f_X(x) \, dx.$$

Interchange the order:
$$= \int_{-\infty}^{\infty} y \int_{-\infty}^{\infty} f_{XY}(x,y) \, dx \, dy$$
$$= \int_{-\infty}^{\infty} y f_Y(y) \, dy$$
$$= E\{Y\}.$$

For the general case, write $Y = Y^+ - Y^-$ and apply this to both. □

The continuity assumptions on the densities are just there so that we can define the improper Riemann integrals and change their order. Since the integrands are positive, this holds even if the integral is infinite. (If it is infinite in one order, it is infinite on the other, too.) With a little more care, and one theorem (Fubini's Theorem) from integration theory, we could dispense with the continuity assumptions.

Exercise 3.31. Assume the hypotheses of Proposition 3.41.

(a) Define the conditional probabilities of Y given X: $P\{Y \in A \mid X\}$.

(b) Extend this to functions g: if $g(Y)$ is integrable, define $E\{g(Y) \mid X\}$ and prove that $E\{E\{g(Y) \mid X\}\} = E\{g(Y)\}$.

(c) If f and g are bounded continuous functions, show that
$$E\{f(X)g(Y)\} = E\{f(X)E\{g(Y) \mid X\}\}.$$

Remark 3.42. We derived the conditional density in a cavalier fashion. We should justify it. The conditional density is, after all, a density. Does it work? That is, does it correctly compute the probabilities that a conditional density should? Let us see. A conditional density should give us conditional probabilities, so that, for example, we should have $P\{c \leq Y < d | X = x\} = \int_c^d f_{Y|X}(y|x)\, dy$. This, in turn, should give us the unconditional probability, say, $P\{a \leq X < b,\, c \leq Y < d\} = \int_a^b P\{c \leq Y < d \mid X = x\} f_X(x)\, dx$. Let us see if it does. Put the two above integrals together, then substitute for $f_{Y|X}$:

$$\int_a^b \left(\int_c^d f_{Y|X}(y|x)\, dy\right) f_X(x)\, dx = \int_a^b \int_c^d \frac{f_{XY}(x,y)}{f_X(x)} f_X(x)\, dy\, dx$$
$$= \int_a^b \int_c^d f_{XY}(x,y)\, dy\, dx$$
$$= P\{a \leq X < b,\, c \leq Y < d\}.$$

So $f_{X|Y}$ does indeed act exactly as expected. This justifies the definition we gave[9].

Remark 3.43. Suppose X and Y are independent. Then the conditional densities of Y given X should be independent of X. Let us check this. We have $f_{XY}(x,y) = f_X(x)f_Y(y)$, so that
$$f_{Y|X}(y \mid x) = \frac{f_{XY}(y,x)}{f_X(x)} = f_Y(y),$$
as expected.

---

[9]When we define the general conditional expectation, we will use a very similar property. Mathematicians quite often define an object simply by giving its principal properties. This could be called the duck principle of definition, following the old proverb: "If it looks like a duck, acts like a duck and quacks like a duck, it *is* a duck."

3.7.1. Sums of Random Variables.
Let X and Y have the joint density $f_{XY}(x,y)$. What is the density of $X+Y$?

Proposition 3.44. *Let X and Y have joint density $f_{XY}(x,y)$. Then $Z \equiv X+Y$ has a density $f_Z(z)$ given by*

$$f_Z(z) = \int_{-\infty}^{\infty} f_{XY}(x, z-x)\, dx = \int_{-\infty}^{\infty} f_{XY}(z-y, y)\, dy\,.$$

Proof. $X + Y \leq z \iff (X,Y) \in \{(x,y): x+y \leq z\}$. Then the density F_{X+Y} is

$$F_{X+Y}(z) = P\{X+Y \leq z\} = \iint_{\{u+v \leq z\}} f_{XY}(u,v)\, du\, dv\,.$$

Let $x=u$, $y = u+v$, a transformation with Jacobian 1. Then this is

$$= \iint_{\{(x,y): y \leq z\}} f_{XY}(x, y-x)\, dx\, dy = \int_{-\infty}^{z} \left[\int_{-\infty}^{\infty} f_{XY}(x, y-x)\, dx\right] dy\,.$$

But $F_{X+Y}(z) = \int_{-\infty}^{z} f_{X+Y}(y)\, dy$, so, comparing the two integrals for F_{X+Y}, we see that $f_{X+Y}(y)$ must equal the term in square brackets. Add X and Y in the opposite order to get the second formula. □

Corollary 3.45. *If X and Y are independent, then*

$$f_{X+Y}(z) = \int_{-\infty}^{\infty} f_X(x) f_Y(z-x)\, dx = f_X * f_Y(z)\,.$$

Notation. $f_X * f_Y$ is the **convolution** of X and Y. Note that, since $X+Y = Y+X$, we could have integrated in different orders to see that $f_X * f_Y(z) = f_Y * f_X(z)$.

3.7.2. Bivariate and Multivariate Gaussian.

Definition 3.46. The **standard bivariate Gaussian density** with parameter ρ, $-1 < \rho < 1$, is

(3.23) $$f(x,y) = \frac{1}{2\pi\sqrt{1-\rho^2}} e^{-\frac{x^2 - 2\rho xy + y^2}{2(1-\rho^2)}}\,;$$

ρ is the **correlation** of X and Y.

Remark 3.47. If $\rho = 0$, this is the density of a pair of independent $N(0,1)$ random variables.

Proposition 3.48. *If X and Y have a standard bivariate Gaussian joint density, then both X and Y have $N(0,1)$ distributions, and $E\{XY\} = \rho$. In particular, if $\rho = 0$, X and Y are independent.*

The proof follows from Exercise 3.32 below.

Exercise 3.32. Let U and V be independent $N(0,1)$ random variables. Put $X = aU + bV$, $Y = cU + dV$ where $a^2 + b^2 = 1$, $c^2 + d^2 = 1$, and $ac + bd = \rho$. Show that both X and Y have $N(0,1)$ distributions, that $E\{XY\} = \rho$, and that their joint density is given by (3.23).

The bivariate Gaussian is a special case of the multivariate Gaussian density (3.28). The calculations are actually nicer in the general case, since vector and matrix notation makes evident some things which are hidden in the bivariate formula. Above all, it makes clear the intimate connection with geometry.

Let us establish some notation. "Vector" means "column vector" unless we say otherwise. If A is a vector or matrix, then A^T is its transpose, and $|A|$ is its determinant. In this notation, the usual vector product of A and B is $A \cdot B \equiv A^T B$. Expectations are computed coordinatewise.

Definition 3.49. If $U = (U_1, U_2, \ldots, U_n)^T$, where the U_i are integrable random variables, then $E\{U\} = (E\{U_1\}, E\{U_2\}, \ldots, E\{U_n\})^T$. Similarly, if $W = (W_{ij})$ is a matrix whose entries are random variables, then $E\{W\} = (E\{W_{ij}\})$.

The expectation of a vector is a vector, and the expectation of a matrix is a matrix.

Exercise 3.33. Show that expectations of random matrices and vectors are linear. That is, if A and B are constant $m \times n$ matrices and X and Y are $n \times p$ matrices whose elements are integrable random variables, show that $E\{AX + BY\} = AE\{X\} + BE\{Y\}$.

We can extend the idea of covariance to vector-valued variables. Let $X_1, X_2, \ldots, X_n$ be square-integrable random variables. Let

$$X \stackrel{def}{=} \begin{pmatrix} X_1 \\ X_2 \\ \vdots \\ X_n \end{pmatrix} \qquad \mu \stackrel{def}{=} \begin{pmatrix} E\{X_1\} \\ E\{X_2\} \\ \vdots \\ E\{X_n\} \end{pmatrix} \stackrel{def}{=} \begin{pmatrix} \mu_1 \\ \mu_2 \\ \vdots \\ \mu_n \end{pmatrix}.$$

Definition 3.50. The **covariance matrix** of $X_1, \ldots, X_n$ (or, equivalently, of the vector X) is the $n \times n$ matrix $V = (v_{ij})$ where $v_{ij} = E\{(X_i - \mu_i)(X_j - \mu_j)\}$. In vector terms:

$$V = E\{(X - \mu)(X - \mu)^T\}.$$

Some facts about covariance matrices:

$1°$ A covariance matrix V is symmetric and positive definite.

3.7. Conditional Distributions, Densities, and Expectations

Symmetry is clear: $v_{ij} = E\{(X_i-\mu_i)(X_j-\mu_j)\} = E\{(X_j-\mu_j)(X_i-\mu_i)\} = v_{ji}$. Let λ be an n-vector. Consider $\lambda^T V \lambda$. Write it as

$$\lambda^T V \lambda = \lambda^T E\{(X-\mu)(X-\mu)^T\}\lambda = E\{\lambda^T(X-\mu)(X-\mu)^T \lambda\}.$$

But $\lambda^T(X-\mu)$ is a scalar, so this equals $E\{(\lambda^T(X-\mu))^2\} \geq 0$. Thus V is positive definite.

2° A covariance matrix has a square root.

Indeed, any symmetric positive-definite matrix has a square root: if V is a symmetric $n \times n$ positive definite matrix, then there is an $n \times n$ matrix A such that $V = AA^T$. A can even be taken to be symmetric.

Let us treat the multivariate Gaussian case. Define it as follows:

Definition 3.51. a. A family $X_1, \ldots, X_n$ of random variables is **jointly,** or **multivariate Gaussian** if there are real numbers μ_i and a_{ij}, $i = 1, \ldots, n$, $j = 1, \ldots, m$ and i.i.d. $N(0,1)$ random variables $\xi_1, \ldots, \xi_m$ such that

$$(3.24) \qquad X_i = a_{i1}\xi_1 + a_{i2}\xi_2 + \cdots + a_{im}\xi_m + \mu_i, \quad i = 1, \ldots, n.$$

b. A family $\{X_\alpha, \alpha \in I\}$ of random variables is **jointly Gaussian** if every finite sub-family is jointly Gaussian.

To put this in vector/matrix form, let

$$(3.25) \qquad \xi \stackrel{\text{def}}{=} \begin{pmatrix} \xi_1 \\ \xi_2 \\ \vdots \\ \xi_m \end{pmatrix} \qquad X \stackrel{\text{def}}{=} \begin{pmatrix} X_1 \\ X_2 \\ \vdots \\ X_n \end{pmatrix} \qquad A \stackrel{\text{def}}{=} \begin{pmatrix} a_{11} & \cdots & a_{1m} \\ \vdots & \vdots & \vdots \\ a_{n1} & \cdots & a_{nm} \end{pmatrix}.$$

Then the vector form of (3.24) is

$$(3.26) \qquad X = A\xi + \mu.$$

Note. This definition may appear contrived. Theorem 6.9 below gives an alternative approach which is more elegant, but harder to verify.

To simplify notation, let us assume that $\mu \equiv 0$, i.e., that our random variables have mean zero. If necessary, we can add the means in at the end.

Remark 3.52. Any linear combination of multivariate Gaussian random variables is Gaussian.

This is immediate—a linear combination of linear combinations of the ξ_i is still a linear combination of the ξ_i.

The covariance matrix of ξ is the $m \times m$ identity I. Indeed, the ij^{th} entry of $\xi\xi^T$ is $E\{\xi_i\xi_j\} = \delta_{ij}$, which is the ij^{th} entry of I. (δ_{ij} is the Kronecker delta.) Let us compute the covariance matrix of X, remembering that $\mu = 0$.

We will only treat the case $m = n$ here. We will see in Example 6.8.2 that this is actually the general case.

$$(3.27) \qquad V \stackrel{\text{def}}{=} E\{XX^T\} = E\{A\xi\xi^T A^T\} = AE\{\xi\xi^T\}A^T = AA^T.$$

This has a striking consequence: *any* symmetric, positive definite matrix is the covariance matrix of jointly Gaussian random variables. Conversely, for jointly Gaussian random variables, the mean and the covariance matrix determine the distribution.

Theorem 3.53. (*i*) *Let V be an $n \times n$ symmetric positive-definite matrix, and let μ be an n-vector. Then there exist jointly Gaussian random variables $X_1, \ldots, X_n$ with mean μ and covariance matrix V.*

(*ii*) *Conversely, the mean and covariance function uniquely determine the distribution of multivariate Gaussian random variables.*

Proof. (*i*) Let A, ξ and X be as in (3.25) and (3.26). Then $E\{A\xi\} = 0$ so that X has mean μ and, by (3.27), its covariance matrix is V.

(*ii*) This is immediate if V is non-singular, since V and μ determine the density (see (3.28) below) and hence the distribution. If V is singular, let $\eta_1, \ldots, \eta_n$ be i.i.d. $N(0,1)$ random variables which are independent of X and Y. Then for any $\varepsilon > 0$, $X + \varepsilon\eta$ has covariance matrix $V + \varepsilon^2 I$, where I is the $n \times n$ identity. This is non-singular, for, if a is an n-vector, $a^T(V+\varepsilon^2 I)a = a^T V a + \varepsilon^2 a^T a$. But V is positive definite, so that $a^T V a \geq 0$, and $a^T a > 0$ unless $a = 0$. Thus $X + \varepsilon\eta$ and $Y + \varepsilon\eta$ have the same mean and non-singular covariance matrix, so they have the same distribution. Let $\varepsilon \to 0$ to see that X and Y have the same distribution. □

Example 3.53.1. Multivariate Gaussian Density★. Let X be a multivariate Gaussian vector of the form (3.26), where A is an $n \times n$ matrix. We will derive its density. Note first that if its covariance matrix V is singular, X does not have a density. Indeed, if V is singular, there exists a vector t such that $Vt = 0$. Then $t \cdot X = 0$, for $E\{(t \cdot X)^2\} = E\{(t^T X)(X^T)t\} = t^T E\{XX^T\}t = t^T V t = 0$. This means that X lies on the $n-1$-dimensional hyperplane $t \cdot x = 0$ in $\mathbb{R}^n$, which has zero volume. Hence X cannot have a density. Thus, suppose V is non-singular. Then $V = AA^T$, where A is a non-singular $n \times n$ matrix. Let $\Lambda = V^{-1} = (A^{-1})^T A^{-1}$.

Note. $y = Ax$ is a linear transformation of $\mathbb{R}^n \mapsto \mathbb{R}^n$. Its Jacobian is

$$\left|\frac{\partial(y_1, \ldots, y_n)}{\partial(x_1, \ldots, x_n)}\right| = |A|,$$

which is constant.

Let $\phi(x_1, \ldots, x_n)$ be a smooth bounded function on $\mathbb{R}^n$. Let f_X be the density of X, i.e., the n-dimensional joint density of $X_1, \ldots, X_n$. The

3.7. Conditional Distributions, Densities, and Expectations

expectation of $\phi(X)$ is
$$E\{\phi(X)\} = \int_{\mathcal{R}^n} \phi(y)\, f_X(y)\, dy\,.$$

Change variables: $y = Ax$:
$$= \int_{\mathcal{R}^n} \phi(Ax)\, f_X(Ax)\, |A|\, dy\,.$$

But $X = A\xi$, so this is also
$$= E\{\phi(A\xi)\} = \int_{\mathcal{R}^n} \phi(Ax)\, f_\xi(x)\, dx\,.$$

This holds for all such ϕ, so, comparing the two, we see that $f_X(Ax) = \frac{1}{|A|} f_\xi(x)$ or
$$f_X(y) = \frac{1}{|A|} f_\xi(A^{-1}y)\,.$$

But we know that $f_\xi(x) = \frac{1}{\sqrt{2\pi}} e^{-\frac{x^2}{2}} \ldots \frac{1}{\sqrt{2\pi}} e^{-\frac{x^2}{2}} = \frac{1}{(2\pi)^{\frac{n}{2}}} e^{-\frac{x^T x}{2}}$.

Now $x = A^{-1}y$ so $x^T x = y^T (A^{-1})^T A^{-1} y = y^T \Lambda y$. Moreover, $1/|A| = \sqrt{|\Lambda|}$. Thus
$$f_X(y) = \frac{|\Lambda|^{\frac{1}{2}}}{(2\pi)^{\frac{n}{2}}} e^{-\frac{1}{2} y^T \Lambda y}\,.$$

We assumed that the random variables had mean zero. This is harmless, since, if they are not, we simply subtract their means. Thus, in terms of the mean μ and covariance function V:

Definition 3.54. The $N(\mu, V)$ **multivariate Gaussian density** on $\mathbb{R}^n$ is

(3.28) $$f(y) = \frac{|\Lambda|^{\frac{1}{2}}}{(2\pi)^{\frac{n}{2}}} e^{-\frac{1}{2}(y-\mu)^T \Lambda (y-\mu)}\,,$$

where V is a $n \times n$ non-singular covariance matrix, $\Lambda = V^{-1}$, and y and μ are n−vectors.

Remark 3.55. (*i*) The quantity $y^T V y = \sum_{i,j} y_i v_{ij} y_j$ is called a **quadratic form**.

(*ii*) If ξ is a vector whose coordinates are i.i.d. $N(0,1)$ random variables, and if A is a non-singular $n \times n$ matrix, then $X = A\xi$ has the density (3.28). Conversely, if V is a symmetric, non-singular positive-definite matrix, there is a non-singular matrix A such that $V = AA^T$, and $Y \stackrel{\text{def}}{=} A\xi$ has the given density.

(*iii*) If $X_1, \ldots, X_n$ are uncorrelated multivariate Gaussian random variables, they are independent, for their joint density factors. Indeed, if they are uncorrelated, then both V and its inverse Λ are diagonal, so the quadratic form in (3.28) equals $\sum_i \frac{1}{2}\lambda_{ii}(y_i - \mu_i)^2$ and the joint density factors into const $\cdot \prod_i \exp(\lambda_{ii}(y_i - \mu_i)^2/2)$.

(iv) In case $n = 2$, X and Y are $N(0,1)$, and $E\{XY\} = \rho$, the covariance matrix is
$$V = \begin{pmatrix} 1 & \rho \\ \rho & 1 \end{pmatrix}$$
and its inverse is
$$\Lambda = V^{-1} = \frac{1}{1-\rho^2}\begin{pmatrix} 1 & -\rho \\ -\rho & 1 \end{pmatrix}.$$

Plug this into (3.28) to get the bivariate normal density (3.23).

(v) The entire distribution of a family of jointly Gaussian random variables is determined by (a) their means and (b) their covariances.

Example. The geometry of random variables⋆. Let $\xi_1, \ldots, \xi_n$ be i.i.d. $N(0,1)$ random variables. Write them as a column vector $\xi = (\xi_1, \ldots, \xi_n)^T$. Let $\mathcal{X}$ be the space of all linear combinations of the ξ_i. Then $\mathcal{X}$ is a vector space: it contains sums and linear combinations of its elements. Since linear combinations of Gaussians are Gaussian, $\mathcal{X}$ is a space of mean-zero Gaussian random variables[10]. Moreover, we can define an inner product in $\mathcal{X}$ by $X \cdot Y \stackrel{\text{def}}{=} E\{XY\}$, and a norm by $\|X\|^2 = E\{X^2\}$—we use the double bars to distinguish it from the absolute value of X, which is a random variable.

In fact, it mirrors the n-dimensional vector space $\mathbb{R}^n$. Suppose we represent elements of $\mathbb{R}^n$ as column vectors. Then map $\mathbb{R}^n$ to $\mathcal{X}$ by $A \mapsto A^T\xi$. We claim that this map is an isomorphism which preserves inner products. First, the map is linear, and any linear combination of the ξ can be written in the form $A^T\xi$ for some $A \in \mathbb{R}^n$—A is just the vector of coefficients— so the map is onto. If $X = A^T\xi$ and $Y = B^T\xi$, then $E\{XY\} = E\{A^T\xi B^T\xi\} = E\{A^T\xi\xi^T B\} = A^T E\{\xi\xi^T\} B$. But the ξ are i.i.d. mean zero and variance one, so $E\{\xi\xi^T\} = I$, the identity. Thus the above equals $A^T B$, which is just the inner product in $\mathbb{R}^n$. So the map preserves inner products, and is an isometry. In short, $\mathcal{X}$ is isometric to $\mathbb{R}^n$, and has inner product $X \cdot Y = E\{XY\}$ and norm $\|X\| = \sqrt{E\{X^2\}}$.

The correlation has an interesting interpretation. If A and B are unit vectors in $\mathbb{R}^n$, then $A \cdot B = \cos\theta$, where θ is the angle between A and B. Then, if $Y = A \cdot \xi$ and $Z = B \cdot \xi$, they have unit variance, and $\text{Cor}(Y, Z) = \text{Cov}(A \cdot \xi, B \cdot \xi) = A^T B = \cos\theta$. So θ can be interpreted as the angle

---

[10]The geometric interpretation of Gaussian random variables and the fact that conditional expectations correspond to projections are heavily used in statistics, adorning such diverse fields as the analysis of variance [32] and Wiener's prediction theory. When infinitely many variables are involved, the natural setting is indeed a Hilbert space.

between the random variables Y and Z (!) and the correlation is the cosine of this angle.

This holds for non-unit vectors as well: in general, just apply the above to $Y/\sqrt{\text{Var}(Y)}$ and $Z/\sqrt{\text{Var}(Z)}$ instead of Y and Z.

So we can think of each random variable as a vector. The whole geometry of Euclidean space carries over. With one extra intriguing fact: if two jointly Gaussian random variables are orthogonal (i.e., their covariance vanishes) they are independent. So, of the angle θ between any two is ninety degrees, they are independent.

Let us close this example with a question. In probability, we often deal with sequences of random variables. Suppose that $\xi_1, \xi_2, \ldots$ is an infinite sequence of i.i.d. $N(0,1)$ random variables. If we let them generate a vector space, would it have infinite dimensions? Does this lead us to something new and interesting?

Hint: In three words, "Yes. Hilbert space."

Problems 3.7

3.34. Let U and X be random variables, where U is $U(0,1)$, and, given $U = u$, X has an exponential distribution with parameter $= 1/u$.

(a) Find the joint density of U and X.

(b) Find $E\{X\}$ and $E\{UX\}$.

(c) Are U and X independent?

3.35. Let X be uniform on $[0,1]$ and, given X, let Y be uniform on $[0, X^2]$. Find the marginal density of Y, and find $E\{X \mid Y\}$.

3.36. Suppose X has an exponential distribution with parameter λ and, given X, suppose that Y has an exponential distribution with mean X. What is the joint density of X and Y? Find $E\{Y\}$.

3.37. Two light fixtures hang in different hallways of an apartment building. For each fixture, a given light bulb lasts an exponential time with mean $\lambda > 0$ before it has to be replaced. (λ is the same for both.) There is a stack of n replacement bulbs in a closet in hallway A, and k replacement bulbs in the corresponding closet in hallway B. What is the probability that hallway A runs out of bulbs before B does?

3.38. Let X and Y be independent Gaussian random variables with the same variance. Show that X+Y and X-Y are independent.

3.39. (Continuation.) Let U and V be independent $N(0,1)$ random variables, let $\phi, \theta \in [0, 2\pi)$, and put $X = U \sin\theta + V \cos\theta$ and $Y = U \sin\phi + V \cos\phi$. Express the correlation ρ of X and Y in terms of θ and ϕ, and interpret this in terms of "the angle between X and Y."

3.40. Let X and Y be bivariate mean zero Gaussian random variables with variance one and correlation ρ. Show that there are independent Gaussian random variables U and V, and reals a, b, c and d such that $X = aU + bV$ and $Y = cU + dV$.

3.41. Construct random variables X and Y which have $N(0,1)$ distributions such that $E\{XY\} = 0$ but X and Y are **not** independent.

3.42. Let Z be a complex-valued random variable. Show that $|E\{Z\}| \leq E\{|Z|\}$.

3.43. (Generalized Chebyshev inequality.) Let Φ be a strictly positive increasing function, X a random variable. Prove that
$$P\{\Phi(X) \geq \lambda\} \leq \frac{1}{\Phi(\lambda)} E\{\Phi(X)\}.$$

3.44. The **Cantor set** C is the subset of $[0,1]$ formed by removing successive "open middle thirds". It can be succinctly defined using the ternary expansion: write $x \in [0,1]$ in the base 3: $x = \sum_{n=1}^{\infty} a_n 3^{-n}$, where each a_n equals zero, one or two. The Cantor set is the set of all x for which $a_n \neq 1$ for all n. (If x has two expansions, put it in the Cantor set if *one* of them has no ones.)

(a) Show that C is a closed set and that it is obtained by first removing the open middle third $(1/3, 2/3)$, then the two middle thirds of the remaining intervals $(1/9, 2/9)$ and $(7/9, 8/9)$, and so on.

(b) Show that the Cantor set can be put into one-to-one correspondence with $[0,1]$, and is therefore uncountable.

(c) Let X be a random variable on the probability space $([0,1], \mathcal{B}([0,1]), dx)$ (see Example 1.14.2) defined by $X(x) = 1$ if $x \in C$, $X = 0$ otherwise. Show that $P\{X = 1\} = 0$.

3.45. Not all continuous distribution functions have densities. Here is an example, called the **Cantor function**. It is defined on $[0,1]$ as follows. Let $x \in [0,1]$ and let (a_n) be the coefficients in its ternary expansion: $x = \sum_n a_n 3^{-n}$. Let $N = \infty$ if none of the $a_n = 1$, and otherwise, let N be the smallest n for which $a_n = 1$. Set $b_n = a_n/2$ if $n < N$, and set $b_N = 1$. Then set
$$F(x) = \sum_{n=1}^{N} b_n 2^{-n}.$$

(a) Show that F is constant on each interval in the complement of the Cantor set (Exercise 3.44), and that it maps the Cantor set *onto* $[0,1]$.

(b) Deduce that F (extended to $\mathbb{R}$ in the obvious way) is a continuous distribution function, and the corresponding measure puts all its mass on the Cantor set.

(c) Conclude that F has no density, i.e. there is no function $f(x)$ for which $F(x) = \int_0^x f(y)\, dy$ for all $x \in [0,1]$. F is a *purely singular* distribution function.

3.46. Let $X_1, X_2, \ldots, X_n$ be independent random variables with mean zero and variance one. Let Y be a square-integrable random variable. Let $a_j = E\{YX_j\}$. Prove **Bessel's inequality**: $E\{Y^2\} \geq \sum_{j=1}^n a_j^2$. Show that Bessel's inequality holds if the X_i are only uncorrelated, not necessarily independent. Show that this extends to an infinite sequence of X_n.

[Hint: Take the expectation of $\left(Y - \sum_{j=1}^n E\{YX_j\} X_j\right)^2$.]

3.7. Conditional Distributions, Densities, and Expectations

3.47. Let $X_1, X_2, \ldots$ be independent mean zero, variance one random variables. Let Y be a square integrable random variable. Show that $E\{YX_n\} \longrightarrow 0$ as $n \to \infty$.

[Hint: Use the previous problem.]

3.48. Let X and Y be independent random variables. Suppose that X has an absolutely continuous distribution. Show that $X + Y$ also has an absolutely continuous distribution.

[Hint: The interchange of order of integration holds for iterated Stieltjes integrals, too.]

3.49. A family decides that they will continue to have children until the first boy is born, then they will stop. Assume that boys and girls are equally likely, and that successive births are independent. What is the expected total number of girls born in the family? What is the expected number of children?

3.50. Flip a coin with $P\{\text{heads}\} = p$. Let X_n be the number of flips it takes to get a run of n successive heads. (If the first flip is tail and the next n flips are heads, then $X_n = n + 1$.)

    (a) Consider $E\{X_{n+1} \mid X_n = k\}$ and relate $E\{X_n\}$ and $E\{X_{n+1}\}$.
    (b) Show that $E\{X_n\} = \sum_{k=1}^{n} \frac{1}{p^k}$.

Chapter 4

Convergence

4.1. Convergence of Random Variables

Let $(\Omega, \mathcal{F}, P)$ be a probability space, and let $X, X_1, X_2, \ldots$ be a sequence of random variables. What does it mean to say that "X_n converges to X?"

The X_n are random variables, which are functions on Ω, so this involves convergence of functions[1]. There are a number of possibilities.

Definition 4.1. (X_n) **converges pointwise** to X if $X_n(\omega) \longrightarrow X(\omega)$ for all $\omega \in \Omega$.

This is the usual definition of convergence in calculus, but, strangely enough, it is seldom used for random variables. It is too strong. We need to look at some weaker forms of convergence.

Definition 4.2. *(i)* (X_n) converges to X **in probability** if for all $\varepsilon > 0$,
$$P\{|X_n - X| > \varepsilon\} \longrightarrow 0 \text{ as } n \to \infty.$$

(ii) (X_n) converges to X **almost everywhere** or **with probability one** if there exists an event Λ of probability zero such that if $\omega \notin \Lambda$, then $X_n(\omega) \longrightarrow X(\omega)$ as $n \to \infty$.

(iii) (X_n) converges to X **in the mean** or **in L^1** if X is integrable and
$$E\{|X_n - X|\} \longrightarrow 0 \text{ as } n \to \infty.$$

(iv) Let $p \geq 1$. (X_n) converges to X **in L^p** if $|X|^p$ is integrable and
$$E\{|X_n - X|^p\} \longrightarrow 0 \text{ as } n \to \infty.$$

---

[1] The convergence theorems in this chapter are due to Henri Lebesgue. They transformed analysis, and probabilists are quite happy to borrow them.

(v) Let F_n and F be the distribution functions of X_n and X, respectively. (X_n) converges to X **in distribution** if $F_n(x) \longrightarrow F(x)$ for all x which are continuity points of F.

Remark 4.3. This puts the main types of convergence on the table. We will not use convergence in distribution until §6.2, but the others will soon come into play.

The word "almost" is used in a specific sense in probability. When we say something occurs "**almost everywhere**" or "**almost surely**" we mean that it occurs *except* possibly on an event of probability zero. They are abbreviated "a.e." and "a.s.", respectively. The two are used interchangeably.

In particular, almost everywhere convergence is like pointwise convergence except that there can be a set of probability zero on which the sequence fails to converge. Intuitively, what happens on a set of probability zero does not matter, so this is "pointwise convergence to all good intents and purposes." We will see a lot of it. Convergence in probability is closely connected with a.e. convergence, but turns out to be weaker. The other two brands of convergence involve convergence in an "average" sense. Let us examine some of the relations between the different modes of convergence.

Proposition 4.4. *If $X_n \longrightarrow X$ in L^p for some $p > 0$, then $X_n \longrightarrow X$ in probability.*

Proof. This is immediate from Chebyshev's inequality, since, if $\varepsilon > 0$,

$$P\{|X_n - X| > \varepsilon\} \leq \frac{1}{\varepsilon^p} E\{|X_n - X|^p\} \longrightarrow 0$$

since $X_n \longrightarrow X$ in L^p. □

Careful: convergence in probability does not imply convergence in L^1 or in L^p for any $p > 0$.

Proposition 4.5. *Suppose X_n converges to X a.e. Then X_n converges to X in probability.*

Proof. There is a set Λ of probability zero such that if $\omega \notin \Lambda$, $X_n(\omega)$ converges to $X(\omega)$. Consider $Z_n \stackrel{def}{=} \sup\{|X_k - X| : k \geq n\}$. Note that $X_n(\omega) \longrightarrow X(\omega) \iff Z_n(\omega) \longrightarrow 0$. Let $\varepsilon > 0$ and let $\Gamma_n^\varepsilon = \{Z_n \geq \varepsilon\}$. Notice that if $\omega \in \bigcap_n \Gamma_n^\varepsilon$, then $Z_n(\omega)$ does not converge to zero, so that $\bigcap_n \Gamma_n^\varepsilon \subset \Lambda$. Now $\Gamma_n^\varepsilon \supset \Gamma_{n+1}^\varepsilon \supset \ldots$, so $P\{\Gamma_n^\varepsilon\} \longrightarrow P\{\bigcap_n \Gamma_n^\varepsilon\} \leq P\{\Lambda\} = 0$. Thus, as $|X_n - X| \leq Z_n$, $P\{|X_n - X| \geq \varepsilon\} \leq P\{\Gamma_n^\varepsilon\} \longrightarrow 0$. This holds for all $\varepsilon > 0$, so $X_n \longrightarrow 0$ in probability. □

Example 4.5.1. The converse of Proposition 4.5 is false. Convergence in probability does not imply convergence a.e. Indeed, consider the probability

4.1. Convergence of Random Variables

space $([0,1], \mathcal{B}, dx)$, with $dx =$ Lebesgue measure, and random variables $X_1, X_2, \ldots$ defined by

$$X_1 \equiv 1,$$
$$X_2 = I_{[0,1/2]},$$
$$X_3 = I_{(1/2,1]},$$
$$X_4 = I_{[0,1/4]},$$
$$X_5 = I_{(1/4,1/2]},$$
$$\vdots$$
$$X_8 = I_{[0,1/8]},$$
$$\vdots$$

Note that the X_n converge to zero in probability, since for $0 < \varepsilon < 1$, $P\{|X_n - 0| \geq \varepsilon\} = P\{X_n \neq 0\} \to 0$. On the other hand, there is no x for which $X_n(x) \to 0$, since for all x, there are infinitely many n for which $X_n(x) = 1$, and infinitely many n for which $X_n(x) = 0$. Indeed, for each x and each k there is exactly one n between 2^k and $2^{k+1} - 1$ for which $X_n(x) = 1$. Therefore, the sequence $X_n(x)$ does not converge for any x.

Let us check that the limit of a sequence of random variables is a random variable. This is not quite trivial, since a random variable on $(\Omega, \mathcal{F}, P)$ is a function on Ω—no trouble there, since the limit of functions is a function—which satisfies one extra condition, namely that it is measurable, or—if we denote it by X—that $\{X < x\} \in \mathcal{F}$ for all x. Keep in mind that once we admit limiting operations, we have to admit infinite values. So the random variables in the following proposition may take on the values $\pm\infty$.

Proposition 4.6. *Let* $(\Omega, \mathcal{F}, P)$ *be a probability space and let* $X_1, X_2, \ldots$ *be extended real-valued random variables. Then the following are also random variables:*

(i) $M(\omega) = \sup\{X_j(\omega), j = 1, 2, \ldots\}$;
(ii) $L(\omega) = \inf\{X_j(\omega), j = 1, 2, \ldots\}$;
(iii) $m(\omega) = \limsup_{n \to \infty} X_n(\omega)$;
(iv) $\ell(\omega) = \liminf_{n \to \infty} X_n(\omega)$;
(v) $X_\infty(\omega) = \begin{cases} \lim_n X_n(\omega) & \text{if the limit exists,} \\ 0 & \text{otherwise.} \end{cases}$

Proof. (i) $\{\omega : M(\omega) \leq x\} = \bigcap_{j=1}^{\infty} \{\omega : X_j(\omega) \leq x\}$. But $\mathcal{F}$ is a sigma-field and $\{X_j \leq x\} \in \mathcal{F}$, (for X_j is a random variable) so the intersection is also in $\mathcal{F}$. Thus $\{M \leq x\} \in \mathcal{F}$, hence M is a random variable.

(ii) This follows directly from (i) since $\inf_n X_n = -\sup_n\{-X_n\}$.

(iii) Let $m_n = \sup_{j \geq n} X_j$. This is a random variable by (i), and therefore $m = \inf_n m_n$ is a random variable by (ii).

(iv) This follows from (iii) since $\liminf_n X_n = -\limsup_n(-X_n)$.

(v) The limit exists on the set $\Lambda_\infty \stackrel{\text{def}}{=} \{\omega : m(\omega) = \ell(\omega)\}$. This is in $\mathcal{F}$ since both ℓ and m, and hence $m - \ell$, are random variables. Define $Y_n = X_n I_{\Lambda_\infty}$, which, being the product of random variables, is a random variable. Then $\lim_n Y_n(\omega)$ exists for every ω and equals $X_\infty(\omega)$. Thus $X_\infty \equiv \limsup_n Y_n$, which is a random variable by (iii). □

Proposition 4.7. *Let X, X_n be a sequence of random variables. Then (X_n) converges to X in probability if and only if*

$$\lim_{n \to \infty} E\left\{\frac{|X_n - X|}{1 + |X_n - X|}\right\} = 0.$$

We leave the proof as an exercise.

The following result is called a lemma for historical reasons, but it deserves to be called a theorem. It is so useful that people have referred to a certain type of proof as "Borel-Cantellizing". First, we need to introduce the lim inf and the lim sup of a sequence of sets.

Let $\Lambda_1, \Lambda_2, \ldots$ be a sequence of subsets of Ω.

Definition 4.8. $\limsup_n \Lambda_n \stackrel{\text{def}}{=} \{\omega : \omega \in \Lambda_n \text{ for infinitely many } n\}$.

$$\liminf_n \Lambda_n \stackrel{\text{def}}{=} \{\omega : \omega \in \Lambda_n \text{ for all but finitely many } n\}.$$

Another way of stating it is that $\liminf_n \Lambda_n$ is the set of ω for which there exists N such that $\omega \in \Lambda_n$ for all $n \geq N$, and $\limsup_n \Lambda_n$ is the set of ω such that, for all N, there exists $n \geq N$ for which $\omega \in \Lambda_n$. These can be expressed in terms of unions and intersections of the Λ_n.

Exercise 4.1. Show that
(a) $\liminf_n \Lambda_n = \bigcup_{n=m}^\infty \bigcap_{j=m}^\infty \Lambda_j$;
(b) $\limsup_n \Lambda_n = \bigcap_{n=m}^\infty \bigcup_{j=m}^\infty \Lambda_j$;
(c) $\limsup_n \Lambda_n \supset \liminf_n \Lambda_n$;
(d) $\limsup_n \Lambda_n = (\liminf \Lambda_n^c)^c$;
(e) $\liminf_n I_{\Lambda_n} = I_{\{\liminf_n \Lambda_n\}}$;
(f) $\limsup_n I_{\Lambda_n} = I_{\{\limsup_n \Lambda_n\}}$.

We gave the definitions of the lim sup and lim inf in their most intuitive form, and we should think of them that way. When we come to proofs,

4.1. Convergence of Random Variables

we will fall back on the formal statements (a) and (b). We will abbreviate "infinitely often" by "i.o." and refer to lim sup Λ_n as the set "$\{\Lambda_n \text{ i.o.}\}$". The statements (e) and (f) justify the names "lim inf" and "lim sup".

The Borel-Cantelli Lemma has two parts. The second is a partial converse to the first, and is often called "The converse Borel-Cantelli Lemma."

Theorem 4.9 (Borel-Cantelli Lemma). *Let $\Lambda_1, \Lambda_2, \ldots$ be a sequence of events. Suppose that $\sum_{n=1}^{\infty} P\{\Lambda_n\} < \infty$. Then*
$$P\{\limsup_n \Lambda_n\} = 0.$$
Conversely, if the Λ_n are independent, there are only two possibilities:
$$P\{\limsup_n \Lambda_n\} = \begin{cases} 0 & \iff \sum_{n=1}^{\infty} P\{\Lambda_n\} < \infty, \\ 1 & \iff \sum_{n=1}^{\infty} P\{\Lambda_n\} = \infty. \end{cases}$$

Proof. The lim sup of the Λ_n can be written as $\lim_{m \to \infty} \bigcup_{n=m}^{\infty} \Lambda_n$. Notice that the union decreases as m increases, so
$$P\{\limsup_n \Lambda_n\} = \lim_{m \to \infty} P\Big\{ \bigcup_{n=m}^{\infty} \Lambda_n \Big\}$$
$$\leq \lim_{m \to \infty} \sum_{n=m}^{\infty} P\{\Lambda_n\}.$$

But this is the tail of a convergent series (the series $\sum_n P\{\Lambda_n\}$ converges by hypothesis) so it converges to zero.

For the converse, consider
$$P\{\limsup_n \Lambda_n\} = 1 - P\{\liminf_n \Lambda_n^c\}$$
$$= 1 - \lim_{m \to \infty} P\Big\{ \bigcap_{n=m}^{\infty} \Lambda_n^c \Big\}$$
$$= 1 - \lim_{m \to \infty} \prod_{n=m}^{\infty} P\{\Lambda_n^c\}.$$

But if $0 < x_n < 1$, $\prod_n (1 - x_n)$ is either zero or strictly positive according to whether $\sum_n x_n = \infty$ or $\sum_n x_n < \infty$, respectively. If the sum is finite, $\lim_{m \to \infty} \prod_{n=m}^{\infty} (1 - x_n) = 1$. Thus, the probability of the lim sup equals
$$1 - \lim_{m \to \infty} \prod_{n=m}^{\infty} (1 - P\{\Lambda_n\}) = \begin{cases} 1 & \text{if } \sum_n P\{\Lambda_n\} = \infty, \\ 0 & \text{if } \sum_n P\{\Lambda_n\} < \infty. \end{cases}$$
$\square$

The Borel-Cantelli Lemma gives us a good way of showing that something does not occur infinitely often. One merely has to check that the sum of the probabilities is finite. Here is an application. While it is not true that convergence in probability implies convergence a.e., something almost as good is true: there exists an a.e.-convergent subsequence. So, while the original series may fail to converge a.e., we can always find a subsequence which does.

Theorem 4.10. *Let X and $X_1, X_2, \ldots$ be random variables, and suppose that $X_n \to X$ in probability. Then there exists a subsequence (n_k) such that $X_{n_k} \longrightarrow X$ a.e.*

Proof. For each m, $P\{X_n - X| > \frac{1}{m}\} \longrightarrow 0$ as $n \to \infty$. Choose n_1 large enough so that $P\{X_{n_1} > 1\} < 1/2$. We can choose a sequence (n_m), $m = 2, 3, \ldots$ by induction: suppose $n_1 < n_2 < \cdots < n_{m-1}$ have been chosen. Then choose n_m so that $n_m > n_{m-1}$ and

$$P\left\{|X_{n_m} - X| > \frac{1}{m}\right\} < \frac{1}{2^m},$$

Let $\Lambda_m = \{|X_{n_m} - X| > 1/m\}$. Then $\sum_m P\{\Lambda_m\} < \infty$, so by the Borel-Cantelli Lemma, $P\{\limsup_m \Lambda_m\} = 0$. In other words, $P\{\Lambda_m \text{ i.o.}\} = 0$. If something does not happen infinitely often, it happens only finitely often, and there must be a last time. So, if $\omega \in \Lambda_n$ only finitely often, there is an N (depending on ω) such that for all $m \geq N$, $\omega \notin \Lambda_m$, hence $|X_{n_m}(\omega) - X(\omega)| \leq 1/m$ for all $m > N$. This means that $X_{n_m}(\omega) \to X(\omega)$ as $m \to \infty$. This is true for a.e. ω, hence the subsequence converges a.e. $\square$

4.2. Convergence Theorems for Expectations

One recurring question in analysis is this: suppose a sequence of functions converges. Do their integrals converge to the integral of the limit? In probability, the question becomes: suppose X_n is a sequence of random variables which converges to X in some sense. Does $E\{X_n\} \to E\{X\}$? In other words, is $\lim_n E\{X_n\} = E\{\lim_n X_n\}$?

The answer is, "Not always." Take, for example, the functions $f_n(x) = nI_{(0,1/n)}(x)$. Then $f_n(x) \to f(x) \equiv 0$ for each x, while $\int_0^1 f_n(x)\,dx = 1$ for all n. So the limit of the integrals is one, while the integral of the limit is zero.

We need theorems to tell us when we can go to the limit under the expectation. Notation: write $x \wedge y = \min(x, y)$, $x \vee y = \max(x, y)$.

Lemma 4.11. *Let $X \geq 0$ be a random variable with values in the extended half-line $[0, \infty]$. Then*

$$\lim_{n \to \infty} E\{X \wedge n\} = E\{X\}.$$

4.2. Convergence Theorems for Expectations

Proof. Since X is positive, its expectation exists, but may be infinite, so the two sides of the above equation may be infinite.

If $P\{X = \infty\} > 0$, then $X \wedge n = n$ on $\{X = \infty\}$, and $E\{X \wedge n\} = nP\{X = \infty\} \to \infty = E\{X\}$. Thus we may assume $X < \infty$ a.s., and, as sets of probability zero do not affect expectations, we may even assume X is finite everywhere.

It is enough to show that for each k, $\lim_{n \to \infty} E\{X \wedge n\} \geq E\{\underline{X}_k\}$, where $\underline{X}_k$ is the lower dyadic approximation of §3.1, since then $E\{X\} \geq \lim_{n \to \infty} E\{X \wedge n\} \geq E\{\underline{X}_k\} \to E\{X\}$ as $k \to \infty$. But now

$$E\{X \wedge n\} \geq E\{\underline{X}_k \wedge n\}$$
$$\geq E\{\underline{X}_k I_{\{\underline{X}_k \leq n\}}\}$$
$$= \sum_{j=1}^{2^k n} \frac{j}{2^n} P\left\{\underline{X}_k = \frac{j}{2^n}\right\}.$$

Let $n \to \infty$. By the definition of the sum of a series, this has the limit

$$\sum_{j=1}^{\infty} \frac{j}{2^n} P\left\{\underline{X}_k = \frac{j}{2^n}\right\} = E\{\underline{X}_k\},$$

whether or not the sum is finite. □

This brings us to three important theorems (though, typically, one is called a lemma): the Monotone Convergence Theorem, Fatou's Lemma, and the Dominated Convergence Theorem. (The capitalization is deserved. They are important.) The first is the Monotone Convergence Theorem.

Theorem 4.12 (Monotone Convergence Theorem). *Let X, $X_1, X_2, \ldots$ be positive random variables such that for each n, $X_n \leq X_{n+1}$ a.s. and $\lim X_n = X$ a.s. Then*

$$\lim_{n \to \infty} E\{X_n\} = E\{X\}.$$

Remark 4.13. This holds even if $E\{X\} = \infty$.

Proof. Let us first get rid of the "almost everywheres". Let $\Lambda_1 = \{\omega : \exists n \ni X_n(\omega) > X_{n+1}(\omega)\}$; $\Lambda_2 = \{\omega : \exists n \ni X_n < 0\}$; and $\Lambda_3 = \{\omega : \lim X_n(\omega) \neq X(\omega)\}$, and let $\Lambda = \Lambda_1 \cup \Lambda_2 \cup \Lambda_3$.

$$P\{\Lambda_1\} \leq \sum_n P\{X_n > X_{n+1}\} = 0,$$
$$P\{\Lambda_2\} \leq \sum_n P\{X_n < 0\} = 0,$$
$$P\{\Lambda_3\} = 0.$$

Thus $P\{\Lambda\} = 0$. We can now replace X and X_n by $X' = XI_{\Lambda^c}$ and $X'_n = X_n I_{\Lambda^c}$, respectively. The X''s satisfy the conditions identically—no exceptional ω's—and, as they equal the original random variables a.e., they have the same expectations. Consequently, we can assume that the original variables satisfy the conditions identically.

Now $0 \leq E\{X_1\} \leq E\{X_2\} \leq \cdots \leq E\{X\}$ so that $\lim E\{X_n\}$ exists, finite or infinite, and $\lim_n E\{X_n\} \leq E\{X\}$. We will be done if we show that $\lim_n E\{X_n\} \geq E\{X\}$. In fact, $E\{X\} = \lim_N E\{X \wedge N\}$ by Lemma 4.11, so it is enough to show that $\lim_n E\{X_n\} \geq E\{X \wedge N\}$ for each N, or even that $\lim_n E\{X_n \wedge N\} \geq E\{X \wedge N\}$. Thus, by replacing X_n and X by $X_n \wedge N$ and $X \wedge N$, respectively, it is sufficient to show it for the case where $0 \leq X_n \leq X \leq N$ and $X_n(\omega) \to X(\omega)$ for all ω.

Let $\varepsilon > 0$ and put $A_{\varepsilon,n} = \{\omega : X_n(\omega) < X(\omega) - \varepsilon\}$. Then $A_{\varepsilon,n} \supset A_{\varepsilon,n+1}$, and $\bigcap_n A_{\varepsilon,n} = \emptyset$. Therefore, $P\{A_{\varepsilon,n}\} \to 0$ as $n \to \infty$; but $X - X_n$ is at most N on the set $A_{\varepsilon,n}$, and is bounded by ε elsewhere. Thus $\lim_n(E\{X\} - E\{X_n\}) \leq \lim_n(NP\{A_{\varepsilon,n}\} + \varepsilon) = \varepsilon$, hence $\lim_n E\{X_n\} \geq E\{X\} - \varepsilon$. Since ε is arbitrary, we are done. $\square$

Remark 4.14. Let Λ be an event of probability zero, and let $X = 0$ on Λ^c, $X = \infty$ on Λ. We saw in §2.3 that $E\{X\} = 0$. We can confirm this with the monotone convergence theorem. Let $X_n = nI_\Lambda$. Then $X_n \uparrow X \Rightarrow E\{X_n\} \uparrow E\{X\}$. But $E\{X_n\} = 0$ for all n, so $E\{X\} = 0$ as claimed.

The monotone convergence theorem does not require integrability, which turns out to be very useful. There is a corollary for positive decreasing sequences, but for this, we do need integrability.

Corollary 4.15. *Let X and $X_1, X_2, \ldots$ be random variables such that for each n, $X_n \geq X_{n+1} \geq 0$ a.e. and $\lim_n X_n = X$ a.e. Suppose that $E\{X_1\}$ is finite. Then*

$$\lim_{n \to \infty} E\{X_n\} = E\{X\}.$$

Proof. $X_1 - X_n$ increases to $X_1 - X$, and satisfies the hypotheses of the monotone convergence theorem. Since X_1 and therefore X and all the X_n are integrable, we have $E\{X_1\} - E\{X_n\} \longrightarrow E\{X_1 - X\} = E\{X_1\} - E\{X\}$, so $E\{X_n\} \to E\{X\}$. $\square$

Example 4.15.1. To see that the integrability condition is necessary, consider $X(x) = 1/nx$ on the probability space $\bigl((0,1), \mathcal{B}, P\bigr)$, where $dP = dx$ is Lebesgue measure. Then $X_n(x) \downarrow 0$ for all $x \in (0,1)$, but $E\{X_n\} = \infty$ for all n, so it does not converge to 0.

4.2. Convergence Theorems for Expectations

Theorem 4.16 (Fatou's Lemma). *Let X_n be a sequence of positive random variables. Then*
$$\liminf_{n \to \infty} E\{X_n\} \geq E\{\liminf_{n \to \infty} X_n\}.$$

Proof. Recall that for a sequence of reals, $\liminf_n x_n = \lim_{n\to\infty} \left(\inf_{j \geq n} x_j\right)$, so that $\liminf_n x_n \geq \inf_{j \geq n} x_j$ and $\inf_{j \geq n} x_j \uparrow \liminf_n x_n$ as $n \to \infty$.

Apply this to the random variables X_n. Certainly,
$$E\{X_n\} \geq E\{\inf_{j \geq n} X_j\}.$$
Let $n \to \infty$. Then
$$\liminf_n E\{X_n\} \geq \liminf_n E\{\inf_{j \geq n} X_j\}.$$
But as $n \to \infty$, $\inf_{j \geq n} X_j \uparrow \liminf_n X_n$, so the lim inf is actually an increasing limit. By the monotone convergence theorem, it is
$$= E\{\liminf_n X_n\}.$$
$\square$

Remark 4.17. Fatou's Lemma does not assume that the X_n converge in any way.

This brings us to Lebesgue's dominated convergence theorem.

Theorem 4.18 (Dominated Convergence Theorem). *Let X and X_n, $n = 1, 2, \ldots$ be random variables. Suppose that there is a random variable Y such that*
  (i) $|X_n| \leq Y$ a.s. for all n;
  (ii) $\lim_{n \to \infty} X_n = X$ either a.e. or in probability.
  (iii) $E\{Y\} < \infty$.
Then
$$\lim_{n \to \infty} E\{X_n\} = E\{X\}.$$

Proof. First suppose that $X_n \longrightarrow X$ a.e. Then $Y - X_n \geq 0$ a.e., and $Y - X_n \longrightarrow Y - X$ a.e. By Fatou's Lemma,
$$\liminf_n E\{Y - X_n\} \geq E\{\liminf_n (Y - X_n)\} = E\{Y - X\}.$$
But the left-hand side equals $E\{Y\} - \limsup_n E\{X_n\}$ and the right-hand side equals $E\{Y\} - E\{X\}$. Since $E\{Y\}$ is finite, we can subtract it from both sides to see that
$$\limsup_n E\{X_n\} \leq E\{X\}.$$

On the other hand, $Y + X_n$ is also positive, and applying Fatou's Lemma to it gives
$$\liminf_n E\{Y + X_n\} \geq E\{\liminf_n (Y + X_n)\} = E\{Y + X\}.$$
The left-hand side equals $E\{Y\} + \liminf_n E\{X_n\}$. Subtract $E\{Y\}$ from both sides, to see that
$$\liminf_n E\{X_n\} \geq E\{X\}.$$
But $\liminf_n E\{X_n\} \leq \limsup_n E\{X_n\}$, so both lim inf and lim sup must be equal, and equal to $E\{X\}$. This proves the theorem for a.e. convergence.

Now suppose $X_n \to X$ in probability. Since $E\{|X_n|\} \leq E\{Y\}$, the sequence of expectations is bounded. If it does not converge to $E\{X\}$, there exists a subsequence (X_{n_k}) such that $E\{X_{n_k}\}$ converges to a limit $L \neq E\{X\}$. (X_{n_k}) converges to X in probability so there exists a *further* subsequence (X_{n_j}) which converges a.e. to X. By what we have just proved, $E\{X_{n_j}\} \to E\{X\}$. This is a contradiction, since we supposed that the original subsequence, and therefore the further subsequence, converged to $L \neq E\{X\}$. □

Corollary 4.19 (Bounded Convergence Theorem). *Let X and X_n, $n = 1, 2, \ldots$ be random variables and let $M \geq 0$ be a real number such that for each n, $|X_n| \leq M$ a.e. Suppose that (X_n) converges to X either a.e. or in probability. Then*
$$\lim_{n \to \infty} E\{X_n\} = E\{X\}.$$

Proof. Let $Y = M$ and apply the Dominated Convergence Theorem. □

Corollary 4.20. *Let X and X_n, $n = 1, 2, \ldots$ be random variables such that $X_n \to X$ either a.e. or in probability. Let Y be an integrable random variable, such that $|X_n| \leq Y$ a.e. for each n. Then X_n converges to X in L^1. Conversely, if X_n converges to X in L^1, $E\{X_n\} \to E\{X\}$.*

Proof. $|X_n - X| \leq 2Y$ and $|X_n - X| \to 0$. Thus $E\{|X_n - X|\} \to 0$ by the Dominated Convergence Theorem, so the sequence converges in L^1.

Conversely, if X_n converges to X in L^1, then
$$|E\{X\} - E\{X_n\}| \leq E\{|X - X_n|\} \longrightarrow 0.$$
□

Exercise 4.2. Let $1 \leq p < q$. Show that if a sequence of random variables X_n converges to a random variable X in L^q, that it also converges to X in L^p.

Exercise 4.3. Let X be an integrable random variable, and let Λ_n be a sequence of events with $P\{\Lambda_n\} \to 0$. Prove that $\int_{\Lambda_n} X \, dP \to 0$.

4.3. Applications

Example 4.20.1. Differentiation under the Expectation. Let X be a random variable with moment generating function $M(\theta)$. If $M(\theta)$ exists for θ in some neighborhood of zero, we claimed in Theorem 2.38 that $M'(0) = E\{X\}$. This was based on differentiating the formula for $M(\theta)$ *inside* the expectation. We could not justify it at the time, but now we can.

Proposition 4.21. *Suppose $M(\theta)$ exists in a neighborhood of the origin. Then the following hold in a neighborhood of the origin.*

(i) $M'(\theta) = E\{Xe^{\theta X}\}$. *More generally,*

(ii) $M^{(k)}(\theta) = E\{X^k e^{\theta X}\}$.

Proof. By the definition of the derivative,

$$M'(\theta) = \lim_{h \to 0} \frac{M(\theta+h) - M(\theta)}{h}$$

(4.1)
$$= \lim_{h \to 0} E\left\{\frac{e^{(\theta+h)X} - e^{\theta X}}{h}\right\}.$$

To show this, we will find an integrable random variable Y which dominates the integrand in (4.1), and apply the Dominated Convergence Theorem.

Choose $b > 0$ small enough that both $M(b)$ and $M(-b)$ exist, and let $0 < \varepsilon < b$. Assume θ and h are small enough that θ and $\theta \pm h$ are in $(-b+\varepsilon, b-\varepsilon)$. By the mean value theorem, for any x, there is θ' (which depends on x) between θ and $\theta + h$ such that

$$\frac{e^{(\theta+h)x} - e^{\theta x}}{h} = xe^{\theta' x}.$$

Note that

$$|xe^{\theta' x}| \leq |x|e^{|\theta' x|} \leq |x|e^{(b-\varepsilon)|x|} \leq Ce^{b|x|},$$

where C is the maximum value of $x \mapsto |x|e^{-\varepsilon|x|}$. Now $e^{b|x|} \leq e^{bx} + e^{-bx}$ so

$$|Xe^{\theta' X}| \leq C(e^{bX} + e^{-bX}) \stackrel{\text{def}}{=} Y.$$

But Y is integrable, since $E\{Y\} = C(M(b) + M(-b)) < \infty$. According to the Dominated Convergence Theorem, we can go to the limit under the integral in (4.1) to see that

$$M'(\theta) = E\{Xe^{\theta X}\}.$$

Set $\theta = 0$ to get the formula for the first moment in Theorem 2.38. We leave it to the reader to prove by induction that one can differentiate any number of times under the expectation. $\square$

Problems 4.3

4.4. Let (X_n) and (Y_n) be sequences of random variables and let a be a constant. Show that the following are true for convergence in probability and convergence a.e. Show that (a) and (b) hold for convergence in L^1, too. Give examples in which (c) and (d) do not hold for L^1 convergence.

Suppose that $X_n \longrightarrow X$ and $Y_n \longrightarrow Y$. Then
  (a) $X_n + Y_n \longrightarrow X + Y$.
  (b) $aX_n \longrightarrow aX$.
  (c) $X_n Y_n \longrightarrow XY$.
  (d) If $P\{X = 0\} = 0$, $\frac{Y_n}{X_n} \longrightarrow \frac{Y}{X}$.

4.5. Let $r > 1$. Show that if $X_n \longrightarrow X$ in L^r, then $X_n \longrightarrow X$ in L^p for all $1 \leq p \leq r$.

4.6. Show that $\rho(X, Y) \stackrel{def}{=} E\left\{\frac{|Y-X|}{1+|Y-X|}\right\}$ is a metric on the space of all random variables. (Regard two random variables as equal if they are equal almost surely.) Prove Proposition 4.7 by showing that for a sequence (X_n) of random variables, $X_n \longrightarrow X$ in probability if and only if $\rho(X_n, X) \longrightarrow 0$. (This shows that convergence in probability is metrizable.)

4.7. Show that if $X_n \longrightarrow X$ in probability and if $X_n \longrightarrow Y$ in probability, that $X = Y$ a.s.

4.8. Show that it is possible to have $X_n \longrightarrow X$ in probability, but not to have $P\{X_n \in (a, b)\} \longrightarrow P\{X \in (a, b)\}$ for all intervals (a, b).

4.9. Let B_n be a countable collection of Borel subsets of $[0, 1]$. Suppose that there is $\delta > 0$ such that the Lebesgue measure of each B_n is greater than δ. Show that there is at least one point which belongs to infinitely many B_n.

4.10. Let $X_1, X_2, \ldots$ be i.i.d. Cauchy random variables. Show that $\limsup_{n \to \infty} \frac{\log |X_n|}{\log n} = 1$ a.s.

4.11. Let (X_n) be a sequence of random variables. Show that

$$X_n \longrightarrow \infty \text{ a.s.} \iff P\{X_n < M \text{ i.o.}\} = 0, \quad \forall M.$$

4.12. The following statements are either true or false. Say which. If true, give a (short) proof. If false, give a counterexample. In all cases, X and $X_1, X_2, \ldots$ are random variables on a given probability space, and C is a constant.
  (a) If $X_n \longrightarrow C$ in probability, then $X_n^2 \longrightarrow C^2$ in distribution.
  (b) If $X_n \longrightarrow 0$ in distribution, then $X_n \longrightarrow 0$ in probability.
  (c) If $X_n \longrightarrow X$ in distribution, then $X_n \longrightarrow X$ in probability.
  (d) $\limsup_{n \to \infty} X_n \leq \limsup_{n \to \infty} E\{X_n\}$.
  (e) If every subsequence of (X_n) contains a further subsequence which converges a.s., then (X_n) converges in probability.

4.13. Show that (X_n) converges in probability to X if and only if every subsequence (X_{n_k}) contains a further subsequence that converges a.e. to X.

4.3. Applications

4.14. Let (X_n) be a sequence of random variables which converge in probability to X. Show that if f is a continuous function on $\mathbb{R}$, then $f(X_n) \longrightarrow f(X)$ in probability. (Careful: f may not be uniformly continuous, or even bounded. Try truncating at $\pm A$ for large A.)

4.15. Show by example that it is possible to have $\sup_n X_n = \infty$ a.e. but have no subsequence X_{n_k} such that $\lim_k X_{n_k} = \infty$ in probability.

4.16. Show that convergence in probability for arbitrary random variables may be reduced to convergence in probability for bounded random variables by the transformation $X' = \arctan X$.

4.17. Let (X_n) be a sequence of integrable random variables which increases to a limit X, and $\sup_n E\{X_n\} < \infty$. Show that if $X = \lim_n X_n$, that $X < \infty$ a.s.

4.18. Show that the following is a metric on the space of random variables, and that a sequence of random variables converges in this metric if and only if it converges in probability.
$$\rho(X, Y) = \inf\{\varepsilon > 0 : P\{|Y - X| > \varepsilon\} \le \varepsilon\}.$$

4.19. While convergence in probability and convergence in L^p are metrizable, convergence a.e. is not, in general. Show that there does not exist a metric $\rho(X, Y)$ on random variables such that $\rho(X_n, X) \longrightarrow 0 \iff X_n \longrightarrow X$ a.e.
[Hint: If X_n did *not* converge to X a.s., there would be a subsequence (n_k) and $\varepsilon > 0$ such that $\rho(X_{n_k}, X) \ge \varepsilon$ for all k.]

4.20. A sequence (X_n) of random variables is said to be **Cauchy in probability** if for each $\varepsilon > 0$ there is N such that for all $m, n > N$, $P\{|X_m - X_n| > \varepsilon\} < \varepsilon$.

Show that if (X_n) is Cauchy in probability, there exists a random variable X such that $X_n \longrightarrow X$ in probability.
[Hint: Show there is a subsequence (n_k) for which $P\{|X_{n_{k+1}} - X_{n_k}| > 2^{-k}\} < 2^{-k}$. Show that $\sum(X_{n_{k+1}} - X_{n_k})$ converges to some Y. Let $X = Y + X_{n_1}$ and show that X works.]

4.21. Show that for any sequence (X_n) of random variables, there exists a sequence of constants (A_n) such that $X_n/A_n \longrightarrow 0$ a.s.

4.22. (Egorov's Theorem.) Suppose a sequence (X_n) of random variables converges on a set Λ of strictly positive probability. Show that for any $\varepsilon > 0$ there exists a set $\Lambda_\varepsilon \subset \Lambda$ such that $P\{\Lambda - \Lambda_\varepsilon\} < \varepsilon$ and (X_n) converges uniformly to X on Λ_ε.
[Hint: Show that it suffices to show it when $\Lambda = \Omega$ and the limit is zero. Let $E_{mk} = \bigcap_{n=k}^{\infty}\{|X_n| \le 1/m\}$. Show that for each m, there is a k such that $P\{E_{mk}\} > 1 - \varepsilon/2^m$. Take Λ_ε to be the intersection of the E_{mk}.]

4.23. Let $X_1, X_2, \ldots$ be a sequence of independent random variables, and let X be a random variable. Show that if $X_n \longrightarrow X$ a.e., then X is a.s. constant. If $X_n \longrightarrow X$ in distribution, is X necessarily constant?

4.24. Suppose that $X_1, X_2, \ldots$ is a sequence of integrable random variables which decreases to a random variable X. Show that if $\inf_n E\{X_n\} > -\infty$, that $X \in L^1$ and X_n converges to X in L^1.

4.25. Let $X_1 \leq X_2 \leq \ldots$ be an increasing sequence of integrable random variables. Suppose there is M such that $E\{X_n\} \leq M$ for all n. Show that $X \stackrel{\text{def}}{=} \lim_{n\to\infty} X_n$ exists and is a.s. finite.

4.26. Let (X_n) be a sequence of positive integrable random variables which converges a.e. to an integrable random variable X. Suppose $E\{X_n\} \longrightarrow E\{X\}$. Show that for all bounded random variables Y, $E\{YX_n\} \longrightarrow E\{YX\}$. Show by example that the conclusion may be false if the X_n are not positive.

4.27. The following extends Fatou's Lemma. Let (X_n) be a sequence of random variables and let Y be an integrable random variable such that $X_n \geq Y$ a.s. for all n. Show that $\liminf_n E\{X_n\} \geq E\{\liminf_n X_n\}$.

4.28. The following extends the dominated convergence theorem. Let (X_n) be a sequence of random variables which converges a.e. or in probability to a random variable X. Suppose (X_n) is dominated by a sequence (Y_n): $|X_n| \leq Y_n$ a.s. for all n. Suppose that (Y_n) converges a.e. to a random variable Y, and that $\lim_n E\{Y_n\} = E\{Y\}$. Show that $E\{X_n\} \longrightarrow E\{X\}$.

4.29. Suppose (X_n) is a sequence of integrable random variables which converges a.s. to an integrable random variable X. Prove that a necessary and sufficient condition that $E\{|X - X_n|\} \longrightarrow 0$ is that $E\{|X_n|\} \longrightarrow E\{|X|\}$.

[Hint: Apply Fatou's Lemma to the positive and negative parts of X_n.]

4.30. Let $X_0, X_1, X_2, \ldots$ be a sequence of random variables with $\sum_{n=0}^{\infty} E\{|X_n|\} < \infty$. Show that $X \stackrel{\text{def}}{=} \sum_{n=0}^{\infty} X_n$ exists a.s. and that $E\{X\} + \sum_{n=0}^{\infty} E\{X_n\}$.

4.31. Let $R = [a,b] \times [c,d]$ be a rectangle, and let f be a real-valued function on R. Use the dominated convergence theorem to prove the following standard theorem from advanced calculus which justifies interchanging the order of differentiation and integration.

Theorem. *Suppose f and $\frac{\partial f}{\partial x}$ are continuous on R. Define $G(x) \stackrel{\text{def}}{=} \int_c^d f(x,y)\,dy$. Then G is differentiable, and $G'(x) = \int_c^d \frac{\partial f}{\partial x}(y)\,dy$.*

4.32. Let (X_n) be a sequence of random variables which converges in distribution to a random variable X. Show that there exists a probability space $(\Omega, \mathcal{F}, P)$, a sequence of random variables (Y_n) and a random variable Y on $(\Omega, \mathcal{F}, P)$ such that $Y_n \longrightarrow Y$ a.s.

[Hint: Concentrate on the distribution functions. Try $Y_n = G_{F_n}(\text{something})$. The proof is relatively straightforward if all distribution functions F_n and F are continuous and strictly increasing. The general case is more delicate.]

4.33. Let (X_n) be a sequence of random variables. Suppose that for every $a < b$, $P\{X_n < a \text{ i.o. and } X_n > b \text{ i.o.}\} = 0$. Show that $\lim_{n\to\infty} X_n$ exists a.e., but may be infinite.

[Hint: Consider rational a and b first.]

4.34. Let $X_n \leq Y_n \leq Z_n$, $n = 1, 2, 3, \ldots$. Suppose that with probability one, $X_n \longrightarrow X$, $Y_n \longrightarrow Y$ and $Z_n \longrightarrow Z$. Show that if $E\{X_n\} \longrightarrow E\{X\}$ and $E\{Z_n\} \longrightarrow E\{Z\}$, then $E\{Y_n\} \longrightarrow E\{Y\}$.

4.3. Applications

4.35. Let (X_n) be a sequence of random variables which converges a.s. to a random variable X. Suppose $E\{|X_n|\} \leq \tan^{-1} n$. Show that X is integrable, and find upper and lower bounds for its expectation.

Chapter 5

Laws of Large Numbers

This chapter covers several theorems, called laws of large numbers. Their prototype is the law of averages. Suppose an experiment is repeated independently n times. Then

$$\frac{1}{n}\# \text{ successes in } n \text{ trials} \longrightarrow P\{ \text{ success } \}$$

as $n \to \infty$. That is, if an experiment is repeated independently, the average number of successes tends to the probability of success in each experiment.

This is so deeply ingrained that it is almost folklore, so much so that there have been serious attempts to use it to define probability. We ourselves are not above using it to motivate things. But, in the mathematical theory, at least, it is a theorem, not a definition. We will prove it here. The fact that our theorems describe things that actually happen in the real world is immensely reassuring!

Rephrase the law of averages: let $X_i = 1$ if the ith trial is a success, $X_i = 0$ if not. Let $\mu = P\{X_i = 1\} = P\{ \text{ success } \}$. Then

(5.1) $$\frac{1}{n}\sum_{i=1}^{n} X_i \longrightarrow \mu.$$

This is now a statement about sums of random variables. Noting that $\mu = E\{X_i\}$, it says that the average of a sum of i.i.d random variables tends towards their expected value. In this form, (5.1) is called the Law of Large Numbers[1]. This is the subject of this chapter.

5.1. The Weak and Strong Laws

At the cost of replacing X_i by $X_i - \mu$, we can suppose the X_i have mean zero: $\frac{1}{n}\sum_{i=1}^n X_i \longrightarrow \mu \iff \frac{1}{n}\sum_{i=1}^n (X_i - \mu) \longrightarrow 0$.

We should also remark that "$\longrightarrow$" can have different meanings, such as convergence a.e., or convergence in probability, so (5.1) can stand for several different theorems according to the exact hypotheses on the X_i and the exact form of convergence. Our first theorem deals with convergence in probability and is called the Weak Law of Large Numbers.

Notation. $S_n = X_1 + \cdots + X_n$.

Theorem 5.1 (Weak Law of Large Numbers). *Let X_1, X_2, ... be a sequence of independent, mean zero random variables. Suppose the variance of X_i is less than or equal to σ^2. Then*

$$\lim_{n\to\infty} \frac{1}{n} S_n = 0 \quad \text{in probability}.$$

Proof. $E\left\{\left(\frac{1}{n}\sum_{i=1}^n X_i\right)^2\right\} = \frac{1}{n^2}\sum_{i=1}^n E\{X_i^2\} \leq \frac{1}{n^2} n\sigma^2 = \frac{\sigma^2}{n}$.

Thus by Chebyshev's inequality, if $\varepsilon > 0$, then

$$P\left\{\frac{1}{n}|S_n| \geq \varepsilon\right\} \leq \frac{\sigma^2}{n\varepsilon^2} \longrightarrow 0.$$

Thus $\lim_{n\to\infty} \frac{1}{n} S_n = 0$ in probability, as claimed. □

This version of the Law of Large Numbers is called "weak" because it only guarantees convergence in probability, not almost everywhere. Almost everywhere theorems are called "strong laws". Here is one, called Cantelli's Strong Law. It requires fourth moments.

Theorem 5.2 (Cantelli's Strong Law). *Let X_1, X_2, ... be a sequence of independent, mean zero random variables, such that for each i, $E\{X_i^4\} \leq M$, where $M > 0$ is a constant. Then*

$$\lim_{n\to\infty} \frac{1}{n} S_n = 0 \quad a.e.$$

Proof. $E\left\{\left(\frac{1}{n}S_n\right)^4\right\} = \frac{1}{n^4}\sum_{i,j,k,\ell=1}^n E\{X_i X_j X_k X_\ell\}$.

If $i \neq j, k$, or ℓ, then $E\{X_i X_j X_k X_\ell\} = E\{X_i\}E\{X_j X_k X_\ell\} = 0$, since $E\{X_i\} = 0$. Therefore there are only two types of non-zero terms:

---

[1]Statements of the law of large numbers can be traced as far back as Gerolamo Cardano in the 14th century, but the first proof was in Jacob Bernoulli's Ars Conjectandi in 1713.

5.1. The Weak and Strong Laws

- $n(n-1)$ terms of the form $E\{X_i^2 X_j^2\} = E\{X_i^2\}E\{X_j^2\}$, $i \neq j$.
- n terms of the form $E\{X_i^4\}$.

By hypothesis, $E\{X_i^4\} \leq M$, and, by the Schwarz inequality, $E\{X_i^2\} = E\{1 \cdot X_i^2\} \leq E\{1^2\}^{1/2} E\{X_i^4\}^{1/2} \leq M^{1/2}$. Thus $E\{X_i^2\}E\{X_j^2\} \leq M$. Thus

$$E\left\{\left(\frac{1}{n}S_n\right)^4\right\} = \frac{1}{n^4}\sum_{i \neq j} E\{X_i^2 X_j^2\} + \frac{1}{n^4}\sum_i E\{X_i^4\}$$
$$\leq \frac{1}{n^4}\left[n(n-1)M + nM\right]$$
$$= \frac{M}{n^2}.$$

By Chebyshev's inequality

$$P\left\{\frac{|S_n|}{n} > \frac{1}{n^{1/8}}\right\} \leq n^{1/2}\frac{M}{n^2} = \frac{M}{n^{3/2}}.$$

But now,

$$\sum_n P\left\{\frac{|S_n|}{n} > \frac{1}{n^{1/8}}\right\} \leq \sum_n \frac{M}{n^{3/2}} < \infty.$$

By the Borel-Cantelli Lemma, with probability one, $|S_n|/n > 1/n^{1/8}$ for only finitely many n. In other words, with probability one, $|S_n|/n \leq 1/n^{1/8}$ for all large enough n. In particular, S_n/n converges to zero a.s. □

Remark 5.3. (i) These theorems all require higher moments, while their conclusions involve only the mean of the random variables, i.e., the first moment. Ideally, its hypotheses should only involve first moments. Indeed, there is one law of large numbers which does not require higher moments. Kolmogorov's strong law states that if the X_i are i.i.d. random variables with mean zero, then $S_n/n \to 0$ a.s. We will prove this later.

(ii) There is actually a.e. convergence under the hypotheses of the weak law. (See Exercise 5.2.)

(iii) The proof of Theorem 5.2 gives an estimate of the speed of convergence to zero: $|S_n/n| \leq n^{-1/8}$ as $n \to \infty$. The true speed of convergence is faster, though: it is very nearly $1/\sqrt{n}$. In fact, $S_n/n = O(1/\sqrt{n})$, which is the content of the Central Limit Theorem, our next subject of study.

Exercise 5.1. (Kronecker's Lemma.) Show that if (a_n) is a sequence of numbers such that $\sum_n a_n/n$ converges, then $\frac{1}{n}(a_1 + \cdots + a_n) \longrightarrow 0$ as $n \to \infty$.
[Hint: Let $b_n = \sum_{j=1}^n a_j/j$. Let $a_0 = b_0 = 0$. Then $a_n = n(b_n - b_{n-1})$. Note that $(1/n)(a_1 + \cdots + a_n) = (1/n)\sum_{j=1}^n j(b_j - b_{j-1})$. Use Abel summation to show that this equals $b_n - (1/n)\sum_{j=0}^{n-1} b_j$. Use the fact that b_n converges to a finite number to conclude that this quantity goes to zero.]

Exercise 5.2. Use Exercise 5.12 and Kronecker's Lemma to show that there is a.e. convergence under the hypotheses of the weak law of large numbers. (Theorem 5.1.)

Example 5.3.1 (Glivenko-Cantelli Theorem). *Let $X_1, X_2, \ldots$ be a sequence of i.i.d. random variables with common distribution function F. Define the* **empirical distribution function $F_n(x)$** *by*

$$F_n(x) = \frac{1}{n}\big\{\#j \leq n : X_j \leq x\big\}.$$

Then F_n converges uniformly to F a.s. as $n \to \infty$.

Proof. This depends less on F than it might appear. We will prove it for the uniform distribution; this implies the general case.

Let $U_1, U_2, \ldots$ be a sequence of independent $U(0,1)$ random variables. Let F^0 be the $U(0,1)$ distribution function, and let F_n^0 be the empirical distribution function of the first n U_i.

Fix x and let $Y_i = I_{\{U_i \leq x\}}$. Then $Y_1, Y_2, \ldots$ are i.i.d. Bernoulli random variables with expectation $P\{U_1 \leq x\} = x$. Moreover, $\frac{1}{n}\sum_{j=1}^n Y_j = F_n^0(x)$, so by the Cantelli Strong Law, $F_n^0(x) \to F^0(x)$ a.s. for each fixed x. This is true simultaneously for all rational x. Furthermore, F^0 and the F_n^0 are monotone and F^0 is continuous, so by Dini's theorem, the sequence (F_n^0) not only converges to F^0 for every x, it converges uniformly. Thus the theorem holds for uniform random variables.

Now consider the general case, where the X_i have distribution function F. Let G_F be the inverse of F, defined in Section 2.2. By Corollary 2.17, $X_i \stackrel{\text{def}}{=} G_F(U_i)$, $i = 1, 2, \ldots$ are i.i.d. random variables with distribution function F. Moreover, from Lemma 2.16, $U_i \leq F(x) \iff X_i \leq x$.

Thus

$$F_n(x) = \frac{1}{n}\#\{i : X_i \leq x\} = \frac{1}{n}\#\{i : U_i \leq F(x)\} = F_n^0(F(x)).$$

Therefore

$$\sup_x |F(x) - F_n(x)| = \sup_x |F_0(F(x)) - F_n^0(F(x))| \leq \sup_x |F^0(x) - F_n^0(x)| \to 0,$$

as we have just proved. Therefore F_n converges uniformly to F. $\square$

Example 5.3.2. Bernstein's of Weierstrass' Theorem. *Let f be continuous on $[0,1]$. Then there exists a sequence $p_n(x)$ of polynomials such that*

$$\lim_{n \to \infty} \Big[\sup_{0 \leq x \leq 1} |f(x) - p_n(x)|\Big] = 0.$$

In other words, the polynomials are dense in $C[0,1]$.

Proof. (Serge Bernstein.) Let $0 \leq p \leq 1$. Let $X_1, X_2, \ldots$ be i.i.d. Bernoulli (p) random variables, and let $S_n = X_1 + \cdots + X_n$. Then S_n is $B(n,p)$, so $E\{S_n\} = np$, $\text{Var}(S_n) = npq$, where $q = 1 - p$. Moreover, $P\{S_n = k\} = \binom{n}{k} p^k q^{n-k}$. By the weak law of large numbers, S_n/n converges to p in probability. Let $\delta > 0$. By Chebyshev's inequality, noting that $pq \leq 1/4$,

$$P\left\{\left|\frac{1}{n}S_n - p\right| > \delta\right\} \leq \frac{\text{Var}(S_n)}{n^2 \delta^2} = \frac{npq}{n^2 \delta^2} \leq \frac{1}{4n\delta^2}.$$

Now let $f \in C[0,1]$. $\frac{1}{n}S_n \to p \Longrightarrow f\left(\frac{S_n}{n}\right) \to f(p)$ in probability. By the bounded convergence theorem,

$$E_p\left\{f\left(\frac{S_n}{n}\right)\right\} - f(p) \to 0,$$

where we use E_p to remind us that the expectation depends on p. Now f is continuous, and therefore uniformly continuous on $[0,1]$. Let $\varepsilon > 0$ and choose $\delta > 0$ such that $|y - x| < \delta \Longrightarrow |f(y) - f(x)| \leq \varepsilon$. Then

$$\left|E_p\left\{f\left(\frac{S_n}{n}\right)\right\} - f(p)\right| \leq \varepsilon P\left\{\left|\frac{S_n}{n} - p\right| < \delta\right\} + 2\|f\|_\infty P\left\{\left|\frac{S_n}{n} - p\right| \geq \delta\right\}$$

$$\leq \varepsilon + \frac{\|f\|_\infty}{2n\delta^2}.$$

The bound is independent of p, so $E_p\{f(S_n/n)\} \longrightarrow f(p)$ uniformly in p, $0 \leq p \leq 1$. Now put $x = p$ and use the fact that S_n is binomial $B(n,p)$:

$$E_x\left\{f\left(\frac{S_n}{n}\right)\right\} = \sum_{k=0}^n f\left(\frac{k}{n}\right) \binom{n}{k} x^k (1-x)^{n-k} \stackrel{\text{def}}{=} p_n(x).$$

The p_n are polynomials in x, and we just proved that they converge uniformly to f. □

Notice that the proof explicitly constructs the approximating polynomials. They are called **Bernstein polynomials**.

5.2. Normal Numbers

Take a sequence of real numbers. Ask: "Is it a random sequence?" Or, rather, since it is deterministic and therefore not random, "Does it act like a random sequence?"

There are a number of ways to test this. One is to check that the digits in the sequence have the right limiting frequencies. Those that do are called "normal".

The problem of normal numbers[2] comes up when we apply these ideas to the sequence of digits in the binary expansions of real numbers.

---

[2]Émile Borel introduced normal numbers in 1909 [6]. However, we still do not know if π, e or $\sqrt{2}$ are normal. According to Theorem 5.6, the odds are strongly in favor.

Let $x \in [0,1]$. The binary expansion of x is $.x_1 x_2 x_3 \ldots$, where the x_i are zeros and ones.
$$x = \sum_{n=1}^{\infty} \frac{x_n}{2^n}.$$

Remark 5.4. (i) We can do the same with the decimal expansion, or, indeed, with the expansion to the base b for any integer $b \geq 2$: if $x = .b_1 b_2 b_3 \cdots$, then $x = \sum_{n=1}^{\infty} \frac{x_n}{2^n}$, where the b_i are integers from 0 to $b-1$. We will do the binary case $b = 2$ in detail, but, with the obvious changes, they carry over directly to the other bases.

(ii) The expansion is not quite unique: for example, $.0111111\cdots = .100000\ldots$. In fact, any number of the form $a/2^n$, where $a \leq 2^n$ is an integer, has two expansions, one ending in zeros, the other ending in ones. All other numbers have a unique expansion. We can make the expansion unique by agreeing to take the one ending in zeros. In any case, there are only countably many such x.

Does the sequence of x_i's act like a random sequence? Let us test the frequency of the different digits. If the sequence is random, the frequency of 0's and 1's should both be $1/2$.

Let $\nu_k(x, n) = \#\{j \leq n : x_j = k\}$, $k = 0, 1$. Then $\nu_k(x, n)$ is the frequency of k's in the first n places of the expansion of x.

Definition 5.5. (i) x is **normal to the base 2** if $\lim_{n \to \infty} \nu_k(x, n) = 1/2$, $k = 0, 1$.

(ii) Similarly, x is **normal to the base b** if its expansion to the base b satisfies $\lim_{n \to \infty} \nu_k(x, n) = 1/b$, $k = 0, \ldots, b-1$.

(iii) x is **normal** if it is normal to all bases $b = 2, 3, \ldots$.

Most numbers one can think of are not normal. For instance, $1/3 = .33333\ldots$ is not normal to the base 10. On the other hand, the number $.12345678910111213141 5\ldots$ is. But it is not clear if it is normal to any other base, and it is not known if particular numbers like π, e, and $\sqrt{2}$ are normal. In fact, one might ask, "Are there any normal numbers?" This question was answered by Émile Borel.

Theorem 5.6 (Borel). *Almost all numbers are normal. That is, there is a Borel set Λ of Lebesgue measure zero, such that every $x \in [0,1] - \Lambda$ is normal.*

The proof uses the unit probability space $([0,1], \mathcal{B}(0,1), P)$, where P is Lebesgue measure. Define $X_n(x) = x_n$, where x_n is the nth digit of the binary expansion of x. Then $X_1, X_2, \ldots$ are random variables. Moreover:

5.2. Normal Numbers

Proposition 5.7. *The successive digits $X_1, X_2, \ldots$ of the binary expansion form a sequence of i.i.d random variables. The same is true if the X_i are obtained from the expansion to the base b; in this case the X_i are uniform on $0, \ldots, b-1$.*

Proof. We will prove this for $b=2$. This already introduces all the ideas. We leave it to the interested reader to modify the proof for general b.

Notice that $X_1(x)$ equals zero on $[0, 1/2)$ and equals one on $[1/2, 1)$; $X_2(x)$ is zero on $[0, 1/4) \cup [1/2, 3/4)$. In each case, $P\{X_i = 0\} = 1/2$.

In general, if $X_1(x) = x_1, \ldots, X_n(x) = x_n$, where the x_i are zero or one, let $t_n = \sum_{i=1}^{n} x_i 2^{-i}$. Notice that x is necessarily in $[t_n, t_n + 2^{-n})$, so that for each $x_1, \ldots, x_n$, $P\{x : X_i(x) = x_i, i = 1, \ldots, n\} = 2^{-n}$. To find $P\{X_n = x_n\}$, sum over the 2^{n-1} possible values of $X_1, \ldots X_{n-1}$ to see that $P\{X_n = 0\} = P\{X_n = 1\} = 2^{n-1} 2^{-n} = 1/2$. Thus, $P\{X_n = x_i \mid X_1 = x_1, \ldots, X_{n-1} = x_{n-1}\} = 1/2$. Since this holds for all possible $x_1, \ldots, x_{n-1}$,

$$P\{X_n = 0 \mid X_1, \ldots, X_{n-1}\} = P\{X_n = 1 \mid X_1, \ldots, X_{n-1}\} = \frac{1}{2}.$$

It follows that X_n is independent of $(X_1, \ldots, X_{n-1})$. Apply the following exercise to finish the proof. □

Exercise 5.3. Let $X_1, X_2, \ldots$ be a sequence of random variables. Suppose that X_n is independent of $(X_1, \ldots, X_{n-1})$ for each $n \geq 2$. Show that $X_1, X_2, \ldots$ are independent.

Proof. (of Theorem 5.6). Let us do the calculations for base 2. The other bases are similar. The X_i are random variables on the probability space $([0,1], \mathcal{B}(0,1), P)$. Then ν_k, for $\nu_1(x,n) = (X_1(x) + \cdots + X_n(x))/n$. The X_i are bounded and independent, with mean $1/2$, so the Cantelli strong law implies that $\nu_1(x,n) \to 1/2$ for a.e. x.

In other words, there is a set Λ_2 of x such that $P\{\Lambda_2\} = 0$, and if $x \in [0,1] - \Lambda_2$, then $\nu_1(x,n) \to 1/2$. Since $\nu_0(x,n) = 1 - \nu_1(x,n)$, it follows that $\nu_0(x,n) \to 1/2$ as well. Thus, if $x \notin \Lambda_2$, x is normal to the base 2.

If we start with an expansion to the base b, then the X_n take on values $0, 1, \ldots, b-1$, and the same argument we used for the binary expansion tells us that they are i.i.d. and uniform on $0, 1, \ldots, b-1$: $P\{X_n = i\} = 1/b$, $i = 0, 1, \ldots, b-1$. Let $Y_{k,n} = 1$ if $X_n = k$, and $Y_{k,n} = 0$ otherwise. For each k, $Y_{i,1}, Y_{i,2}, \ldots$ are independent Bernoulli $(1/b)$, and $\nu_k(x,n) = (Y_{k,1}(x) + \cdots + Y_{k,n}(x))/n$, so, once again, the Cantelli strong law implies that for a.e. x, $\nu_k(x,n) \to 1/b$. That is, there is a set $\Lambda_{b,k} \subset [0,1]$ such that $P\{\Lambda_{b,k}\} = 0$ and if $x \in [0,1] - \Lambda_{b,k}$, $\nu_k(x,n) \to 1/b$. If $x \notin \Lambda_b \stackrel{\text{def}}{=} \Lambda_{b,0} \cup \cdots \cup \Lambda_{b,b-1}$, $\nu_k(x,n) \to 1/b$ for all k, and x is normal to the base b. Note that $P\{\Lambda_b\} \leq P\{\Lambda_{b,0}\} + \cdots + P\{\Lambda_{b,b-1}\} = 0$.

This is true for all b. Let $\Lambda = \bigcup_{b=2}^{\infty} \Lambda_b$ and note that $P\{\Lambda\} \leq \sum_b P\{\Lambda_b\} = 0$. If $x \in [0,1] - \Lambda$, then x is normal to the base b for all b, and therefore normal. □

Remark 5.8. We can pose further tests of normality, such as as the frequency of blocks of k numerals. We can also test the frequency along subsequences $n_1, n_2, n_3, \ldots$ As long as we impose only countably many tests, almost every x will pass them. However, for each specific x, there are subsequences along which the test fails.

5.3. Sequences of Random Variables: Existence*

This is a detour, but we want to take advantage of the work we just did. We have already encountered infinite sequences of random variables, and we will soon meet stochastic processes and random functions which involve uncountable families of random variables. The problem is: do they really exist? That is, is there a probability space and a family of random variables on it which really behave the way we claim that they do? The answer is "Yes" for single random variables (Corollary 2.17). It is not hard to extend that to a finite family of random variables. But the extension to infinitely many random variables is less obvious. We must deal with them and we would like some assurance that we are dealing with something solid, not something imaginary, non-existent.

The most general existence theorem is beyond our reach, but we can prove something which is sufficient for our needs: there exist infinite sequences of independent random variables with any given distributions. Almost all the processes in this book can be constructed in one way or another from such sequences.

In fact, we just did the hard part while proving the normal number theorem, for we constructed a sequence of independent Bernoulli (1/2) random variables (X_n) on the unit probability space $([0,1], \mathcal{B}(0,1), P)$. We also saw that the random variable $X(x) = x$ is $U(0,1)$ and satisfies $X = \sum_n X_n 2^{-n}$.

The same will be true for any other sequence of independent Bernoulli (1/2) random variables: the corresponding sum will have a uniform distribution. Thus we have:

Lemma 5.9. *If $Y_1, Y_2, \ldots$ is a sequence of independent Bernoulli (1/2) random variables, then*

$$U \stackrel{def}{=} \sum_{n=1}^{\infty} \frac{Y_n}{2^n}$$

is a $U(0,1)$ random variable.

5.3. Sequences of Random Variables: Existence*

Remark 5.10. The $X_n(x)$ are well-known in analysis: the functions $r_n(x) \stackrel{\text{def}}{=} 1 - 2X_n(x)$ are called **Rademacher functions**.

We will use them to construct a sequence of independent random variables with arbitrary distributions. Let $F_1, F_2, \ldots$ be a sequence of distribution functions.

Theorem 5.11. *There exists a sequence of independent random variables $Y_1, Y_2, \ldots$ such that Y_n has distribution function F_n.*

Proof. We will do this in two stages. First we will use Cantor's diagonal argument to construct a sequence of i.i.d. uniform random variables. Write the Bernoulli variables X_n in an array, filling up the diagonals as follows:

(5.2)
$$\begin{array}{ccccc} X_1 & X_3 & X_6 & X_{10} & X_{15} & \cdots \\ X_2 & X_5 & X_9 & X_{14} & \cdots \\ X_4 & X_8 & X_{13} & \cdots \\ X_7 & X_{12} & \cdots \\ X_{11} & \cdots \\ \vdots \end{array}$$

Let $U_1 = X_1/2 + X_3/2^2 + X_6/2^3 + \ldots$. In other words, the first row is the binary expansion of U_1: $U_1 = .X_1 X_3 X_6 X_{10} \ldots$. Similarly, let $U_2 = .X_2 X_5 X_6 \ldots$ and, in general, let the nth row be the binary expansion of U_n, $n = 1, 2, \ldots$.

The array consists of independent Bernoulli $(1/2)$ random variables. Moreover, if $n \neq m$, the nth and m^{th} rows are disjoint families of the X_i, so the rows are independent.

(*i*) Each U_n is a $U(0,1)$ random variable by Lemma 5.9.

(*ii*) The U_n are independent, since they are sums of disjoint independent families of random variables.

Recall the inverses G_{F_n} of the distribution functions F_n. Define $Y_n \stackrel{\text{def}}{=} G_{F_n}(U_n)$. By Corollary 2.17, Y_n is a random variable with distribution function F_n, and the desired sequence is $Y_1, Y_2, \ldots$. $\square$

Exercise 5.4. (**Monte Carlo Integration.**) Let f be a square-integrable Borel function on $[0,1]$. Let $U_1, U_2, \ldots$ be independent $U(0,1)$ random variables.

(a) Show that with probability one, $\frac{1}{n} \sum_{k=1}^{n} f(U_k) \longrightarrow \int_0^1 f(x) \, dx$ as $n \to \infty$. The sum is called a **Monte Carlo estimate** of the integral.

(b) Let I_n be the Monte Carlo estimate. How large must n be in order that $E\{(I_n - \int_0^1 f(x) \, dx)^2\} < 1/10$? (The answer depends on f.)

(c) Monte Carlo integration is often used for high-dimensional integrals, which can be hard to compute numerically. Let $f(x_1, x_2, \ldots, x_n)$ be a bounded Borel function of n variables, defined on $[0,1]^n$. Find a Monte Carlo estimate of the integral $\int_0^1 \ldots \int_0^1 f(x_1, \ldots, x_n)\, dx_1 \ldots dx_n$. Answer question (b) for this estimate, and compare the one and n-dimensional cases.

5.4. Sigma Fields as Information

Sigma fields are information[3]. To phrase it more carefully, or at least with more verbosity: in probability, σ-fields are used to encode information[4]. It takes practice to learn to tease the information from the encoding, but it is worth it.

Let Ω be a set and let $\mathcal{G}$ be a σ-field of subsets of Ω. Let X be a real-valued function on Ω.

Definition 5.12. X is **measurable with respect to $\mathcal{G}$** if, for all $x \in \mathbb{R}$, $\{X \leq x\} \in \mathcal{G}$. We also say that X is **$\mathcal{G}$-measurable**.

The following proposition makes this definition useful.

Proposition 5.13. *Let X be a $\mathcal{G}$-measurable function. Then for all Borel sets B, $\{X \in B\} \in \mathcal{G}$.*

Proof. This is close to the proof for the case $\mathcal{G} = \mathcal{F}$ in Proposition 2.8. Let $\mathcal{H}$ be the class of sets $A \subset \mathbb{R}$ for which $\{X \in A\} \in \mathcal{G}$. To finish the proof, we only have to show that $\mathcal{B} \subset \mathcal{H}$.

From the definition of measurability, $(-\infty, x] \in \mathcal{H}$ for all x. The empty set is clearly in $\mathcal{H}$. If $A \in \mathcal{H}$, then $\{X \in A^c\} = \{X \in A\}^c$ which is in $\mathcal{G}$ since $\mathcal{G}$ is closed under complementation. If $A_1, A_2, \ldots$ are in $\mathcal{H}$, then $\{X \in \bigcup_n A_n\} = \bigcup_n \{X \in A_n\}$. But $\{X \in A_n\} \in \mathcal{G}$ and $\mathcal{G}$ is a σ-field, so it is closed under countable unions, hence $\{X \in \bigcup_n A_n\} \in \mathcal{G} \Rightarrow \bigcup_n A_n \in \mathcal{H}$. Thus $\mathcal{H}$ is a σ-field which contains all the intervals $(-\infty, x]$. But the Borel sets $\mathcal{B}$ are the smallest such σ-field, so $\mathcal{B} \subset \mathcal{H}$. □

Measurability is new only in name; we have seen it before. Borel functions on $\mathbb{R}$ are $\mathcal{B}$-measurable. A random variable on $(\Omega, \mathcal{F}, P)$ is an $\mathcal{F}$-measurable function on Ω. Much of what we have proved for random variables carries over directly to $\mathcal{G}$-measurable functions. The following easy exercise shows why.

Exercise 5.5. Let $\mathcal{G}$ be a σ-field of subsets of a non-empty set Ω. Show that

(a) there exists a probability measure P on $(\Omega, \mathcal{G})$.

---

[3]Warning: There is no probability in this section. It is purely about sets and functions.
[4]Sigma fields were originally introduced to get around the fact that not all subsets of $\mathbb{R}$ have a length. However, Kolmogorov's definition of a conditional expectation [21], and their use in martingale theory showed how well they describe information in the raw. Take a close look at Proposition 5.16 and the remark following it for some intuition on this.

5.4. Sigma Fields as Information

(b) Suppose X is a $\mathcal{G}$-measurable function. Then X is a random variable on the probability space $(\Omega, \mathcal{G}, P)$.

This tells us, for instance, that the sup, lim sup, limit, etc. of a sequence of $\mathcal{G}$-measurable functions are $\mathcal{G}$-measurable—just substitute the words "random variable" for "$\mathcal{G}$-measurable" and point to Proposition 4.6.

Recall that $\sigma(\mathcal{C})$ is the σ-field generated by the class $\mathcal{C}$ of sets. Random variables can also generate σ-fields.

Definition 5.14. Let X be a function on Ω. Then the **σ-field generated by X**, denoted $\sigma(X)$, is the smallest σ-field containing the sets $\{X \leq x\}$. Similarly, if $\{X_\alpha, \alpha \in I\}$ is a family of functions on Ω, the **σ-field generated by $\{X_\alpha, \alpha \in I\}$**, denoted $\sigma(\{X_\alpha, \alpha \in I\})$, is the smallest σ-field containing all the sets $\{X_\alpha \leq x\}$, $\alpha \in I$, $x \in \mathbb{R}$.

Here is an exercise which gives some practice using the definitions.

Exercise 5.6. Show that X is $\sigma(X)$-measurable, and that $\sigma(X)$ is the smallest σ-field $\mathcal{G}$ such that X is $\mathcal{G}$-measurable.

The following lemma and proposition should help explain why σ-fields contain information. Intuitively, $\sigma(X)$ contains all the information in the function X. We will expand on this after we have proved them.

Lemma 5.15. *Let X be a function on Ω. Then $\sigma(X) = X^{-1}(\mathcal{B})$, i.e., each set $\Lambda \in \sigma(X)$ is of the form $\Lambda = X^{-1}(B)$ for some Borel set B.*

Proof. Let $\mathcal{H}$ be the class of sets of the form $\Lambda = X^{-1}(B) = \{\omega : X(\omega) \in B\}$ for some $B \in \mathcal{B}$. By the Exercise 5.6 and Proposition 5.13, $\mathcal{H} \subset \sigma(X)$. On the other hand, $\mathcal{H}$ contains $(-\infty, x]$, since, as X is $\sigma(X)$-measurable, $\{X \leq x\} \in \sigma(X)$; but $\mathcal{H}$ is a σ-field. (This is a general fact: inverse images preserve set-operations, so the inverse image of a σ-field is a σ-field.) Thus, $\mathcal{H}$ is a σ-field of subsets of $\mathbb{R}$ which contains $\{X \leq x\}$ for all x. Therefore X is $\mathcal{H}$-measurable, so evidently $\mathcal{H}$ contains $\sigma(X)$, which is the smallest such σ-field. Thus the two σ-fields must be equal. $\square$

Proposition 5.16. *Let X and Y be real-valued functions on Ω. Then Y is measurable with respect to $\sigma(X)$ if and only if there exists a Borel function f such that $Y = f(X)$, i.e., $Y(\omega) = f(X(\omega))$ for all $\omega \in \Omega$.*

Proof. Suppose that $Y = f(X)$ for some Borel function f. Let $B \in \mathcal{B}$. Then $\{Y \in B\} = \{f(X) \in B\} = \{X \in f^{-1}(B)\}$. But as f is Borel, $f^{-1}(B) \in \mathcal{B}$, and this last set is in $\sigma(X)$. In particular, if $B = (-\infty, x]$, $\{Y \leq x\} \in \sigma(X)$, so Y is $\sigma(X)$-measurable.

This brings us to the nub of the proof, the converse. Suppose that Y is measurable with respect to $\sigma(X)$. We claim that there is a Borel function f such that $Y = f(X)$. Fix n for the moment. For each k, $\{Y \in (k2^{-n}, (k+1)2^{-n}]\} \in \sigma(X)$. By Lemma 5.15 there exists a Borel set B_{kn} such that $\{Y \in (k2^{-n}, (k+1)2^{-n}]\} = \{X \in B_{kn}\}$. The sets $\{X \in B_{kn}\}$ are necessarily disjoint. Define
$$f_n(x) = \sum_k I_{B_{kn}}(x).$$
Then $f_n(X) \leq Y \leq f_n(X) + 2^{-n}$. (Indeed, in the notation of Chapter 3, $f_n(X) = \underline{X}_n$.) Now for each x, $f_n(x) \leq f_{n+1}(x) \leq f_n(x) + 2^{-n}$, so that the sequence (f_n) converges monotonically to a function $f(x)$. Since the f_n are Borel, so is f, and clearly $Y = f(X)$. □

Remark 5.17. This gives a remarkable answer to the question, "Is Y a function of X?" It does so without having to construct the function. We merely have to check that Y is $\sigma(X)$-measurable.

This is the sense in which $\sigma(X)$ contains "all the information in X." $\sigma(X)$ not only gives us the values of X, it gives us the values of any function of X. (We are probably attributing more generosity to the σ-field than it deserves. By "gives us the values" we mean that it allows us to reconstruct the values from its sets.)

5.5. Another Look at Independence

Let us recast the idea of independence in terms of σ-fields.

Definition 5.18. Two σ-fields $\mathcal{F}$ and $\mathcal{G}$ are *independent* if, whenever $\Lambda \in \mathcal{F}$ and $\Gamma \in \mathcal{G}$, then Λ and Γ are independent.

More generally:

Definition 5.19. (*i*) A finite number of σ-fields $\mathcal{F}_1, \mathcal{F}_2, \ldots, \mathcal{F}_n$ are **independent** if whenever $\Lambda_1 \in \mathcal{F}_1, \ldots, \Lambda_n \in \mathcal{F}_n$, then
$$P\{\Lambda_1 \cap \cdots \cap \Lambda_n\} = \Pi_{j=1}^n P\{\Lambda_j\}.$$

(*ii*) An arbitrarily family $\{\mathcal{F}_\alpha, \alpha \in I\}$ of σ-fields is **independent** if each finite subfamily is.

Sigma fields are independent if their elements are independent. The reader might want to see why the problem of "pairwise independence" vs. "joint independence" of more than two σ-fields (see §1.5) does not come up.

We can describe the independence of random variables in terms of the σ-fields they generate.

5.6. Zero-one Laws

Proposition 5.20. (*i*) *Two random variables X and Y are* **independent** *if and only if $\sigma(X)$ and $\sigma(Y)$ are independent.*

(*ii*) *More generally, a family of random variables $\{X_\alpha,\ \alpha \in I\}$ is* **independent** *if the family of σ-fields $\{\sigma(X_\alpha),\ \alpha \in I\}$ is independent.*

Proof. Suppose X and Y are independent, and let $\Lambda \in \sigma(X)$ and $\Gamma \in \sigma(Y)$. Then there are Borel sets A and B such that $\Lambda = \{X \in A\}$ and $\Gamma = \{Y \in B\}$, and we have

$$\begin{aligned}P\{\Lambda \cap \Gamma\} &= P\{X \in A,\ Y \in B\} \\ &= P\{X \in A\}P\{Y \in B\} \\ &= P\{\Lambda\}P\{\Gamma\}.\end{aligned}$$

This shows that Λ and Γ are independent, and hence that $\sigma(X)$ and $\sigma(Y)$ are independent.

Conversely, if $\sigma(X)$ and $\sigma(Y)$ are independent, then for any Borel A and B, the sets $\{X \in A\}$ and $\{Y \in B\}$ are independent, since they are in $\sigma(X)$ and $\sigma(Y)$, respectively. Thus

$$P\{X \in A,\ Y \in B\} = P\{X \in A\}P\{Y \in B\},$$

showing that X and Y are independent.

We leave (*ii*) as an exercise. $\square$

Here is an (easy) exercise which shows how σ-fields can be used to draw unexpected conclusions.

Exercise 5.7. Let X be a random variable whose range is all of $\mathbb{R}$, let f be a Borel function, and let $Y = f(X)$. Show that f is invertible if and only if $\sigma(Y) = \sigma(X)$.

Exercise 5.8. Let $(\Omega, \mathcal{F}, P)$ be a probability space, and let X be a random variable. We say a σ-field $\mathcal{G} \subset \mathcal{F}$ is **trivial** if $\Lambda \in \mathcal{G} \Rightarrow$ either $P\{\Lambda\} = 0$ or $P\{\Lambda\} = 1$. Prove that X is a.e. constant if and only if $\sigma(X)$ is trivial. ("X is a.e. constant" means, of course, there is a constant c such that $P\{X = c\} = 1$.)

5.6. Zero-one Laws

Example 5.20.1. Let $X_1,\ X_2,\ldots$ be independent random variables. Question: what do the following three examples have in common?

(*i*) Suppose the X_i are i.i.d. and that X_1 has mean zero. Then with probability one, $\lim_{n \to \infty} S_n/n = 0$.

(*ii*) Suppose the X_i are i.i.d. and that X_1 is not integrable. Then the probability that S_n/n converges is zero. In fact, with probability one, $\limsup_{n \to \infty} |X_n|/n = \infty$.

(iii) Suppose that X_n is Bernoulli with parameter p_n. Then with probability one,
$$\limsup_{n \to \infty} X_n = \begin{cases} 0 & \text{if } \sum_n p_n < \infty, \\ 1 & \text{if } \sum_n p_n = \infty. \end{cases}$$

Indeed, (i) follows from Kolmogorov's Law of Large Numbers, (ii) was shown in Exercise 5.23 and (iii) is a consequence of the converse Borel-Cantelli Lemma.

Did your answer include the following? If so, bravo.

(a) They all involve sequences of independent random variables.

(b) They all involve limiting behavior of the sequence and (this last one is harder)

(c) all the probabilities are zero or one. Nothing in between. No probabilities of 1/2 or 1/3. Just 0 and 1.

There are many, many examples of this type. When we see such a curious fact, we have to ask if perhaps it is not a coincidence, and if there is something deeper going on; and there is. In fact, (a) and (b) together imply (c). The tricky thing is to find a way to state this mathematically so that we can actually prove it. How can we define something so all-inclusive as "limiting behavior"? For this, we turn to σ-fields.

Definition 5.21. Let $X_1, X_2, \ldots$ be random variables and let $\mathcal{F}_n^*$ be the σ-field $\sigma(X_{n+1}, X_{n+2}, \ldots)$. The **tail field** of the sequence is the σ-field $\mathcal{F}_\infty^* = \bigcap_n \mathcal{F}_n^*$. Elements in $\mathcal{F}_\infty^*$ are **tail events.**

Being the intersection of σ-fields, $\mathcal{F}_\infty^*$ is a σ-field. It is sometimes called the **remote field**. It contains interesting events: for instance, the set on which the sequence (X_n) converges is a tail event. Indeed, for all n, the convergence is determined by $X_{n+1}, X_{n+2}, X_{n+3}, \ldots$, so the set on which the series converges is an element of $\mathcal{F}_n^*$. This is true for all n, so it is in the intersection, $\mathcal{F}_\infty^*$. In fact, almost anything to do with the limiting behavior of the X_i will be in $\mathcal{F}_\infty^*$.

Here is a theorem which covers all the examples above, called the Borel Zero-One Law[5].

Theorem 5.22 (Borel Zero-One Law)**.** *Let $X_1, X_2, \ldots$ be a sequence of independent random variables. Then its tail field is trivial, that is, any set in the tail field has probability zero or one.*

---

[5]It is more often called the Kolmogorov zero-one law. The attribution to Borel—which we follow out of habit—probably dates to the converse Borel-Cantelli lemma, which is a special case.

5.6. Zero-one Laws

Proof. Let $\Lambda \in \mathcal{F}_\infty^*$. Let $\mathcal{H}$ be the class of all events Γ such that
$$P\{\Gamma \cap \Lambda\} = P\{\Gamma\}P\{\Lambda\}. \tag{5.3}$$

Let $\mathcal{F}_n = \sigma(X_1, \ldots, X_n)$, and put $\mathcal{F}_\infty = \sigma(X_1, X_2, \ldots)$. Intuitively, $\mathcal{F}_n$ contains all the information in the first n random variables, and $\mathcal{F}_\infty$ contains the information in all of them. Note that $\mathcal{F}_n$ and $\mathcal{F}_n^*$ are generated by disjoint families of independent random variables, and are therefore independent. Moreover, $\Lambda \in \mathcal{F}_n^*$, so any $\Gamma \in \mathcal{F}_n$ satisfies (5.3). Thus $\mathcal{F}_n \subset \mathcal{H}$. Consequently, $\bigcup_n \mathcal{F}_n \subset \mathcal{H}$. Now $\bigcup_n \mathcal{F}_n$ is a field, but it may not be a σ-field. Let $\mathcal{F}_\infty = \sigma(\bigcup_n \mathcal{F}_n)$. We claim that $\mathcal{F}_\infty \subset \mathcal{H}$. To see this, note that $\mathcal{H}$ contains the field $\bigcup_n \mathcal{F}_n$. Then, by the Monotone Class Theorem (Theorem 1.5), it is sufficient to prove that $\mathcal{H}$ is a monotone class.

Suppose that $\Gamma_1, \Gamma_2, \ldots$ are in $\mathcal{H}$: (5.3) holds for each Γ_n. If either $\Gamma_1 \subset \Gamma_2 \subset \ldots$ or $\Gamma_1 \supset \Gamma_2 \ldots$, we can go to the limit on both sides to see that (5.3) holds for $\Gamma = \lim_n \Gamma_n$. Thus $\mathcal{H}$ is a monotone class, and therefore it contains $\mathcal{F}_\infty$. But certainly $\Lambda \in \mathcal{F}_\infty$, so we can take $\Gamma = \Lambda$: $P\{\Lambda\} = P\{\Lambda\}P\{\Lambda\}$. That is, Λ is independent of itself! But there are only two solutions of $x = x^2$, zero and one. Therefore, $P\{\Lambda\} = 0$ or 1. Done! □

So, any event in the tail field has probability either 0 or 1, and any tail random variable (i.e., $\mathcal{F}_\infty^*$-measurable) is constant a.e. It goes without saying that the hypothesis of independence is essential. For instance, if all the X_n are equal, the tail field is $\sigma(X_1)$.

Example 5.22.1. Theorems like the Borel Zero-One Law are valuable, for they give us a lot of information almost for free: no calculation is required. Let $X_1, X_2, X_3, \ldots$ be independent random variables, let $S_n = X_1 + \cdots + X_n$ and let A be a Borel set. The following are tail events and therefore have probability zero or one: $\{\limsup X_n \in A\}$, $\{\liminf_n S_n/n \in A\}$, $\{(X_n) \text{ converges }\}$, $\{(S_n) \text{ converges }\}$, $\{X_n > 0 \text{ i.o.}\}$. But $\{S_n > 0 \text{ i.o.}\}$ and $\{|S_n| < 1 \text{ i.o.}\}$ are not tail events, although they do actually have probability either zero or one. (Indeed, changing the value of X_1 may change either of these events.)

Random variables like $\limsup_n S_n/n$ and $\limsup_n S_n/\sqrt{n}$, are measurable with respect to the tail field, and are a.e. constant. However, while the convergence/divergence of the series $\sum_n X_n$ is a tail event, and therefore has probability zero or one, the sum itself is not: it depends on all the X_n.

Problems 5

5.9. Consider an experiment with probability p of success, probability $1 - p$ of failure. Suppose it is repeated independently, infinitely often. Use the law of large number to show that, not only will the experiment eventually produce a success, but with probability one, it will produce infinitely many of them.

5.10. Let (X_n) be random variables and let $S_n = X_1 + \cdots + X_n$. Suppose that $X_n \to 0$ a.e. Show that $S_n/n \to 0$. Show that this is no longer true for convergence in probability: $X_n \to 0$ in probability does not imply $S_n/n \to 0$ in probability, even if the X_n are independent.
[Hint: Let X_n equal 0 and 2^n with probabilities $1 - 1/n$ and $1/n$ respectively.]

5.11. Let $X_1, X_2, \ldots$ be independent random variables with mean zero and finite variance. Put $S_n = X_1 + \cdots + X_n$.

(a) Let $\varepsilon > 0$ and set $\Lambda = \{\omega : \max_{1 \leq k \leq n} |S_n(\omega)| \geq \varepsilon\}$. Define $\Lambda_k = \{\omega : \max_{1 \leq j \leq k-1} |S_j(\omega)| < \varepsilon, |S_k(\omega)| \geq \varepsilon\}$, $k = 1, \ldots, n$. Show that the Λ_k are disjoint and $\Lambda = \bigcup_{k=1}^n \Lambda_k$.

(b) Show that $S_n - S_k$ is independent of S_k.

(c) Show that $\int_{\Lambda_k} S_n^2 \, dP \geq \int_{\Lambda_k} S_k^2 \, dP$.
[Hint: $S_n = S_k + (S_n - S_k)$.]

(d) Deduce that $E\{S_n^2\} \geq \sum_{k=1}^n \int_{\Lambda_k} X_k^2 \, dP \geq \varepsilon^2 P\{\Lambda\}$.

(e) Conclude **Kolmogorov's Inequality:**

(5.4) $$P\{\max_{1 \leq k \leq n} |S_n| \geq \varepsilon\} \leq \frac{1}{\varepsilon^2} E\{S_n^2\}.$$

[Note that if $\max_{1 \leq k \leq n} |S_n|$ is replaced by $|S_n|$ itself, this is just Chebyshev's inequality. So Kolmogorov's inequality is a significant extension of Chebyshev's inequality. Inequalities like this will reappear in Chapter 9.]

5.12. (Convergence of Series.) Let $X_1, X_2, \ldots$ be independent random variables with mean zero and finite variance. Put $S_n = X_1 + \cdots + X_n$. Show that if $\sum_{n=1}^\infty E\{X_n^2\} < \infty$, that $\sum_{n=1}^\infty X_n$ converges a.s.
[Hint: Use Kolmogorov's inequality (Exercise 5.11.)]

5.13. (Random Signs Problem.) Let (a_n) be a sequence of real numbers. Choose the sign, positive or negative, of a_n at random. Does $\sum_n \pm a_n$ converge? To answer this, let $X_1, X_2, \ldots$ be i.i.d. with $P\{X_n = 1\} = P\{X_n = -1\} = 1/2$. Let (a_n) be a sequence of real numbers. Show that if $\sum_n a_n^2 < \infty$, then $\sum_n a_n X_n$ converges a.s.
[Note: The random series actually converges if and only if the sum of squares converges, but the converse is slightly harder. The definitive result on the convergence of random series is Kolmogorov's Three Series Theorem.]

5.14. Let $X_1, \ldots, X_n$ and $Y_1, \ldots, Y_n$ be random variables. Suppose that for any bounded continuous fumctions $f_1(x), \ldots, f_n(x)$, $E\{\prod_{i=1}^n f(X_i)\} = E\{\prod_{i=1}^n f(Y_i)\}$. Show that $(X_1, \ldots, X_n)$ and $(Y_1, \ldots, Y_n)$ have the same joint distribution.
[Hint: $I_{(-\infty, x]}(t)$ is an increasing limit of continuous functions.]

5.15. (Normal Numbers.) Let $x \in [0, 1]$ and let $.x_1 x_2 x_3 \ldots$ be its expansion to the base b, $b \geq 2$. Let $k \geq 1$ be an integer, and let $\xi_1, \xi_2, \ldots$ be successive blocks of k digits: $\xi_n \stackrel{\text{def}}{=} (x_{nk+1}, \ldots, x_{(n+1)k})$. Let η be a fixed k-tuple of integers from 0 to $b-1$. Show that for a.e. (Lebesgue) $x \in [0, 1]$, that $\lim_{n \to \infty} \frac{1}{n} \#\{k \leq n : \xi_k = \eta\} = b^{-k}$. Show, moreover, that this is true simultaneously for all $b = 2, 3, \ldots$ for all $k \geq 1$, and for all such k-tuples η. (Note that the case $k = 1$ has been done: it is the normal number theorem.)

5.6. Zero-one Laws

5.16. (Continuation.) Let $n_1, n_2, \ldots$ be a fixed subsequence of $1, 2, 3, \ldots$ and let and $x = .x_1 x_2 x_3 \cdots$ be the expansion of x to the base 2. Show that for a.e. $x \in [0,1]$, $\lim_{m \to \infty} \frac{1}{m} \#\{k \leq m : x_{n_k} = 1\} = \lim_{m \to \infty} \frac{1}{m} \#\{k \leq m : x_{n_k} = 0\} = \frac{1}{2}$. Show, however, that this does not hold simultaneously for all subsequences, and that in fact, there is *no* $x \in [0,1]$ for which this holds for all possible subsequences.

5.17. The Cantor set C (see Exercise 3.44) is defined to be the set of all $x \in [0,1]$ whose trinary expansion contains no 1's. Use the law of large numbers to show that the Lebesgue measure of C is zero.

5.18. Use probability to show that the set of $x \in [0,1]$ whose decimal expansion contains no 3's has zero Lebesgue measure. Deduce that there exist two sets A and B of Lebesgue measure zero such that $\mathbb{R} = A + B$, that is, every $x \in \mathbb{R}$ can be written as a sum $x = y + z$ where $y \in A$ and $z \in B$.
[Hint: Law of Large Numbers.]

5.19. Let X be a random variable with an absolutely continuous distribution. Show that with probability one, the fractional part of X is a normal number.

5.20. Is the sum of two normal numbers, modulo one, a normal number?
[Hint: If α is a non-normal number, the set of differences between α and all normal numbers (mod 1) has probability one.]

5.21. Show that under the hypotheses of the Cantelli strong law, with probability one, for any $\varepsilon > 0$, $|S_n|/n \leq 1/n^{1/4-\varepsilon}$ for all sufficiently large n.

5.22. A sequence of real numbers $x_1, x_2, \ldots$ with values in $[0,1]$ is said to be **uniformly distributed** if for each pair (a,b) of real numbers with $0 \leq a < b \leq 1$,
$$\lim_{n \to \infty} \frac{1}{n} \#\{j \leq n : a < x_j < b\} = b - a.$$
Prove that a uniformly distributed sequence exists.

5.23. Let $X_1, X_2, \ldots$ be a sequence of i.i.d. random variables with $E\{|X_1|\} = \infty$.
  (a) Show that $\limsup_{n \to \infty} |X_n|/n = \infty$ a.s.
  (b) Deduce that $\limsup_{n \to \infty} |X_1 + \cdots + X_n|/n = \infty$.

5.24. Let $X_1, X_2, \ldots$ be i.i.d. exponential (λ). Show that there exists C_0 such that $C < C_0$ implies $P\{X_n \geq C \log n \text{ i.o.}\} = 1$ and $C > C_0$ implies $P\{X_n \geq C \log n \text{ i.o.}\} = 0$. What if $C = C_0$? What is the value of C_0?

5.25. Let $X_1, X_2, \ldots$ be independent but not necessarily identically distributed random variables. Consider the power series
$$\sum_{n=0}^{\infty} X_n z^n.$$

  (a) Show that for each z, the series either converges a.s. or diverges a.s.
  (b) Show that the radius of convergence of the series is deterministic, i.e., almost surely constant.

Chapter 6

Convergence in Distribution and the CLT

6.1. Characteristic Functions

We remarked in §3.2 that the moment generating function $M(t)$ is a form of Laplace transform, and that it can be inverted to get the distribution. That is, when the moment generating function exists. It often doesn't. In fact, the mere existence of a moment generating function is a strong condition. For example, the moment generating function of the Cauchy distribution only exists at $t = 0$.

We want something valid for all distributions. The solution is to replace t by it, where $i = \sqrt{-1}$. (This connects it with the Fourier transform instead of the Laplace transform.) The result is called the characteristic function[1].

Definition 6.1. The **characteristic function** of a random variable X is the function
$$\phi(t) \stackrel{\text{def}}{=} E\{e^{itX}\}.$$

Remark 6.2. (i) This is the expectation of a complex-valued random variable. It can be handled by writing the random variable in terms of its real and imaginary parts. If U and V are real-valued random variables, then $Z \stackrel{\text{def}}{=}$

---

[1]The characteristic function, along with the continuity and inversion theorems, was introduced by Paul Lévy. It will be our principle tool for proving the central limit theorem. While there are shorter proofs, the characteristic-function approach is so flexible and useful that it is worthwhile for its own sake. See [17] for a nice proof using Trotter's Gaussian-convolution approach.

$U + iV$ is a complex random variable, and we define $E\{Z\} = E\{U\} + iE\{V\}$. The expectation has complex values. We leave it to the reader to verify the usual properties. In particular, if $|Z| = \sqrt{Z\bar{Z}}$ is the modulus of Z, then $|E\{Z\}| \leq E\{|Z|\}$.

(ii) $\phi(t) = E\{\cos tX\} + iE\{\sin tX\}$.

(iii) The characteristic function exists for all random variables. Indeed, both the sine and cosine are bounded, so the expectations of $\sin(tX)$ and $\cos(tX)$ exist.

(iv) If X has a density f, then

$$\phi(t) = \int_{-\infty}^{\infty} e^{itx} f(x)\, dx,$$

which is the Fourier transform of f.

More generally, we can write ϕ in terms of the distribution function:

$$\phi(t) = \int_{-\infty}^{\infty} e^{itx}\, dF(x).$$

Think of the characteristic function as the Fourier transform of the distribution of X. Thus the characteristic function extends the classical Fourier transform. We will assume the classical Fourier inversion theorem below.

(v) If X has a density f, then f may be recovered from the characteristic function ϕ. Indeed, by the Fourier inversion theorem, for a.e. x,

$$f(x) = \lim_{T \to \infty} \frac{1}{2\pi} \int_{-T}^{T} \phi(t) e^{-itx}\, dt.$$

The Cauchy principal value of the integral is necessary because ϕ need not be integrable, even if f is. (For example, if $f = I_{(-1,1)}/2$, $\phi(t) = \sin(t)/t$.) If ϕ is integrable, there is no need for the Cauchy principal value.

The characteristic function is due to Paul Lévy, who proved two fundamental theorems about it, both named after him: Lévy's Inversion Formula, and Lévy's Continuity Theorem. It is particularly well-suited to handling sums of independent random variables, and it was the basic tool in his deep study of sums of independent random variables.

Let us begin our story with some elementary properties of the characteristic function.

Theorem 6.3. (i) $\phi(0) = 1$, and $|\phi(t)| \leq 1$ for all $t \in \mathbb{R}$.

(ii) $\phi(t)$ is a uniformly continuous function of t.

6.1. Characteristic Functions

(iii) $\phi(t)$ is positive definite, that is, for all n, real $t_1, \ldots, t_n$ and complex $z_1, \ldots, z_n$,

$$\sum_{k,j=1}^{n} \phi(t_k - t_j) z_k \bar{z}_j \geq 0.$$

Proof. (i) $\phi(0) = E\{e^{i0X}\} = 1$, $\quad |\phi(t)| \leq E\{|e^{itX}|\} = E\{1\} = 1$.

(ii) $|\phi(t+s) - \phi(t)| = |E\{e^{i(t+s)X} - e^{itX}\}| \leq E\{|e^{itX}(e^{isX} - 1)|\} = E\{|e^{isX} - 1|\}$.

Let $\delta(s) = E\{|e^{isX} - 1|\}$. Then $|e^{isX} - 1| \leq 2$ and $|e^{isX} - 1| \to 0$, as $s \to 0$, so by the bounded convergence theorem, $\delta(s) \to 0$ as $s \to 0$. Thus, for all t, $|\phi(s+t) - \phi(s)| \leq \delta(s)$. The bound is independent of t, so that ϕ is not only continuous, but uniformly continuous.

(iii) To show that ϕ is positive definite, choose real $t_1, \ldots, t_n$ and complex $z_1, \ldots, z_n$. Then

$$\sum_{k,j=1}^{n} \phi(t_k - t_j) z_k \bar{z}_j = E\left\{\sum_{k,j=1}^{n} e^{(t_k - t_j)X} z_k \bar{z}_j\right\}$$

$$= E\left\{\sum_{k,j=1}^{n} e^{it_k X} z_k e^{-it_j X} \bar{z}_j\right\}$$

$$= E\left\{\left(\sum_{k=1}^{n} z_k e^{it_k X}\right) \overline{\left(\sum_{j=1}^{n} z_j e^{it_j X}\right)}\right\}$$

$$= E\left\{\left|\sum_{k=1}^{n} z_k e^{it_k X}\right|^2\right\} \geq 0.$$

$\square$

Like the moment generating function, the characteristic function, generates the moments of a random variable by differentiation.

Theorem 6.4. *Let ϕ be the characteristic function of a random variable X.*

(i) *If $E\{|X|^k\} < \infty$, then ϕ is k times differentiable, and*

(6.1) $$\phi(t) = \sum_{j=0}^{k} \frac{1}{j!} E\{X^j\} (it)^j + o(t^k) \quad \text{as } t \to 0.$$

In particular, for $n = 1, \ldots, k$,

(6.2) $$\phi^{(n)}(0) = i^n E\{X^n\}.$$

(ii) *Conversely, if $\phi^{(2k)}(0)$ exists and is finite, then $E\{|X|^{2k}\} < \infty$.*

Proof. (*i*) We first show that if X has k moments, then the characteristic function is k times differentiable, and we derive the formula for the derivatives. The key question is this: can we differentiate the formula $\phi(t) = E\{e^{itX}\}$ under the expectation? If we can, the formula is immediate. While the proof is technically by induction, the main ideas are apparent in the case $k = 1$. Suppose $E\{|X|\} < \infty$. Let $h > 0$, and notice that ϕ' is the limit as $h \to 0$ of the following.

$$(6.3) \quad \frac{1}{h}(\phi(t+h) - \phi(t)) = E\left\{e^{itX}\frac{e^{ihX} - 1}{h}\right\}$$

$$= E\left\{e^{itX}\frac{\cos(hX) - 1}{h} + ie^{itX}\frac{\sin(hX)}{h}\right\}.$$

By the mean value theorem, there are θ and θ'—which may depend on ω—between 0 and h such that

$$\frac{1}{h}(e^{ihX} - 1) = -X\sin(\theta X) + iX\cos(\theta' X)).$$

But the sine and cosine are bounded by one, so

$$(6.4) \quad \left|\frac{1}{h}(e^{ihX} - 1)\right| \leq \sqrt{2}|X|.$$

Since the complex exponential has modulus 1, the integrand in (6.3) is bounded in modulus by $\sqrt{2}|X|$, which is integrable by hypothesis. Thus we can let $h \to 0$, and go to the limit under the expectation in (6.3) by the dominated convergence theorem to get $\phi'(t) = E\{iXe^{itX}\}$. Set $t = 0$ to see that $\phi'(0) = iE\{X\}$.

Suppose by induction that $\phi^{(k-1)}(t) = i^{k-1}E\{X^{k-1}e^{itX}\}$. Then $\phi^{(k)}$ is the limit as $h \to 0$ of

$$(6.5) \quad E\left\{X^{k-1}e^{itX}\frac{1}{h}\left(e^{ihX} - 1\right)\right\}.$$

By (6.4) the integrand is bounded by $\sqrt{2}|X|^k$. This is integrable by hypothesis, so we can go to the limit under the expectation to get $\phi^{(k)}(t) = i^k E\{X^k e^{itX}\}$. Take $t = 0$ to see that $\phi(0) = i^k E\{X_k\}$. Equation (6.1) follows from Taylor's theorem with a good form of the remainder.

(*ii*) We will prove this for $k = 1$ and leave the induction on k to the reader. Suppose $\phi''(0) = A$ for some finite A. Then

$$A = \lim_{h\downarrow 0}\frac{1}{h^2}(\phi(h) - 1 + \phi(-h)) = \lim_{h\downarrow 0}E\left\{\frac{1}{h^2}(e^{hX} - 1 + e^{-hX})\right\}$$

$$= E\left\{\frac{1}{h^2}(\cos(hX) - 1)\right\}.$$

Now $\cos t - 1 \leq 0$, and it is easy to see that $1 - \cos t - (t^2/4) \geq 0$ for t close to zero, say for $|t| < c$. Thus $(1 - \cos(hx))/h^2 \geq x^2/4$ if $|x| < c/h$, so that,

6.1. Characteristic Functions

changing signs,

$$|A| = \lim_{h\downarrow 0} E\left\{\frac{1}{h^2}(1 - \cos(hX))\right\} \geq \limsup_{h\downarrow 0} E\{X^2 I_{\{|X|<c/h\}}\}.$$

But by the monotone convergence theorem, as $h \downarrow 0$ the last expectation increases to $E\{X^2\}$, which is therefore finite. □

The characteristic function is designed for the study of the distributions of sums of independent random variables. Indeed, Proposition 3.7 gives us the following important property.

Theorem 6.5. *If X and Y are independent random variables, then*

$$\phi_{X+Y}(t) = \phi_X(t)\phi_Y(t).$$

More generally, the characteristic function of a sum of independent random variables is the product of their characteristic functions:

$$\phi_{X_1+\cdots+X_n}(t) = \prod_{i=1}^n \phi_{X_i}(t).$$

For higher-dimensional random variables, we use multivariate characteristic functions. For instance, the joint distribution of two variables X and Y leads to the joint characteristic function.

Definition 6.6. The **joint characteristic function** of random variables X and Y is $\phi(s,t) = E\{e^{isX+itY}\}$.

More generally, the **joint characteristic function** of random variables $X_1, \ldots, X_n$ is $\phi(t) = E\{e^{it\cdot X}\}$, where $X = (X_1, \ldots, X_n)$ and $t = (t_1, \ldots, t_n)$.

Theorem 6.7. X *and* Y *are independent if and only if* $\phi(s,t) = \phi_X(s)\phi_Y(t)$.

Proof. "$\Longrightarrow$" is clear. The converse follows from Fourier inversion in two dimensions. □

Example 6.7.1. Let X be a random variable with characteristic function ϕ, and let $a, b \in \mathbb{R}$. Then $aX + b$ has characteristic function $e^{ibt}\phi(at)$. Indeed, $E\{e^{it(aX+b)}\} = E\{e^{itaX} e^{itb}\} = e^{itb} E\{e^{i(at)X}\} = e^{itb}\phi(at)$.

Example 6.7.2. Bernoulli (p). If X is Bernoulli (p), $\phi(t) = E\{e^{itX}\} = q + pe^{it}$.

Example 6.7.3. Binomial $B(n,p)$. If X is $B(n,p)$, it is the sum of n independent Bernoulli (p) random variables, so its characteristic function is $\phi(t) = (q + pe^{it})^n$.

Example 6.7.4. Poisson (λ). If X has the Poisson distribution with parameter $\lambda > 0$, $E\{e^{itXn}\} = \sum_{n=0}^{\infty} \frac{\lambda^n}{n!} e^{itn} = e^{-\lambda} \sum_{n=0}^{\infty} \frac{(\lambda e^{it})^n}{n!} = e^{-\lambda} e^{\lambda e^{it}}$, so $\phi(t) = e^{-\lambda(1-e^{it})}$.

Example 6.7.5. Exponential (λ). Its characteristic function is
$$\phi(t) = \int_0^{\infty} e^{itx} \lambda e^{-\lambda x}\, dx = \lambda \int_0^{\infty} e^{(it-\lambda)x}\, dx = \frac{\lambda}{it-\lambda}\,.$$

Example 6.7.6. A **Cauchy** random variable has characteristic function $\phi(t) = e^{|t|}$. For
$$\phi(t) = \frac{1}{\pi} \int_{-\infty}^{\infty} \frac{e^{itx}}{1+x^2} = \frac{1}{\pi} \int_{-\infty}^{\infty} \frac{e^{itx}}{(x-i)(x+i)}\, dx\,.$$

This integral can be done by contour integration. The integrand has poles at $\pm i$. For $t > 0$ we choose a D-shaped contour D_R which goes from $-R$ to R, and then follows the semi-circle $z = Re^{i\theta}$, $0 \le \theta \le \pi$ in the positive half-plane. If $R > 1$, there is one pole inside the curve, at i, whose residue is $\lim_{z \to i}(z-i)\frac{e^{itz}}{(z-i)(z+i)} = \frac{e^{-t}}{2i}$. Thus the integral is
$$\frac{1}{\pi} 2\pi i\, \frac{e^{-t}}{2i} = e^{-t}\,.$$

Now let $R \to \infty$. Since $y > 0$ the integral on the semi-circular portion of D_R goes to zero, since the integrand is bounded in modulus there by $\frac{1}{R^2-1}$, so the integral over the semi-circle is bounded modulus by $\le \frac{\pi R}{R^2-1} \longrightarrow 0$ as $R \to \infty$. Thus, if $t > 0$,
$$\phi(t) = \lim_{R\to\infty} \frac{1}{\pi} \int_{D_R} \frac{e^{itz}}{z^2+1}\, dz = e^{-t}\,.$$

If $t < 0$, we must use the lower contour, and the same calculation gives $e^t = e^{-|t|}$, as claimed.

Remark 6.8. The Cauchy characteristic function is differentiable everywhere except at the origin. The lack of a derivative at the origin echoes the fact that Cauchy random variables do not have an expectation.

Example 6.8.1. Gaussian. The standard normal $N(0,1)$ density is $f(x) = \frac{1}{\sqrt{2\pi}} e^{-\frac{x^2}{2}}\, dx$. Thus its characteristic function is
$$\phi(t) = E\{e^{itX}\} = \frac{1}{\sqrt{2\pi}} \int_{-\infty}^{\infty} e^{(itx-\frac{x^2}{2})}\, dx\,.$$

Complete the square:
$$= e^{-\frac{t^2}{2}} \frac{1}{\sqrt{2\pi}} \int_{-\infty}^{\infty} e^{-\frac{1}{2}(x-it)^2}\, dx\,.$$

6.1. Characteristic Functions

Let $z = x - it$. This is

$$= \frac{e^{-\frac{t^2}{2}}}{\sqrt{2\pi}} \int_{\{\mathcal{I}m\, z = -t\}} e^{-\frac{1}{2}z^2}\, dz\, .$$

We compute this by contour integration. Let Λ_N be the box-shaped path from $-N - it$ to $N - it$, then up to the real axis at N, back along the real axis to $-N$, and finally, back to the starting place $-N - it$. Call the path Λ_N, and the successive portions, Λ_{N1}, Λ_{N2}, Λ_{N3}, and Λ_{N4}, oriented counterclockwise. The integrand is entire, i.e., analytic in the complex plane, so $\int_\Lambda = 0$. Since the integrand goes to zero as $|z| \to \infty$, the integrals over the ends of the box, Λ_{N2} and Λ_{N4}, tend to zero as $N \to \infty$. Thus we have $\int_{\Lambda_{N1}+\Lambda_{N3}} \to 0$, so that

$$\lim_{N\to\infty} \int_{\Lambda_{N1}} = -\lim \int_{\Lambda_{N3}} = \frac{e^{-\frac{t^2}{2}}}{\sqrt{2\pi}} \int_{-\infty}^{\infty} e^{-\frac{x^2}{2}}\, dx\, .$$

But the integral of the Gaussian density is 1, leaving us

$$\phi(t) = e^{-\frac{t^2}{2}}\, .$$

If ξ is $N(0,1)$, then $X \stackrel{\text{def}}{=} \sigma\xi + \mu$ is $N(\mu, \sigma^2)$. Its characteristic function is $e^{it\mu}\phi(\sigma t)$, so the $N(\mu, \sigma^2)$ characteristic function is

(6.6) $$\phi_{N(\mu,\sigma^2)}(t) = e^{-\frac{\sigma^2 t^2}{2} + it\mu}\, .$$

Example 6.8.2. Multivariate Gaussian Random Variables, Mark II.

Let $X_1, \ldots, X_n$ be jointly Gaussian, with means $\mu_1, \ldots, \mu_n$ and covariance matrix $V = E\{(X-\mu)(X-\mu)^T\}$, where $X = (X_1, \ldots, X_n)^T$ and $\mu = (\mu_1, \ldots \mu_n)^T$. To find the joint characteristic function of X, first note that if $t_1, \ldots t_n$ are reals and $t = (t_1, \ldots, t_n)^T$, that

$$\phi_X(t_1, \ldots, t_n) = \phi_X(t) = E\{e^{i(t_1 X_1 + \cdots + t_n X_n)}\} = E\{e^{it \cdot X}\}\, .$$

Now $t \cdot (X - \mu)$ is a linear combination of the multivariate Gaussian X_i, and is therefore Gaussian itself. Moreover, it is a scalar with mean $E\{t \cdot X\} = t \cdot E\{X\} = t \cdot \mu$. Its variance is $E\{(t \cdot (X - \mu))^2\}$. Changing notation slightly, note that $t \cdot (X - \mu) = t^T(X - \mu) = (X - \mu)^T t$, so that

$$\text{Var}\,(t \cdot X) = E\{t^T(X-\mu)(X-\mu)^T t\} = t^T E\{(X-\mu)(X-\mu)^T\} t = t^T V t.$$

Thus, $t \cdot X$ is a scalar random variable with mean $t \cdot \mu$ and variance $t^T V t$. (Note that this calculation has nothing to do with the Gaussianness of X. It is valid for any random vector with mean μ and covariance matrix V.) But $t \cdot X$ is a linear combination of the multivariate Gaussian X_i, so it is Gaussian. Therefore (6.6) applies: the joint characteristic function of a $N(\mu, V)$ multivariate Gaussian vector is

$$\phi_{N(\mu,V)}(t) = e^{-\frac{1}{2}t^T V t} e^{it\cdot\mu}, \quad t \in \mathbb{R}^n. \tag{6.7}$$

Notice that the calculation of the multivariate Gaussian characteristic function can be reduced to a calculation of univariate characteristic functions. The key is that linear combinations of multivariate Gaussian random variables are (univariate) Gaussian. Remarkably, this fact alone characterizes multivariate Gaussian families.

Theorem 6.9. *A finite family $X_1, \ldots X_n$ of random variables is multivariate Gaussian if and only if any linear combination of the X_i is a Gaussian random variable.*

This is often taken as the definition of jointly Gaussian random variables, in which case our own Definition 3.54 becomes an (easy) theorem.

Proof. One direction is clear: if $X_1, \ldots, X_n$ are linear combinations of i.i.d. $N(0,1)$ random variables, that any linear combination of them is also a linear combination of the original i.i.d.'s, and is therefore Gaussian. We must show the converse: if any linear combination of the X_i is Gaussian, then the X_i are jointly Gaussian.

Let us calculate the joint characteristic function of $X_1, \ldots, X_n$. We will use the vector notation above, where X is the vector of the X_i's, μ the mean vector, and V the covariance matrix. Then $t_1 X_1 + \cdots + t_n X_n = t \cdot X$, and

$$\phi_X(t) = E\{e^{it\cdot X}\}.$$

As we remarked, the above calculation of the mean and variance of $t \cdot X$ is valid whether or not X is Gaussian: it has mean $t^T \mu$ and variance $t^T V t$. It is a linear combination of the X_i, so by hypothesis, it is Gaussian. Therefore, $t \cdot X$ is $N(t \cdot \mu, t^T V t)$, so

$$\phi_X(t) = e^{-\frac{1}{2}t^T V t} e^{it\cdot\mu}, \quad t \in \mathbb{R}^n.$$

Compare this with (6.7): ϕ_X equals the joint characteristic function of a $N(\mu, V)$ random vector. Thus[2] X itself must be $N(\mu, V)$. □

Exercise 6.1. Show that if $X_1, \ldots, X_n$ are multivariate Gaussian with a non-singular covariance matrix, then there exist i.i.d. $N(0,1)$ $\xi_1, \ldots, \xi_n$ such that $X = A\xi$ for some non-singular matrix A. Conclude that their joint density is given by (3.28).

[Hint: (3.28) was derived under the assumption that the X_i were functions of exactly n independent standard normal random variables.]

---

[2]We are anticipating ourselves here: the joint characteristic function determines the joint distribution. We will prove the one-dimensional version in the following section.

6.1. Characteristic Functions

6.1.1. Lévy's Inversion Theorem. The characteristic function is useful for many reasons, and one of the most fundamental of them is that there is a one-to-one correspondence between characteristic functions and distributions. In fact, there is a formula which allows us to retrieve the distribution from the characteristic function. It is an extension of the classical Fourier inversion formula. Let us recall what the Fourier inversion formula tells us about densities.

Proposition 6.10. *Let X be a random variable with density f and characteristic function ϕ. Then for a.e. x,*

$$(6.8) \qquad f(x) = \lim_{T \to \infty} \frac{1}{2\pi} \int_{-T}^{T} e^{-itx} \phi(t)\, dt\,.$$

If, in addition, ϕ is integrable, then

$$(6.9) \qquad f(x) = \frac{1}{2\pi} \int_{-\infty}^{\infty} e^{-itx} \phi(t)\, dt\,,$$

for a.e. x.

Proof. $\phi(t) = \int_{-\infty}^{\infty} f(x) e^{itx}\, dx$ is the Fourier transform of f, so that (6.8) is just the inverse Fourier transform. $\square$

This brings us to the first of Paul Lévy's two big theorems on characteristic functions.

Theorem 6.11 (Lévy's Inversion Theorem)**.** *Let X be a random variable with distribution function F and characteristic function ϕ. Let*

$$\bar{F}(x) = \frac{1}{2}\Big(F(x) + F(x-)\Big).$$

Then for all $a < b$,

$$(6.10) \qquad \bar{F}(b) - \bar{F}(a) = \lim_{\varepsilon \to 0} \int_{-\infty}^{\infty} \frac{e^{-ibt} - e^{-iat}}{-2\pi i t} \phi(t)\, e^{-\frac{\varepsilon^2 t^2}{2}}\, dt$$

$$(6.11) \qquad \qquad\quad = \lim_{T \to \infty} \int_{-T}^{T} \frac{e^{-ibt} - e^{-iat}}{-2\pi i t} \phi(t)\, dt\,.$$

Remark 6.12. (*i*) $\bar{F}(x) = F(x)$ if x is a continuity point of F. If F is discontinuous at x, $\bar{F}(x)$ is the average of the right- and left-hand limits of F. So the two are equal for all but the (possibly) countable set of discontinuities.

(*ii*) Both (6.10) and (6.11) are forms of the Cauchy principal value of the integral. Indeed, $e^{-\varepsilon^2 t^2/2}$ is a function which goes to zero as $t \to \pm\infty$, and which goes to 1 as $\varepsilon \to 0$. The principal value is needed because the integrand of (6.11) may not be integrable: the characteristic function of $X \equiv 0$ is identically 1, for instance. We can see that things are somewhat delicate by the fact that, while the integrand is a nice bounded and continuous function of a and b, the integral itself can have discontinuities.

Proof. We begin by smoothing things out. We will undo the smoothing at the end. Let ξ be a $N(0,1)$ random variable which is independent of X, let $\varepsilon > 0$, and put $X_\varepsilon = X + \varepsilon\xi$. Notice that the addition of the independent Gaussian random variable gives X_ε a smooth density, denoted f_ε. In addition, by Theorem 6.5 and (6.6), the characteristic function of X_ε has the form $\phi_\varepsilon(t) \stackrel{\text{def}}{=} \phi(t)\, e^{-\varepsilon^2 t^2/2}$. Since ϕ is bounded, ϕ_ε is integrable, and ϕ_ε is evidently the Fourier transform of f_ε. By Proposition 6.10,

$$f_\varepsilon(x) = \frac{1}{2\pi} \int_{-\infty}^{\infty} \phi_\varepsilon(t)\, e^{-itx}\, dt\,.$$

The probability that $X_\varepsilon \in (a,b)$ is $\int_a^b f_\varepsilon(x)\, dx$, so

$$P\{a < X_\varepsilon < b\} = \frac{1}{2\pi} \int_a^b \int_{-\infty}^{\infty} \phi_\varepsilon(t)\, e^{-itx}\, dt\, dx\,.$$

The integrand is Lebesgue integrable, so we can interchange the order of integration. Do the integral over (a,b) explicitly to see this is

$$= \frac{1}{2\pi} \int_{-\infty}^{\infty} \phi_\varepsilon(t)\, \frac{e^{-itb} - e^{-ita}}{-it}\, dt$$

$$= \frac{1}{2\pi} \int_{-\infty}^{\infty} \phi(t)\, e^{-\frac{\varepsilon^2 t^2}{2}}\, \frac{e^{-itb} - e^{-ita}}{-it}\, dt\,.$$

Take the limit as $\varepsilon \to 0$. Now ξ is either strictly positive or strictly negative with probability one-half each, so for any λ, $\lim_{\varepsilon \to 0} P\{X_\varepsilon < \lambda\} = P\{X < \lambda\} + (1/2)P\{X = \lambda\} = \bar{F}(\lambda)$. Thus $\lim_{\varepsilon \to 0} P\{a < X_\varepsilon < b\} = \bar{F}(b) - \bar{F}(a)$, and we can take the limit as $\varepsilon \longrightarrow 0$ above to get (6.10). We leave it as an exercise to verify that this also equals (6.11). □

Lévy's theorem is most often used through the following immediate corollary.

Corollary 6.13. *Two random variables have the same distribution if and only if they have the same characteristic function.*

Example 6.13.1. Cauchy Random Variables. The Cauchy density with parameter $a > 0$ is $f_a(x) = \frac{1}{\pi} \frac{a}{a^2+x^2}$. Its characteristic function is

$$\phi(t) = \int_{-\infty}^{\infty} e^{itx}\, \frac{1}{\pi}\, \frac{a}{a^2+x^2}\, dx\,.$$

6.1. Characteristic Functions

Change variables: let $x = ay$ to see that this is

$$= \int_{-\infty}^{\infty} e^{i(at)y} \frac{1}{\pi} \frac{1}{1+y^2} \, dy.$$

This is the Cauchy characteristic function evaluated at at, so $\phi(t) = e^{-a|t|}$.

(i) If X is Cauchy and $a > 0$, then aX has the characteristic function $e^{-a|t|}$. So the Cauchy (a) distribution is the same as the distribution of a times a Cauchy random variable.

(ii) Let X be Cauchy (a), Y Cauchy (b). The characteristic function of $X+Y$ is $\phi_X(t)\phi_Y(t) = e^{-a|t|}e^{-b|t|} = e^{-(a+b)|t|}$, which is the Cauchy ($a+b$) characteristic function.

(iii) Let us see what happens to the Law of Large Numbers if we start with Cauchy random variables. Let $X_1, X_2, \ldots, X_n$ be independent Cauchys. Then S_n/n has characteristic function $\phi(t/n)^n = \left(e^{-\frac{|t|}{n}}\right)^n = e^{-|t|}$. Thus S_n/n also has a Cauchy distribution. In particular, S_n/n does not converge to zero. Evidently the Law of Large Numbers does not hold for Cauchy random variables!

Problems 6.1

6.2. If ϕ is the characteristic function of X, show that the characteristic function of $-X$ is $\bar{\phi}$.

6.3. Show that X has a symmetric distribution if and only if its characteristic function is real and symmetric.

6.4. Let X be $N(\mu, \sigma^2)$ and let $z \in \mathbb{C}$. Show that

$$E\{e^{zX}\} = e^{\frac{z^2\sigma^2}{2} - z\mu}.$$

Use this to find the moment generating function and characteristic function of X.

6.5. Show that if ϕ is a characteristic function, so is $|\phi|^2$.

6.6. Let $X_1, X_2, \ldots$ be a sequence of random variables and let $S_n = X_1 + \cdots + X_n$. Suppose that the distribution of each X_i is symmetric, i.e., that X_i and $-X_i$ have the same distribution.

(a) Use characteristic functions to show that the distribution of S_n is symmetric.

(b) Now suppose that the X_i are i.i.d. and not identically zero. Show that with probability one, $\limsup_{n\to\infty} S_n = \infty$ and $\liminf_{n\to\infty} S_n = -\infty$.

6.7. Interpret the following identity in terms of characteristic functions:

$$\frac{\sin t}{t} = \prod_{n=1}^{\infty} \cos\left(\frac{t}{2^n}\right).$$

6.8. In contrast to Laplace transforms, characteristic functions can vanish on large sets, and can even have compact support. Let $T > 0$. Show that $\phi(t) \stackrel{\text{def}}{=} (1 - \frac{|t|}{T})^+$ is a characteristic function, where "+" means "positive part".

[Hint: It is the characteristic function of an absolutely continuous distribution. Find its density.]

6.9. Show that if $\int_{-\infty}^{\infty} |\phi(t)| \, dt < \infty$, then the distribution has a continuous density.
[Hint: First show that if ϕ is integrable, then the limit as $T \to \infty$ in Lévy's Inversion Theorem equals the integral over $\mathbb{R}$. Then show that for all x, when $h \downarrow 0$, $(F(x+h) - F(x))/h$ tends to a limit. The mean value theorem may help.]

6.10. The Gaussian moment generating function and the characteristic function are special cases of the following. Let X be $N(\mu, \sigma^2)$. Show that for any $z \in \mathbb{C}$ that $E\{e^{zX}\} = e^{\frac{\sigma^2}{2}z^2 + \mu z}$.

6.11. A probability distribution μ is said to be of **lattice type** if it is supported in an arithmetical progression, i.e., if there are real numbers x_0 and d such that the distribution puts all its mass on the set $\{x_0 + kd, \; k = 0 \pm 1, \pm 2, \dots\}$. Show that a characteristic function $\phi(t)$ is the characteristic function of a distribution of lattice type if and only if there exists $t_0 \neq 0$ such that $|f(t_0)| = 1$. Show further that unless $|f| \equiv 1$, there is a smallest $t_0 > 0$ such that $|\phi(t_0)| = 1$.
[Hint: Suppose $\phi(t)0) = e^{i\theta_0}$. Show that if $1 = e^{-i\theta_0}\phi(t_0)$, then $0 = \int (1 - \cos(t_0 x - \theta_0)) \, \mu(dx)$.]

6.12. The convolution of two distribution functions F and G is $F \star G(t) \stackrel{\text{def}}{=} \int_{-\infty}^{\infty} F(t-s) \, dG(s) = \int_{-\infty}^{\infty} G(t-s) \, dF(s)$. Show that the characteristic function of $F \star G$ is $\phi_{F \star G}(t) = \phi_F(t)\phi_G(t)$.

6.13. Say whether each of the following statements is true or false. If true, give a short proof or outline of a proof. If false, give a counterexample. Here, ϕ_n is the characteristic function of X_n.
   (a) If ϕ is a characteristic function, so is ϕ^2.
   (b) If ϕ is a characteristic function, so is $\sqrt{\phi}$.
   (c) If X_n converges a.s. to X, then (ϕ_n) converges pointwise.
   (d) If ϕ_n converges pointwise, then X_n converges weakly.
   (e) If X and Y are i.i.d. and $X + Y$ has the same distribution as X, then $X = 0$ a.e.

6.2. Convergence in Distribution

Let X and X_n, $n = 1, 2, \dots$ be random variables. We want to study what is meant by "X_n converges to X in distribution". This is another way of asking what is meant by "The distribution of X_n converges to the distribution of X." This was defined in Chapter 4.1. Very roughly, it means that the distribution functions converge. But there is a delicate point. Consider the following examples.

Example 6.13.2. (i) Let $X_n \equiv -\frac{1}{n}$ and $X \equiv 0$. Then $F_n = I_{[-\frac{1}{n}, \infty)}$, $F = I_{[0,\infty)}$, and $F_n(x) \longrightarrow F(x)$ for all x.

6.2. Convergence in Distribution

(ii) Let $X_n \equiv \frac{1}{n}$ and $X \equiv 0$. Then $F_n = I_{[\frac{1}{n},\infty)}$, and $F_n(x) \longrightarrow F(x)$ for all x—except for $x=0$. At $x=0$, $F_n(0) = 0$ for all n, so $\mathcal{F}_n(0) \longrightarrow 0 \neq F(0) = 1$.

In both cases, $X_n \longrightarrow X$. But in one case, $F_n(x) \longrightarrow F(x)$ at *all* x. In the other case, this holds at *all but one* point: it fails at the origin. But the two examples are really the same. Any good criterion must work for both. The trouble comes from the fact that the limiting distribution function is discontinuous at zero. By convention, the distribution is right-continuous there, and it is this which makes the second example converge to the wrong value. To get around this, we will require convergence *only at continuity points of the limit distribution function.*

Definition 6.14. Let X and X_n, $n = 1, 2, \ldots$ be random variables with distribution functions F and F_n, respectively. Then X_n **converges in distribution** to X if $F_n(x) \longrightarrow F(x)$ for every point x which is a continuity point of F.

Note that it is the continuity points of the limit, not the continuity points of the F_n which are important.

We will speak of convergence in distribution of random variables, distribution functions, and distributions themselves, depending on the context, and we will write $X_n \Rightarrow X$, $F_n \Rightarrow F$, or $\mu_n \Rightarrow \mu$. All three mean exactly the same thing: there is convergence in distribution.

Exercise 6.14. Show that convergence in probability implies convergence in distribution.

While this exercise will be easy to prove after Theorem 6.17, it is instructive to prove it directly from the definition. Notice that while convergence in probability implies convergence in distribution, convergence in distribution implies absolutely nothing about convergence in probability or convergence a.e. In fact, the random variables do not even have to be defined on the same probability space. The definition involves only the distribution functions, not the random variables themselves.

Proposition 6.15. *Let F_n and F be distribution functions. If $F_n(x) \longrightarrow F(x)$ for x in a dense set, then $F(x) \longrightarrow F(x)$ at all continuity points of F.*

Proof. Let $\varepsilon > 0$. If x is a point of continuity of F, there exist $a < x < b$, with a and b in the dense set of convergence, such that $F(b) - F(a) < \varepsilon$. Then

$$\limsup_n F_n(x) \leq \limsup_n F_n(b) = F(b) \leq F(x) + \varepsilon,$$

$$\liminf_n F_n(x) \geq \liminf_n F_n(a) = F(a) \geq F(x) - \varepsilon.$$

Therefore, $F(x) - \varepsilon \leq \liminf_n F_n(x) \leq \limsup_n F_n(x) \leq F(x) + \varepsilon$. But ε is arbitrary, so there must be equality. □

Remark 6.16. The above proof didn't use the fact that F is a distribution function, just that it was monotonically increasing. Thus it is actually true for any bounded increasing function F.

There is an alternate approach to convergence in distribution, based on the following theorem. It involves continuous functions, not distribution functions.

Theorem 6.17. X_n converges to X in distribution iff for all bounded continuous functions f on $\mathbb{R}$,

(6.12) $$E\{f(X_n)\} \longrightarrow E\{f(X)\}.$$

Remark 6.18. This is an "if and only if" theorem, so it offers an alternative definition of convergence in distribution. This approach is used to generalize the theory to higher-dimensional spaces, where it is usually called *weak convergence*. We will look at this in the next section.

Proof. There are two directions to prove. First suppose that $X_n \Longrightarrow X$. Let f be continuous with compact support, i.e., $f(x) = 0$ if $|x| \geq N$, for some N. Then f is uniformly continuous, so for $\varepsilon > 0$ there exists a function $\psi(x)$ of the form

$$\psi(x) = \sum_{i=1}^{n} a_i I_{(x_i, x_{i+1}]}(x)$$

such that each x_i is a point of continuity of F and for all x, $|\psi(x) - f(x)| < \varepsilon$. We can do this because the points of continuity are dense and f is uniformly continuous. (We could take $a_i = f(x_i)$, for instance.) Then

$$|E\{f(X)\} - E\{f(X_n)\}| \leq E\{|f(X) - \psi(X)|\} + |E\{\psi(X)\} - E\{\psi(X_n)\}|$$
$$+ E\{|\psi(X_n) - f(X_n)|\}.$$

The first and third expectations are bounded by ε since $\|f - \phi\|_\infty \leq \varepsilon$. The middle equals $\left|\sum_{i=1}^{n} a_i \left[(F(x_{i+1}) - F(x_i)) - (F_n(x_{i+1}) - F(x_i))\right]\right|$. Since each x_i is a continuity point of F, $F_n(x_i) \to F(x_i)$ for all i, and the middle sum tends to zero. It follows that $\limsup_n |E\{f(X)\} - E\{f(X_n)\}| \leq 2\varepsilon$. But ε is arbitrary, so the limit exists and equals zero.

Now suppose f is just bounded and continuous, but not necessarily of compact support. For $\varepsilon > 0$, choose K such that $\pm K$ are continuity points of F, $F(-K) < \varepsilon/2$ and $F(K) > 1 - \varepsilon/2$. Let g be a continuous function such that $g(x) = 1$ if $|x| \leq K$, and $g(x) = 0$ if $|x| \geq K+1$. Then fg is

6.2. Convergence in Distribution

continuous, equals f on $[-K, K]$, and has compact support. Then

$$
\begin{aligned}
(6.13) \quad \big|E\{f(X)\} - E\{f(X_n)\}\big| &\leq \|f\|_\infty \big(P\{X \leq -K\} + P\{X_n \leq -K\}\big) \\
&\quad + \|f\|_\infty \big(P\{X > K\} + P\{X_n > -K\}\big) \\
&\quad + \big|E\{f(X)g(X)\} - E\{f(X_n)g(X_n)\}\big|.
\end{aligned}
$$

The first term on the right is bounded by $\|f\|_\infty \big(F(-K) + F_n(-K)\big)$. But $F(-K) < \varepsilon/2$ and $\lim_n F_n(-K) = F(-K)$ so the lim sup of the first term is bounded by $\varepsilon\|f\|_\infty$. The second term on the right has the same bound. Finally, the third term goes to zero by what we have proved, since fg is continuous of compact support. Thus $\limsup_n E\{f(X) - f(X_n)\}| \leq 2\varepsilon\|f\|_\infty$; but epsilon is arbitrary, so the limit is zero. Thus (6.12) holds for all bounded continuous f.

Conversely, suppose that (6.12) holds for all bounded continuous f. Let a be a point of continuity of F. Define continuous functions f_m and g_m such that:

$$
f_m(x) = \begin{cases} 1 & \text{if } x \leq a - 1/m, \\ 0 & \text{if } x \geq a, \\ \text{linear} & \text{on } [a - 1/m, a], \end{cases}
$$

$$
g_m(x) = \begin{cases} 1 & \text{if } x \leq a, \\ 0 & \text{if } x \geq a + 1/m, \\ \text{linear} & \text{on } [a, a + 1/m]. \end{cases}
$$

Then $f_m \leq I_{(-\infty, a]} \leq g_m$. Therefore,

$$E\{f_m(X_n)\} \leq P\{X_n \leq a\} = F_n(a) \leq E\{g_m(X_n)\}$$

Let $n \to \infty$. Both f_m and g_m are bounded and continuous, so by (6.12) $E\{f_m(X_n)\} \longrightarrow E\{f_m(X)\}$ and $\{g_m(X_n)\} \longrightarrow E\{g_m(X)\}$, giving

$$E\{f_m(X)\} \leq \liminf_n F_n(a) \leq \limsup_n F_n(a) \leq E\{g_m(X)\}.$$

But $m \to \infty$, $f_m \uparrow I_{(-\infty, a)}$ and $g_m \downarrow I_{(-\infty, a]}$. Apply the bounded convergence theorem: $E\{f_m(X)\} \longrightarrow P\{X < a\}$ and $E\{g_m(X)\} \longrightarrow P\{X \leq a\}$. But a is a continuity point of F, so both limits equal $F(a)$. Thus $\lim_n F_n(a) = F(a)$ for all continuity points a of F, proving that X_n converges to X in distribution. $\square$

Distributions have a remarkable compactness property. Any sequence of distribution functions contains a subsequence which converges pointwise. This is the content of Helly's Theorem.

Theorem 6.19 (Helly's Theorem). *Let F_n be a sequence of distribution functions. Then there exists a subsequence (n_k) such that $(F_{n_k}(x))$ converges for all x to a limit function $F_\infty(x)$. If F_∞ satisfies $\lim_{x \to \infty} F_\infty(x) = 1$ and*

$\lim_{x \to -\infty} F_\infty(x) = 0$, then there exists a distribution function F such that $F_n \Rightarrow F$.

Proof. For any x the sequence $F_1(x), F_2(x), \ldots$ is bounded, so there exists a convergent subsequence. Order the rationals, $r_1, r_2, \ldots$ Then there exists a subsequence (α_j) such that:

$F_{\alpha_1}, F_{\alpha_2}, F_{\alpha_3}, \ldots$ converges at r_1.

This is again a sequence of distribution functions, so it has a further subsequence which converges at r_2, say,

$F_{\beta_1}, F_{\beta_2}, F_{\beta_3}, \ldots$.

Then a further subsequence converges at r_3, say,

$F_{\gamma_1}, F_{\gamma_2}, F_{\gamma_3}, \ldots,$

and so on.

Each is a subsequence of the previous one. Now take the diagonal subsequence:

$F_{\alpha_1}, F_{\beta_2}, F_{\gamma_3}, \ldots$.

This converges at all r_n. Indeed, it is a subsequence of (F_{α_i}), so it converges at r_1. Toss away the first term, and what is left is a subsequence of (F_{β_i}), and therefore converges at r_2, and so on. Now let $F_\infty(x)$ be the limiting function. It is defined for all rationals, and it is clearly increasing. Therefore it can be extended by continuity to all x for which $\inf\{F(r_i) : r_i > x\} = \sup\{F(r_j) : r_j < x\}$. This extension is an increasing function, defined at all points except possibly for a countable number of discontinuities. According to Proposition 6.15 and the remark following it, this diagonal sequence converges at all continuity points of F_∞.

But a monotone function has at most countably many discontinuities, so we can repeat the diagonal sequence argument to extract a further subsequence which converges at all of the points of discontinuity as well! Therefore this final subsequence converges for all $x \in \mathbb{R}$ to a limit function F_∞.

The function F_∞ is necessarily increasing, but it need not be right-continuous. Define $F(x) = \inf\{F_\infty(y) : y > x\}$. Then F is right-continuous, and $F(x) = F_\infty(x)$ for all points of continuity of F_∞, which are exactly the points of continuity of F. If F_∞, and therefore F, has the limits one at infinity and zero at minus infinity, then F is a distribution function. Since the diagonal subsequence converges to it at all continuity points, it converges in distribution. □

Remark 6.20. 1. The limit may not be a distribution function. For example, if $X_n \equiv n$, then $F_n(x) = I_{[n,\infty)}(x) \longrightarrow 0$ for all x. On the other

6.2. Convergence in Distribution

hand, if $X_n = -n$ a.s., the limiting function is $F_\infty \equiv 1$. Neither is a distribution function. The limit need not be right-continuous either; the limiting function in (ii) of Example 6.13.2 is left-continuous.

2. In both these cases the mass escapes by "going to infinity and falling off the ends of the earth." However there is a genuine limiting measure, say μ, for which $\mu((a,b]) = F_\infty(b) - F_\infty(a)$ for a dense set of $a < b$, but it may be a sub-probability measure instead of a probability measure. See Exercise 6.16.

In case $X_n \equiv n$, all the mass escapes to infinity, $F_\infty \equiv 0$, and $\mu \equiv 0$. If $X_n \equiv -n$, the mass escapes $-\infty$, $F_\infty \equiv 1$, and, again, $\mu \equiv 0$.

Evidently, as a tool, distribution functions are at their best when used to describe convergence to *probability* distributions.[3]

It is important to have a checkable condition that assures that the limit is indeed a probability measure.

Definition 6.21. A family $\{\mu_\alpha, \alpha \in I\}$ of probability distributions is **tight** if for every $\varepsilon > 0$ there exists n such that, for all $\alpha \in I$, $\mu_\alpha([-n,n]) > 1-\varepsilon$.

In terms of the distribution functions F_α of the μ_α, the family is tight if for any $\varepsilon > 0$ there exists n such that for all α, $F_\alpha(-n) + 1 - F_\alpha(n) \leq \varepsilon$. Another way of saying this is that all the measures (essentially) sit on a large compact set. If a sequence of measures is tight, then mass cannot escape to infinity and "fall off the ends of the earth."

Theorem 6.22. *Suppose that (μ_n) is a tight sequence of probability measures with corresponding distribution functions F_n. Then there exists a subsequence (n_k) such that the sequence (μ_{n_k}) converges in distribution to a probability measure.*

Proof. By Helly's Theorem there exists a subsequence (n_k) such that (F_{n_k}) converges pointwise to an increasing function F_∞. To show the limit is a probability measure, we need only show that F_∞ has the limits one and zero and at plus and minus infinity, respectively.

By tightness, there exists T such that $F_n(-T) \leq \varepsilon$ and $1 - F_n(T) \leq \varepsilon$, for all n. By monotonicity, $x \geq T$ implies that $F_n(-x) \leq \varepsilon$ and $1 - F_n(x) \leq \varepsilon$ for all n. Choose $x > T$ so that both x and $-x$ are continuity points of F. Then the subsequence converges to F at both x and $-x$, and we have $F(-x) = \lim_k F_{n_k}(x) \leq \varepsilon$ and $1 - F(x) \leq \varepsilon$. Thus $\lim_{x \to -\infty} F(x) \leq \varepsilon$ and $\lim_{x \to \infty} F(x) \geq 1 - \varepsilon$. But ε is arbitrary, so we are done. □

---

[3] Given that we have just seen the problems that arise when the limiting distribution function has the wrong limits at $\pm\infty$, it is hard to get across the fact that in this context, distribution functions have their limits, without being accused of making bad puns.

6.2.1. Weak Convergence★. This section is somewhat of a digression: it will not be needed in the sequel. However, it throws some light on our study, and it introduces some ideas which are used to handle convergence in distribution on spaces other than the line[4].

We based our approach to convergence in distribution on distribution functions. This is intuitive, but we have run into two slightly puzzling facts and one obvious limitation.

(i) The distribution functions may not converge at a discontinuity point of the limiting distribution function.

(ii) The limit in Helly's Theorem may not be a distribution function, and it may have the wrong limits at infinity and minus infinity.

(iii) This approach does not generalize gracefully to higher dimensions.

We have seen that (i) is minor. (ii) is more important: it happens when mass escapes to plus or minus infinity. The point (iii) is not obvious, it is merely true. But, along with (i) and (ii), it suggests looking for a more graceful approach. There is one, and in fact, we have already seen it.

Definition 6.23. A sequence of sub-probability measures (μ_n) on $\mathbb{R}$ converges **weakly** to a measure μ_∞ if, for all bounded continuous functions f on $\mathbb{R}$,

$$(6.14) \qquad \lim_{n \to \infty} \int f \, d\mu_n = \int f \, d\mu_\infty.$$

The sequence converges **vaguely** if (6.14) holds for all continuous f of compact support on $\mathbb{R}$.

Remark 6.24. (a) The continuous functions of compact support are a subset of the bounded continuous functions, so that weak convergence implies vague convergence. The converse is not true, however.

(b) If μ_n converges to μ weakly, then $\lim \mu_n(\mathbb{R}) = \mu(\mathbb{R})$. (To see this, just take $f \equiv 1$ in the definition.) Thus weak convergence of probability measures is just convergence in distribution. If μ_n only converges vaguely, however, $\mu(\mathbb{R})$ can be strictly smaller than $\liminf \mu_n(\mathbb{R})$. In that case, the sequence does not converge weakly. This is the case in (ii) above.

Example 6.24.1. Let μ_n be the unit mass at n. Then (μ_n) does not converge weakly, but it converges vaguely to the zero measure.

Indeed, if f is a continuous and has compact support, $f(n) = 0$ for all large enough n, so $\lim_n \int f \, d\mu_n = \lim_n f(n) = 0$. This is exactly the integral of f with respect to the zero measure.

---

[4]See Billingsley [**4**] for a lucid general treatment of weak convergence.

6.2. Convergence in Distribution

Lemma 6.25. *Let $A \subset \mathbb{R}$ be an open set and suppose μ_n converges vaguely to a measure μ. Then $\mu(A) \leq \liminf_n \mu_n(A)$.*

Proof. If A is open, there exists a sequence g_k of non-negative continuous functions of compact support which increases to I_A. (See Exercise 6.18.) Then
$$\mu(A) = \lim_{k \to \infty} \int g_k \, d\mu$$
$$= \lim_{k \to \infty} \left(\lim_{n \to \infty} \int g_k \, d\mu_n \right).$$
But $\int g_k \, d\mu_n \leq \mu_n(A)$, so this is
$$\leq \lim_{k \to \infty} \left(\liminf_{n \to \infty} \mu_n(A) \right) = \liminf_{n \to \infty} \mu_n(A).$$
$\square$

Taking $A = \mathbb{R}$, we see that the vague limit of sub-probability measures is also a sub-probability measure.

In case the μ_n are all probability measures, then by Theorem 6.17, weak convergence is equivalent to convergence in distribution. Vague convergence, however, is not; the limit may well be a sub-probability measure, as we just saw in Example 6.24.1.

Let us restate the definition of tightness, in order to emphasize the role of compactness.

Definition 6.26. A family $\{\mu_\alpha, \alpha \in I\}$ of sub-probability measures is **tight** if for each $\varepsilon > 0$ there exists a compact K_ε such that for all α, $\mu_\alpha(K_\varepsilon^c) \leq \varepsilon$.

Helly's theorem and a modification of Theorem 6.17 tell us the following.

Theorem 6.27 (Helly-Bray Theorem). *Consider a sequence (μ_n) of sub-probability measures. Then there exists a subsequence (μ_{n_k}) which converges vaguely to a sub-probability measure μ. If the μ_n are tight, then μ_{n_k} converges weakly to μ.*

Proof. (Outline). Define $F_n(x) = \mu_n((-\infty, x])$. By Helly's Theorem and the remark following it, there exists a subsequence which converges pointwise to an increasing function $\bar{F}_\infty$. Let $F_\infty(x)$ be the right-continuous version of $x \mapsto \bar{F}_\infty(x) - \inf_y \bar{F}_\infty(y)$. Show that F_∞ is the distribution function of a sub-probability measure μ_∞. Then $F_{n_k}(b) - F_{n_k}(a) \longrightarrow F_\infty(b) - F_\infty(a)$ for each continuity interval $(a, b]$ of μ_∞. Now the first part of the proof of Theorem 6.17, with the obvious modifications, shows that μ_{n_k} converges vaguely to μ_∞. Finally, if the (μ_n) are tight, then the μ_n-integrals of a bounded continuous function can be approximated uniformly in n by the integrals of a continuous function of compact support. Thus for a tight sequence, vague convergence implies weak convergence. □

Remark 6.28. (*i*) The weak convergence approach is not limited to measures on the line; it makes sense for measures on general topological spaces. The above results can be extended to Euclidean spaces of any dimension, and, more generally, to spaces such as locally compact separable metric spaces.

(*ii*) The above are really compactness results. The Helly-Bray theorem, for instance, says that the set of sub-probability measures is compact in the topology of vague convergence, and that a tight family of sub-probability measures is relatively compact in the topology of weak convergence[5]. In this generality, the relative compactness of tight families of measures (and its converse) is called Prokhorov's Theorem.

6.3. Lévy's Continuity Theorem

Let μ and μ_n be a sequence of probability measures on $\mathbb{R}$ with characteristic functions ϕ and ϕ_n, respectively. Let us first establish that convergence in distribution implies the convergence of characteristic functions.

Proposition 6.29. *Suppose $\mu_n \Rightarrow \mu$. Then for all t, $\phi_n(t) \longrightarrow \phi(t)$.*

Proof. The function e^{itx} is bounded and continuous, so—with the obvious extension to complex integrals—Theorem 6.17 implies that $\phi_n(t) = \int e^{itx} \mu_n(dx) \longrightarrow \int e^{itx} \mu(dx) = \phi(t)$, which shows pointwise convergence. □

The big result is the converse: convergence of characteristic functions implies convergence in distribution of probability measures. First, we need a lemma which shows how to use the characteristic function to get bounds on the tails of the distribution.

Lemma 6.30. *Let μ be a probability measure with characteristic function ϕ, and let $A > 0$. Then*

$$(6.15) \qquad \mu\{[-2A, 2A]\} \geq A \left| \int_{-\frac{1}{A}}^{\frac{1}{A}} \phi(t) \, dt \right| - 1.$$

Proof. We claim that

$$(6.16) \qquad \frac{1}{2T} \int_{-T}^{T} \phi(t) \, dt = \int_{-\infty}^{\infty} \frac{\sin Tx}{Tx} \mu(dx).$$

---

[5]Those familiar with Banach space theory might recognize the kinship with the fact that the closed unit ball in the dual space of $C[0,1]$ is compact in the weak-star topology.

6.3. Lévy's Continuity Theorem

Indeed, the left-hand side is $\int_{-T}^{T} \int_{-\infty}^{\infty} e^{itx} \mu(dx)\, dt$. Interchange the order of integration, and do the integral over t: it equals $(e^{iTx} - e^{-iTx})/ix = \sin(Tx)/x$. This gives (6.16).

The integrand in the right-hand side of (6.16) is bounded by 1 for all x, and is bounded by $1/2TA$ for $|x| > 2A$. Therefore,

$$\left|\frac{1}{2T}\int_{-T}^{T} \phi(t)\, dt\right| \leq \mu([-2A, 2A]) + \frac{1}{2TA}\big(1 - \mu([-2A, 2A])\big)$$

$$= \left(1 - \frac{1}{2TA}\right)\mu([-2A, 2A]) + \frac{1}{2TA}.$$

Take $T = 1/A$ to see that

$$\frac{A}{2}\left|\int_{-\frac{1}{A}}^{\frac{1}{A}} \phi(t)\, dt\right| \leq \frac{1}{2}\mu([-2A, 2A]) + \frac{1}{2}.$$

This gives the desired result. □

Remark 6.31. This gives us a criterion for tightness in terms of characteristic functions. A sequence μ_n of probability measures is tight if for all $\varepsilon > 0$ there is a $\delta > 0$ such that

$$\frac{1}{2\delta}\int_{-\delta}^{\delta} \phi_n(t)\, dt > 1 - \varepsilon.$$

This brings us to

Theorem 6.32 (Lévy's Continuity Theorem). *Let (μ_n) be a sequence of probability measures with characteristic functions (ϕ_n). Suppose that*

 (a) *(ϕ_n) converges pointwise to a function ϕ_∞.*

 (b) *ϕ_∞ is continuous at $t = 0$.*

Then

 (i) *There exists a probability measure μ_∞ such that $\mu_n \Rightarrow \mu_\infty$.*

 (ii) *ϕ_∞ is the characteristic function of μ_∞.*

Remark 6.33. $\phi_n(0) = 1$ for all n, so that $\phi_\infty(0) = 1$. Thus (b) $\iff \lim_{t \to 0} \phi_\infty(t) = 1$.

Keep in mind that we do not know a priori that ϕ_∞ is a characteristic function. That is part of the conclusion of the theorem.

Proof. The plan of the proof is simple: first use the lemma to show that the sequence (μ_n) is tight. Then use tightness and Helly's theorem to pull out the limiting distribution. By the lemma, if $\delta > 0$, then

(6.17) $$\mu_n\left(\left[-\frac{2}{\delta}, \frac{2}{\delta}\right]\right) \geq \frac{1}{\delta}\left|\int_{-\delta}^{\delta} \phi_n(t)\, dt\right| - 1.$$

Now $\phi_n \longrightarrow \phi_\infty$ boundedly, so by the bounded convergence theorem, the right-hand side converges to

$$\left|\frac{1}{\delta}\int_{-\delta}^{\delta}\phi_\infty(t)dt\right| - 1.$$

Moreover, ϕ_∞ is continuous at zero, so we can choose δ small enough so that $|\phi_\infty(t) - 1| < \varepsilon/2$ if $|t| < \delta$. Now choose N large enough so that

$$\frac{1}{\delta}\int_{-\delta}^{\delta}|\phi_n(t) - \phi_\infty(t)|\,dt < \frac{\varepsilon}{2} \quad \text{if } n \geq N.$$

Note that $(1/2\delta)\int_{-\delta}^{\delta}\phi_\infty(t)\,dt$ is an average, and if ϕ_∞ lies in the disk $\{z : |z - 1| < \varepsilon/4\}$ in the complex plane, so does its average. Therefore, the modulus of its average is at least $1 - \varepsilon/4$. Thus,

$$\left|\frac{1}{\delta}\int_{-\delta}^{\delta}\phi_n(t)\,dt\right| \geq \left|\frac{1}{\delta}\int_{-\delta}^{\delta}\phi_\infty(t)\,dt\right| - \frac{2}{\delta}\int_{-\delta}^{\delta}|\phi_n(t) - \phi_\infty(t)|\,dt.$$

If $n > N$, this is

$$\geq \left(2 - \frac{\varepsilon}{2}\right) - \frac{\varepsilon}{2} = 2 - \varepsilon.$$

From (6.17), for $n > N$, $\mu_n\big([-2/\delta, 2/\delta]\big) \geq 1 - \varepsilon$. Thus the sequence (μ_n) is tight.

Therefore, by Helly's Theorem, there exists a subsequence (n_k) and a probability measure μ_∞ such that $\mu_{n_k} \Rightarrow \mu_\infty$. Let $\tilde{\phi}$ be the characteristic function of μ_∞. Then by Proposition 6.29, $\phi_{n_k}(t) \longrightarrow \tilde{\phi}(t)$ for all t. But $\phi_{n_k} \longrightarrow \phi_\infty$ by hypothesis, so evidently ϕ_∞ is the characteristic function of the probability measure μ_∞.

This shows that every subsequence of (μ_n) has a further subsequence which converges to a probability measure... which has to be μ_∞. But convergence in distribution is metrizable (see the problems) so this implies that the whole sequence converges in distribution to μ_∞. □

To summarize: probability measures converge if and only if their characteristic functions converge. Even better, if the characteristic functions converge to a limit function which is continuous at zero, that function is necessarily the characteristic function of a probability measure, and the sequence converges in distribution to that measure. So to show convergence in distribution, we merely have to verify that the characteristic functions converge pointwise, and that their limit is continuous at the origin.

Example 6.33.1. Poisson Limits. We stated the Poisson limit theorem in Theorem 2.42. While it is easily proved directly, it is even easier to prove it by characteristic functions. Recall the result: if $0 < p_n < 1$ and

6.3. Lévy's Continuity Theorem

$np_n \longrightarrow \lambda > 0$ as $n \to \infty$, a sequence of $B(n, p_n)$ random variables converges in distribution to a Poisson (λ) distribution.

To prove this, we need an extension of one of the fundamental limits in calculus: $(1+x/n)^n \longrightarrow e^x$. It is a good exercise in using the complex log, so we leave its proof to the reader.

Lemma 6.34. *Let z and z_n be complex variables such that $z_n \longrightarrow z$ as $n \to \infty$. Then*
$$\lim_{n\to\infty} \left(1 + \frac{z_n}{n}\right)^n = e^z.$$

To prove the Poisson limit theorem from this, note that the $B(n,p)$ characteristic function is $(1-p+pe^{it})^n$. Rewrite this, let $n \to \infty$, and apply the lemma:
$$\left(1 + \frac{np_n(e^{it}-1)}{n}\right)^n \longrightarrow e^{\lambda(e^{it}-1)} \quad \text{as } np_n \to \lambda.$$

The characteristic functions converge and the limit is continuous at zero—in fact, it is the Poisson (λ) characteristic function—so the $B(n, p_n)$ distributions do indeed converge to the Poisson (λ) distribution. $\square$

Example 6.34.1. Poisson Characterization. There are numerous characterizations of the Poisson distribution. The following one can often identify the Poisson distribution when it occurs in nature.

Proposition 6.35. *Let X be a positive integer-valued random variable with the following properties:*

(i) For each n there exist i.i.d. random variables $Y_1^n, \ldots, Y_n^n$ such that $Y_1^n + \cdots + Y_n^n$ has the same distribution as X;

(ii) $P\{Y_n \geq 2\} = o\left(P\{Y_n = 1\}\right)$ as $n \to \infty$.

Then X is Poisson (λ), where $\lambda = -\log\left(P\{X=0\}\right)$.

Remark 6.36. In the usual terminology, (i) says that the distribution of X is **infinitely divisible**. Condition (ii) rules out the possibility that, for example, the Y_i^n, and therefore X, take on only even values.

Proof. Note first that $P\{X = 0\} > 0$, since for any n, $P\{X = 0\} = P\{Y_1^n = 0\}^n$, and if $P\{Y_1^n = 0\} = 0$, then $Y_1^n + \cdots + Y_n^n \geq n$, implying that X is infinite. Therefore, we can let $P\{X = 0\} = e^{-\lambda}$ for some $\lambda > 0$. Then $P\{Y_1^n = 0\} = e^{-\lambda/n}$, and $P\{Y_1^n = 1\} = 1 - e^{-\lambda/n} - P\{Y_1^n \geq 2\} = (1-\varepsilon)(1-e^{-\lambda/n})$, where $\varepsilon \to 0$ as $n \to \infty$ by (ii). Therefore, the characteristic function of Y_1^n can be written as
$$\phi_n(t) = e^{-\lambda/n} + (1-\varepsilon)(1-e^{-\lambda/n})e^{it} + \sum_{k\geq 2} P\{Y_1^n = k\}e^{itk}$$
$$= e^{-\lambda/n} + (1-\eta)(1-e^{-\lambda/n})e^{it},$$

where η is a complex number depending on t whose modulus is bounded by ε. Indeed, the sum over k above—call it S—is bounded in modulus by $P\{Y_1^n \geq 2\} = \varepsilon(1 - e^{-\lambda/n})$, so we just set $\eta = e^{-it}S/(1 - e^{-\lambda/n})$.) Thus the characteristic function ϕ_X of X is given by

$$\phi_X(t) = e^{-\lambda}\big(1 + (1-\eta)e^{\lambda/n}(1 - e^{-\lambda/n})e^{it}\big)^n$$

$$= e^{-\lambda}\left(1 + \frac{1}{n}n(1 - e^{-\lambda/n})e^{-\lambda/n}(1-\eta)e^{it}\right)^n$$

But $\eta \to 0$ and $e^{-\lambda/n} \to 1$ as $n \to \infty$, so by Lemma 6.34, the right side of the equation converges to $\exp(-\lambda(1 - e^{it}))$, which is the Poisson (λ) characteristic function. Thus X is indeed Poisson (λ) as claimed. □

This characterization can be used to show that the distribution of, say, the number of Geiger-counter clicks in one minute from a weak radioactive source is Poisson. (To apply (i) for $n = 60$, say, divide the minute into sixty one-second intervals. Any physicist should be able to explain why the clicks in different sub-intervals are independent and identically distributed. To apply (ii), one may have to divide the minute into even shorter sub-intervals to assure that the probability p of one click is small, in which case the probability of two or more will be $O(p^2)$.)

For another example, consider the number of stars seen by a telescope in a typical one-degree solid angle portion of the night sky. If we avoid star-clusters and the Milky Way, the stars will be more-or-less randomly distributed. Again, to get (i), divide the area into n equal sub-areas, and let Y_i^n be the number of stars in the ith one. Actually, (ii) is questionable since an appreciable fraction of stars are double-star systems, so it would not hold if the telescope could separate double-stars.

Problems 6.3

6.15. Let X_n and Y_n be i.i.d. binomial random variables with a $B(n, \frac{\lambda}{n})$ distribution, where $\lambda > 0$. Let $Z_n = X_n - Y_n$. Show that (Z_n) converges in distribution. What can you say about its limit?

6.16. Show that the limit function F_∞ in Helly's theorem corresponds to a sub-probability measure μ in the sense that if $a < b$ are continuity points of F_∞, then $\mu((a,b]) = F_\infty(b) - F_\infty(a)$.

6.17. A **tent function** is a continuous function $f(x)$ which vanishes if x is not in some interval (a, b), which is strictly positive at $x = (a+b)/2$, and which is linear in the two intervals $(a, (a+b)/2)$ and $((a+b)/2, b)$. (It is so-called because its graph looks like a tent.) Show that a sequence of probability measures μ_n converges vaguely to a measure μ if and only if $\int f(t)\,\mu_n(dt) \longrightarrow \int f(t)\,\mu(dt)$ for every tent function f.

[Hint: First show that every continuous function of compact support can be uniformly approximated by a finite linear combination of tent functions.]

6.3. Lévy's Continuity Theorem

6.18. Show that the indicator function of an open set is an increasing limit of a sequence of continuous functions of compact support.

6.19. Let μ_n be a sequence of probability measures on $\mathbb{R}$ that converges weakly to a probability measure μ. Let ∂A be the boundary of A. Show that:

(a) if G is open, then $\liminf_n \mu_n\{G\} \geq \mu\{G\}$.

(b) If K is closed, then $\limsup_n \mu(K) \leq \mu\{K\}$.

(c) If $\mu(\partial A) = 0$, then $\lim_n \mu_n(A) = \mu(A)$.

Show by example that the inequalities in (a) and (b) can be strict.
[Hint: Exercise 6.18 may help.]

6.20. (Continuation). Let μ and μ_n be probability measures on $\mathbb{R}$. Suppose that for each set A whose boundary has zero μ measure, $\lim_n \mu_n(A) = \mu(A)$. Show that (μ_n) converges to μ in distribution.
[Call the above condition (d). Note that (a), (b), (c) and (d) are each *equivalent* to $\mu_n \Longrightarrow \mu$. This is sometimes called the Portmanteau Theorem.]

6.21. Suppose that X_n converges weakly to X. Then $E\{f(X_n)\}$ converges to $E\{f(X)\}$ for all bounded continuous functions f on $\mathbb{R}$.

(a) Show by example that this is not true for all bounded Borel functions.

(b) Give an example of a sequence which converges vaguely but not weakly.

6.22. (a) Show that if (X_n) and (Y_n) converge in probability to X and Y, respectively, that $(X_n + Y_n)$ converges in distribution to $X + Y$.

(b) Give an example of random variables (X_n), (Y_n), X and Y on the same probability space such that X_n and Y_n converge in distribution to X and Y, respectively, but $X_n + Y_n$ does not converge in distribution to $X + Y$.

6.23. Let X, (X_n), (Y_n) be random variables on the same probability space, such that (X_n) converges in distribution to X, and (Y_n) converges in distribution to a constant c. Show that

(a) $X_n + Y_n \Longrightarrow X + c$ and

(b) $Y_n X_n \Longrightarrow cX$.

[Hint: Show first that Y_n converges in probability.]

6.24. (Skorokhod Representation Theorem.) Suppose that X_n converges to X in distribution. Show the following, proved by Skorokhod in 1956.

> **Theorem** (Skorokhod). There is a probability space $(\Omega, \mathcal{F}, P)$ and random variables $\hat{X}_n$ and $\hat{X}$ defined on it such that for all n, $\hat{X}_n$ and $\hat{X}$ have the same distributions as X_n and X, respectively, and $(\hat{X}_n)$ converges a.s. to $\hat{X}$.

[Hint: Let U be $U(0,1)$ and let $\hat{X}_n = G_{F_n}(U)$ and $\hat{X} = G_F(U)$, where G_F is the left-continuous inverse of F (see §2.2). Then show that G_{F_n} converges to G_F at most points.]

6.4. The Central Limit Theorem

The Central Limit Theorem has a distinguished history, and it was long considered the crown jewel of classical probability[6]. It dates to the very birth of the subject: Abraham deMoivre proved it for Bernoulli random variables in 1733. Laplace [23] fleshed it out in 1812, to what is now called the deMoivre-Laplace theorem[7].

It received attention from many of the most distinguished mathematicians of the next century, including Gauss, who lent his name to Gaussian random variables. It only reached a definitive form in 1922, when Lindeberg [27] proved the theorem which is now named for him, and Feller showed that it was indeed definitive by showing that Lindeberg's condition was necessary as well as sufficient.

In mathematics, though, the designation "definitive" may have a short lifetime. Answering one question can lead to many more, and new open questions, consequences, and extensions of the central limit theorem are appearing, even to this day[8].

The first motivation for looking at it comes from the law of large numbers. If $X_1, X_2, \ldots$ is an i.i.d. sequence of random variables with mean m, then the Law of Large numbers tells us that

$$\frac{1}{n} \sum_1^n X_j \longrightarrow m \quad \text{a.s.}$$

But how fast does this sequence converge? Look at the difference:

$$\left(\frac{1}{n} \sum_1^n X_{j=1}^n\right) - m = \frac{1}{n} \sum_{j=1}^n (X_j - m).$$

How fast does this go to zero? Multiply by $\sqrt{n}$:

$$\sqrt{n}\left[\frac{1}{n} \sum_{j=1}^n (X_j - m)\right].$$

The central limit theorem not only tells us that this is of moderate size, it tells us its distribution, for any (large) value of n.

Recall that if X is a random variable with mean μ and variance σ^2, then its characteristic function ϕ_X has the Taylor expansion

$$\phi_X(t) = 1 + i\mu t - \frac{1}{2}(\sigma^2 + \mu^2)t^2 + o(t^2) \quad \text{as } t \to 0.$$

---

[6]In fact, the word "central" was originally used because it was considered to be at the very center of the subject.

[7]See [15] for a discussion of this.

[8]For example, the invariance principle (Chapter 10) is an application of the central limit theorem applied to stochastic processes.

6.4. The Central Limit Theorem

Here is our first version of the Central Limit Theorem.

Theorem 6.37. *Let $X_1, X_2, \ldots$ be i.i.d. random variables with mean μ and variance σ^2. Let $S_n = X_1 + \cdots + X_n$. Then*
$$\frac{S_n - \mu}{\sqrt{n\sigma^2}}$$
converges in distribution to an $N(0,1)$ random variable.

Proof. Replace X_n by $X'_n = (X_n - \mu)/\sigma$. Then X_n has mean zero and variance one. Its characteristic function is $\phi(t) = E\{e^{itX'}\} = 1 - t^2/2 + e(t^2)$, where e is a function such that $e(t)/t \to 0$ as $t \to 0$. Then $S'_n \stackrel{\text{def}}{=} (X'_1 + \cdots + X'_n)/\sqrt{n}$ has characteristic function
$$\phi_{S'_n}(t) = \phi\left(\frac{t}{\sqrt{n}}\right)^n = \left(1 + \frac{t^2}{2n} + e\left(\frac{t^2}{n}\right)\right)^n.$$

Take $z_n = -(t^2/2) + ne(t^2/n)$ in the lemma and note that $z_n \to -t^2/2$ to see that for all t,
$$\phi_{S_n}(t) \longrightarrow e^{-\frac{1}{2}t^2}.$$
This is the characteristic function of a $N(0,1)$ random variable. By Lévy's Continuity Theorem, S'_n converges in distribution to a standard normal. □

However, this theorem doesn't indicate how widely the phenomenon occurs. While nature supplies us with endless numbers of random variables, it is not always obliging enough to give them identical distributions. In practice, we may have only a vague idea of their distributions, but still, the Gaussian distribution comes up repeatedly. The following, which we might call the "central limit principle" is the essence of the Central Limit Theorem.

> The sum of a large number of small independent random variables is nearly normal.

That is, if $X_1, X_2, \ldots, X_n$ are independent random variables, if n is large, if the X_j are all small, and if $S_n = X_1 + \cdots + X_n$, then
$$\frac{S_n - E\{S_n\}}{\sqrt{\text{Var}(S_n)}} \text{ is nearly } N(0,1).$$

This is a rule of thumb, not a theorem. To make it into mathematics, we have to define exactly what is meant by "large", "small" and "nearly". With rigorous definitions, it becomes a theorem.

To see why this situation occurs so often, consider, for example, the measurement error in a scientific experiment. Typically, the final measurement depends on many things: the alignment, manufacture and calibration of the experimental apparatus, the ambient temperature and other weather conditions, and many more. Each of those is subject to its own errors. The experimenters take immense care to minimize them, but they can't eliminate them entirely. So, while the parameter values should be $x_1, x_2, \ldots, x_n$, their actual values are $x_1 + \varepsilon_1, \ldots, x_n + \varepsilon_n$, where the ε_j are random variables, probably independent or nearly so. Without errors, the final measurement would be some function of the parameters, say $f(x_1, \ldots, x_n)$. But the errors are there, and the result actually is $f(x_1 + \varepsilon_1, \ldots, x_n + \varepsilon_n)$. If the function is at all smooth, we can expand f by Taylor's formula to see that

$$(6.18) \quad f(x_1 + \varepsilon_1, \ldots, x_n + \varepsilon_n) - f(x_1, \ldots, x_n) = \sum_{j=1}^{n} \frac{\partial f}{\partial x_j}(x_1, \ldots, x_n)\varepsilon_j$$
$$+ \text{higher order terms.}$$

If the ε_j are small—as they should be—the higher order terms will be nearly negligible, and the error is effectively the sum of the small independent random variables. The central limit principle tells us that the distribution of the error must be approximately normal. Its mean will hopefully be close to zero, and its variance will be on the order of $\sum_j (\partial f/\partial x_j)^2 \text{Var}(\varepsilon_j)$.

Note that (6.18) can come up in many situations having nothing to do with experimental error. It just describes a situation which depends smoothly on many different quantities, and not too strongly on any one of them. So, in fact, the Gaussian distribution can arise in almost any sufficiently complicated situation. Here are a few standard examples:

(1) Exam scores in a large class. There are many students, many problems in the exam, and each problem has a few points to win or lose. One usually finds that the scores are approximately normal, which leads to the common practice of "grading on a curve," the curve being, of course, the (in)famous "bell curve".

(2) Heights in a large homogeneous population of people. Height is influenced by many genes, as well as other factors like nutrition. It is reasonable to assume that the effect of a single gene is comparatively small, and to expect that the heights follow an approximate normal distribution.

(3) Errors in a fine Swiss wristwatch over a week. The errors come from small errors in the size of the many mechanical parts, from dust and impurities in the lubrication, from random temperature changes, and from the effects of the movement of the watch, among others. Generations of watchmakers have worked to minimize all these sources of error, so that the total error is a sum of many small random contributions, and therefore has a nearly normal distribution.

6.4. The Central Limit Theorem

We could continue the list, but it is already clear that the normal distribution can arise simply from complexity, and that we can anticipate it in almost any sufficiently complex situation.

We will give one version of the central limit theorem, due to Lyapounov, which involves nearly arbitrary distributions. There is a trade off: if we abandon the assumption of identical distribution, it no longer suffices just to have finite means and variances; we need a little bit[9] more. In this case, we assume the existence of third moments.

Theorem 6.38 (Lyapounov's Central Limit Theorem). *Let $X_1, X_2, \ldots$ be independent mean zero random variables with $\operatorname{Var}(X_j) = \sigma_j^2$ and $\gamma_j \stackrel{\text{def}}{=} E\{|X_j|^3\} < \infty$. Let $S_n = \sum_{j=1}^n X_j$ and $s_n^2 = \sum_{j=1}^n \sigma_j^2$. Suppose that $(1/s_n^3) \sum_{j=1}^n \gamma_j \longrightarrow 0$ as $n \to \infty$. Then*

$$\frac{S_n}{s_n} \Rightarrow N(0,1) \quad \text{as } n \to \infty.$$

Proof. This proof is typical of the subject: the characteristic function of the sum is the product of the characteristic functions of the summands, so its log is the sum of the logs. It is easy to see that the principal term goes to the right limit, and then we use the third moment condition to show that the error terms go to zero.

Let $\hat{\phi}_j$ be the characteristic function of X_j. By Theorem 6.4, $\hat{\phi}_j$ has three derivatives, and its third derivative is bounded in modulus by γ_j, so a three-term Taylor expansion is

$$\hat{\phi}_j\left(\frac{t}{s_n}\right) = 1 - \frac{t^2}{2s_n^2}\sigma_j^2 + \Lambda_j \gamma_j \left(\frac{t}{s_n}\right)^3,$$

where Λ_j is a function of t; all we need know is that it is bounded by $1/6$. The X_k are independent, so the characteristic function ϕ_n of S_n/s_n is

$$\phi_n(t) = \prod_{j=1}^n \left(1 - \frac{t^2}{2s_n^2}\sigma_j^2 + \Lambda_j \gamma_j \left(\frac{t}{s_n}\right)^3\right).$$

Let $\theta_{jn} = -\frac{\sigma_j^2 t^2}{2s_n^2} + \Lambda_j \gamma_j \left(\frac{t}{s_n}\right)^3$, so that

$$\phi_n = \prod_{j=1}^n (1 - \theta_{jn}).$$

---

[9] For years, the quest for the definitive central limit theorem involved the search for that "little bit". That was answered in 1922 by Lindeberg, who introduced the Lindeberg condition and Feller in 1935, who showed that Lindeberg's condition was indeed that "little bit".

Fix t in the following. Take the logarithm of ϕ_n. (The complex log is multi-valued, so we have to be careful.)

$$\log \phi_n = \sum_{j=1}^{n} \log(1 + \theta_{jn}).$$

(a) We claim $\sup_{j \leq n} |\theta_{jn}| \to 0$ as $n \to \infty$.

Indeed, by Lyapounov's inequality (Corollary 3.23) our hypotheses imply that

$$\sup_{j \leq n} \left(\frac{\sigma_j^2}{s_n^2}\right) \leq \sup_{j \leq n} \left(\frac{\gamma_j}{s_n^3}\right)^{\frac{2}{3}} \to 0.$$

(b) Moreover, $\sum_{j=1}^{n} |\theta_{jn}|$ is bounded and $\sum_{j=1}^{n} \theta_{jn} \to -\frac{t^2}{2}$ as $n \to \infty$.

Indeed $s_n^{-2} \sum_{j=1}^{n} \sigma_j^2 = 1$, so $\left|\frac{t^2}{2} + \sum_{j=1}^{n} \theta_{jn}\right| \leq |\Lambda_j| t^3 s_n^{-3} \sum_{j=1}^{n} \gamma_j$, and, as $|\Lambda_j| \leq 1/6$, this goes to zero by hypothesis. Boundedness is clear since the only term which might change sign with j is Λ_j, and $|\Lambda_j| \leq 1/6$.

(c) $\sum_{j=1}^{n} |\theta_{jn}^2| \leq \left(\sup_{i \leq n} |\theta_{in}|\right) \sum_{j=1}^{n} |\theta_{jn}| \to 0$ by (a) and (b).

By (a) there is N such that $|\theta_{jn}| < 1/2$ for all $n \geq N$, and all $j \leq n$. Recall the expansion of the logarithm: for small z, $\log(1+z) = z + \Gamma z^2$, where Γ is a function of z, and, as long as $|z| \leq 1/2$, $|\Gamma| \leq 1$. Thus

$$\log \phi_n = \sum_{j=1}^{n} \log(1 + \theta_{jn})$$

$$= \sum_{j=1}^{n} \theta_{jn} + \sum_{j=1}^{n} \Gamma \Lambda_j \theta_{jn}^2.$$

The first sum converges to $-t^2/2$ by (b), and, as $|\Gamma \Lambda_j| \leq 1/6$, the second converges to zero by (c). Therefore, $\log \phi_n(t) \to -t^2/2$.

It follows that $\phi_n(t) \to e^{-t^2/2}$, which is the characteristic function of a standard normal. This is true for every t, and Lévy's continuity theorem tells us that the sum converges in distribution to a standard normal. □

Remark 6.39. We cannot apply the Central Limit Theorem blindly. For instance, if $X_1, X_2, \ldots$ are independent random variables with a Cauchy distribution, which is symmetric and has characteristic function $e^{-|t|}$, then $(X_1 + \cdots + X_n)/\sqrt{n}$ has characteristic function $e^{-t\sqrt{n}}$. As $n \to \infty$, this converges to zero for $t \neq 0$. This is the characteristic function of the zero measure, so the distributions converge to the zero measure: the central limit principle does not apply. The reason it doesn't is that the random variables X_j are not small. It is quite possible that the maximum of $X_1, X_2, \ldots, X_n$ is larger than the sum of the rest—and even larger than the sum of their absolute values.

6.4. The Central Limit Theorem

6.4.1. Some Consequences and Extensions. The central limit theorem is an approximation theorem at heart. This is clear from the central limit principle, which claims that certain distributions are *nearly* Gaussian. But, as often happens, the approximation is stated as a limit theorem. The limit theorem tells us that there *is* an approximation theorem, but it does not tell us how good the approximation is. For example, the law of large numbers tells us that the average of sums of i.i.d. random variables tends to the expectations, but not how fast. The central limit theorem then tells us that the error is on the order of $1/\sqrt{n}$. But can we refine that "order of"? What about the speed of convergence in the central limit theorem itself? Here are two different answers.

The Berry-Esseen Theorem.

The first is the Berry-Esseen Theorem. This tells us that the distribution functions converge at a rate of about $1/\sqrt{n}$, and it gives the "about" in terms of the third moment of the underlying distribution.

Theorem 6.40. *Let $X_1, X_2, \ldots$ be i.i.d. with mean zero and variance σ^2. Then there exists a universal constant C_0 such that*

$$(6.19) \qquad \sup_x \left| P\left\{ \frac{S_n}{\sigma\sqrt{n}} \leq x \right\} - \Phi(x) \right| \leq C_0 \frac{E\{|X|^3\}}{\sigma^3 \sqrt{n}}.$$

When we say the constant C_0 is universal, we mean that it holds for all distributions. We will not prove this. See, for instance, [9] Theorem 7.4.1 for a proof.

The Law of the Iterated Logarithm.

The Central Limit Theorem tells us that $\sqrt{n}$ times the error in the law of large numbers converges to zero *in distribution,* but doesn't tell us whether the *sequence* converges or not. In fact, it does not. The sequence $(S_n - E\{S_n\})/\sqrt{n}$ not only fails to converge a.e., its lim sup and lim inf are infinite. (See Exercise 6.29.)

This leads us to ask: what, exactly, is the asymptotic behavior of the sequence $S_n - E\{S_n\}$, $n = 1, 2, \ldots$?

The original impetus for this question came from number theory, not probability. It involved Borel's normal numbers. But this was soon seen to be just the Bernoulli case of a more general phenomenon. There were a number of partial results on the way to the final answer. Evidently, the sequence is not $O(1/\sqrt{n})$. Hausdorff (1913) showed that it behaved as $O(\sqrt{n^{1+\varepsilon}})$ and Hardy and Littlewood (1914) improved that to $O(\sqrt{n \log n})$. Then Khintchine (1922) improved that to $O(\sqrt{n \log \log n})$, and then finally, (1924) completed this by showing that this estimate couldn't be improved. This was so striking that Khinchine's theorem received the euphonious title "Law of the Iterated Logarithm."

Theorem 6.41 (Law of the Iterated Logarithm). *Let $X_1, X_2, \ldots$ be i.i.d. random variables with mean zero and variance σ^2. Let S_n be the sum of the first n of them. Then*

$$\limsup_{n \to \infty} \frac{S_n}{\sqrt{2\sigma^2 n \log(\log n)}} = 1 \quad a.e. \tag{6.20}$$

So the rate of convergence is slightly—very slightly—slower than $1/\sqrt{n}$: it is exactly $1/\sqrt{n \log \log n}$. To understand why this theorem is so celebrated, one must see how tiny, how miniscule, that loglog term is. "Tiny" and "miniscule" might seem to be strange descriptions of a function which tends to infinity, but observe how slowly it does so. For example, physicists estimate that there are between 10^{80} and 10^{100} sub-atomic particles in this universe. But $\log \log(10^{100})$ is less than five and a half. So, even if we had been flipping coins once a second since the birth of the universe—about 4×10^{17} flips by the latest estimates—we would barely be able to see the effect of that loglog term.

We will prove this later, when we discuss Brownian motion. (See Theorems 10.38 and 10.62.)

6.5. Stable Laws*

If $X_1, X_2, \ldots$ are i.i.d. with mean zero, variance one, then by the central limit theorem,

$$\frac{X_1 + \cdots + X_n}{\sqrt{n}} \Rightarrow N(0, 1).$$

What happens if we drop the hypothesis that the common variance is finite? We can no longer expect the central limit theorem to hold, so there is no reason to expect that the limiting distribution is normal. In addition, the norming constant $\sqrt{n}$ is too specific, so let us norm by a constant, say η_n. While we are at it, let us drop the hypothesis that the mean be zero, and center the sum by, say, β_n. This leads to the following situation.

Let $X_1, X_2, \ldots$ be i.i.d., and suppose that there exist norming constants $\eta_1, \eta_2, \ldots$ and $\beta_1, \beta_2, \ldots$ such that

$$\frac{X_1 + \cdots + X_n - \beta_n}{\eta_n} \Longrightarrow F$$

for some distribution F.

6.5. Stable Laws*

What can we say about the limiting distribution F? We will see that we can say quite a lot. The class of possible F is exactly the class of **stable laws**[10][11].

The Holzmark Distribution.

Let us consider a physical example, Holzmark's distribution. Suppose that stars are randomly distributed in space. What is the force of gravity that one feels from them? To be specific, suppose the stars are distributed uniformly along the x-axis.[12].

Thus, suppose n stars of mass m are uniformly distributed on $(-n, n)$. Let X_i be the position of the ith star. Gravity is an inverse-square force, so the force[13] G_n at the origin is

$$G_n = \sum_{i=1}^n mg \frac{\operatorname{sgn} X_i}{X_i^2},$$

where g is the constant of gravity. Let us compute the characteristic function of G_n:

$$E\{e^{itmg\frac{\operatorname{sgn} X_i}{X_i^2}}\} = \frac{1}{2n} \int_{-n}^{n} e^{itmg\frac{\operatorname{sgn} x}{x^2}} dx$$

$$= \frac{1}{n} \int_0^n \cos \frac{mgt}{x^2} dx$$

$$= 1 - \frac{1}{n} \int_0^n \left(1 - \cos \frac{mgt}{x^2}\right) dx.$$

The characteristic function of G_n is the nth power of this, so

$$\phi_n(t) = \left(1 - \frac{1}{n} \int_0^n \left(1 - \cos \frac{tmg}{x^2}\right) dx\right)^n.$$

Now for large x, the integrand is $\sim (tmg)^2/x^4$ so the integral from n to infinity is small, certainly $o(1/n)$, and we can write it as

$$= \left(1 - \frac{1}{n} \int_0^\infty \left(1 - \cos \frac{tmg}{x^2}\right) dx + o(1/n)\right)^n \longrightarrow e^{-\int_0^\infty \left(1 - \cos \frac{tmg}{x^2}\right) dx}.$$

To see how this integral depends on t, substitute $y = x/\sqrt{|t|mg}$. It becomes $\sqrt{|t|mg} \int_0^\infty (1 - \cos(1/y^2)) \, dy$. Let $c = \sqrt{mg} \int_0^\infty (1 - \cos(1/y^2)) \, dy$. Then for all t,

$$\phi_n(t) \longrightarrow \phi(t) = e^{-c|t|^{1/2}}.$$

---

[10]In this context, "law" just means "distribution". It has even been suggested to call convergence in distribution "legal convergence".

[11]Stable laws were introduced by Paul Lévy, along with the weak limit theorem which applies to them [25].

[12]If the stars are uniformly distributed in all of 3-space, the question of stellar gravity fields becomes singular—infinities have to cancel—and open questions remain.

[13]We only consider the x-component of the force; the other components vanish.

By Levy's Continuity Theorem, ϕ is the characteristic function of a probability measure and G_n converges in distribution to it.

Note that if X and Y are independent random variables with characteristic function ϕ, then $X+Y$ has the characteristic function $\phi(t)^2 = e^{-2c|t|^{1/2}} = \phi(4t)$. Thus $(X+Y)/2^2$ has the same distribution function as X. We will see that this means that X has a stable law of exponent $1/2$.

Notation. $X \stackrel{d}{\sim} Y$ means that X and Y have the same distribution.

Definition 6.42. A distribution is **stable** if it is not concentrated at a single point and if, whenever X and Y are i.i.d. with that distribution, then for any strictly positive constants A and B, there exist η and β such that

(6.21) $$\frac{AX + BY - \beta}{\eta} \sim X.$$

By induction, we see that if $X_1, \ldots, X_n$ are i.i.d. stable random variables, there are η_n and β_n such that

(6.22) $$\frac{X_1 + \cdots + X_n - \beta_n}{\eta_n} \stackrel{d}{\sim} X_1.$$

Lemma 6.43. *Let X and Y be i.i.d. with a stable law. Then $X - Y$ has a symmetric stable law with the same norming constant η and $\beta = 0$ in (6.21).*

Proof. Let ϕ be the common characteristic function of X and Y. By (6.21), $\phi(At/\eta)\phi(Bt/\eta)e^{-it\beta} = \phi(t)$. Let X_i and Y_i be independent copies of X, $i = 1, 2$. The characteristic function of $-X$ is $\bar{\phi}$, so the characteristic function of $A(X_1 - Y_1)/\eta + B(X_2 - Y_2)/\eta$ is $\phi(At/\eta)\phi(Bt/\eta)\bar{\phi}(At/\eta)\bar{\phi}(Bt/\eta) = \phi(t)\bar{\phi}(t) = |\phi(t)|^2$, which is the characteristic function of $X - Y$. □

Exercise 6.25. Let X and Y be i.i.d. random variables. Show that if $X + Y$ and X have the same distribution, then $X = 0$ a.s. Conclude that $\eta \neq 1$ in (6.21).

Theorem 6.44. *The only possible norming constants in (6.22) are $\eta_n = n^{1/\alpha}$, where $\alpha > 0$ is a constant.*

Remark 6.45. We will see later that $0 < \alpha \leq 2$.

Proof. Let $X_1, \ldots, X_n$ be i.i.d. stable random variables.. By Lemma 6.43 we can suppose the X_i are symmetric.

6.5. Stable Laws*

Let $n = rk$, where r and k are integers, and break up the sum into r blocks of k:

$$\eta_{rk} X_1 \stackrel{d}{\sim} \underbrace{(X_1 + \cdots + X_k)}_{\stackrel{d}{\sim}\, \eta_k X_1} + \underbrace{(X_{k+1} + \cdots + X_{2k})}_{\stackrel{d}{\sim}\, \eta_k X_2} + \cdots$$

$$+ \underbrace{(X_{(r-1)k+1} + \cdots + X_{rk})}_{\stackrel{d}{\sim}\, \eta_k X_r} .$$

By (6.22) the sum of the terms in the ith parenthesesis has the same distribution as $\eta_k X_i$, $i = 1, \ldots, r$, so, again by (6.22), this sum is

$$\stackrel{d}{\sim} \eta_k (X_1 + \cdots + X_r) \stackrel{d}{\sim} \eta_k \eta_r X_1 .$$

Thus, for positive integers r and k we have

(6.23) $$\eta_{rk} = \eta_k \eta_r .$$

We claim that η_n increases with n. First, since $\eta_{m+n} X_1 \stackrel{d}{\sim} \eta_m X_1 + \eta_n X_2$, for any $x > 0$, $P\{\eta_{m+n} X_1 \geq \eta_m x\} \geq \frac{1}{2} P\{\eta_m X_1 > \eta_m x\} = \frac{1}{2} P\{X_1 > x\}$, where we have used the fact that X_2 is symmetric. Thus

$$P\left\{X_1 \geq \frac{\eta_m}{\eta_{m+n}} x\right\} \geq \frac{1}{2} P\{X_1 > x\} > 0 .$$

This implies that η_m / η_{m+n} is bounded above in both m and n. Let $r > 1$ and $p \geq 1$ be integers. Take $m = r^p$ and $m + n = (r+1)^p$. By (6.23), $\eta_{r^p} / \eta_{(r+1)^p} = (\eta_r / \eta_{r+1})^p$. Now the left-hand side is bounded, so this must remain bounded as $p \to \infty$. But this implies that $\eta_r / \eta_{r+1} \leq 1$, which proves the claim.

Now let $j, k \geq 1$ be integers. For $p \geq 1$, there is an integer λ for which $j^\lambda \leq k^p < j^{\lambda+1}$. Since η_n increases in n and $\eta_{j^\lambda} = \eta_j^\lambda$, this implies that $\eta_j^\lambda \leq \eta_k^p \leq \eta_j^{\lambda+1}$.

Now the η_n increase in n, and clearly $\eta_1 = 1$. Moreover, $\eta_2 \neq 1$ by Exercise 6.25, so that $\eta_n > 1$ for $n \geq 2$. Take logs of the two double-inequalities above to see that

$$\lambda \frac{\log j}{\log k} \leq p \leq (\lambda + 1) \frac{\log j}{\log k} ,$$

$$\lambda \log \eta_j \leq p \log \eta_k \leq (\lambda + 1) \log \eta_j ,$$

or, noting that $\eta_j > 1$, so $\log \eta_j > 0$,

$$\frac{\lambda}{p} \leq \frac{\log k}{\log j} \leq \frac{\lambda + 1}{p} ,$$

$$\frac{\lambda}{p} \leq \frac{\log \eta_k}{\log \eta_j} \leq \frac{\lambda + 1}{p} .$$

Eliminate p by using the first inequalities in the second to get
$$\frac{\lambda}{\lambda+1}\frac{\log k}{\log j} \leq \frac{\log \eta_k}{\log \eta_j} \leq \frac{\lambda+1}{\lambda}\frac{\log k}{\log j}.$$

Now let $\lambda \longrightarrow \infty$ to see that
$$\frac{\log \eta_k}{\log \eta_j} = \frac{\log k}{\log j} \Longrightarrow \frac{\log \eta_k}{\log k} = \frac{\log \eta_j}{\log j}.$$

As j and k are arbitrary, the ratio must be a positive constant, so there is an $\alpha > 0$ such that
$$\frac{\log \eta_k}{\log k} = \frac{1}{\alpha}.$$

Consequently, $\eta_k = k^{\frac{1}{\alpha}}$, $k = 1, 2, \ldots$. $\square$

We can now complete our description of symmetric stable laws by giving their characteristic functions.

Theorem 6.46. *Let ϕ be the characteristic function of a symmetric stable law. Then there exists α, $0 < \alpha \leq 2$, and $c > 0$ such that*

(6.24) $$\phi(t) = e^{-c|t|^\alpha}.$$

Proof. We have seen that ϕ is real and symmetric, and Theorem 6.44 tells us that there exists $\alpha > 0$ such that $\phi(t) = \phi(t/n^{1/\alpha})^n$, $n = 1, 2, \ldots$, or, if we replace t by $n^{1/\alpha}t$,

(6.25) $$\phi(n^{1/\alpha}t) = \phi(t)^n, \ n = 1, 2, \ldots$$

We claim that $\phi(t) > 0$ for all t.

To see this, suppose $\phi(t)$ does vanish somewhere, and let t_0 be the smallest t for which $\phi(t) = 0$. Then $t_0 > 0$ since ϕ is continuous and $\phi(0) = 1$. Then by (6.25) with $n = 2$, $\phi(t_0/2^{1/\alpha}) = \phi(t_0)^2 = 0$; but $t_0/2^{1/\alpha} < t_0$, a contradiction. Done!

Thus, ϕ is strictly positive, so we can take its logarithm: let $\psi(t) = \log \phi(t)$. From (6.25) ψ satisfies

(6.26) $$\psi(t) = n\psi\left(\frac{t}{n^{1/\alpha}}\right).$$

Note that for any $c > 0$, $\psi(t) = -ct^{1/\alpha}$ satisfies (6.26) for $t \geq 0$. We must show that these are the only solutions. Note that ϕ is an even function, so it is enough to consider positive t. Replacing t by $n^{1/\alpha}t$ in (6.26), we have $\psi(n^{1/\alpha}t) = n\psi(t)$ for every integer $n \geq 1$. Combining the two,
$$\psi\left(\frac{m^{1/\alpha}}{n^{1/\alpha}}\right) = \frac{m}{n}\psi(1).$$

6.5. Stable Laws*

or, letting $t = (m/n)^{1/\alpha}$, $\psi(t) = t^\alpha \psi(1)$. This holds for a dense set of t, and ψ is continuous, so it holds for all positive t:
$$\psi(t) = t^\alpha \psi(1),\ t \geq 0\,.$$

Just set $c = -\psi(1)$ to see that all solutions of (6.26) are of the form $\phi(t) = e^{-c|t|^\alpha}$, where $c = -\log \phi(1)$.

To see that $\alpha \leq 2$, suppose the contrary, that $\alpha > 2$. We claim that $\phi(t) = e^{-c|t|^\alpha}$ is not a characteristic function for any $c > 0$. Indeed, suppose ϕ is the characteristic function of a random variable X. Now ϕ is non-constant, twice differentiable and $\phi''(0) = 0$, so that X has a finite second moment given by $E\{X^2\} = -\phi''(0) = 0$. Therefore, $X \equiv 0$, so its characteristic function is identically one; but ϕ is non-constant, a contradiction. $\square$

Remark 6.47. If $\alpha = 2$, the stable (α) distribution is Gaussian. If $\alpha = 1$, it is Cauchy. If $\alpha < 2$, $\phi''(0)$ fails to exist, so the distribution has no variance. If $\alpha \leq 1$, $\phi'(0)$ fails to exist, and the distribution has no mean.

For each α, $0 < \alpha \leq 2$, there is a weak convergence result much like the central limit theorem. For $\alpha = 2$, it *is* the central limit theorem. For each α, $0 < \alpha \leq 2$, there is a class of distributions, called the **domain of attraction** of stable (α) and denoted $\mathcal{D}(\alpha)$, which contains the distributions with the property that if $X_1, X_2, \ldots$ are independent with that distribution, and if $S_n = X_1 + \cdots + X_n$, then
$$\frac{S_n}{n^{\frac{1}{\alpha}}} \Longrightarrow \text{stable}\,(\alpha)\,.$$

The stable (2) distributions are Gaussian, so $\mathcal{D}(2)$ contains all distributions with finite variance. For other values of α, membership in $\mathcal{D}(\alpha)$ depends on the tails of the distribution. For instance, it can be shown that $\mathcal{D}(\alpha)$ contains a distribution with distribution function F if
$$F(x) \sim \frac{c}{(1+|x|)^{\frac{1}{\alpha}}} \text{ as } t \to -\infty \qquad 1 - F(x) \sim \frac{d}{(1+|x|)^{\frac{1}{\alpha}}} \text{ as } t \to \infty\,,$$
where c and d are positive and need not be equal; if they are equal, the limit is symmetric stable, and if not, it is unsymmetric stable. A distribution with density of the form $f(x) = c(1+|x|^{1+\alpha})$ would be in $\mathcal{D}(\alpha)$, for instance. Note that this settles the question of existence of stable laws, since a posteriori they are weak limits of known probability distributions.

There is a more direct way to see that the stable distributions exist, at least for $\alpha \leq 1$. G. Pólya proved the following.

Theorem 6.48 (G. Polya). *If a real-valued function ϕ is symmetric ($\phi(t) = \phi(-t)$) and $\phi(|t|)$ is decreasing, convex and $\phi(0) = 1$, then ϕ is the characteristic function of a probability distribution. If $\lim_{t \to \infty} \phi(|t|) = 0$, this distribution has a density.*

Since $e^{-c|t|^\alpha}$ satisfies this for $\alpha \le 1$, it must be the characteristic function of an absolutely continuous probability distribution. While we know the densities for the Cauchy and the Gaussian distributions, we do not, in general, know them for other stable distributions.

Problems 6

6.26. Ten numbers are rounded to the nearest integer and then summed. Use the central limit theorem to determine the approximate probability that the sum of the unrounded numbers equals the rounded sum of the original numbers. (Assume the fractional part of each number is uniformly destributed.)

6.27. Let f be a bounded Borel function on the unit hypercube $C \stackrel{\text{def}}{=} [0,1]^d$ in $\mathbb{R}^d$. Let f be a bounded Borel function defined on C, and compute $I \stackrel{\text{def}}{=} \int_C f(x)\,dx$ by Monte Carlo integration as follows: let $X_1, \ldots, X_n$ be independent and uniform on C. Then $\bar{f}_n \stackrel{\text{def}}{=} \frac{1}{n} \sum_{j=1}^n f(X_j)$ converges a.s. to I. Let $\varepsilon > 0$, and find an approximate value of n such $P\{|\bar{f}_n - I| > \varepsilon\} < .05$. (Give the answer in terms of some integrals of f.) Find the smallest n which works for all functions f such that $\sup\{|f(x)| : x \in C\} \le 1$. Would the value of n be smaller if the functions were continuous?

6.28. Let $X_1, X_2, \ldots$ be a sequence of i.i.d. random variables with common characteristic function
$$f(t) = e^{-t^2 - \sqrt{|t|}}.$$

(a) Find real numbers η_n such that $\frac{X_1 + \cdots + X_n}{\eta_n}$ converges in distribution to a non-degenerate probability distribution as $n \to \infty$.

(b) What possible distributions can the limit have?

(c) Prove or disprove: $E\{|X_1|\} < \infty$.

6.29. Let $S_n = \left(\sum_1^n X_i\right)/\sqrt{n}$, where the X_i are i.i.d. mean zero and variance one. Show that the lim sup and lim inf of the sequence (S_n) are infinite.

6.30. Lazzerini determined the value of π from Buffon's needle problem; the numbers were given in Chapter 2. Use these and the central limit theorem to estimate the probability that a result could be as good as his. Compare this with the true binomial probabilities. Hint: You won't have to calculate many...]

6.31. Let (X_n) be a sequence of i.i.d. random variables. Show there is no sequence of constants c_n such that $\sum(X_n - c_n)$ converges in distribution to a random variable, unless the common distribution of the X_n is a point mass.
[Hint: Exercise 6.11 may help.]

6.32. Show that the Y_i^n in Proposition 6.35 are themselves Poisson.

6.33. Let F and F_n, $n = 1, 2 \ldots$ be distribution functions. Show that if $F_n(x) \to F(x)$ and $F_n(x-) \to F(x-)$ as $n \to \infty$ for all x, then the convergence is uniform.

6.34. (Lévy 45-degree distance.) If F is a distribution function, let $\bar{F}$ be the filled-in graph of F; if x is a discontinuity point of F, add the line segment connecting the points $(x, F(x-))$ and $(x, F(x))$. Then $\bar{F}$ is the union of the graph of F with all these line segments. (See Figure 1.) Note that a line of slope minus 1 intersects $\bar{F}$ in exactly one point. Define the distance between distribution functions F and G as the supremum of the distances between the intersections of lines of slope minus 1 with $\bar{F}$ and $\bar{G}$. Show that this is a metric on the set of all distribution functions, and that convergence in this metric corresponds to convergence in distribution.

Figure 1. The distribution function F and the distance between $\bar{F}$ and $\bar{G}$.

6.35. An astronomer wishes to measure the distance in light years from her observatory to a distant star. Because of atmospheric disturbances and experimental error, each measurement will not be accurate, but will only be approximate. As a result, she will make a series of independent measurements, and use the average value as the final estimate. She believes that she can account for all systematic errors, so that her measurements have a mean d (the true distance) and a standard deviation of four light years. If she wishes to be 95% sure that her estimate is correct to within ± 0.5 light years, approximately how many measurements should she make?
[Hint: $\Phi(1.64) = .9495$, $\Phi(1.65) = .9505$, $\Phi(1.95) = .9744$, $\Phi(1.96) = .9750$.]

Supplementary Problems

6.36. The following three statements are incorrect as they stand. Add a hypothesis to each which corrects it. [Some hypotheses are better than others: "Assume zero equals one" may be logically correct but it won't get any credit.]

(a) Let (X_n) be a sequence of integrable random variables, and let X be a random variable. Suppose that (X_n) converges in probability to X as $n \to \infty$. Then $E\{X_n\} \longrightarrow E\{X\}$ as $n \to \infty$.

(b) Let (f_n) be a sequence of characteristic functions and let f be a complex-valued function on $\mathbb{R}$. Suppose that for each t, $f_n(t)$ converges to $f(t)$ as $n \to \infty$. Then f is the characteristic function of a probability distribution.

(c) Let (X_n) be a sequence of independent random variables of mean zero and variance one. Then $(X_1 + \cdots + X_n)/\sqrt{n}$ converges in distribution to a $N(0,1)$ random variable.

6.37. A gambler plays a series of independent games. The probability of winning or losing each game is $1/2$. The gambler bets \$k on the kth game.

Let S_n be the total amount of money the gambler has won in the first n games. (S_n may be negative if the gambler has lost.)

(a) Let $Z_n = n^{-p} S_n$. Find p such that the variance of Z_n has a non-zero finite limit as $n \to \infty$.

(b) Show that for this p, the sequence (Z_n) converges in distribution to a random variable Z. What is the distribution of Z?

[Hint: Your guess as to the limit is correct, but... can you prove it?]

6.38. Let (X_n) and (Y_n) be sequences of random variables such that as $n \to \infty$, (X_n) converges in distribution to a random variable X and $Y_n - X_n \to 0$ in probability. Show that Y_n converges in distribution to X.

6.39. (Close Elections.) In the American presidential election in the year 2000, 5,962,657 votes were cast in the state of Florida. According to the official tally, 2,912,790 votes were cast for the Republican candidate, Mr. George W. Bush, and 2,912,253 votes were cast for the Democratic candidate, Mr. Al Gore. The other 137,614 votes were cast for other candidates. So, out of almost six million votes, the difference between the top two candidates was less than 550 votes, less than one one-hundredth of one percent of the votes cast. This led to a wrangle in the courts over recounts, allegations of voting irregularities and disputes over "hanging chads" and "butterfly ballots", but that is another story. The interesting question is this: suppose that, in some sense, the voters are equally divided. There are still random elements at work: the rate of voter participation in that election was only 51.3%, and events can cause many people to change their minds. How unlikely is this result?

Suppose that each of the 5,825,043 voters who voted either Republican or Democratic decides independently to vote Republican with probability 1/2, and Democratic with probability 1/2. Let X be the number of Republican votes, and $N-X$ the number of Democratic votes, where $N = 5,825,043$. Find the probability that the absolute value of the difference between the Republican and Democratic votes is 537 or less.

6.40. (Continuation: another election model.) In the 2000 federal election, 51.3% of the eligible voters nationwide voted. Suppose there are 12,000,000 eligible voters in Florida, and suppose that exactly 6,000,000 of these would vote for Bush and exactly 6,000,000 would vote for Gore... *if* they vote. But each voter chooses independently whether or not to vote with probability 1/2 each. Let B be the number of Bush voters and G the number of Gore voters. Find the approximate probability that $|B - G| \leq 537$.

6.41. (Continuation. Getting out the vote: the effect of small biases.) On the day of an important election, political parties do what is called "getting out the vote." They make sure that voters from their own party actually do go to the polls and vote. Consider the situation of the preceding problem, with one difference. Suppose that one of the two parties does nothing, but the other gets out the vote and assures that 6,000 of their voters (or 0.1%) go to the polls—for sure. The other 5,994,000 of their voters—and all 6,000,000 of the opposing party's voters—choose at random between voting or not voting at all. Estimate the probability that the first party wins the election.

6.42. Let $\phi_{X,Y}(s,t)$ be the joint characteristic function of random variables X and Y. If X and Y are independent, then $\phi_{XY}(s,t) = \phi_X(s)\phi_Y(t)$. But there exist non-independent X and Y such that for all t, $\phi_{XY}(t,t) = \phi_X(t)\phi_Y(t)$. (Note that this is the characteristic function of $X + Y$.) Find such a pair.

Chapter 7

Markov Chains and Random Walks

This chapter marks a change of viewpoint, and a transition from what might be called classical probability to the study of stochastic processes. Some of the same subjects—sums of independent random variables, weak convergence—will return in different guises. But we will shift from a static to a dynamic viewpoint. The emphasis changes from single random variables and limit theorems to processes which evolve in time.

7.1. Stochastic Processes

Definition 7.1. A **stochastic process** is a family $\{X_t,\ t \in T\}$ of random variables defined on the same probability space.

Here, T is a parameter set. A priori, it can be anything: a stochastic[1] process is just a collection of random variables. The definition is so general as to be almost all-encompassing, but we usually have something more specific in mind. We think of a stochastic process as something evolving randomly in time. Common choices for T are the positive integers and the positive half-line. If $T = \{0, 1, 2, \dots\}$, the process is $X_0, X_1, X_2, \dots$. It is called a *discrete-parameter* process. If $T = [0, \infty)$, the process is $\{X_t,\ t \geq 0\}$, a *continuous-parameter* process.

We have already seen important examples: a sequence $X_1, X_2, \dots$ of independent random variables is a discrete-time stochastic process. So is the sequence $S_1, S_2, \dots$ of partial sums of the X_i. The gambler's ruin (Section

---
[1] "Stochastic" is a Greek word for "random."

1.7) is another example. It describes a gambler's fortune as the game of chance progresses. A typical example of a continuous-time stochastic process would be $\{D(t), t \geq 0\}$, where $D(t)$ is the Dow-Jones stock-market average at time t. But there are other stochastic processes with larger parameter sets, such as $\{T(x,y,t) : t \geq 0,\ 0 \leq x \leq 360,\ 0 \leq y \leq 180\}$, where $T(x,y,t)$ is the temperature at latitude x, longitude y, and time t on the surface of the world.

We need one fact from measure theory. Let $\{X_t,\ t \in T\}$ be a stochastic process on $(\Omega, \mathcal{F}, P)$. Its finite-dimensional distributions are of the form $P\{(X_{t_1}, \ldots, X_{t_n}) \in A\}$, where $n = 1, 2, \ldots$, $t_1, \ldots t_n \in T$, and A is a Borel set in $\mathbb{R}^n$. Then:

Theorem 7.2. *The entire distribution of a stochastic process is determined by its finite-dimensional distributions.*

Exercise 7.1. Let $\{X_t,\ t \in T\}$ be a stochastic process on $(\Omega, \mathcal{F}, P)$. Let $\mathcal{F}_{t_1,\ldots,t_n}$ be the σ-field of sets of the form $\{(X_{t_1}, \ldots, X_{t_n}) \in A\}$, $t_1, \ldots, t_n \in T$, $n = 1, 2, \ldots$, and A is a Borel set in $\mathbb{R}^n$. These sets are called **cylinder sets**. The probabilities of the cylinder sets are the finite dimensional distributions of X.

(a) Show that the cylinder sets form a field $\mathcal{F}_0$, but not necessarily a sigma-field. Let $\mathcal{F} = \sigma(\mathcal{F}_0)$.

(b) Show that if P and Q are probability measures on $\mathcal{F}$ which agree on $\mathcal{F}_0$, then $P = Q$. Thus prove Theorem 7.2.

[Hint: Use Theorem 1.7.]

A consequence of this is:

Corollary 7.3. *Two stochastic processes with the same finite-dimensional distributions have the same distribution.*

7.2. Markov Chains

Let $X_1, X_2, \ldots$ be a stochastic process. Suppose the X_n take values in some finite or countable space, say $S = \{s_n,\ n = 1, 2, \ldots\}$. We call S the **state space**. If $X_n = s_i$, we say "X_n is in state s_i." To simplify notation, write i instead of s_i, so that the state space is a subset of the integers, and the X_n take integer values.

Definition 7.4. The process $\{X_n,\ n = 0, 1, 2, \ldots\}$ is a **Markov process** if, for each n, i and $i_0, \ldots, i_n$,

(7.1) $P\{X_{n+1} = i \mid X_0 = i_0, \ldots, X_n = i_n\} = P\{X_{n+1} = i \mid X_n = i_n\}$.

(7.1) is called the **Markov property.**

Think of the parameter as time. If n is the present—i.e., "now"—then $\{1, 2, \ldots n-1\}$ is the past, and $\{n+1, n+2, \ldots\}$ is the future. According to

7.2. Markov Chains

(7.1), the distribution of X_{n+1} depends only on the present—i.e., X_n—not on the past. In other words, only the present value is important, not how the process got here[2].

Exercise 7.2. Extend (7.1) by showing that if (X_k) is a Markov chain, $n > 0$ and $A \subset \mathbb{R}^{n+1}$, then

(7.2) $\quad P\{X_{n+1} = j \mid X_n = i, (X_0, \ldots, X_n) \in A\} = P\{X_{n+1} = j \mid X_n = i\}$.

Definition 7.5. A Markov chain X has **stationary transition probabilities** if for all i, j, and n,

$$P\{X_{n+1} = j \mid X_n = i\} = P\{X_1 = j \mid X_0 = i\}.$$

In this case, the quantities

$$P_{ij} \stackrel{def}{=} P\{X_1 = j \mid X_0 = i\}$$

are called the **transition probabilities** of X, and the matrix

$$\mathbb{P} \stackrel{def}{=} (P_{ij})$$

is called the **transition probability matrix.**

The transition probabilities tell us how the process evolves. The laws of nature do not change with time, so if our Markov chain describes a physical process, and if the environment is not changing either, we would expect the process to evolve in the same way, regardless of what time the clock reads. In other words, we would expect its transition probabilities to be stationary. This is why processes with stationary transition probabilities are so common.

We will assume from now on that our Markov chains have stationary transition probabilities, unless we specifically say otherwise. It is possible—though not always useful—to transform any Markov chain into a Markov chain with stationary transition probabilities. See Exercise 7.3.

Exercise 7.3. Let $X = \{X_n, n = 0, 1, 2, \ldots\}$ be a Markov chain, which may not have stationary transition probabilities. Define a new process $Y = \{Y_n, n = 0, 1, 2, \ldots\}$ by

$$Y_n = (X_0, \ldots, X_n).$$

(So Y_n takes values in $\mathbb{Z}^n$.) Show that Y is a Markov chain with stationary transition probabilities.

[Hint: The state space is $\bigcup_n \mathbb{Z}^n$, not $\mathbb{Z}$, and you must extend the transition probabilities to $\bigcup_n \mathbb{Z}^n$ to use the definition of stationarity.]

---

[2] Markov chains, processes, and the Markov property are named after the Russian mathematician A. A. Markov, who introduced them and proved the ergodic theorem [**28**]. "Stopping times" were originally called "Markov times".

Remark 7.6. Let $\mathbb{P} = (P_{ij})$ be a transition probability matrix. Then
   (i) $\forall i, j$: $P_{ij} \geq 0$;
   (ii) $\forall i$: $\sum_j P_{ij} = 1$.

Indeed, P_{ij} is a probability, which implies (i), and (ii) follows since $1 = P\{X_{n+1} \in \mathbb{R} \mid X_n = i\} = \sum_j P\{X_{n+1} = j \mid X_n = i\}$.

Definition 7.7. The distribution of X_0 is called the **initial distribution**, denoted $\mu_i = P\{X_0 = i\}$.

Proposition 7.8. *The distribution of a Markov chain is determined by its initial distribution and its transition probability matrix. In particular, for any $n \geq 0$ and integers $i_0, \ldots, i_n$,*

(7.3) $$P\{X_0 = i_0, \ldots, X_n = i_n\} = \mu_{i_0} P_{i_0 i_1} P_{i_1 i_2} \ldots, P_{i_{n-1} i_n}.$$

Proof. By Theorem 7.2 it is enough to prove (7.3). By the Markov property,

$$P\{X_0 = i_0, \ldots, X_n = i_n\} = P\{X_n = i_n \mid X_0 = i_0, \ldots, X_{n-1} = i_{n-1}\}$$
$$\times P\{X_0 = i_0, \ldots, X_{n-1} = i_{n-1}\}$$
$$= P\{X_n = i_n \mid X_{n-1} = i_{n-1}\} P\{X_0 = i_0, \ldots, X_{n-1} = i_{n-1}\}$$
$$= P_{i_{n-1} i_n} P\{X_0 = i_0, \ldots, X_n = i_{n-1}\}.$$

By induction, this is

$$= P_{i_{n-1} i_n} P_{i_{n-2} i_{n-1}} P\{X_0 = i_0, \ldots, X_{n-2} = i_{n-2}\}$$
$$\vdots$$
$$= P_{i_{n-1} i_n} \ldots P_{i_0 i_1} P\{X_0 = i_0\}$$
$$= P_{i_{n-1} i_n} \ldots P_{i_0 i_1} \mu_{i_0}.$$

$\square$

Markov chains have the "restarting property". A chain can restart at each time n, and the post-n process is again a Markov chain with the same transition probabilities, as the following exercise shows.

Exercise 7.4. Show that for positive integers n and k and states $j_0, \ldots j_k$, that

(7.4) $$P\{X_n = j_0, \ldots, X_{n+k} = j_k\} = P\{X_n = j_0\} P_{j_0 j_1} P_{j_1 j_2} \ldots P_{j_{k-1} j_k},$$

and compare these with the probabilities for $X_0, \ldots, X_k$. Conclude that $X_0, X_1, \ldots$ and $X_k, X_{k+1}, \ldots$ are Markov chains with the same transition probabilities.

The n-step Transition Probabilities. Let $P_{ij}^{(0)} \overset{\text{def}}{=} \delta_{ij}$, $P_{ij}^{(1)} = P_{ij}$, and, in general, $P_{ij}^{(n)} = P\{X_n = j \mid X_0 = i\}$. Let $\mathbb{P}^{(n)} = (P_{ij}^{(n)})$ be the matrix of **n-step transition probabilities.** (Note that $\mathbb{P}^{(0)}$ is the identity matrix.)

7.2. Markov Chains

Let us calculate $P_{ij}^{(2)}$:

$$\begin{aligned} P_{ij}^{(2)} &= P\{X_2 = j \mid X_0 = i\} \\ &= \sum_k \frac{P\{X_2 = j,\, X_1 = k,\, X_0 = i\}}{P\{X_0 = i\}} \\ &= \sum_j \frac{\mu_i P_{ik} P_{kj}}{\mu_i} \\ &= \sum_j P_{ik} P_{kj}\,. \end{aligned}$$

But this is just the ij^{th} entry of the product $\mathbb{P} \times \mathbb{P}$. Thus $\mathbb{P}^{(2)} = \mathbb{P}^2$. We can proceed by induction to see that the n-step probability transition matrix is the nth power of $\mathbb{P}$:

$$\mathbb{P}^{(n)} = \mathbb{P}^n\,.$$

It follows that for any positive integers m and n, $\mathbb{P}^{(m+n)} = \mathbb{P}^{(m)} \mathbb{P}^{(n)}$. The formula of matrix multiplication leads us to the Chapman-Kolmogorov equations:

Theorem 7.9 (Chapman-Kolmogorov Equations). *Let m, n be positive integers. Then for all states i and j,*

(7.5) $$P_{ij}^{(m+n)} = \sum_k P_{ik}^{(m)} P_{kj}^{(n)}\,.$$

The Distribution of X_n. Let μ_0 be the initial distribution of the process: $\mu_0(i) = P\{X_0 = i\}$. More generally let μ_n be the distribution of X_n: $\mu_n(i) = P\{X_n = i\}$. Then μ_1 is given by

$$\mu_1(j) = P\{X_1 = j\} = \sum_i P\{X_0 = i, X_1 = j\} = \sum_i \mu_0(i) P_{ij}\,.$$

More generally, the probability $\mu_n(j)$ that $X_n = j$ can be given in terms of the n-step transition probabilities:

(7.6) $$\mu_n(j) = \sum_i \mu_0(i) P_{ij}^{(n)}\,.$$

A close look reveals a hidden matrix multiplication. Indeed, if we consider the initial distribution μ_0 and the distributions μ_n as as row vectors, e.g., $\mu_0 = (\mu_0(0), \mu_0(1), \dots)$, then (7.6) can be written in terms of the n-step transition matrices, or equivalently, in terms of the powers of the transition matrix:

$$\mu_1 = \mu_0 \mathbb{P}\,, \qquad \mu_n = \mu_0 \mathbb{P}^n\,.$$

Examples. Here is a scattering of examples of Markov chains from various branches of science and mathematics, along with an idea of the kind of questions that one might ask about them.

1. Gambler's Ruin, $N = 5$. (See §1.7.) Let X_n be the gambler's fortune after the nth play. At each stage, the fortune moves up or down by one, according to whether the gambler wins or loses that bet, until it reaches either 0 or 5. When it reaches 0 or 5, it stays there. So that for $i = 1, \ldots, 4$, $P_{ii+1} = P_{ii-1} = 1/2$, and $P_{ij} = 0$ if $|j-i| \neq 1$, while $P_{00} = 1$, $P_{0j} = 0$, $j \neq 0$, and $P_{55} = 1$, $P_{5j} = 0$ if $j \neq 5$. The transition probability matrix is

$$\mathbb{P} = \begin{pmatrix} 1 & 0 & 0 & 0 & 0 \\ \frac{1}{2} & 0 & \frac{1}{2} & 0 & 0 \\ 0 & \frac{1}{2} & 0 & \frac{1}{2} & 0 \\ 0 & 0 & \frac{1}{2} & 0 & \frac{1}{2} \\ 0 & 0 & 0 & 0 & 1 \end{pmatrix}.$$

The main questions we would ask of this model are, "What is the probability that the gambler goes broke, i.e., the probability that the chain reaches zero?" and "How long does it take"? (See §1.9.)

2. Simple symmetric random walk[3]. Consider a person walking, taking steps at random, one step to the left, one to the right, independently and with equal probability. The state space is $\mathbb{Z}$, the space of all integers, and state i represents i steps from the origin. The walker goes from state i to state $i \pm 1$, with probability 1/2 each. Thus $P_{ii+1} = P_{ii-1} = 1/2$, and $P_{ij} = 0$ if $|i - j| \neq 1$. The transition probability matrix $\mathbb{P}$ is an infinite matrix, but it is a relatively simple one, for it is tri-diagonal. It has zeros on the main diagonal, 1/2 on the diagonals above and below, and zeros on all other diagonals,

$$\mathbb{P} = \begin{pmatrix} \ddots & \vdots & \vdots & \vdots & \vdots & \vdots & \\ \cdots & 0 & \frac{1}{2} & 0 & 0 & 0 & \cdots \\ \cdots & \frac{1}{2} & 0 & \frac{1}{2} & 0 & 0 & \cdots \\ \cdots & 0 & \frac{1}{2} & 0 & \frac{1}{2} & 0 & \cdots \\ \cdots & 0 & 0 & \frac{1}{2} & 0 & \frac{1}{2} & \cdots \\ \cdots & 0 & 0 & 0 & \frac{1}{2} & 0 & \cdots \\ & \vdots & \vdots & \vdots & \vdots & \vdots & \ddots \end{pmatrix}.$$

Questions one might ask about this: Does this walk reach every possible point? Does it tend to equilibrium in some sense?

3. A reflecting random walk on $\{0, 1, \ldots, n\}$. Let $0 \leq p, q, r$ with $p + q + r = 1$. Suppose the probability of a step to the right is p, to the left

---

[3]This is also called the Drunkard's walk, for obvious reasons. But it comes up in many other situations, such as, for instance, the Gambler's Ruin.

7.2. Markov Chains

is q, and the probability of staying at the state is r. Then for $0 < i < n$, $P_{ii+1} = p$, $P_{ii-1} = q$, $P_{ii} = r$, and $P_{ij} = 0$ if $|j - i| > 1$. If it is at 0 or n, it moves away on the next step: $P_{01} = P_{nn-1} = 1$. Thus,

$$\mathbb{P} = \begin{pmatrix} 0 & 1 & 0 & 0 & 0 & \cdots & 0 \\ q & r & p & 0 & 0 & \cdots & 0 \\ 0 & q & r & p & 0 & \cdots & 0 \\ 0 & 0 & q & r & p & \cdots & 0 \\ & & \ddots & \ddots & \ddots & & \\ 0 & 0 & 0 & \cdots & q & r & p \\ 0 & 0 & 0 & \cdots & 0 & 1 & 0 \end{pmatrix}.$$

The primary question here is whether or not it tends to equilibrium.

4. Branching processes. Suppose an organism produces a random number ξ of offspring at the end of its life. We count by generations, so the lifetime of each individual is 1. Suppose that the family sizes for different times and different individuals are i.i.d., and that $P\{\xi = i\} = a_i$, $i = 0, 1, 2, \ldots$, where $a_i \geq 0$ and $\sum a_i = 1$. Let X_n be the population size after the nth generation. Thus, if there are i individuals at time n, each is replaced by an independent family whose size has the same distribution as ξ. Thus

$$P_{ij} = P\{X_{n+1} = j \mid X_n = i\} = P\{\xi_1 + \cdots + \xi_i = j\},$$

where the ξ_j are independent copies of ξ.

The main question here is, "What is the probability that the population dies out?"; i.e., what is the probability that the chain ever reaches 0?

5. A Queuing Model. This models the behavior at, say, the check-out counter in a supermarket. Customers arrive for service and take their place in the waiting line. Time is divided into periods. If the line is not empty at the start of a period, a single customer is served, and then new customers arrive. The number of arrivals in the nth period is ξ_n, where ξ_i, $i = 1, 2, \ldots$ are i.i.d. with $P\{X_1 = k\} = a_k$, $k = 0, 1, 2, \ldots$. The state of the system is the number of customers waiting in line *before* the service is performed. The state space is $S = \{0, 1, 2, \ldots\}$.

If the state of the system is $i \geq 1$, then the number of customers at the next stage is the present number plus that number of new arrivals minus one (for the customer served), so the new state is $i + \xi_n - 1$. If $i = 0$, no service is performed, so the next state is ξ_n. In either case, $X_{n+1} = (X_n - 1)^+ + \xi_n$.

The transition probabilities are

$$\mathbb{P} = \begin{pmatrix} a_0 & a_1 & a_2 & \cdots & & \\ a_0 & a_1 & a_2 & \cdots & & \\ 0 & a_0 & a_1 & a_2 & \cdots & \\ 0 & 0 & a_0 & a_1 & a_2 & \cdots \\ \cdots & & & & & \end{pmatrix}.$$

Some questions one might want to ask about this: does $X_n \to \infty$? Or does the system approach equilibrium? If so, what is the equilibrium distribution? Intuitively, $X_n \to \infty$ if $E\{\xi_n\} > 1$. (You can prove that with what you already know!) Otherwise the system tends to equilibrium.

6. Genetics. Sewall Wright introduced the following, called a "simple haploid" model. The population consists of N individuals. Each individual has two genes at a certain locus on a certain chromosome, so there is a fixed population of $2N$ genes associated with that locus. There are two types of gene, a and A. The type of the gene may change from generation to generation. At each generation, each gene "chooses a parent at random" independently from the existing population, and inherits its new type from that parent. That is, if there are j type a genes in the population and $2N-j$ type A genes, then, in the next generation, a given gene becomes type a or type A with probabilities p_j and q_j, respectively, independently of the other genes, where p_j and q_j are given by[4]

$$P\{\text{becomes type } a\} = \frac{j}{2N} \stackrel{\text{def}}{=} p_j, \quad P\{\text{becomes type } A\} = 1 - \frac{j}{2N} \stackrel{\text{def}}{=} q_j.$$

At the nth generation, let the state be X_n, the number of type a genes in that generation. The state space is $S = \{0, 1, 2, \ldots, 2N\}$.

There are $2N$ genes, so that if $X_n = j$, X_{n+1} will be binomial:

(7.7) $$P_{jk} = P\{X_{n+1} = k \mid X_n = j\} = \binom{2N}{k} p_j^k q_j^{2N-k}.$$

Note that states 0 and $2N$ are absorbing: indeed $X_n = 0 \Rightarrow j = 0 \Rightarrow X_{n+1} = 0$, and $X_n = 2N \Rightarrow j = 2N \Rightarrow X_{n+1} = 2N$.

Questions: What is the probability of absorption in the states 0 and $2N$? Is absorption sure? If so, how long does it take?

It is possible to introduce mutation and selection into this by changing the p_j and q_j. To include mutation, for example, suppose that at each stage, type a mutates into type A with probability α_1, and type A mutates to type a with probability α_2. Assume the gene first chooses its parent,

---
[4] Yes, it itself is included among the parents it might choose. This might bother philosophers and biologists, but not mathematicians.

then mutates. Then $p_j = \frac{j}{2N}(1-\alpha_1) + \left(1 - \frac{j}{2N}\right)\alpha_2$, and $q_j = 1 - p_j$. The transition probabilities are again given by (7.7).

If both α_1 and α_2 are strictly positive, then 0 and $2N$ are no longer absorbing states. In that case, we would expect the chain to tend to an equilibrium. This brings up another question: what is the equilibrium distribution?

7. Success Runs. Consider an experiment which has probability p of success, and $q = 1 - p$ of failure. The experiment is repeated independently. We want to know how many times in a row it succeeds. So at time n, we let X_n be the number of successes in a row up to and including the nth repetition. That is, the state of the system is the length of the success run in progress at the moment. For example the sequence

$$S\ F\ S\ S\ F\ S\ F\ S\ S\ \underbrace{S}_{X_n=3}\ \underbrace{F}_{X_{n+1}=0}$$

leads to successive states of the system: $1, 0, 1, 2, 0, 1, 0, 1, 2, 3, 0$.

If $X_n = i$, then $X_{n+1} = i + 1$ if the next experiment is a success (probability p), and $X_{n+1} = 0$ if not, (probability q.) Thus

$$\mathbb{P} = \begin{pmatrix} q & p & 0 & 0 & \cdots \\ q & 0 & p & 0 & \cdots \\ q & 0 & 0 & p & 0 & \cdots \\ \vdots & & \vdots & & \end{pmatrix}.$$

Typical questions: How long do we have to wait for a run of length n? What is the probability that a run of length m occurs during the first n steps? Is there a typical length of run?

7.2.1. Conditional Independence and the Markov Property.

Definition 7.10. Events A and B are **conditionally independent** given C if

$$P\{A \cap B \mid C\} = P\{A \mid C\}P\{B \mid C\}.$$

If X, Y and Z are random variables, and Z is discrete, then X and Y are **conditionally independent** given Z if for all Borel sets A and B,

(7.8) $\quad P\{X \in A,\ Y \in B \mid Z\} = P\{X \in A \mid Z\}P\{Y \in B \mid Z\}.$

Exercise 7.5. Show that if $P\{A \mid B \cap C\} = P\{A \mid C\}$, then A and B are conditionally independent given C.

Exercise 7.6. Suppose A, B, and C are independent events. Show that $A \cap C$ and $B \cap C$ are conditionally independent given C.

Remark 7.11. This also defines the conditional independence of families $X_1, \ldots, X_n$ and $Y_1, \ldots, Y_m$ of random variables: simply take X and Y to be vector-valued: $X = (X_1, \ldots, X_n)$ and $Y = (Y_1, \ldots, Y_m)$.

Recall that (7.8) is equivalent to the following for all possible z:

$$P\{X \in A, Y \in B \mid Z = z\} = P\{X \in A \mid Z = z\} P\{Y \in B \mid Z = z\}.$$

Here is another statement of the Markov property in terms of the restarting property.

Theorem 7.12. *Let $\{X_n, n = 0, 1, 2, \ldots\}$ be a Markov chain. Then for each n:*

(i) $\{X_{n+j}, j = 0, 1, \ldots\}$ is a Markov chain with transition probability matrix $\mathbb{P}$;

(ii) given X_n, it is conditionally independent of $X_0, \ldots, X_n$.

Proof. By Exercise 7.4, $X_n, X_{n+1}, \ldots$ has the same finite-dimensional probabilities as $X_1, X_2, \ldots$. Therefore the two processes have the same distribution. Since one is a Markov chain with transition probabilities $\mathbb{P}$, so is the other. To see that the pre- and post-n chains are conditionally independent, let $\Lambda = \{X_0 = i_0, \ldots, X_n = i_n\}$ and $\Gamma = \{X_n = i_n, X_{n+1} = j_1, \ldots, X_{n+k} = j_k\}$. Then,

$$P\{\Lambda \cap \Gamma \mid X_n = i_n\} = \frac{P\{\Lambda \cap \Gamma\}}{P\{X_n = i_n\}}.$$

By (7.3) this is

$$= \frac{1}{P\{X_n = i_n\}} P\{X_0 = i_0\} P_{i_0 i_1} \ldots P_{i_{n-1} i_n} P_{i_n j_1} P_{j_1 j_2} \ldots P_{j_{k-1} j_k}$$

$$= \frac{P\{X_0 = i_0\} P_{i_0 i_1} \ldots P_{i_{n-1} i_n}}{P\{X_n = i_n\}} \frac{P\{X_n = i_n\} P_{i_n j_1} P_{j_1 j_2} \ldots P_{j_{k-1} j_k}}{P\{X_n = i_n\}}$$

$$= P\{\Lambda \mid X_n = i_n\} P\{\Gamma \mid X_n = i_n\}.$$

This is true for all k and all choices of the i_p and j_q, which implies that the processes are conditionally independent given X_n. □

Remark 7.13. (i) In one sense, Markov chains generalize sequences of independent random variables, but instead of being independent, $X_0, \ldots, X_n$ and X_{n+1} are *conditionally* independent given X_n.

(ii) Think of X_n as the present state, $X_0, \ldots, X_{n-1}$ as the past, and $X_{n+1}, X_{n+2}, \ldots$ as the future. Then Theorem 7.12 can be interpreted as a satisfyingly symmetric statement of the Markov property: "The past and the future are conditionally independent given the present."

Problems 7.2

7.7. Let $\{X_n,\ n=0,1,2,\dots\}$ be a Markov chain. Show that the process of pairs $\{(X_n, X_{n+1}),\ n=0,1,2,\dots\}$ is also a Markov chain.

7.8. Let $\{X_n,\ n=0,1,2,\dots\}$ be a Markov chain whose transition probabilities may not be stationary. Define $\hat{X}_n$ to be the n-tuple $\hat{X}_n \stackrel{\text{def}}{=} (X_0,\dots,X_n)$. Show that $(\hat{X}_n)$ is a Markov chain with stationary transition probabilities. (What is its state space?)

7.9. Let $Z_1, Z_2, \dots$ be i.i.d. random variables with values ± 1, probability one-half each. Which of the following are Markov chains? Give the transition probabilities of those that are:

(a) $X_n = Z_n Z_{n+1}$, (b) $X_n = Z_n + Z_{n+1}$, (c) $X_n = Z_n + \max(Z_1, \dots, Z_n)$,
(d) $X_n = (Z_n, Z_{n+1})$, (e) $X_n = (Z_1 + \cdots + Z_n)^2$, (f) $X_n = 10 Z_n + Z_{n+1}$.

7.10. (Random Walks on Groups.) Let $\mathfrak{G}$ be a finite or countable group. Let $M_1, M_2, \dots$ be a sequence of i.i.d. random variables with values in $\mathfrak{G}$. Define $X_0 = g$ for some element $g \in \mathfrak{G}$, and define $X_{n+1} = X_n M_{n+1}$. Then (X_n) is called a **right random walk on** $\mathfrak{G}$. (The "right" refers to multiplication by M_n on the right. There is also a left random walk.)

(a) Show that (X_n) is a Markov chain and find its transition probabilities.

(b) Show that if $g \in \mathfrak{G}$ and if (X_n) is a right random walk from the identity, then (gX_n) is a random walk from g.

(c) Let $Y_n = X_n^{-1}$. Show that (Y_n) is a left random walk.

7.11. A transition matrix $\mathbb{P} = (p_{ij})$ is **stochastic**, i.e., $p_{ij} \geq 0$ for all i, j and $\sum_j p_{ij} = 1$ for all i. It is **bi-stochastic** if, in addition, $\sum_i p_{ij} = 1$ for all j, i.e., if both row and column sums equal one. Show that if a Markov chain has a bi-stochastic transition probability matrix, then all of its n-step transition matrices are bi-stochastic.

7.3. Classification of States

We asked, "What do Markov chains do?" Let's start by answering a simpler question: "Where do Markov chains go?"

Definition 7.14. A state j is **accessible** from a state i if there exists $n \geq 0$ such that $P_{ij}^{(n)} > 0$. We write $i \longrightarrow j$.

The states i and j **communicate** if $i \longrightarrow j$ and $j \longrightarrow i$. We write $i \longleftrightarrow j$.

Remark 7.15. If i and j do *not* communicate, then either $P_{ij}^{(n)} = 0$ for all n, or $P_{ji}^{(n)} = 0$ for all n.

Proposition 7.16. *Communication is an equivalence relation.*

Proof. We must show that communication is reflexive, symmetric, and transitive. Take $n = 0$; then $P_{ii}^{(0)} = 1$, which implies $i \longleftrightarrow i$.
(Reflexivity)

$i \longleftrightarrow j \iff j \longleftrightarrow i$ since the definition is symmetric in i and j.
(Symmetry)

$i \longrightarrow j$ and $j \longrightarrow k \implies i \longrightarrow k$. (Transitivity.)

Indeed, $i \longrightarrow j \implies \exists\, m \ni P_{ij}^{(m)} > 0$ and $j \longrightarrow k \implies \exists\, n \ni P_{jk}^{(n)} > 0$. Then by (7.5), $P_{ik}^{(m+n)} = \sum_\ell P_{i\ell}^{(m)} P_{\ell k}^{(n)} \geq P_{ij}^{(m)} P_{jk}^{(n)} > 0$, hence $i \longrightarrow k$. Therefore, "$\longleftrightarrow$" is an equivalence relation. $\square$

Definition 7.17. A set C of states is an **equivalence class** if:

(i) $i, j \in C \Rightarrow i \longleftrightarrow j$,

(ii) $i \in C$ and $i \longleftrightarrow j \Rightarrow j \in C$.

Communication, being an equivalence relation, divides the state space into equivalence classes $C_1, C_2, \ldots$. There may be a finite or infinite number of them. A Markov chain is **irreducible** if it is a single equivalence class, that is, if all states communicate.

If two states are in the same equivalence class, the chain can get from either one to the other. It may be possible for the chain to get from one equivalence class to another, but if it does, it cannot return.

Example 7.17.1. 1. Consider the chain on states 1, 2, 3, 4, and 5 with transition probability matrix

$$\mathbb{P} = \begin{pmatrix} \frac{1}{2} & \frac{1}{2} & 0 & 0 & 0 \\ \frac{1}{4} & \frac{3}{4} & 0 & 0 & 0 \\ 0 & 0 & 0 & 1 & 0 \\ 0 & 0 & \frac{1}{2} & 0 & \frac{1}{2} \\ 0 & 0 & 0 & 1 & 0 \end{pmatrix}.$$

There are two equivalence classes, $C_1 \stackrel{\text{def}}{=} \{1, 2\}$ and $C_2 \stackrel{\text{def}}{=} \{3, 4, 5\}$. It is not possible to get from one to the other, so if the process starts in one of the classes, it stays there forever.

2. Now let

$$\mathbb{P} = \begin{pmatrix} \frac{1}{2} & \frac{1}{2} & 0 & 0 & 0 \\ \frac{1}{4} & 0 & \frac{3}{4} & 0 & 0 \\ 0 & 0 & 0 & 1 & 0 \\ 0 & 0 & \frac{1}{2} & 0 & \frac{1}{2} \\ 0 & 0 & 0 & 1 & 0 \end{pmatrix}.$$

Once again, $C_1 \stackrel{\text{def}}{=} \{1, 2\}$ and $C_2 \stackrel{\text{def}}{=} \{3, 4, 5\}$ are equivalence classes. But now it is possible to go from C_1 to C_2, since state 3 is accessible from state

7.3. Classification of States

2 : $P_{23} = 3/4 > 0$. However, it is not possible to go from states 2, 3 or 4 to either 1 or 2, since all those transition probabilities vanish.

3. Consider uniform motion to the right: if $X_n = i$, then $X_{n+1} = i + 1$, for all i. Then $P_{ii+1} = 1$, $P_{ij} = 0$ if $j \neq i+1$. In this chain, every state is its own equivalence class.

Periodicity.

Definition 7.18. The **period of state i** is the greatest common divisor of the set $\{n \geq 1 : P_{ii}^{(n)} > 0\}$. We denote the period of i by $d(i)$. If $d(i) = 1$, we say that i is **aperiodic**.

If m and n are integers, and m divides n, we write $m \mid n$.

In a simple random walk, the process goes from a state i to either $i+1$ or $i-1$ at each step, and each state has period 2. Indeed, at each step, the process goes between the set of even numbers and the set of odd numbers, so it can only return to its starting place after an even number of steps. And it is possible to go from i to $i+1$ and back in two steps. Thus the g.c.d. of the possible return times is two.

Consider the chain on $\{1, \ldots, n\}$ with transition probability matrix $\mathbb{P}$ given by

$$\mathbb{P} = \begin{pmatrix} 0 & 1 & 0 & 0 & \ldots & 0 \\ 0 & 0 & 1 & 0 & \ldots & 0 \\ 0 & 0 & 0 & 1 & \ldots & 0 \\ \vdots & & & \vdots & & \vdots \\ 0 & 0 & 0 & 0 & \ldots & 1 \\ 1 & 0 & 0 & 0 & \ldots & 0 \end{pmatrix}.$$

In this case the process goes from $1 \longrightarrow 2 \longrightarrow 3 \longrightarrow 4 \ldots \longrightarrow n$ and then back to 1. It returns to its original starting place after n steps, so $d(1) = n$. The same is true for any other state, so $d(i) = n$ for $i = 1, \ldots, n$.

Proposition 7.19. *(i) if $i \longleftrightarrow j$, then $d(i) = d(j)$.*

(ii) If state i has period $d(i)$, then there is an integer N such that

$$P_{ii}^{(nd(i))} > 0 \text{ for all } n \geq N.$$

(iii) $P_{ji}^{(m)} > 0 \implies P_{ji}^{(m+nd(i))} > 0$ for all large enough n.

Proof. If $i \longleftrightarrow j$, then there are m and n, such that $P_{ij}^{(m)} > 0$ and $P_{ji}^{(n)} > 0$. Thus, $P_{ii}^{(m+n)} \geq P_{ij}^{(m)} P_{ji}^{(n)} > 0$. This implies that $d(i) \mid m+n$.

By the same reasoning, if $P_{jj}^{(r)} > 0$ for some r, then $P_{ii}^{(r+m+n)} > 0$, so $d(i) \mid r+m+n$. But we just saw that $d(i) \mid m+n$, so evidently $d(i) \mid r$. Thus, $d(i) \mid d(j)$. By symmetry, $d(j) \mid d(i)$, so they are equal.

(ii) is an exercise in number theory which we leave to the reader, and (iii) is a consequence of (ii). □

7.4. Stopping Times

Consider a gambler playing roulette at a casino. He has decided that he will place one single bet on his lucky number seventeen, but he has not decided when to make it. We have no idea exactly how he will decide when to bet, but we can say a few general things. The first is that he cannot look into the future: certainly, he would like to place a bet on the $n + 1^{\text{st}}$ spin of the wheel only if seventeen will come up then, but he has to bet *before* the wheel is spun, so he cannot know whether or not his number will come up. All he knows when he makes the bet are the results of previous plays, and he has to base his decision on this. Of course, he may have other information, too, but let us ignore that for the sake of discussion.

Let $X_1, X_2, \ldots$ be the results of the plays. The gambler has to base his decision whether or not to bet on the $n + 1^{\text{st}}$ spin on the observed values of $X_1, \ldots, X_n$. That is all the information he has. So his strategy for betting must be this: there is a set $A \subset \mathbb{R}^n$ such that he will bet on the $n + 1^{\text{st}}$ spin if the observed values of the n-tuple $(X_1, \ldots, X_n)$ fall into the set A. Otherwise he will wait till later to bet.

The particular choice of A is the gambler's business, not ours. Our interest is this: it is the most general rule for deciding when to bet.

Let us focus on the time T that he places his bet. Then T is a random variable, and $T = n \iff (X_1, \ldots, X_n) \in A$.

Note that strategies like "bet on seventeen immediately after the zero or double zero appears" and "bet on the fifth turn of the wheel" are of this form.

It is convenient to allow the possibility that the gambler never bets. If he never bets, we say $T = \infty$. Thus the betting time T is characterized by the following:

1° T is a random variable taking values in $\{1, 2, \ldots, \infty\}$;

2° for each n there is $A_n \subset \mathbb{R}^n$ such that $T = n \iff (X_1, \ldots, X_n) \in A_n$.

We can streamline this criterion by introducing σ-fields. Let $\mathcal{F}_n$ be the class of all events of the form $\{\omega : (X_1(\omega), \ldots, X_n(\omega)) \in B\}$, where B is a Borel set in $\mathbb{R}^n$. (That is, $\mathcal{F}_n = \sigma\{X_1, \ldots, X_n\}$.)

Then T, the time of the bet, is random and satisfies:

- T takes values in $\{1, 2, \ldots, \infty\}$ and
- for each n, $\{T = n\} \in \mathcal{F}_n$.

7.4. Stopping Times

Remark 7.20. $\mathcal{F}_n$ is a σ-field and for each n, $\mathcal{F}_n \subset \mathcal{F}_{n+1}$.

Indeed, we already know that $\mathcal{F}_n$ is a σ-field, and that if $\Lambda \in \mathcal{F}_n$, then there is a Borel set A for which $\Lambda = \{(X_1, \ldots, X_n) \in A\}$. We can also write Λ in terms of $(X_1, X_2, \ldots, X_{n+1})$, for $\Lambda = \{(X_1, \ldots, X_n, X_{n+1}) \in A \times \mathbb{R}\}$, which is in $\mathcal{F}_{n+1}$. Therefore, $\mathcal{F}_n \subset \mathcal{F}_{n+1}$.

Such times are called *stopping times*[5].

The gambler may also have extra information. Recalling that σ-fields are information in the raw, we can take this into account by adding the extra information to the σ-fields $\mathcal{F}_n$. Thus $\mathcal{F}_n$ may be larger than $\sigma\{X_1, \ldots, X_n\}$.

Definition 7.21. Let $(\Omega, \mathcal{F}, P)$ be a probability space. Let $\mathcal{F}_0 \subset \mathcal{F}_1 \subset \mathcal{F}_2 \subset \ldots$ be a sequence of sub-sigma-fields of $\mathcal{F}$. A random variable T with values in $0, 1, 2, \ldots, \infty$ is a **stopping time** if for every $n = 0, 1, 2, \ldots$, the set $\{T = n\} \in \mathcal{F}_n$.

Here is an equivalent definition which is often used.

Proposition 7.22. *A random variable with values in $0, 1, 2, \ldots, \infty$ is a stopping time relative to the sigma-fields $\mathcal{F}_0 \subset \mathcal{F}_1 \subset \mathcal{F}_2 \subset \ldots$ if and only if $\{T \leq n\} \in \mathcal{F}_n$ for $n = 0, 1, 2, \ldots$.*

Proof. If T is a stopping time, then $\{T \leq n\} = \bigcup_{m=0}^{n} \{T = m\}$. If $m \leq n$, then $\{T = m\} \in \mathcal{F}_m \subset \mathcal{F}_n$. Thus $\{T \leq n\}$ is a finite union of elements of the σ-field $\mathcal{F}_n$, and is therefore in $\mathcal{F}_n$ itself. Conversely, if $\{T \leq n\} \in \mathcal{F}_n$ for all n, then $\{T = n\} = \{T \leq n\} - \{T \leq n = 1\}$. By hypothesis, $\{T \leq n\} \in \mathcal{F}_n$ and $\{T \leq n - 1\} \in \mathcal{F}_{n-1} \subset \mathcal{F}_n$, so the difference, $\{T = n\} \in \mathcal{F}_n$, and T is a stopping time. $\square$

Stopping times are important: they are times when interesting things can happen. And they will appear frequently in the sequel. Let us say more about the intuition behind them.

In the example above, the stopping time is a time at which the gambler can make a decision. The important thing is that the gambler can recognize the time when it arrives. That is exactly what the condition "$\{T = n\} \in \mathcal{F}_n$" means. For the gambler's total knowledge at time n is $\mathcal{F}_n$, and, since $\{T = n\}$ (and $\{T \neq n\}$ as well) are part of that knowledge, he knows whether or not T arrives then.

---

[5] These were introduced by A. A. Markov for Markov chains, and they are often called Markov times. They came up in more general settings, e.g., to show that gambling systems do not work and in particular, to answer the question of whether a gambler could improve the odds by opting to stop playing the game at a certain time, hence the name "stopping time". They are also called "optional times".

So, intuitively, a stopping time is simply a (possibly random) time that can be recognized when it arrives. It is not necessary to wait, or to look into the future to divine it. We can recognize it in real time.

In fact, stopping times are so pervasive that even times which seem to be fixed are actually random. For instance, the time of my 21st birthday—which seemed important at the time—depends on the date of my birth, which is definitely a random time. (A glance at the biology of reproduction makes it clear how frighteningly random that time was[6].)

Exercise 7.12. Show that if T is a stopping time, then for each n, the sets $\{T = n\}$, $\{T < n\}$, $\{T \geq n\}$ and $\{T > n\}$ are in $\mathcal{F}_n$.

Example 7.22.1. (i) The constant random variable $T \equiv n$ is a stopping time.

Indeed $\{T \leq m\}$ is either Ω or $\emptyset$, depending on whether $n \leq m$ or not. Both are in $\mathcal{F}_m$.

(ii) Let $\mathcal{F}_n = \sigma(X_0, \ldots, X_n)$, $n = 0, 1, 2, \ldots$, and let B be a Borel subset of the state space. Let $T_B = \inf\{n \geq 0 : X_n \in B\}$, where we make the convention that the infimum of the empty set is ∞. Then T_B is a stopping time, called the **first hitting time of B**.

For, $\{T_B = n\} = \{X_0 \in B^c, \ldots, X_{n-1} \in B^c, X_n \in B\} \in \mathcal{F}_n$.

(iii) If T is a stopping time and $m \geq 0$ is an integer, then $T + m$ is a stopping time.

Indeed, if $n \geq m$, $\{T + m = n\} = \{T = n - m\} \in \mathcal{F}_{n-m} \subset \mathcal{F}_n$, because the σ-fields increase. And if $n < m$, $\{T + m = n\} = \emptyset$, which is also in $\mathcal{F}_n$.

(iv) If S and T are stopping times, so are $S \wedge T = \min\{S, T\}$ and $S \vee T = \max\{S, T\}$. In particular, if m is a positive integer, $T \wedge m$ and $T \vee n$ are stopping times.

To see this, note that $\{S \wedge T \leq n\} = \{S \leq n\} \cup \{T \leq n\}$ and $S \vee T \leq n\} = \{S \leq n\} \cap \{T \leq n\}$. By Exercise 7.12, all four of these sets are in $\mathcal{F}_n$, hence so are $\{S \wedge T \leq n\}$ and $\{S \vee T = n\}$.

We will need to define the process at a stopping time, or, more generally, at any random time.

Definition 7.23. Let R be a random variable taking values in $0, 1, 2 \ldots$. Then X_R is the random variable defined by

$$X_R(\omega) = X_{R(\omega)}(\omega).$$

---

[6] Even the dates of the Gregorian calendar are random: biology aside, the official date of Christ's birth was not decided until 525 A.D., and was based on the best estimate they could make at the time. Given the incomplete and occasionally contradictory historical records available then, it could easily be off by several years. This raises an interesting point: 1 A.D. was not a stopping time. 526 A.D. was.

7.4. Stopping Times

That is, X_R is defined by

$$X_R(\omega) = X_n(\omega) \quad \text{if } \omega \in \{R = n\}.$$

There is something to verify about this definition, namely that X_R actually is a random variable, i.e., that for any Borel set A, $\{X_R \in A\} \in \mathcal{F}$. But this is clear since $\{X_R \in A\} = \bigcup_{n=0}^{\infty} \{R = n\} \cap \{X_n \in A\}$, and both $\{R = n\}$ and $\{X_n \in A\}$ are in $\mathcal{F}$, hence so is their intersection.

Note: In particular, we will want to use X_T for a stopping time T, but there is a potential problem: we allow infinite values for stopping times, but X_T is not defined on the set $\{T = \infty\}$. However, we will be careful to confine ourselves to sets on which T is finite. (Notice that if $X_T = i$, then T must be finite[7].)

We have just argued that stopping times are really like fixed times, even though they are random. The σ-field $\mathcal{F}_n$ represents the information available—that is, everything we know—up to the present time n. We need to extend this idea to stopping times. That is, we need a σ-field to represent the history of the process up to a stopping time T. We are tempted to say that it is all the events determined by $X_0, \ldots, X_T$, but... T is random, so the number of elements in the sequence is not fixed. However, if $T = n$, the sequence is $X_0, \ldots, X_n$, and so the event is determined by $X_0, \ldots, X_n$. This leads us to define the past before T as follows.

Definition 7.24. Let T be a stopping time relative to the σ-fields $\mathcal{F}_0 \subset \mathcal{F}_1 \subset \mathcal{F}_2 \subset \ldots$. We define $\boldsymbol{\mathcal{F}_T}$ by

$$\mathcal{F}_T \stackrel{\text{def}}{=} \{\Lambda \in \mathcal{F} : \Lambda \cap \{T \leq n\} \in \mathcal{F}_n, \ n = 0, 1, 2, \ldots\}.$$

We call this "the past before T".

Remark 7.25. We can replace "$\{T \leq n\}$" by "$\{T = n\}$" in the definition; that is,

$$\mathcal{F}_T \stackrel{\text{def}}{=} \{\Lambda \in \mathcal{F} : \Lambda \cap \{T = n\} \in \mathcal{F}_n, \ n = 0, 1, 2, \ldots\}.$$

The equivalence of these two characterizations of $\mathcal{F}_T$ follows from the equations

$$\begin{cases} \Lambda \cap \{T \leq n\} = \bigcup_{j=0}^{n} \Lambda \cap \{T = j\}, \\ \Lambda \cap \{T = n\} = \Lambda \cap \{T \leq n\} - \Lambda \cap \{T \leq n-1\}. \end{cases}$$

---

[7]If we want to be formal, we can take a point ∂ which is *not* in the state space, and arbitrarily define $X_\infty = \partial$. With the obvious extension of the transition probabilities (e.g., set $P_{\partial\partial} = 1$), X becomes a Markov chain on $\mathbb{N} \cup \{\partial\}$, and X_T is well-defined even on $\{T = \infty\}$.

This gives us a choice of conditions to verify when we want to show[8] something is in $\mathcal{F}_T$.

The following proposition collects some elementary properties of $\mathcal{F}_T$.

Proposition 7.26. *Let S and T be stopping times. Then:*

(i) $\mathcal{F}_T$ is a σ-field.

(ii) Both T and $X_T I_{\{T<\infty\}}$ are $\mathcal{F}_T$-measurable.

(iii) If $S \leq T$, then $\mathcal{F}_S \subset \mathcal{F}_T$.

Proof. We leave (*i*) for the exercises.

(*ii*) Note that X_T is not defined on $\{T = \infty\}$, which is why we restrict ourselves to the set where T is finite. It is enough to prove that $\{X_T = j\} \in \mathcal{F}_T$. To see this, just test it with $\{T = n\}$: for $n = 0, 1, 2,, \ldots$

$$\{X_T = j\} \cap \{T = n\} = \underbrace{\{X_n = j\}}_{\in \mathcal{F}_n} \cap \underbrace{\{T = n\}}_{\in \mathcal{F}_n} \in \mathcal{F}_n.$$

To see T is $\mathcal{F}_T$-measurable, it is enough to show $\{T \leq m\} \in \mathcal{F}_T$ for all m. But for $m, n = 0, 1, 2, \ldots$

$$\{T \leq m\} \cap \{T \leq n\} = \{T \leq m \wedge n\} \in \mathcal{F}_{m \wedge n} \subset \mathcal{F}_n,$$

so $\{T \leq m\}$ is indeed in $\mathcal{F}_T\}$.

(*iii*) Let $\Lambda \in \mathcal{F}_S$. To show $\Lambda \in \mathcal{F}_T$, test it with $\{T \leq n\}$. As $S \leq T$,

$$\Lambda \cap \{T \leq n\} = \underbrace{\Lambda \cap \{S \leq n\}}_{\in \mathcal{F}_n \text{ (def of } \mathcal{F}_S)} \cap \underbrace{\{T \leq n\}}_{\in \mathcal{F}_n} \in \mathcal{F}_n,$$

for $n = 0, 1, 2, \ldots$, hence $\Lambda \in \mathcal{F}_T$. □

7.5. The Strong Markov Property

The Markov property of a Markov chain $X_0, X_1, \ldots$ can be expressed: "For each n, the process $X_n, X_{n+1}, \ldots$ is a Markov chain with the same transition probabilities as $X_0, X_1, \ldots$; moreover, given X_n, it is conditionally independent of $X_0, \ldots, X_n$."

It is remarkable that this remains true if we replace the fixed time n by any finite stopping time T whatsoever.

Let $\{X_n,\ n = 0, 1, 2, \ldots\}$ be a Markov chain with transition probability matrix $\mathbb{P} = (P_{ij})$, and let $\mathcal{F}_n = \sigma(X_0, \ldots, X_n)$. We can rephrase the Markov property in terms of $\mathcal{F}_n$: for $n \geq 1$, $\Lambda \in \mathcal{F}_n$, and states i, j, (7.1) becomes:

(7.9) $\qquad \Lambda \in \mathcal{F}_n \implies P\{X_{n+1} = j \mid X_n = i, \Lambda\} = P_{ij}\,,$

---

[8]This is a minor point in the present setting. It will be more important when we look at continuous parameter processes, where the situation is more delicate.

7.5. The Strong Markov Property

or, equivalently,

(7.10) $$P\{\Lambda;\, X_n = i, X_{n+1} = j\} = P\{\Lambda;\, X_n = i\} P_{ij}\,.$$

Theorem 7.27 (Strong Markov Property, discrete case[9]). *Let T be a finite stopping time and i a state. Given that $X_T = i$, the process $\{X_{T+n},\, n = 0, 1, 2, \dots\}$ is a Markov chain with transition probability matrix $\mathbb{P}$, which is conditionally independent of the past before T.*

Proof. Let S be a finite stopping time and let $\Lambda \in \mathcal{F}_S$. Then for states i and j,

$$P\{\Lambda;\, X_S = i, X_{S+1} = j\} = \sum_{k=0}^{\infty} P\{\Lambda \cap \{S = k\};\, X_k = i, X_{k+1} = j\}\,.$$

Now $\Lambda \in \mathcal{F}_S$, so $\Lambda \cap \{S = k\} \in \mathcal{F}_k$. By (7.10), this is

$$= \sum_{k=1}^{\infty} P\{\Lambda \cap \{S = k\};\, X_k = i\} P_{ij}\,.$$

Factor out P_{ij} and note the sum equals $P\{\Lambda;\, X_S = i\}$, so that

(7.11) $$P\{\Lambda;\, X_S = i, X_{S+1} = j\} = P\{\Lambda;\, X_S = i\} P_{ij}\,.$$

Divide both sides by $P\{\Lambda;\, X_S = i\}$ to get

(7.12) $$P\{X_{S+1} = j \mid X_S = i;\, \Lambda\} = P_{ij}\,.$$

This is the main step. To see that the post-T process is a Markov chain, just take $S = T + n$ and $\Lambda = \{X_T = i_0, \dots, X_{T+n} = i_n\}$. Since S is a stopping time and $\Lambda \in \mathcal{F}_S$, (7.12) says that

$$P\{X_{T+n+1} = i_{n+1} \mid X_{T+n} = i_n, \dots, X_T = i_0\} = P_{i_0 i_1}\,.$$

This implies that (X_{T_n}) is a Markov chain with transition matrix $\mathbb{P}$.

To see that the post-T chain is conditionally independent of $\mathcal{F}_T$, it is enough to show the finite-dimensional distributions are. Let $\Lambda \in \mathcal{F}_T$, put $\Gamma_0 = \{X_T = i_0\}$, ..., $\Gamma_n = \{X_T = i_0, \dots, X_{T+n} = i_n\}$. Now for each m,

---

[9] It is not hard to work around the strong Markov property in the discrete parameter case, but it is simply too useful and too intuitive to ignore. Moreover, it is powerful and decidedly non-trivial in the continuous parameter case—see Blumenthal's zero-one law, for instance—so we will introduce it now; then we will be familiar with it when we really need it.

$\Lambda \cap \Gamma_{m+1} = \Lambda \cap \Gamma_m \cap \{X_{m+1} = i_{m+1}\}$, so we can apply (7.10) repeatedly:

$$P\{\Lambda \cap \Gamma_n\} = P\{\Lambda \cap \Gamma_{n-1}; X_{T+n-1} = i_{n-1}, X_{T+n} = i_n\}$$
$$= P\{\Lambda \cap \Gamma_{n-1}; X_{T+n-1} = i_{n-1},\} P_{i_{n-1} i_n}$$
$$= P\{\Lambda \cap \Gamma_{n-2}; X_{T+n-2} = i_{n-2},\} P_{i_{n-2} i_{n-1}} P_{i_{n-1} i_n}$$
$$\vdots$$
$$= P\{\Lambda; X_T = i_0\} P_{i_0 i_1} \ldots P_{i_{n-1} i_n}.$$

Divide both sides by $P\{X_T = i_0\}$ to see that

$$\frac{P\{\Lambda \cap \Gamma_n\}}{P\{X_T = i_0\}} = \frac{P\{\Lambda; X_T = i_0\}}{P\{X_T = i_0\}} \frac{P\{X_T = i_0\} P_{i_0 i_1} \ldots P_{i_{n-1} i_n}}{P_{i_0 i_1} \ldots P_{i_{n-1} i_n}},$$

or, as $P\{X_T = i_0\} P_{i_0 i_1} \ldots P_{i_{n-1} i_n} = P\{\Gamma_n\}$,

$$P\{\Lambda \cap \Gamma_n \mid X_T = i_0\} = P\{\Lambda \mid X_T = i_0\} P\{\Gamma_n \mid X_T = i_0\}.$$

This holds for each choice of n, $i_0, \ldots, i_n$, and $\Lambda \in \mathcal{F}_t$; and it implies that the post-T process is conditionally independent of $\mathcal{F}_T$ given X_T. □

It may not be clear in this setting whether the strong Markov property is deep or whether it is just a triviality tricked out in fancy mathematical trappings. In fact, it is somewhere between: the heart of the proof is the derivation of (7.11), which is only a few lines, and, until one actually tries to prove it, it seems so intuitively obvious that many use it without even noticing. However, intuitive as it may seem, it is subtle. For instance, here is a consequence to make one think.

Corollary 7.28 (Mark I). *Gambling systems don't work.*

This statement is melodrama, not mathematics, but it has a point. By "gambling system" we mean a strategy for playing a game like roulette, which offers a series of i.i.d. plays, and which may involve changing the size of the bets or not betting at all, depending perhaps on the previous results. The point is that any strategy must involve stopping times, and the strong Markov property tells us that stopping times do not change the transition probabilities. Let us state this more rigorously.

Corollary 7.29 (Mark II). *Let $X_0, X_1, \ldots$ be i.i.d. random variables with a discrete distribution. Let T be a finite stopping time. Then X_{T+1} has the same distribution as X_0, independent of $\mathcal{F}_T$.*

Proof. $X_0, X_1, \ldots$ is a Markov chain with transition probabilities $P_{ij} = P\{X_{n+1} = x_j \mid X_n = x_i\} = P\{X_{n+1} = x_j\} \stackrel{\text{def}}{=} p_j$. So P_{ij} does not depend on i. By the strong Markov property, $X_T, X_{T+1}, \ldots$ is again a Markov chain

with the same transition probabilities, so that if $\Lambda \in \mathcal{F}_T$,
$$P\{X_{T+1} = x_j \mid \Lambda, X_T = x_i\} = P_{ij} = p_j = P\{X_0 = x_j\}.$$
Thus X_{T+1} has the same distribution as X_0, regardless of Λ and the value of X_T. Therefore X_{T+1} is independent of $\Lambda \cap \{X_T = i\}$ for all i, and therefore of Λ. But Λ is an arbitrary element of $\mathcal{F}_T$, hence X_{T+1} is independent of $\mathcal{F}_T$, and has the same distribution as X_0. □

We will take up the question of gambling systems again when we study martingales in Chapter 9.

Problems 7.5

7.13. Let $X_0, X_1, X_2, \ldots$ be a simple symmetric random walk with $X_0 = 0$. Let $T_n = \inf\{k : X_k = n\}$. Show that $T_1, T_2 - T_1, T_3 - T_2, \ldots$ are i.i.d. random variables.

7.14. Express $T_{A \cup B}$ in terms of T_A and T_B. What about $T_{A \cap B}$?

7.15. Let S and T be stopping times. Show that
  (a) $\mathcal{F}_T$ is a σ-field.
  (b) $\Gamma \in \mathcal{F}_S \Longrightarrow \Gamma \cap \{S \leq T\} \in \mathcal{F}_T$.
  (c) $\{S \leq T\} \in \mathcal{F}_S \cap \mathcal{F}_T$.

7.16. Show that the time of the third entry into a set B is a stopping time. Do the same for the nth entry.

7.17. Let (X_n) be a Markov chain and let A be a subset of its state space. Let $S_1 = \inf\{j \geq 1 : X_j \in A\}$, and, for $n \geq 2$, define S_n by induction: $S_n = \inf\{j > S_{n-1} : X_j \in A\}$. Assume $S_n < \infty$ for all n. Let $Y_n = X_{S_n}$, $n = 1, 2, \ldots$. Show that Y_n is a Markov chain with state space A.

7.18. Let T be a stopping time for a Markov chain (X_n), and put $Y_n = X_{T \wedge n}$. Show that (Y_n) is a Markov chain if and only if $T = T_A$ for some subset A of the state space.

7.19. A coin with probability p of heads is flipped successively. Find the distribution of the number of flips it takes for the pattern "HT" to appear. Do the same for HHT.

7.6. Recurrence and Transience

We will learn more in this section about what a Markov chain does. The key question is: does the process ever return to its initial state?

Notation. Let P^i and E^i denote the probabilities and expectations respectively when $P\{X_0 = i\} = 1$. So, for example, $P^i\{\Lambda\} = P\{\Lambda \mid X_0 = i\}$ and $E^i\{Y\} = E\{Y \mid X_0 = i\}$.

Definition 7.30. The **first-hitting time of i** is $T_i = \inf\{n \geq 1 : X_n = i\}$.

If $X_0 \neq i$, then T_i is the first time the chain reaches i, but if $X_0 = i$, it is the first time the chain *returns* to i. In either case it is a stopping time.

(It is important that the inf in the definition is over strictly positive n.) We use the convention that the inf of the empty set is plus infinity, so that if the chain never hits i, $T_i = \infty$; but if $T_i < \infty$, then $X_{T_i} = i$. Accessibility has an intuitive interpretation in terms of the T_i.

Exercise 7.20. Let $i \neq j$. Show that $i \longrightarrow j \iff P\{T_j < \infty \mid X_0 = i\} > 0$, i.e., j is accessible from i iff it is possible for the chain to get from i to j.

Definition 7.31. A state i is **recurrent** if $P\{T_i < \infty \mid X_0 = i\} = 1$. It is **transient** otherwise.

If the process starts at i, its first return is at time T_i. It may return again several times. Let T_i^n be the time of the nth return to i, which can be defined by induction: if $T_i^1, \ldots, T_i^n$ have been defined, set

$$T_i^{n+1} = \inf\{k > T_i^n : X_k = i\}.$$

Definition 7.32. The **total number of returns to** i is

$$N_i \stackrel{\text{def}}{=} \#\{n : T_i^n < \infty\} = \#\{n \geq 1 : X_n = i\}.$$

Let $p = P^i\{T_i < \infty\}$.

Proposition 7.33. *For* $n = 0, 1, 2, \ldots$, $P^i\{N_i \geq n\} = p^n$. *In particular, if* $X_0 = i$ *and* i *is recurrent,* $N_i = \infty$ *a.s., and if* i *is transient, then* N_i *is a geometric random variable with parameter* p*, and*

(7.13) $$E^i\{N_i\} = \frac{p}{1-p}.$$

Proof. Let $p^{(n)} = P^i\{N_i \geq n\}$. Then $p^{(1)} = p$ by definition. Suppose $p^{(k)} = p^k$ for $k = 1, 2, \ldots, n$. We claim this also holds for $n+1$. Note that $N_i \geq n \iff T_i^n < \infty$. Now if $T_i^n < \infty$, $X_{T_i^n} = i$, so the post-T_i^n chain starts from i. Apply the Markov property at T_i^n: the post-T_i^n process is again a Markov chain from i and it has the same distribution as the original process. In particular, the probability that it returns to i is p. But if it returns, it does so at T_i^{n+1}, which means that $T_i^{n+1} < \infty$ and therefore that $N_i \geq n+1$. Since the post-T_i^n process is (conditionally) independent of $\mathcal{F}_{T_i^n}$ we have $P^i\{N_i \geq n+1\} = P^i\{N_i \geq n\}p$, or $p^{(n+1)} = p^{(n)}p$. Thus by the induction hypothesis, $p^{(n+1)} = p^{n+1}$, as claimed. □

Remark 7.34. (*i*) Note that if the chain starts at a recurrent state, it returns to that state infinitely often.

(*ii*) With the obvious interpretation in case $p = 1$, (7.13) holds for all states, and implies that

(7.14) $$\begin{cases} i \text{ is transient} \iff E^i\{N_i\} < \infty, \\ i \text{ is recurrent} \iff E^i\{N_i\} = \infty. \end{cases}$$

7.6. Recurrence and Transience

If we express this in terms of the transition probabilities, we get a useful criterion.

Theorem 7.35. *A state i is recurrent iff $\sum_n P_{ii}^{(n)} = \infty$, transient iff $\sum_n P_{ii}^{(n)} < \infty$.*

Proof. Note that $N_i = \sum_n I_{\{X_n=i\}}$. If $X_0 = i$ a.s., then $E^i\{N_i\} = E^i\{\sum_n I_{\{X_n=i\}}\} = \sum_n E^i\{I_{\{X_n=i\}}\} = \sum_n P_{ii}^{(n)}$, since $P_{ii}^{(n)} = P^i\{X_n = i\}$. The result now follows from (7.14). □

We can now link recurrence and accessibility.

Corollary 7.36. *If $i \longrightarrow j$ and i is recurrent, then $i \longleftrightarrow j$, $P^j\{T_i < \infty\} = 1$ and j is recurrent.*

Proof. Let $X_0 = i$. Since i is recurrent, $N_i = \infty$ a.s., and $i \longrightarrow j$, so $P^i\{T_j < \infty\} > 0$. But in order for N_i to be infinite, the chain *must* return from j to i. Apply the strong Markov property at T_j: $X_{T_j} = j$, so the post-T_j chain has the same distribution as the original chain from j. But the post-T_j chain is sure to reach i, so evidently the original chain from j is also sure to reach it, i.e. $P^j\{T_i < \infty\} = 1$. In particular $i \longleftrightarrow j$.

Thus there are r and s such that $P_{ij}^r > 0$ and $P_{ji}^s > 0$. Then $P_{jj}^{r+n+s} \geq P_{ji}^s P_{ii}^n P_{ij}^r$ for any n. (Indeed, P_{jj}^{r+n+s} is the probability of returning to j in $r+n+s$ steps, while $P_{ji}^s P_{ii}^n P_{ij}^r$ is the probability of doing this by going from j to i in s steps, then going from i to i in the next n steps, and then going back to j in the following r steps.) Thus

$$\sum_n P_{jj}^n \geq \sum_n P_{jj}^{r+n+s} \geq P_{ij}^r P_{ji}^s \sum_n P_{ii}^n.$$

But $P_{ij}^r P_{ji}^s > 0$ and, as i is recurrent, the sum of the P_{ii}^n is infinite. Therefore, so is the sum of the P_{jj}^n, so j is recurrent. □

Remark 7.37. Recurrence and transience are class properties. That is, let C be an equivalence class of states. According to Corollary 7.36, either

(*i*) all states in C are recurrent, in which case it is called a **recurrence class**, or

(*ii*) all states in C are transient, in which case it is a **transience class**.

We asked what a Markov chain does. According to the following, it keeps quite busy.

Theorem 7.38. *Let C be a recurrence class. If $X_0 \in C$, then with probability one X visits every state in C infinitely often. That is,*

$$P\{N_i = \infty, \, \forall i \in C \mid X_0 \in C\} = 1.$$

Proof. Let $i \in C$. By Proposition 7.33, $N_i = \infty$ a.s. on $\{X_0 = i\}$. Moreover, if $j \in C$, Corollary 7.36 implies that $T_j < \infty$ a.s. on $\{X_0 = i\}$. By the strong Markov property, the post-T_j chain has the same distribution as the chain starting at j. Therefore, Proposition 7.33 applies to it as well, and $P^i\{N_j = \infty\} = P^j\{N_j = \infty\} = 1$. This is true for all $j \in C$, so
$$P^i\{\exists j \in C \ni N_j < \infty\} \le \sum_{j \in C} P^j\{N_j < \infty\} = 0,$$
because each summand vanishes. □

Stirling's Formula. Stirling's formula gives a useful approximation to the factorial. Let us sneak up on it by first making a rough guess as to its size. Now $n!$ is the product of the first n integers. We might guess that it is approximately equal to the average, $n/2$, raised to the nth power. That will not be quite correct, so let's put in a correction factor a_n and set
$$n! = a_n \left(\frac{n}{2}\right)^n.$$
Then
$$\frac{a_n}{a_{n-1}} = \frac{n!\left(\frac{n-1}{2}\right)^{n-1}}{(n-1)!\left(\frac{n}{2}\right)^n} = 2\left(\frac{n-1}{n}\right)^{n-1} = 2\left(1 - \frac{1}{n}\right)^{n-1}.$$
This tends to $2e^{-1}$ as $n \to \infty$. Thus a_n should be near $(2/e)^n$. This leads us to try $a_n = 2^n e^{-n} b_n$, or $n! = b_n n^n e^{-n}$ for some b_n. Then
$$\frac{b_n}{b_{n-1}} = \left(1 - \frac{1}{n}\right)^{n-1} e.$$
Take logs and use Taylor's expansion: $\log(1-1/n) = -1/n - 1/2n^2 - 1/3n^3 - \cdots$
$$\log \frac{b_n}{b_{n-1}} = (n-1)\log\left(1 - \frac{1}{n}\right) + 1$$
$$= 1 - 1 - \frac{1}{2n} - \frac{1}{3n^2} - \cdots + \frac{1}{n} + \frac{1}{2n^2} + \cdots.$$
Thus, $\log b_n - \log b_{n-1} = \frac{1}{2n} + \frac{1}{6n^2} + \cdots$. If we just take the first term, this suggests that $\log b_n \sim \sum_1^n 1/(2j) \sim (1/2)\log n$, or $b_n \sim \sqrt{n}$. For the next stage, then, we try $n! = c_n n^{n+1/2} e^{-n}$ for some c_n. We do the same thing with c_n:
$$\frac{c_n}{c_{n-1}} = e\left(1 - \frac{1}{n}\right)^{n-\frac{1}{2}},$$
leading to
$$\log c_n - \log c_{n-1} = -\frac{1}{12n^2}.$$
The c_n decrease, say $c_n \downarrow c > 0$, and this suggests that $\log c_n \sim \log c - \sum_{j=1}^n \frac{1}{12n^2} \sim \log c + \frac{1}{12n}$. This leads to the approximation $n! \sim$

7.6. Recurrence and Transience

$cn^{n+1/2}e^{-n+1/12n}$ for some constant c. It takes some further analysis to find that $c = \sqrt{2\pi}$. This gives Stirling's approximation:

Theorem 7.39 (Stirling's Formula). $n! = \sqrt{2\pi} n^{n+\frac{1}{2}} e^{-n+\frac{\theta_n}{12n}}$, where $0 < \theta_n < 1$.

It is possible to get further terms in the asymptotic expansion, but the approximation $n! \sim \sqrt{2\pi} n^{n+1/2} e^{-n}$ is already quite good for large n. One can see from the error term that the error is already within 10% for $n = 5$, and less than 1% for $n = 10$.

7.6.1. Examples: Random Walks. One Dimension. Let $\{X_n,\, n = 1, 2, \dots\}$ be a simple random walk on the line: if $X_n = i$, then in the next step, X_{n+1} equals either $i+1$ or $i-1$ with probabilities p and $q = 1-p$, respectively, so $P_{ii+1} = p$, $P_{ii-1} = q$. All states have period 2, and, clearly all states communicate, so the random walk is irreducible. We would like to know if it is recurrent or transient. Since all states communicate, it is enough to consider the origin. We will use Theorem 7.35 to determine the recurrence/transience of X.

If $X_0 = 0$, then $P_{00}^n = 0$ for odd values of n, since the random walk can only return to the origin on even n. In order to return in $2n$ steps, it must take exactly n steps to the right, n to the left. The number of ways to place n steps to the right in $2n$ places is $\binom{2n}{n}$, and each particular order has probability $p^n q^n$. Use Stirling's formula in the form $n! \sim \sqrt{2\pi} n^{n+1/2} e^{-n}$:

$$P_{00}^{2n} = \binom{2n}{n}(pq)^n = \frac{(2n)!}{n!n!}(pq)^n \sim \frac{(4pq)^n}{\sqrt{\pi n}}.$$

Now $pq \leq 1/4$ and $pq = 1/4$ only if $p = q = 1/2$. If $p \neq q$, $\sum_{n=1}^{\infty} P_{00}^{2n} \leq \sum_{n=1}^{\infty}(4pq)^n < \infty$. If $p = 1/2$, then $\sum_{n=1}^{\infty} P_{00}^{2n} \sim \sum_n \frac{1}{\sqrt{n\pi}} = \infty$. Thus the random walk is transient if $p \neq 1/2$ and recurrent if $p = 1/2$.

Two Dimensions. Consider a simple symmetric random walk on the integer lattice in the plane. The walk moves from a point (i,j) to one of its four nearest neighbors $(i \pm 1, j)$ $(i, j \pm 1)$ with probability 1/4 each. The chain is irreducible, so we need only check the origin for recurrence. All states have period 2, so again $P_{00}^{2n+1} = 0$. In order to return in $2n$ steps, the random walk must take the same number of steps up and down, and the same number to the left and to the right. Therefore, it can take j steps to the left, j to the right, k steps up and k steps down, where $j + k = n$. By the multinomial formula, there are $\frac{(2n)!}{j!j!k!k!}$ ways to choose them, and each has probability $1/4^{2n}$, so that

$$P_{00}^{2n} = \sum_{j+k=n} \frac{(2n)!}{j!j!k!k!}\left(\frac{1}{4}\right)^{2n}.$$

We need to work on this before applying Stirling's formula. Multiply and divide by $(n!)^2$:

$$= \frac{(2n)!}{n!n!} \sum_{j+k=n} \frac{n!}{j!k!} \frac{n!}{j!k!} \frac{1}{4^{2n}}$$

$$= \binom{2n}{n} \frac{1}{4^{2n}} \sum_{k=0}^{n} \binom{n}{k}\binom{n}{n-k}.$$

Interpret the sum as the number of ways to choose n from $2n$ by first choosing k from the first n, and $n-k$ from the second n. This is $\binom{2n}{n}$, so

$$P_{00}^{2n} = \binom{2n}{n}^2 \frac{1}{4^{2n}} = \frac{((2n)!)^2}{(n!)^4}.$$

By Stirling's formula this is

$$\sim \frac{2^{4n}}{\pi n} \frac{1}{4^{2n}} = \frac{1}{\pi n}.$$

Therefore, $\sum_n P_{00}^{2n} \sim \sum_n \frac{1}{\pi n} = \infty$, and the random walk is recurrent.

Three Dimensions. The symmetric random walk on the three-dimensional integer lattice moves from a point (i,j,k) to one of the six nearest neighbors $(i \pm 1, j, k)$, $(i, j \pm 1, k)$ and $(i, j, k \pm 1)$ with probability $1/6$ each. Again, the chain is irreducible and has period two. It can return to the origin in $2n$ steps by taking i left, i right, j forward, j backward, k up and k down, where $i+j+k = n$. The number of ways of doing this is $\frac{(2n)!}{i!i!j!j!k!k!}$, so

$$P_{00}^{2n} = \sum_{i+j+k=n} \frac{(2n)!}{i!i!j!j!k!k!} \frac{1}{6^{2n}}.$$

Factor out $(2n)!/2^{2n}$ and multiply and divide by $n!^2$ to see that this is

$$= \frac{1}{2^{2n}} \binom{2n}{n} \sum_{i+j+k=n} \left\{ \frac{(n)!}{i!j!k!3^n} \right\}^2.$$

We recognize the term inside the braces as a probability from a trinomial distribution: it is the probability that, for instance, n equally likely moves are divided into i left, j forward, and k up. Thus the probabilities sum to one. Therefore the sum of their squares is less than or equal to the largest single term. (For any probabilities p_n, $\sum_n p_n^2 \leq \max p_n$.) But the largest single term comes when i, j and k are equal, or nearly so. Thus, we apply Stirling's formula with $i = j = k = n/3$ to see that the maximum term is $\sim 3^{3/2}/(2\pi n)$. We saw above that $2^{2n}\binom{2n}{n} \sim 1/\sqrt{\pi n}$. Thus,

$$P_{00}^{2n} \sim \frac{1}{2}\left(\frac{3}{\pi n}\right)^{\frac{3}{2}}.$$

7.6. Recurrence and Transience

But this is summable: $\sum_n P_{00}^{2n} < \infty$. Therefore the random walk is transient. It is not hard to see that random walks in dimensions higher than three are also transient. Intuitively, in order to return to the origin, all coordinates of the random walk must simultaneously return to the origin. Since the random walk in three dimensions is transient, the first three coordinates simultaneously return to zero only finitely often, and adding more coordinates only makes the get-together harder.

To summarize:

Theorem 7.40. *The simple symmetric random walk is recurrent in dimensions one and two, and transient in dimensions three and higher. Non-symmetric simple random walks are transient in all dimensions.*

Problems 7.6

7.21. Discuss the following statement: "The entire transition probability matrix of a Markov chain can be recovered by the observation of a single sample path."

7.22. Show that if C and D are equivalence classes, then the chain may be able to go from C to D, but if it does, it can never return.

7.23. Use the gambler's ruin to show that the simple symmetric random walk in one dimension is recurrent, and that the simple non-symmetric random walk ($p \neq 1/2$) is transient.

7.24. (Reflection Principle for the simple symmetric random walk.) Let (X_n) be a simple symmetric random walk with $X_0 < b$, where b is an integer. Show $P\{\max_{0 \leq j \leq n} X_j \geq b\} = 2P\{X_n \geq b\} - P\{X_n = b\}$. Use this to give another proof that X_n is recurrent.
[Hint: Show that for $a \geq 1$, $\{X_n = b + a\}$ and $\{T_b < \infty, X_n = b - a\}$ have the same probability.]

7.25. Let (X_n) be a simple random walk from the origin, and let p be the probability of a positive step. Let m be a positive integer. Show that the expected number of times that $X_n = m$ **before** the process returns to the origin is $(p/q)^m$ if $p < 1/2$, and 1 if $p = 1/2$. What if $p > 1/2$?

7.26. Let (X_n) be a recurrent Markov chain with $X_0 = j$. Let T_n be the time of the nth return to j. Show that $T_1, T_2 - T_1, T_3 - T_2, \ldots$ are i.i.d.

7.27. Show the estimates in Stirling's formula are good enough to determine convergence. That is, use the error bounds in Stirling's formula to show that the series in Example 7.6.1 actually do converge or diverge as claimed.

7.28. A simple symmetric random walk from zero will eventually reach any state N. Find the expected number of times it is at zero before reaching N.

7.29. Let (X_n) be a simple reflecting random walk on $0, 1, \ldots, N$.
  (a) Find the average amount of time spent in state i, $i = 0., \ldots, N$.
  (b) Find the expected return time to 0.

7.30. The simple symmetric random walk is often called the Drunkard's Walk, because it describes the path of someone whose steps are as likely to be backward as forward, which might describe someone leaving a tavern somewhat the worse for wear. Explain why such a person, starting at the tavern, will eventually reach home. Find the expected length of the path he traverses, i.e., what is the expected number of steps he takes before arriving home?

7.7. Equilibrium and the Ergodic Theorem for Markov Chains

It is natural to ask if a given physical system is in equilibrium, or if it tends to equilibrium as time passes. This is a question of distributions: by tending to equilibrium, we mean that the distribution of the process tends to a limit. (We repeat: this is purely a convergence of distributions, not sample paths, and does not imply in any way that the sample paths themselves converge. The process may be in statistical equilibrium while at the same time its sample paths can vary wildly.)

Let us begin with the basic theorem. We will add the frills later. Recall that if $X_0 = i$, that $T_i = \inf\{n \geq 1 : X_n = i\}$ is the first return time to i.

Theorem 7.41 (Basic Limit Theorem). *Suppose j is a recurrent aperiodic state in an irreducible Markov chain. Then for each i,*

$$\lim_{n \to \infty} P_{ij}^{(n)} = \frac{1}{E^j\{T_j\}} . \tag{7.15}$$

Remark 7.42. Note that the limit is independent of the starting state i.

Proof. Let $f_k = P^j\{T_j = k\}$. As j is recurrent, T_j is finite and $\sum_k f_k = 1$. To simplify the proof[10], suppose that $P_{jj} > 0$, so that $f_1 > 0$. Set $r_0 = 1$, $r_n = P^j\{T_j > n\} = \sum_{k=n+1}^{\infty} f_k$. We claim that

$$\sum_{k=0}^{n} r_k P_{jj}^{(n-k)} = 1 . \tag{7.16}$$

To see this, let $L_n = \sup\{k \leq n : X_k = j\}$ be the last time before n that the chain hits j. (This is *not* a stopping time!) Now $X_0 = j$ so $0 \leq L_n \leq n$,

---

[10]This avoids some number-theoretic complications, but does not change the probabilistic content of the proof.

7.7. Equilibrium and the Ergodic Theorem for Markov Chains

and

$$1 = P^j\{L_n \le n\} = \sum_{k=0}^{n} P^j\{L_n = k\}$$

$$= \sum_{k=0}^{n} P^j\{X_k = j, X_{k+1} \ne j, \ldots, X_n \ne j\}$$

$$= \sum_{k=0}^{n} P^j\{X_k = j\} P^j\{T_j > n-k\}$$

$$= \sum_{k=0}^{n} P_{jj}^{(k)} r_{n-k}.$$

Interchange k and $n-k$ to get (7.16).

Now set $p_n = P_{jj}^{(n)}$ and note that

$$p_n = \sum_{k=1}^{n} P^j\{T_j = k, X_n = j\}$$

or

(7.17) $$p_n = \sum_{k=1}^{n} f_k P_{jj}^{(n-k)},$$

which is the **renewal equation.**

Now define $\lambda = \limsup_{n\to\infty} p_n$ and choose a subsequence (n_k) along which $\lim_k p_{n_k} = \lambda$. We claim that $\lim_k p_{n_k-1} = \lambda$ also. Indeed, use the renewal equation:

$$p_{n_k} = p_{n_k-1} f_1 + p_{n_k-2} f_2 + \cdots + f_{n_k}$$

$$= f_1 p_{n_k-1} + \sum_{i=2}^{n_k} p_{n_k-i} f_{n_k}.$$

Let $k \longrightarrow \infty$. The left-hand side tends to λ. Use $\liminf_k (a_k + b_k) \le \liminf_k a_k + \limsup_k b_k$ on the right to see that

$$\lambda \le f_1 \liminf_k p_{n_k-1} + \sum_{i=2}^{n_k} f_i \limsup_k p_{n_k-i}.$$

But $\limsup_{k\to\infty} p_{n_k-i} \le \lambda$, so

$$\lambda \le f_1 \liminf_k p_{n_k-1} + \lambda \sum_{k=2}^{\infty} f_i = f_1 \liminf_k p_{n_k-1} + \lambda(1 - f_1).$$

Thus, $\lambda f_1 \le f_1 \liminf_k p_{n_k-1}$ or, since $f_1 > 0$ by hypothesis, $\liminf_k p_{n_k-1} \ge \lambda \ge \limsup_k p_{n_k-1}$, which implies equality.

By induction, $\liminf_k p_{n_k-i} = \lambda$, $i = 1, 2, \ldots$. Now we can use (7.16). There are two cases.

Case 1. $E^j\{T_j\} < \infty$. Then $\sum_i r_i = \sum_i P^j\{T_j > i\} = E^j\{T_j\} < \infty$. But the sum in (7.16) is dominated by $\sum_i r_i$, so that we can go to the limit by dominated convergence:

$$1 = \lim_k \sum_{i=0}^{n_k} r_i \underbrace{p_{n_k-i}}_{\to \lambda} = \lambda \sum_{i=0}^{\infty} r_i = \lambda E^j\{T_j\},$$

Since the expectation is finite, we can divide by it to see that $\lambda = 1/E^j\{T_j\}$.

Case 2. $E^j\{T_j\} = \infty$. Then $\sum_i r_i = \infty$, so we cannot use dominated convergence; but fix N. For $n_k > N$:

$$\sum_{i=0}^{n_k} r_i p_{n_k-i} \geq \sum_{i=0}^{N} r_i p_{n_k-i},$$

so

$$1 \geq \liminf_{k \to \infty} \sum_{i=0}^{N} r_i p_{n_k-i}$$

$$\geq \liminf_{k \to \infty} \sum_{i=0}^{N} r_i \underbrace{p_{n_k-i}}_{\to \lambda}$$

$$= \lambda \sum_{i=0}^{N} r_i.$$

Let $N \to \infty$. Since the sum tends to infinity, evidently λ must be zero. Thus, $\limsup p_n = 1/E^j\{T_j\}$. (This is true in both cases.) Now we repeat the argument for the lim inf. Let $\mu = \liminf_n p_n$ and let (n_k) be a sequence along which $p_{n_k} \longrightarrow \mu$. Then show $p_{n_k-i} \longrightarrow \mu$ for $i = 1, 2, \ldots$. Finally, identify $\mu = 1/E^j\{T_j\}$. We leave the details to the reader.

To show that $i \neq j \implies \lim P_{ij}^{(n)} = \lim P_{jj}^{(n)}$, let $a_n = P^i\{T_j = n\}$, where T_j is the first hit of j. Then

(7.18) $$P_{ij}^{(n)} = \sum_{k=1}^{n} a_k \underbrace{P_{jj}^{(n-k)}}_{\to \lambda} \longrightarrow \lambda \sum_{k=1}^{\infty} a_k = \lambda,$$

where we have used the dominated convergence theorem and the fact that, since i and j are communicating recurrent states, $\sum_k a_k = 1$. □

Definition 7.43. A recurrent state j is **positive recurrent** if $E^j\{T_j\} < \infty$. It is **null-recurrent** if $E^j\{T_j\} = \infty$.

7.7. Equilibrium and the Ergodic Theorem for Markov Chains

Proposition 7.44. *Let X be an irreducible aperiodic chain. If $i \longleftrightarrow j$, then i is positive recurrent if and only if j is.*

Proof. Suppose j is positive recurrent. Choose m, n, such that $P_{ij}^{(m)} > 0$ and $P_{ji}^{(n)} > 0$. since j is positive-recurrent, there is $a > 0$ for which $P_{jj}^{(k)} \longrightarrow a$.

$$P_{ii}^{(n+k+m)} \geq P_{ij}^{(m)} \underbrace{P_{jj}^{(k)}}_{\to a} P_{ji}^{(n)} \longrightarrow a P_{ij}^{(m)} P_{ji}^{(n)} > 0.$$

Thus, $\lim_{k \to \infty} P_{ii}^{(k)} = 1/E^i\{T_i\} > 0$, so $E^i\{T_i\} < \infty$, and i is positive recurrent. By symmetry, i positive recurrent $\Rightarrow j$ is positive recurrent. $\square$

Remark 7.45. *Positive-recurrence, null-recurrence and transience are class properties.*

Theorem 7.41 only covers the irreducible aperiodic case. Before extending it to the periodic case, consider the very simple example in the following exercise.

Exercise 7.31. Let X be a Markov chain on the state space $S = \{0, 1\}$, with transition probability matrix $\mathbb{P} = \begin{pmatrix} 0 & 1 \\ 1 & 0 \end{pmatrix}$.

(a) Show that X is irreducible with period 2, but that the chain $Y_k \stackrel{\text{def}}{=} X_{2k}$, $k = 0, 1, 2, \ldots$ is reducible, with two disjoint recurrence classes, S_1 and S_2. Show that the original chain alternates between S_1 and S_2,

(b) Show that X has a unique stationary probability distribution, but that the distribution of the chain does not necessarily tend to it. Show that, in fact, the chain tends to a stationary distribution if and only if it starts with it.

To extend our results to periodic chains, note that a chain with period d can only return to its initial state at multiples of d, so that $P_{jj}^{(k)} = 0$ unless k is a multiple of d. This suggests looking at the limit over multiples of d. Alternatively, we can look at the Césaro limit. This will give us a formula valid for periodic and aperiodic chains alike, and which extends easily to reducible chains.

Exercise 7.32. Let X be an irreducible chain with period d. Show that there exist d disjoint sets of states, $S_1, \ldots, S_d$, such that if the chain starts in S_1, it rotates through $S_2, S_3, \ldots, S_d$ and then back to S_1.

Now we can give the result for general chains. We will state it in terms of the Césaro limits of the $P_{ij}^{(k)}$, since these always exist for periodic states, whereas the ordinary limits may not.

Theorem 7.46. *Let X be a Markov chain. Then for each state j, if j has period $d(j)$*

$$\lim_{n \to \infty} P_{jj}^{(nd(j))} = \frac{d(j)}{E^j\{T_j\}} \tag{7.19}$$

and

$$\lim_{n \to \infty} \frac{1}{n} \sum_{k=1}^{n} P_{jj}^{(k)} = \frac{1}{E^j\{T_j\}}. \tag{7.20}$$

Consequently, if C is the class containing j, then for each state i,

$$\lim_{n \to \infty} \frac{1}{n} \sum_{k=1}^{n} P_{ij}^{(k)} = P^i\{T_C < \infty\} \frac{1}{E^j\{T_j\}}. \tag{7.21}$$

Proof. First note that if j is transient, $E^j\{T_j\} = \infty$, so both sides of (7.19) and (7.21) vanish. Now suppose j is recurrent. If it has period $d(j)$, consider the chain $Y_k \stackrel{\text{def}}{=} X_{kd(j)}$, $k = 0, 1, 2, \ldots$. Y is aperiodic, and if $Y_0 = j$, Y_k stays in the recurrence class containing j. In fact, j must be in one of the S_i of Exercise 7.32, so that Y_k actually stays in S_i. If we consider Y as a chain on S_i, it is both irreducible and aperiodic. Its one-step transition probabilities are $P_{ij}^{(d(j))}$, and its return time to state j is $T_j/d(j)$. Thus (7.19) follows from (7.15).

But $P_{jj}^{(k)}$ vanishes unless k is a multiple of $d(j)$, so that the Césaro sum of the $P_{jj}^{(k)}$ must equal $1/E^j\{T_j\}$, giving (7.20).

Now suppose that $X_0 = i$, where $i \neq j$. Once (X_n) hits the class C, it will eventually hit j, so $P^i\{T_j < \infty\} = P^i\{T_C < \infty\}$. Thus,

$$P^i\{X_k = j\} = \sum_{m=1}^{k} P^i\{T_j = m\} P_{jj}^{(k-m)},$$

and therefore, summing first over k,

$$\frac{1}{n} \sum_{k=1}^{n} P^i\{X_k = j\} = \frac{1}{n} \sum_{m=1}^{n} P^i\{T_j = m\} \sum_{k=m}^{n} P_{jj}^{(k-m)}$$

$$= \sum_{m=1}^{n} P^i\{T_j = m\} \left[\frac{1}{n} \sum_{k=1}^{n-m} P_{jj}^{(k)} \right].$$

But as $n \to \infty$, the term in brackets tends to $1/E^j\{T_j\}$ by (7.20), and $\sum_{m=0}^{n} P^i\{T_j = m\} \to P^i\{T_j < \infty\} = P^i\{T_C < \infty\}$. Then (7.21) follows from the dominated convergence theorem. □

7.7. Equilibrium and the Ergodic Theorem for Markov Chains

7.7.1. Stationary Distributions and Equilibrium.

Definition 7.47. A distribution μ on the state space is **stationary** if $\mu_j = \sum_i \mu_i P_{ij}$, or, in terms of matrix multiplication, if $\mu = \mu \mathbb{P}$.

If μ is a stationary distribution, and if X_0 has distribution μ, then X_1 also has the distribution μ, and, by induction, so do $X_2, X_3, \ldots$. So if the process has a stationary initial distribution, it will always have that distribution.

Let X be an irreducible aperiodic Markov chain. By Theorem 7.41, it has a limiting distribution. Intuitively, the limiting distribution must be stationary, but there is a catch: it may not be a probability distribution. The next theorem shows that if the chain is positive recurrent, it is indeed a stationary probability distribution.

Theorem 7.48. *Let X be an irreducible aperiodic chain. Let $\pi_j = \lim_n P_{jj}^{(n)}$. If X is either transient or null-recurrent, then $\pi_j = 0$ for all j. If it is positive-recurrent, then:*

(i) $\pi_j > 0 \; \forall j$ and $\sum_j \pi_j = 1$;
(ii) $\pi_j = \sum_i \pi_i P_{ij} \; \forall j$;
(iii) (π_j) is the unique probability distribution which satisfies (ii).

Proof. Fix an integer $M > 1$. Note that $P_{ij}^{(n)} \longrightarrow \pi_j$ as $n \to \infty$, so that

$$1 \geq \sum_{j=1}^{M} P_{ij}^{(n)} \longrightarrow \sum_{1}^{M} \pi_j$$

and

$$P_{ij}^{(n+1)} \geq \sum_{k=1}^{M} P_{ik}^{(n)} P_{kj} \implies \pi_j \geq \sum_{k=1}^{M} \pi_k P_{kj}.$$

Let $M \to \infty$ in these inequalities to see that

(7.22) $$\sum_j \pi_j \leq 1,$$

(7.23) $$\pi_j \geq \sum_i \pi_i P_{ij} \; \forall j.$$

Multiply (7.23) by $P_{j\ell}$ and sum over j:

$$\sum_j \pi_j P_{j\ell} \geq \sum_j \sum_k \pi_k P_{kj} P_{j\ell} = \sum_k \pi_k \sum_j P_{kj} P_{j\ell},$$

where we can interchange the order of summation since the summands are positive. Note that the sum over j equals $P_{k\ell}^{(2)}$ and by (7.23) the left-hand side is dominated by π_ℓ. Thus,

$$\pi_\ell \geq \sum_k \pi_k P_{k\ell}^{(2)} \,.$$

By induction,

$$\pi_j \geq \sum_i \pi_i P_{ij}^{(n)}$$

for all n. We claim that there is equality for all n and j. If not, there exist some j and n for which there is strict inequality. Now sum over j:

$$\sum_j \pi_j > \sum_j \sum_i \pi_i P_{ij}^{(n)} = \sum_i \pi_i \sum_j P_{ij}^{(n)} \geq \sum_i \pi_i \,.$$

This is a contradiction. Therefore,

$$\pi_j = \sum_i \pi_i P_{ij} \,, \forall n \,.$$

Thus (π_i) is stationary. We know from (7.22) that it is a sub-probability distribution. We must show it is actually a probability distribution.

Let $\varepsilon > 0$ and choose M so large that $\sum_{M+1}^\infty \pi_i < \varepsilon$. Then

$$\sum_{i=1}^M \pi_i P_{ij}^{(n)} \leq \pi_j \leq \sum_{i=1}^M \pi_i P_{ij}^{(n)} + \varepsilon \,.$$

Let $n \to \infty$:

$$\sum_{i=1}^M \pi_i \pi_j \leq \pi_j \leq \sum_{i=1}^M \pi_i \pi_j + \varepsilon \,.$$

Now let $M \to \infty$ and $\varepsilon \to 0$ to see that for all j,

$$\pi_j = \pi_j \sum_i \pi_i \,.$$

There are two possibilities: either $\pi_j = 0$ for all j, or there exists j for which $\pi_j > 0$. In the latter case, $\sum_i \pi_i = 1$. Since the chain is positive recurrent, we must have $\pi_j > 0$ for all j, and $\sum_j \pi_j = 1$. This proves (i) and (ii).

It remains to show that (ii) determines the π_j. Suppose that (x_j) is a probability distribution satisfying $x_j = \sum_i x_i P_{ij}$. As with the π_j, it follows that $x_j = \sum_i x_i P_{ij}^{(n)}$ for all n. Let $n \to \infty$ to see that

$$x_j \geq \sum_i x_i \pi_j = \pi_j \sum_i x_i = \pi_j \,.$$

Thus $x_j \geq \pi_j$ for all j. But both the x_j and the π_j sum to one, so they must be equal. $\square$

7.7. Equilibrium and the Ergodic Theorem for Markov Chains

Intuitively, the Markov chain tends to equilibrium in the sense that its distribution approaches the stationary distribution π. If the chain is transient, then the stationary distribution is identically zero. This is to be expected, since a transient chain simply goes away. If the chain is null-recurrent, the distribution also tends to zero, but this time it is not because the chain simply goes away—it does not—but because it returns at longer and longer intervals, so that at any fixed time, it is unlikely to be anywhere nearby. If the chain is positive-recurrent, then the process returns regularly, and its distribution tends to a stationary distribution. Notice that the process may be in equilibrium statistically, but it still visits every state infinitely often.

Finally, let us tie this together with the ergodic theorem. Ergodic theorems relate the time averages of a single sample path to the probabilistic averages, i.e., to the averages over ω, which is to say the expectations. For an aperiodic, irreducible, and recurrent Markov chain (X_n), define the number of visits to j before time n by

$$N_j^n = \#\{m : 1 \leq m \leq n, X_m = j\}.$$

Then N_j^n/n is the average amount of time that X spends at j up to time n, and $\lim_{n \to \infty} N_j^n/n$ (if the limit exists) is the long-term average time spent at j.

Let T_j^n be the time of the nth return to j. Then $T_j^n = \inf\{k : N_j^k = n\}$, and if $m = T_j^n$, then $N_j^m = n$. (The chain is recurrent, so all these stopping times are finite, and $N_j^n \to \infty$ as $n \to \infty$.)

By Exercise 7.26, if $X_0 = j$, then $T_j^1, T_j^2 - T_j^1, T_j^3 - T_j^2, \ldots$ are i.i.d. random variables. Therefore, $T_j^n = T_j^1 + (T_j^2 - T_j^1) + \cdots + (T_j^n - T_j^{n-1})$, and, by the law of large numbers

$$\lim_{n \to \infty} \frac{T_j^n}{n} = E^j\{T_j^1\} \text{ for all } j.$$

If j is transient or null-recurrent, this limit is zero. Now if $T_j^m \leq n \leq T_j^{m+1}$, then $m \leq N_j^n \leq m+1$. (This just says that n is between the mth and $m+1^{\text{st}}$ returns to j, so there have been at least m but not more than $m+1$ visits by time n.) This leads to:

(7.24) $$T_j^m \leq n \leq T_j^{m+1} \implies \frac{m}{T_j^{m+1}} \leq \frac{N_j^n}{n} \leq \frac{m+1}{T_j^m}.$$

But now, as $m \to \infty$, so does n, and both sides of (7.24) converge to $1/E^j\{T_j^1\}$. Therefore, so does N_j^n/n. We have proved:

Theorem 7.49. *Let X be an irreducible, aperiodic, and positive-recurrent Markov chain. Let (π_i) be its stationary distribution. Then with probability*

one, for all j,
$$\lim_{n\to\infty} \frac{1}{n} N_j^n = \lim_{n\to\infty} P_{jj}^{(n)} = \pi_j.$$

Remark 7.50. This extends easily to null-recurrent and transient chains, where the limit is zero for all states. The extension to periodic chains is more complicated. As we can see from Exercise 7.31, the distribution may tend to a limit cycle rather than a limit, and this cycle may depend on the initial measure. We will leave this to the problems.

7.8. Finite State Markov Chains

If the chain has only finitely many states, its behavior is simpler.

Theorem 7.51. *Let X be a Markov chain with finitely many states. Then:*

  (i) *there is at least one positive-recurrent state ;*

  (ii) *there are no null-recurrent states;*

  (iii) *regardless of the initial distribution, the chain will eventually be absorbed into a class of recurrent states.*

Proof. (i) For any i and n, $\sum_j P_{ij}^{(n)} = 1$, so
$$1 = \sum_j \left[\frac{1}{n} \sum_{m=1}^{n} P_{ij}^{(m)} \right].$$

Let $n \to \infty$. By Theorem 7.46 the Cesaro average in brackets tends to a limit for each j, say ρ_j. By (7.21), $\rho_j = 0$ if j is transient or null-recurrent. Since the sum over j is finite, we can go to the limit as $n \to \infty$ to conclude that
$$1 = \sum_j \rho_j.$$

Therefore, at least one ρ_j must be strictly positive, so that j is a positive-recurrent state.

(ii) If there exists a null-recurrent state, then (i) applies to the recurrence class containing it: that class must contain a positive-recurrent state. This contradicts the fact that all states in it must be null-recurrent, so there are no null-recurrent states.

(iii) By (ii), all states are either transient or positive recurrent. If the chain ever reaches a recurrence class, it stays there. If $\mathcal{T}$ is the class of all transient states, for any i,
$$P^i\{X_n \in \mathcal{T}\} = \sum_{j \in \mathcal{T}} P_{ij}^{(n)} \longrightarrow 0,$$

since $P_{ij}^{(n)} \longrightarrow 0$ as $n \to \infty$ for any transient state. So the chain eventually leaves the transient states, and is absorbed into a recurrence class. □

According to Theorem 7.51, an irreducible aperiodic Markov chain with finitely many states is necessarily positive recurrent, and therefore has a unique stationary probability distribution. Moreover, the chain will tend to that distribution. Thus, an irreducible aperiodic finite-state chain will always tend to equilibrium. According to Theorem 7.48, the stationary distribution is the unique solution of the system

$$\pi_j \geq 0, \ j = 1, \ldots, n,$$

$$\sum_j \pi_j = 1,$$

$$\pi_j = \sum_i \pi_i P_{ij}, \ j = 1, \ldots, n.$$

Card Shuffling. A deck of cards is shuffled periodically to randomize it. A shuffle is just a remixing of the cards which changes their order in a (hopefully) random way. In practice, a single shuffle is not sufficient, for skilled players can still profit from what they know of the pre-shuffled order, so the deck is usually shuffled several times in succession. But how well does shuffling actually randomize the deck?

Suppose the deck has N cards, numbered from one to N. Let $x^{(1)}$ be the number of the card at the top of the deck, $x^{(2)}$ the number of the second, and so on. If the deck is in random order, then $P\{x^{(1)} = k\} = 1/N$ for $k = 1, \ldots, N$, and the same is true for $x^{(2)}, \ldots, x^{(N)}$. This is enough for dealing a single card, but in general, we need more. In fact, we would like the entire joint distribution of $(x^{(1)}, \ldots, x^{(N)})$ to be uniform.

The deck at any time contains the N cards in some order, and it is this order which is important, so let us represent its state X by a permutation of $1, \ldots, N$:

$$X = \begin{pmatrix} 1 & 2 & \ldots & N \\ x^{(1)} & x^{(2)} & \ldots & x^{(N)} \end{pmatrix}.$$

Let $\mathfrak{S}_N$ be the set of all $N!$ permutations of $1, \ldots, N$. A shuffle just permutes the existing order of the cards, so *definition*: a **shuffle** is a random variable taking values in $\mathfrak{S}_N$. Let Π be a shuffle. If the the deck is in state X, then, after the shuffle, its new state is $\Pi \circ X$. If the deck is perfectly randomized, X has a uniform distribution on $\mathfrak{S}$. (This is equivalent to saying that $(x^{(1)}, \ldots, x^{(N)})$ has a uniform distribution.)

Let $\Pi_1, \Pi_2, \ldots$ be i.i.d. shuffles. If the initial state of the deck is X_0, then its successive states are $X_1 = \Pi_1 X_0$, $X_2 = \Pi_2 \circ X_1, \ldots, X_n = \Pi_n \circ X_{n-1} \ldots$. Note that $X_1, X_2, \ldots$ is a Markov chain on $\mathfrak{S}_N$. In fact, $\mathfrak{S}_N$ is a group under composition, and (X_n) is a random walk on it. Let $\mathbb{P} = (p_{ij})$ be its transition probability matrix. Then $\mathbb{P}$ is *bi-stochastic*, i.e., all elements are positive, and both the column and row sums are one. ($\sum_j p_{ij} = 1$ because the p_{ij} are transition probabilities, and $\sum_i p_{ij} = 1$ because the transpose $\mathbb{P}^T$ is the transition probability matrix generated by the inverse shuffle Π^{-1}.)

Here are three easily-proven facts about finite bi-stochastic chains:

1° $\mathbb{P}$ bi-stochastic $\implies$ so are $\mathbb{P}^2, \mathbb{P}^3, \ldots$.

2° The uniform distribution is stationary.

3° There are no transient states.

If the chain is irreducible, then the uniform distribution is the unique stationary distribution. If it is also aperiodic, then the distribution of X_n tends to a uniform distribution on $\mathfrak{S}_N$, which is to say that the state of the deck approaches a perfectly randomized state. It is known that certain types of shuffle lead to irreducible aperiodic chains. In this case, Markov's theorem (see Exercise 9.26) says that the distribution converges to the uniform density exponentially fast; and it does so in the *total variation distance*: if $P = (p_j)$ on $\mathfrak{S}_N$ the total variation distance of P from the uniform distribution U is $\|P - U\| \stackrel{\text{def}}{=} (1/N!) \sum_{j=1}^{N} |p_j - (1/N!)|$. But the state space is enormous, so even exponential convergence might seem slow.

Bayer and Diaconis [**7**] give a remarkably explicit answer. They claim that "Seven shuffles is enough." They show that using what is called a "riffle shuffle" on a standard 52-card deck that the total variation distance is close to 1 for the first five shuffles, then drops off sharply: It is nearly zero after ten shuffles, and, with a distance of around .3, seven shuffles is a reasonable choice for "enough".

Absorption Probabilities. Let X be a finite-state Markov chain. Let $C_1, C_2, \ldots$ be its recurrence classes. If the chain starts in one of the C_i, it stays there. If it starts from a transient state, it will eventually be absorbed into one of the C_i. The limiting distribution of the chain depends on the probabilities of absorption into the different C_i. (See Theorem 7.46.) These absorption probabilities satisfy a system of linear equations.

Theorem 7.52. *Let X be a finite-state Markov chain with transition probability matrix $\mathbb{P}$. Suppose it has recurrence classes $C_1, \ldots, C_n$. Let $f_j(i)$ be the probability of absorption into the class C_j given that $X_0 = i$, and let f_j be the column vector $(f_j(1), \ldots, f_j(n))^T$. Then each f_j satisfies the equations:*

(7.25) $\quad \begin{cases} f_j(i) = 1, & i \in C_j, \\ f_j(i) = 0, & i \in C_k, \ k \neq j \end{cases}$

7.8. Finite State Markov Chains

and

(7.26)
$$f_j = \mathbb{P} f_j.$$

Proof. If the process starts in C_j, it is already absorbed there, and the absorption probability is one. If it starts in one of the C_k for $k \neq j$, it stays in C_k and never reaches C_j, so the absorption probability vanishes. This accounts for (7.25).

In general, the process will be absorbed in C_j if and only if the hitting time T_{C_j} is finite. Thus, for any i,

$$f_j(i) = P^i\{\text{absorption in } C_j\} = P^i\{T_{C_j} < \infty\}.$$

Let the process take one step, and notice that

$$P^i\{T_{C_j} < \infty\} = \sum_\ell P^i\{X_1 = \ell \text{ and } T_{C_j} < \infty\}$$
$$= \sum_\ell P^i\{X_1 = \ell\} P^i\{T_{C_j} < \infty \mid X_1 = \ell\}.$$

But by the Markov property, $P^i\{T_{C_j} < \infty \mid X_1 = \ell\} = P^\ell\{T_{C_j} < \infty\} = f_j(\ell)$. Note that this is true even if $X_1 \in C_j$, since then absorption in C_j is sure, and $f_j(\ell) = 1$, or if $X_1 \in C_k$, when $f_j(\ell) = 0$. Thus we have

$$f_j(i) = \sum_\ell P_{i\ell} f_j(\ell).$$

This holds for all $j = 1, \ldots, n$ and all states i, which is exactly (7.26). □

Remark 7.53. If there are N states, (7.26) is a system of N linear equations in the N unknowns $f_j(1), \ldots, f_j(N)$. It can be solved by the usual means. (The system is actually smaller than that, since we already know the values for the recurrent states.)

Hitting Probabilities. Theorem 7.52 also gives the probabilities of hitting an arbitrary set A in the state space. Let $S_A = \inf\{n \geq 0 : X_n \in A\}$. Note that if $X_0 \in A$, $S_A = 0$, so that the probability that the chain is ever in A is $P\{S_A < \infty\}$. So let $f_A(i) = P^i\{S_A < \infty\}$. Clearly, $f_A(i) = 1$ if $i \in A$. For $i \in A^c$, consider the Markov chain stopped at S_A: $\{X_{n \wedge S_A}, n = 0, 1, 2, \ldots\}$. This is again a Markov chain (see the problems) whose transition probabilities are $\hat{P}_{ij} = P_{ij}$ if $i \in A^c$, and $\hat{P}_{ij} = \delta_{ij}$ if $i \in A$. Moreover, A is a union of recurrence classes for the stopped chain, and S_A is the time of absorption into A. If we apply Theorem 7.52 to $\hat{\mathbb{P}}$ we get

Theorem 7.54. *The f_A satisfy*

(7.27)
$$f_A(i) = \begin{cases} 1 & \text{if } i \in A, \\ \sum_j P_{ij} f_A(j) & \text{if } i \in A^c. \end{cases}$$

In fact, f_A is the smallest positive solution of the above equations. (See Exercise 9.31.)

Similar reasoning gives the expected value of S_A. Let $e_A(i) = E^i\{S_A\}$. Note that $e_A(i) = 0$ if $i \in A$. If $i \notin A$, then $S_A \geq 1$, so consider the process again after the first step. If it is in A after one step, then $S_A = 1$. If not, we can restart the process again from its new position. If it is at k, say, the conditional expected *remaining* time until S_A is $e_A(k)$. But one step has passed, so the total expected time is $1 + E\{e_A(X_1)\}$. Thus $e_A(i) = 1 + \sum_k P^i\{X_1 = k\}(e_A(k))$. This leads to:

Theorem 7.55. *Let $e_A(i) = E^i\{S_A\}$. Then:*

(7.28) $$e_A(i) = \begin{cases} 0 & \text{if } i \in A, \\ 1 + \sum_{k \in A^c} p_{ik} e_A(k) & \text{if } i \in A^c. \end{cases}$$

There is always a finite solution to (7.27), but there may be none to (7.28); think of the simple symmetric random walk, where the expected hitting times are infinite. Even in finite state chains, it may be impossible to reach A from some state i, so the hitting time would be infinite and $e_A(i) = \infty$. However, if all hitting times have finite expectations, (7.28) holds.

Problems 7.8

7.33. Let X be a Markov chain on states $\{1,2,3,4,5\}$ with transition probability matrix

$$\mathbb{P} = \begin{pmatrix} 1 & 0 & 0 & 0 & 0 \\ \frac{1}{2} & 0 & \frac{1}{2} & 0 & 0 \\ 0 & \frac{1}{2} & 0 & \frac{1}{2} & 0 \\ 0 & 0 & 0 & \frac{1}{3} & \frac{2}{3} \\ 0 & 0 & 0 & \frac{2}{3} & \frac{1}{3} \end{pmatrix}.$$

(a) Find the period of states 1–5.
(b) Classify all states as transient or recurrent.
(c) Find all equivalence classes.
(d) Find $\lim_{n \to \infty} P_{24}^{(n)}$, $\lim_{n \to \infty} P_{54}^{(n)}$.
(e) Let $P\{X_0 = 2\} = P\{X_0 = 3\} = 1/2$. Find the limiting distribution of X_n as $n \to \infty$.

7.34. A single die is thrown repeatedly. Let X_n be the number of spots showing on the first n throws. Find $\lim_{n \to \infty} P\{X_n \text{ is a multiple of } 17\}$.

7.35. (Success Runs.) An experiment is run repeatedly and independently. It has two outcomes: success, or failure, with probabilities p and $q = 1 - p$, respectively, where $0 < p < 1$. Let X_n be the length of the current success run at time n.
(a) Show that the stationary distribution is $\pi_n = qp^n$, $n = 0, 1, 2, \ldots$.

(b) Given $X_0 = 0$, Show that the expected time until the first run of length n is $E^0\{T_n\} = \dfrac{1-p^n}{qp^n}$. (See Exercise 3.50 for another approach.)

7.36. A certain baseball player has a batting average of .357. That is, each time at bat, he has a probability .357 of getting a base hit. Suppose that he gets to bat four times each game, and that the results of different at-bats are independent. What is the probability that he gets at least one base hit per game for 56 straight games? Set this up as a problem in success runs. What is the stationary probability of a 56-game hitting streak? What is the expected waiting time for a streak of that length or longer to appear? Explain exactly what your answers mean. Take the following fact into account. Joe DiMaggio accomplished this in the 1941 season. It has been called the greatest sports feat of all time. His average was .357 in 1941, .325 lifetime. It may help to express time in terms of seasons (154 games per season in DiMaggio's time, excluding playoffs, 162 now). An average player in the major leagues would hit about .260. What would the probabilities and waiting times be for that average?

7.37. A fugitive has two hideouts, A and B, and moves between them. A detective, who knows the hideouts, raids them in hopes of making the arrest. If he raids the hideout while the fugitive is there, he makes the arrest. If he fails, he raids one of the hideouts again the next day. Both change hideouts according to Markov chains. The transition matrices for the detective and for the fugitive are

$$\mathbb{P}_d = \begin{pmatrix} \frac{1}{3} & \frac{2}{3} \\ \frac{1}{2} & \frac{1}{2} \end{pmatrix} \quad \mathbb{P}_f = \begin{pmatrix} \frac{1}{4} & \frac{3}{4} \\ \frac{1}{2} & \frac{1}{2} \end{pmatrix}.$$

(That is, if the detective raids A this time, the probability of raiding A the next time is $1/3$, of B, $2/3$.)

Suppose the detective first raids B while the fugitive is in A.

(a) Find the expected number of steps to catch the fugitive.

(b) Find the probability that the detective catches the fugitive in hideout B.

7.38. Let $\mathbb{P}$ be the transition probability matrix

$$\mathbb{P} = \begin{pmatrix} 0 & u & 1-u & 0 \\ \frac{1}{2} & 0 & \frac{1}{2} & 0 \\ 0 & 0 & 0 & 1 \\ 0 & 0 & 1 & 0 \end{pmatrix}.$$

Show there is exactly one value of u for which $\lim_{n\to\infty} P_{14}^{(n)}$ exists. Find it.

7.39. (Capture-recapture. See Example 2.29.1.) In order to estimate the size of a population of wild animals, scientists capture one of them, tag it, and release the tagged animal. They do this successively until they have a large enough sample for their estimate. Assume that each animal is captured at random and that successive captures are independent. Let N be the total number of animals in the population, and let X_n be the number of different tagged animals after n captures. Show that (X_n) is a Markov chain, and compute its transition probabilities. Find the distribution of the number of untagged animals at the time of the first recapture of a tagged animal.

7.40. A random walk on a graph goes from vertex to adjacent vertex by walking along the connecting edge. It chooses the edge at random: all edges connecting to that vertex are equally likely. Consider a random walk on the graph below.

(a) Find the long-term proportion of time spent at vertex 0.

(b) Find the expected return time to 0.

(c) Find the probability that a random walk from vertex e hits 0 before hitting either a or b.

(d) Suppose the random walk starts at 0. Find the expected time to reach $\{a \cup b\}$.

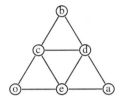

[Hint: Make use of symmetry!]

7.41. Let (X_n) be a right random walk on a finite or countable group $\mathfrak{G}$. (See Exercise 7.10.) Show the following.

(a) Either all states are transient or all states are recurrent. Thus the random walk is recurrent if and only if the identity is recurrent.

(b) If the random walk is recurrent, let $\mathfrak{H}$ be the recurrence class containing the identity. Then $\mathfrak{H}$ is a subgroup.

(c) Any recurrence class is of the form $g\mathfrak{H}$ for some $g \in \mathfrak{G}$.

7.42. Let (X_n) be a right random walk on a finite or countable group $\mathfrak{G}$, and let T be a finite stopping time for (X_n).

(a) Show that $\{X_T^{-1} X_{T+n},\ n = 0, 1, 2, \dots\}$ is a right random walk from the identity on $\mathfrak{G}$.

(b) Let $M_0 = X_0^{-1} X_1$. Let $\mathfrak{H}$ be a positive-recurrence class, and let π be the limiting probability distribution on $\mathfrak{H}$. Show that π is invariant under the law of M_0.

7.43. Consider a finite state Markov chain. Show that if it has a bi-stochastic transition matrix (see Exercise 7.11) then the uniform distribution is stationary. Conclude that it can have no transient states.

7.44. Show that for probabilities (p_n), $\sum_n p_n^2 \leq \max_n p_n$.

The following five exercises concern periodic Markov chains. The simple random walk is a Markov chain with period 2. It alternates between two sets of states, even and odd. The first exercise is to show that the general case is similar: if the period is d, there are d disjoint sets of states, and the chain cycles through them. The subsequent exercises follow from the first one and what we know about aperiodic chains.

7.8. Finite State Markov Chains

7.45. Let X_n be an irreducible recurrent Markov chain of period $d \geq 2$. Let $\mathbb{P}$ be its transition probability matrix and $\mathbb{P}^d$ its d-step transition matrix. Show that the state space is a union of d disjoint sets $E_0, \ldots E_{d-1}$ such that

(a) Each E_i is an equivalence class for $\mathbb{P}^d$.

(b) If $X_0 \in E_0$, then $X_1 \in E_1$, $X_2 \in E_3, \ldots, X_{d-1} \in E_{d-1}$ and $X_d \in E_0$.

7.46. Let (X_n) be an irreducible positive-recurrent Markov chain with period d. Let $E_0, E_1, \ldots, E_{d-1}$ be the $\mathbb{P}^d$-recurrence classes of Exercise 7.45. Show that if (π^k) is the $\mathbb{P}^d$-stationary probability on E_k that $\pi^k = \pi^0 \mathbb{P}^k$, $k = 0, \ldots, d-1$.

7.47. Let (X_n) be an irreducible recurrent Markov chain. Let $\pi_j \stackrel{\text{def}}{=} 1/E\{T_j\}$. Show that the (π_j) form a stationary distribution, even if the chain is periodic.

7.48. Let C be a recurrence class of a Markov chain. Show that if $j \in C$, then $\lim_{n\to\infty}(P_{ij}^{(1)} + \cdots + P_{ij}^{(n)})/n = P^i\{T_j < \infty\}\pi_j$, where $\pi_j = 1/E^j\{T_j\}$.

7.49. Let (X_n) be an irreducible, positive-recurrent Markov chain with period $d > 1$. Show that $\frac{1}{d}(P_{jj}^{(n+1)} + \cdots + P_{jj}^{(n+d)})$ tends to a stationary distribution π. Show also that for each k, as $n \to \infty$, $P_{jj}^{(nd+k)}$ tends to a distribution which depends on k and is stationary under $\mathbb{P}^d$.

7.50. Consider the transition matrix
$$\begin{pmatrix} 1/3 & 2/3 & 0 & 0 & 0 \\ 2/3 & 1/3 & 0 & 0 & 0 \\ 0 & 0 & 1/4 & 0 & 3/4 \\ 1/4 & 0 & 0 & 1/4 & 1/2 \\ 0 & 0 & 1/2 & 0 & 1/2 \end{pmatrix}.$$

(a) Find all classes and classify them as transient or recurrent.

(b) Find $\lim_{n\to\infty} P_{i,5}^{(n)}$ for $i = 1, 2, 3, 4$, and 5.

7.51. A Markov chain has the following transition matrix:
$$\mathbb{P} = \begin{pmatrix} \frac{1}{2} & \frac{1}{4} & \frac{1}{4} \\ \frac{1}{4} & \frac{1}{2} & \frac{1}{4} \\ \frac{1}{2} & \frac{1}{2} & 0 \end{pmatrix}.$$

Find the long-run proportion of times n for which $X_n = X_{n+1}$.

7.52. Coin 1 has $P\{\text{heads}\} = .6$. Coin 2 has $P\{\text{heads}\} = .5$. Start with coin 1, and flip it until the first time tails appears. Then start flipping coin 2 until the first time tails appears, and switch coins again. Continue, switching coins each time tails appears.

(a) If we start with coin 1, what is the probability that the third flip is with coin 2?

(b) What proportion of time, on the average, is coin 1 flipped?

7.53. (Markov's Theorem.) Let $\{X_n, n = 0, 1, 2, \ldots\}$ be a Markov chain on a finite state space S. Suppose that it is irreducible and aperiodic.

(a) Show that there is an integer n_0 such that if $n \geq n_0$, then $P_{ij}^{(n)} > 0$ for all i and j.

(b) Let N be the number of states. Show that for any pair of states i and j, it is possible to get from i to j in $N-1$ or fewer steps.

(c) Show that there is a $\lambda > 0$ and $c > 0$ such that for all $i \in S$ and $n \geq 1$,
$$\sum_{j \in S} |\pi_j - P_{ij}^{(n)}| \leq c e^{-\lambda n}.$$

7.54. There are $2m$ balls, m white and m black. These are originally divided up into two urns, with m balls in each. At each step, a ball is chosen at random from each urn, and the balls are interchanged. Let the state of the system at the nth step be X_n, the number of black balls in the first urn.

(a) Give the transition probabilities. Is the chain irreducible? Recurrent?

(b) Intuitively, what *must* the stationary distribution be? Prove you are right.

[Hint: The identity $\sum_i \binom{m}{i}^2 = \binom{2m}{m}$ might be useful.]

7.9. Branching Processes

Branching processes were first studied by J. Bienaymé in 1845. He proved the basic theorem on extinction, but for some reason it passed unnoticed. They were forgotten until 1874, when they were rediscovered by Galton and Watson, who were concerned about the apparent extinction of family names among the British aristocracy. To study this, they assumed that the family name is carried by the sons[11]: each son carries on the family name, and passes it on to his own sons in the next generation. The name dies out if there are no sons of that name born in some generation. This leads to a mathematical model that has also been used for the evolution of populations, the spread of infectious disease, the survival of genes in a population, and nuclear chain reactions, among others.

Think of a family tree, which lists the members of each generation of an extended family. A branching process simply keeps track of the numbers in the successive generations of the tree. Let Z_0 be the number of individuals in the initial generation, and let Z_n be the number in the nth generation. Each individual in the nth generation is replaced by a new family consisting of his (or her) progeny to form the $n+1^{\text{st}}$ generation. The family sizes—the number of progeny, that is—are random, and all families are independent and identically distributed. Let N be a random variable, representing the number of children born to a given individual. So, for instance, if Z_n is the number of individuals in the nth generation, and if $Z_n = m$, then each of the m individuals has a family; the ith individual has a family of size N_i, where $N_1, \ldots, N_m$ are m independent copies of the random variable N. The

---

[11] This comes from the rules governing inheritance of family names in Britain. The daughters also inherit the family name, but they would change it when they got married, so they would not pass it on to their children. This was probably a better assumption in 1874 than it would be today.

7.9. Branching Processes

number in the next generation is $Z_{n+1} = N_1 + \cdots + N_m$. The population dies out if Z_n is ever zero, since then there are no surviving members to give birth to the next generation: $Z_n = 0 \implies Z_{n+1} = Z_{n+2} = \cdots = 0$.

The population size forms a stochastic process $Z_0, Z_1, Z_2, \ldots$ which takes positive integer values[12].

Question: Is the population sure to die out, or can it possibly survive forever?

This depends on the distribution of N. Clearly, if $P\{N=0\} = 0$ it never dies out, and if $P\{N=0\} = 1$ it dies out immediately.

Each individual gives rise to a line of descendants. Since all the family sizes are independent, two different individuals will give rise to two distinct lines of descendants, *and these lines will be independent*. Thus, if there are m individuals, the ith individual will give rise to, say, $\hat{Z}_1^i$ descendants in the first generation, $\hat{Z}_2^i$ in the second, and so on. The processes $\{\hat{Z}_n^i, n = 0, 1, 2, \ldots\}$ are independent for different i, and they have the same distribution, for each is itself a branching process starting with one individual.

Similarly, each of the individuals alive at time n gives rise to a line of descendants, and these lines are independent branching processes. For instance, if there are m individuals alive at time n, each will give rise to an independent copy of $Z_0, Z_1, Z_2, \ldots$; these are independent not only of each other, but of the entire process[13] before time n. Thus, in particular, a branching process is a Markov chain.

This property of branching processes is worth restating: disjoint families evolve independently. In particular, since the process starting with $Z_0 = m$ is the same as the sum of m independent processes starting with $Z_0 = 1$, it is enough to consider the latter case. Therefore, in the following, we assume $Z_0 = 1$.

The analysis involves probability generating functions. Let $G(s)$ be the probability generating function of N and let $G_n(s)$ be the probability generating function of Z_n:

$$G(s) = E\{s^N\}, \quad s \geq 0,$$
$$G_n(s) = E\{s^{Z_n}\}, \quad s \geq 0.$$

Now N and Z_n are positive-integer valued, so that G and G_n all exist, at least for $0 \leq s \leq 1$, and, for those values of s, $0 \leq G(s) \leq 1$. Moreover, as we saw in Example 2.29.4, $G(0) = P\{N=0\}$, $G(1) = 1$, and $G'(1) = E\{N\}$.

---

[12] As elsewhere in this book, we use "positive" rather than "non-negative" to mean "greater than or equal to zero". When we mean "strictly positive" we will say so. So zero is positive, or, to be pedantic, non-strictly-negative.

[13] Strictly speaking, they are conditionally independent of $Z_0, \ldots, Z_n$ given that $Z_n = m$.

Since $Z_0 = 1$, $Z_1 = N$, hence $G_1(s) = G(s)$. The fundamental result about the G_n is:

Theorem 7.56. $G_{n+1}(s) = G_n(G(s)) = G(G_n(s))$. More generally, for $n, m \geq 1$,

(7.29) $$G_{n+m}(s) = G_n(G_m(s)).$$

Proof. It is enough to prove (7.29). Consider time n, and suppose that $Z_n = k$. Each of the k individuals alive at time n gives rise to its own line of descendants; say the ith individual gives rise to $\hat{Z}_0^i, \hat{Z}_1^i, \ldots$. This has exactly the same distribution as $Z_0, Z_1, \ldots$, and the different $\hat{Z}^i$ are independent. Thus, Z_{n+m} has the same distribution as $\hat{Z}_m^1 + \cdots + \hat{Z}_m^k$. Since the $\hat{Z}_m^i$ are independent,

$$E\{s^{Z_{n+m}} \mid Z_n = k\} = E\left\{s^{\hat{Z}_m^1 + \cdots + \hat{Z}_m^k}\right\}$$
$$= \prod_{i=1}^{k} E\{s^{\hat{Z}_m^i}\}$$
$$= G_m(s)^k,$$

for $\hat{Z}_m^i$ has the same distribution, and therefore the same probability generating function as Z_m. Thus,

$$E\{s^{Z_{n+m}}\} = \sum_{k=0}^{\infty} P\{Z_n = k\} E\{s^{Z_{n+m}} \mid Z_n = k\}$$
$$= \sum_{k=0}^{\infty} P\{Z_n = k\} G_m(s)^k$$
$$= E\{G_m(s)^{Z_n}\}$$
$$= G_n(G_m(s)).$$

Thus, $G_{n+m}(s) = G_n(G_m(s))$ as claimed. $\square$

Let us first find the moments. Let $E\{N\} = \mu$, $\text{Var}\{N\} = \sigma^2$.

Proposition 7.57. $E\{Z_n\} = \mu^n$, $\text{Var}\{Z_n\} = \begin{cases} n\sigma^2 & \text{if } \mu = 1, \\ \frac{\sigma^2(\mu^n - 1)\mu^{n-1}}{\mu - 1} & \text{otherwise}. \end{cases}$

Proof. Recall that if X is a positive integer-valued random variable with probability generating function G_X, then $E\{X\} = G'_X(1)$ and $\text{Var}(X) = G'''_X(1) + G'_X(1) - G'_X(1)^2$. In particular, $G(1) = 1$, $G'(1) = \mu$, and $G''(1) = \sigma^2 + \mu^2 - \mu$.

7.9. Branching Processes

Now $G_{n+1}(x) = G_n(G(s))$ so, differentiating twice,
$$G'_{n+1}(s) = G'_n(G(s))G'(s)$$
$$G''_{n+1}(s) = G''_n(G(s))G'(s)^2 + G'_n(G(s))G''(s).$$

Evaluate the first equation at $s=1$ to see that $G'_{n+1}(1) = G'_n(G(1))G'(1) = G'_n(1)\mu$. Therefore, $E\{Z_{n+1}\} = \mu E\{Z_n\}$. Since $E\{Z_1\} = E\{N\} = \mu$, by induction, $E\{Z_n\} = \mu^n$.

Evaluating the second equation at $s=1$ gives
$$G''_{n+1}(1) = G''_n(1)G'(1)^2 + G'_n(1)G''(1)$$
or, in terms of μ and σ^2,
$$\sigma^2_{n+1} + \mu^{2n+2} - \mu^{n+1} = (\sigma^2_n + \mu^{2n} - \mu^n)\mu^2 + \mu^n(\sigma^2 + \mu^2 - \mu).$$

Solve for σ^2_{n+1}:
$$\sigma^2_{n+1} = \mu^2 \sigma^2_n + \mu^n \sigma^2.$$

Now $\sigma^2_1 = \sigma^2$. If $\mu = 1$, an easy induction gives $\sigma^2_n = n\sigma^2$. If $\mu \neq 1$, the given formula holds for $n = 1$. If it holds for $n - 1$, then
$$\sigma^2_n = \mu^2 \sigma^2_{n-1} + \mu^{n-1}\sigma^2$$
$$= \mu^2 \frac{\sigma^2(\mu^{n-1} - 1)\mu^{n-2}}{\mu - 1} + \mu^{n-1}\sigma^2$$
$$= \frac{\sigma^2(\mu^n - 1)\mu^{n-1}}{\mu - 1}.$$

□

Extinction Probabilities. If $Z_n = 0$, then $Z_{n+1} = Z_{n+2} = \cdots = 0$, so that $P\{Z_n = 0\}$ increases with n. The population will eventually become extinct if $Z_n = 0$ for some n, so the probability of eventual extinction is $\lim_{n \to \infty} P\{Z_n = 0\}$.

Now $P\{Z_n = 0\} = G_n(0)$, so the probability of eventual extinction is $\lim_{n \to \infty} G_n(0)$. If $\mu < 1$, then
$$P\{Z_n > 0\} \leq E\{Z_n\} = \mu^n \longrightarrow 0,$$
so the process will eventually become extinct. However, if $\mu = 1$, $E\{Z_n\} = 1$, and it is not clear whether or not it will become extinct.

Let $v_n = P\{Z_n = 0\}$. Then
$$v_n = P\{X_n = 0\}$$
$$= G_n(0)$$
$$= G(G_{n-1}(0))$$
$$= G(v_{n-1}).$$

Now $P\{Z_n = 0\}$ increases in n, so that $v_n \leq v_{n+1} \leq \cdots \leq 1$. Thus the sequence (v_n) has a limit, say, v. Moreover, G, being a power series with positive coefficients, is continuous on $[0,1]$. Therefore,
$$v \stackrel{\text{def}}{=} \lim v_n = \lim_{n \to \infty} G(v_{n-1}) = G(v).$$
Thus the probability v of eventual extinction is a solution of the equation
$$s = G(s).$$
We must learn just a little more about probability generating functions. Once we do, we will be able to solve this equation by drawing a few pictures.

Lemma 7.58. *Let $0 \leq s \leq 1$. Then $s \mapsto G(s)$ is positive, continuous, increasing, and convex. It is strictly convex if $P\{N \geq 2\} > 0$. Moreover,*
$$\begin{cases} G(0) = P\{N = 0\}, \\ G(1) = 1, \\ G'(1) = \mu, \\ G''(1) = \sigma^2 + \mu^2 - \mu. \end{cases}$$

Proof. G is a power series with radius of convergence at least one, hence it is continuous and continuously differentiable on $[0,1]$. (If its radius of convergence is exactly one, the continuity and differentiability at $s = 1$ follow from Abel's theorem.) The identities involving $G(0)$, $G'(1)$ and $G''(1)$ have already been mentioned. They are included here for ease of reference.

Then $G(s) = E\{s^N\} \geq 0$, $G'(s) = E\{Ns^{N-1}\} \geq 0$, and $G''(s) = E\{N(N-1)s^{N-2}\} \geq 0$. (In the last expectation, for instance, there is no contribution from the set where N equals zero or one, for then $N(N-1) = 0$, and the contribution from the set $N \geq 2$ is positive.) Since G' and G'' are positive, G is increasing and convex. If $P\{N \geq 2\} > 0$, G'' is strictly positive, so G is strictly convex. $\square$

We can see the solutions of $s = G(s)$ by graphing both s and $G(s)$ together. There are four cases, leading to four different pictures, according to the values of $P\{N = 0\}$ and μ.

Case 1: $\mu = 1$, $N \equiv 1$.
Case 2: $\mu \leq 1$, $P\{N = 1\} < 1$.
Case 3: $\mu > 1$, $P\{N = 0\} = 0$.
Case 4: $\mu > 1$, $P\{N = 0\} > 0$.

Theorem 7.59. *The probability of extinction is the smallest root $s_0 \in [0,1]$ of $s = G(s)$. Thus,*

(i) $s_0 = 1$ if $\mu \leq 1$ and $N \not\equiv 1$,
(ii) $s_0 = 0$ if $P\{N = 0\} = 0$,

7.9. Branching Processes

(iii) $0 < s_0 < 1$ if $\mu > 1$ and $P\{N = 0\} > 0$.

Remark 7.60. If we restrict ourselves to non-trivial family sizes, the result is simple to state: extinction is sure if and only if $\mu \leq 1$.

Proof. The probability of extinction is always one of the solutions of $s = G(s)$, and $s = 1$ is always a solution. Question: Are there other solutions, and if so, which is the correct one? If G is strictly convex, there can be at most two solutions in $[0, 1]$, $s = 1$ and another smaller value of s. The possibilities are shown in Figure 1. Of the four cases graphed above, the result is clear from the pictures in the first three. The words below merely justify the pictures.

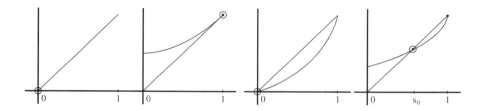

Figure 1. Cases 1–4: The smallest solution of $G(s) = s$ is circled.

Case 1. $N \equiv 1$. This is trivial: $P\{= 1\} = 1$, and each generation has exactly one member, so $Z_n = 1$ for all n, and there is no extinction. Moreover, $G(s) \equiv s$ so every point $s \in [0, 1]$ is a solution of $G(s) = s$. The smallest of these is $s_0 = 0$, which is indeed the probability of extinction.

Case 2. Suppose $\mu \leq 1$ and $p \stackrel{\text{def}}{=} P\{N = 1\} < 1$. If $P\{N \geq 2\} = 0$, then $G(s) = ps + (1 - p)$. As $p < 1$, the only solution of $G(s) = s$ is $s = 1$. If $P\{N \geq 2\} > 0$, then G is strictly convex, so G' is strictly increasing. $G'(s) < \mu \leq 1$ for $s < 1$. Now $G(s) - s = \int_s^1 1 - G'(u)\,du$, $G(s) - s > 0$ if $s < 1$, and again, the only solution of $G(s) = s$ in $[0, 1]$ is $s = 1$, which is therefore the probability of extinction.

Case 3. If $\mu > 1$ and $P\{N = 0\} = 0$, $G(s)$ is strictly convex and there are two solutions to $s = G(s)$, namely $s = 0$ and $s = 1$. But if $P\{N = 0\} = 0$, the population can never die, hence the probability of extinction is zero, the smaller of the two solutions.

Case 4. The case where $\mu > 1$ and $P\{N = 0\} > 0$ is the most interesting one. Evidently $\mu > 1 \implies P\{N \geq 2\} > 0$, which implies that G is strictly convex. Thus there are at most two solutions, one of which is $s = 1$. But $1 < \mu = G'(1)$, so that $G(s) < s$ in some interval $(1 - \varepsilon, 1)$, and $G(0) = P\{N = 0\} > 0$, so $G(s)$ is greater than s at zero, smaller than s near one. By the intermediate value theorem, there is a second solution, $s_0 \in (0, 1)$.

To see that s_0 is the correct solution, let $v_n = P\{Z_n = 0\}$. Figure 2 makes the following argument clear.

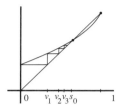

Figure 2. Why s_0 is the smallest solution.

Now $Z_0 = 1$, so $v_0 = 0$. Then $v_1 = G(v_0) \geq v_0$, and $G(v_0) \leq G(s_0) = s_0$. Then $v_2 = G(v_1)$ and $v_1 \leq v_2 \leq s_0$. Proceeding by induction, we find $v_0 \leq v_1 \leq v_2 \leq \cdots \leq v_n \leq \cdots \leq s_0$. The v_n are increasing and bounded above by s_0, so they converge to, say, v. By continuity, $G(v) = \lim_n G(v_n) = \lim_n v_{n+1} = \lim_n v_n = v$. Thus v is a solution of $G(s) = s$ and $v \leq s_0$. Since s_0 is the smallest solution, evidently $v = s_0$. □

Example 7.60.1. In nuclear fission, an energetic neutron encounters the nucleus of an atom of fissionable material. The nucleus may then break up (fission) and, in the process, emit one or several new energetic neutrons. These can strike other nucleii, causing these to decay in turn, leading to a chain reaction. In branching process terminology, a neutron dies out if it fails to encounter another fissionable nucleus. If it encounters another nucleus, however, the neutrons emitted in the resulting fission are its family. Different neutrons are clearly independent, so the total number of neutrons is a branching process.

The two key quantities are the number of energetic neutrons emitted in a fission—which averages about two and a half for ^{235}U—and the probability that a neutron encounters another nucleus and causes it to fission. This depends on several things, notably the amount of fissionable atoms in the neighborhood (leading to the famous "critical mass"), and the amount of non-fissionable material present to absorb neutrons. If the expected number of successful neutrons per fission is less than one, the chain reaction dies out. However, if it is greater than one, the chain reaction may not die out, in which case the number of neutrons increases exponentially. Presto, explosion!

Problems 7.9

7.55. The family-size distribution of a branching process (X_n) is $p_0 = 1/3$, $p_1 = 1/6$ and $p_2 = 1/2$. Suppose $X_0 = 3$.

7.9. Branching Processes

(a) Find $E\{X_{30}\}$.

(b) Find the probability of ultimate extinction.

(c) Find $P\{X_6 = 2 \mid X_5 = 2\}$.

7.56. Let (Z_n) be a branching process with $Z_0 = 1$ and expected family size $\mu < 1$. Let $N = \sum_{n=0}^{\infty} Z_n$ be the total number of individuals in the process over all time. Show that
$$E\{N\} = \frac{1}{1-\mu}.$$

[Hint: Count one for the initial individual, and re-assess the situation after the first family is born.]

7.57. Atoms in a radioactive substance spontaneously fission from time to time. Each fission releases an amount of energy, say q. Nuclear piles use this to provide a steady source of energy. Let r be the average number of atoms which fission spontaneously per unit time. Each atom, when it fissions, releases energetic neutrons. Say a neutron is *successful* if it encounters a nucleus and causes it to fission. Let p_j be the probability that a given fission produces j successful neutrons, and let μ be the expected number of successful neutrons per fission. (To prevent a runaway chain reaction, μ must be strictly less than one, but this can be assured by, for instance, limiting the $^{235}U/^{238}U$ ratio.) Show that the power output, i.e., the energy per unit time produced by the pile, is
$$P = \frac{qr}{1-\mu}.$$

[Hint: Use the previous problem.]

[Note: This ignores the secondary decay of unstable fission products, which increases the power output. It operates on a slower time-scale, so it can be considered as an important additional source of energetic neutrons.]

7.58. (The Spread of Infectious Disease.) In each unit of time, a person infected with a certain disease will either infect one other person ,or exhibit symptoms of the infection and immediately be isolated from the population. The probabilities of these two possibilities are p and $1 - p$, respectively. Each infected person in the population acts independently. Let X_n be the number of undiscovered infected people in the population.

(a) Suppose that a single infected person enters the population at time 0. Find the distribution of the number of undiscovered infected people at the time that the first one is discovered.

(b) Show that X_n is a branching process, and show that there is a critical value p_0 of p such that if $p < p_0$ the infection dies out, and if $p > p_0$ the infection has a strictly positive probability of spreading without limit. What is the critical value p_0?

7.59. In a population of bacteria, each individual lives one day, then either dies with probability $1/4$, or splits into two offspring with probability $3/4$. Suppose that on day 1, the population consists of exactly one bacterium.

(a) Find the expected number of bacteria alive on day 4.

(b) Find the probability the population will eventually die out.

(c) Suppose that the population on day 6 consists of exactly two bacteria. What is the probability this will eventually die out?

7.60. Consider a branching process (Z_n) whose family size is geometric: $P\{N = k\} = 2^{-k-1}$. Assume $Z_0 = 1$.

(a) Show that Z_n has the generating function
$$G_n(s) = \frac{n - (n-1)s}{n + 1 - ns}.$$

(b) Let $T = \min\{n : Z_n = 0\}$ be the extinction time. Find $P\{T = n\}$ for $n \geq 1$ and show that $T < \infty$ a.s.

7.10. The Poisson Process

When we say that an event occurs randomly in some finite interval, we mean that its time of occurrence has a uniform distribution on that interval. But what does it mean for an event to occur randomly in an infinite interval like $[0, \infty)$? Nothing, it turns out. Indeed, a moment's reflection shows that there is no such thing as a random variable distributed uniformly on $(0, \infty)$. (For any number n, such a random variable would be infinitely more likely to be greater than n than less than n!) However, it is possible to have an *infinite* number of events uniformly distributed on the line. Let us see how.

Suppose that events occur at times $T_1 < T_2 < T_3 < \ldots$. If $I \subset [0, \infty)$ is a time interval, let $N(I)$ be the number of events which occur in I and let $|I|$ denote its length. We assume that, for some $\lambda > 0$:

(i) if the intervals $I_1, \ldots, I_n$ are disjoint, $N(I_1), \ldots, N(I_n)$ are independent;

(ii) the distribution of $N(I)$ depends only on the length of I;

(iii) $P\{N(I) = 1\} = \lambda|I| + o(|I|)$ as $|I| \to 0$;

(iv) $P\{N(I) \geq 2\} = o(|I|)$ as $|I| \to 0$.

Proposition 7.61. *For each interval $I \subset \mathbb{R}^+$, $N(I)$ has a Poisson distribution with parameter $\lambda|I|$.*

Proof. First, if I is finite, so is $N(I)$. Indeed, let $I \subset \mathbb{R}^+$ be a finite interval, and let $h = |I|$ be its length. Divide I into n disjoint intervals $I_1, \ldots, I_n$ of length h/n and let $X = N(I)$, $Y_j = N(I_j)$, $j = 1, \ldots, n$. The Y_j are independent, identically distributed (because the I_j all have the same length) and add up to X. Then

$$P\{X = \infty\} = P\{\exists j \leq n \ni Y_j = \infty\} \leq \sum_{j=1}^{n} P\{Y_j > 1\} = n\, o(h/n) \to 0$$

as $n \to \infty$. In particular, for each $n \geq 1$, $N(I)$ is a sum of n i.i.d. random variables Y_j, and $P\{Y_j \geq 2\} = o\big(P\{Y_j = 1\}\big)$ by hypotheses (iii) and (iv)

7.10. The Poisson Process

above, so that Proposition 6.35 implies that $N(I)$ is Poisson. We must check that its parameter is $\lambda|I|$.

Let I, I_j, X and Y_j be as above, and let $h = |I|$. Note that X vanishes iff all the Y_j do, and the Y_j are independent, so that

$$P\{X = 0\} = P\{Y_1 = 0\}^n = (1 - \lambda h/n + o(h/n))^n = \left(1 - \frac{\lambda h + o(1)}{n}\right)^n.$$

Let $n \to \infty$. By Lemma 6.34, the right-hand side tends to $e^{-\lambda h}$. Thus $P\{X = 0\} = e^{-\lambda h}$, so that X is Poisson (λh), as claimed. □

The Poisson process $\{X_t, \ t \geq 0\}$ simply counts the number of events that have happened up to time t, for every t. We define it by $X_0 = 0$ and

$$X_t = N\big((0, t]\big), \quad t \geq 0.$$

Definition 7.62. A process $\{X_t, \ t \geq 0\}$ has **independent increments** if for each n and each $0 \leq s_1 < t_1 \leq s_2 < t_2 \leq \cdots \leq s_n < t_n$, the random variables $X_{t_j} - X_{s_j}$ $j = 1, \ldots, n$ are independent. The increments are **stationary** if for each t, the distribution of $X_{s+t} - X_s$ is independent of s.

Elementary Properties of the Poisson Process. Let us look at some of the basic properties of the Poisson process. First, its distribution is Poisson. More explicitly:

(i) For all $t \geq 0$, X_t has a Poisson (λt) distribution, and if $s < t$, $X_t - X_s$ is Poisson ($\lambda(t-s)$). In particular, $X_t - X_s \overset{d}{\sim} X_{t-s}$.

This holds since $X_t = N\big((0, t]\big)$, which is Poisson (λt), and if $s < t$, $X_t - X_s = N\big((s, t]\big)$, which is Poisson ($\lambda(t-s)$).

Remark 7.63. There are infinitely many events. They are uniformly distributed on $(0, \infty)$ in the sense that the number of events in any time interval has exactly the same distribution as the number in any other interval of the same length. They do not favor any part of the half-line over any other.

(ii) X is a process of stationary independent increments.

Indeed, if $s_1 < t_1 \leq s_2 < t_2 \leq \cdots \leq s_n < t_n$, $X_{t_j} - X_{s_j} = N((s_j, t_j])$, and, as the intervals are disjoint, the $N\big((s_j, t_j]\big)$ are independent. Moreover, $N\big((s, s+t]\big) \overset{d}{\sim} N\big((0, t]\big)$, so the increments are stationary.

Notation. We have been following the usual convention of "omitting the omegas" for random variables. A random variable X is a function on a probability space, so it can be written $X(\omega)$, $\omega \in \Omega$. We commonly drop the ω from the notation: it is understood, so there is no need to put it there. This simplifies the notation, and makes things much clearer... usually. However, we have introduced a second variable: now we are talking about the Poisson process, whose sample paths are real-valued functions on $(0, \infty)$. We will head off some possible confusion by temporarily putting the omegas back where they belong.

If we include the variables, the Poisson process is $\{X_t(\omega),\ t \geq 0,\ \omega \in \Omega\}$. Think of it as a process which evolves in time[14]: for fixed ω, $t \mapsto X_t(\omega)$ is a real-valued function on $[0, \infty)$, called a *sample path*. It is what we would see if we watched the process as it evolved. Since it is a function, we can talk about its continuity, its limits, its differentiability and other properties, just as we can with any other function. Note that $N\big((s, t]\big)$ is also a random variable: $N\big((s, t]\big)(\omega)$. So we will put the ω's in for a (short) while, simply to remind us of this. But only for a short while, since they really do complicate the notation.

(iii) For a.e. ω, $t \mapsto X_t(\omega)$ is an integer-valued, increasing, and right-continuous function.

Indeed, X is a counting process, so it is integer-valued, and if $s < t$, $N\big((0, s]\big)(\omega) \leq N\big((0, t]\big)(\omega)$—every event in $(0, s]$ is also in $(0, t]$—so $t \mapsto X_t(\omega)$ is increasing. There are only finitely many events in any finite interval. Note that if $t_n \downarrow t$, then any event that falls in $(0, t_n]$ for all n is necessarily in $(0, t]$—it is important that the intervals here are closed on the right—so that $N\big((0, t_n]\big)(\omega) \downarrow N\big((0, t]\big)(\omega)$, hence $t \mapsto X_t(\omega)$ is right-continuous.

Since $t \mapsto X_t(\omega)$ is an increasing function, its left limit $X_{t-}(\omega)$ exists for all t. In terms of N, $X_{t-}(\omega) = \lim_{s \to t,\ s < t} N\big((0, s]\big)(\omega) = N\big((0, t)\big)(\omega)$, which is the number of events in the open interval $(0, t)$.

(iv) $t \mapsto X_t(\omega)$ increases only in jumps of size one.

The jump of X at time t is $X_t(\omega) - X_{t-}(\omega)$. Let k be an integer and let us compute the probability that there is a jump of size strictly greater than one for some $t \leq k$. Divide the interval $(0, k]$ into kn equal subdivisions of length $1/n$. Let $t_j = j/n$, $j = 1, \ldots, kn$. Then

$$P\{\omega : \exists t \leq 1 \ni X_t(\omega) - X_{t-}(\omega) \geq 2\} \leq P\{\omega : \exists j \ni X_{t_j}(\omega) - X_{t_{j-1}}(\omega) \geq 2\}$$

$$\leq \sum_{j=1}^{kn} P\{\omega : X_{t_j}(\omega) - X_{t_{j-1}}(\omega) \geq 2\}.$$

The probability is independent of j and equals $P\{X_{1/n} \geq 2\} = o(1/n)$. (Since the random variable is Poisson (λ/n), we could compute the exact probability, but this is good enough.) Therefore the above probability is

$$= kn\, o(1/n) \longrightarrow 0$$

---

[14] In the early days of stochastic process theory, much was made of the fact that a stochastic process is really a random function. This is true, of course, and it is often useful to think of it this way but... this is not one of those times. Keep things simple. Just think of a point evolving randomly as time goes on.

7.10. The Poisson Process

as $n \to \infty$. Thus the probability p_k that there is a jump of size 2 or greater in $(0, k]$ is zero. Therefore the probability of ever having a jump of size 2 or greater is less than or equal to $\sum_k p_k = 0$.

To summarize, the Poisson process is an integer-valued process of stationary independent increments which increases in jumps of size one; and for each time t, X_t is a random variable with a Poisson distribution.

The Poisson process has what we called the "restarting property". It can restart itself at any time t.

Theorem 7.64. *Let X be a Poisson process and $s > 0$. Then the process $\{X_{s+t} - X_s, t \geq 0\}$ is a Poisson process with the same distribution as $\{X_t, t \geq 0\}$, and it is independent of $\{X_t, 0 \leq t \leq s\}$.*

Proof. This is a consequence of the fact that X has stationary independent increments. The increments $X_{s+t} - X_s$ are independent of the increments $X_u - X_0 = X_u$ as long as $u \leq s$, which shows the independence. The fact that $Y_t \stackrel{def}{=} X_{s+t} - X_s$ is a Poisson process follows from the fact that it has the same finite-dimensional distributions as X. This, in turn, follows from the facts that $X_0 = Y_0 = 0$, and for any $0 \leq t_1 \leq \cdots \leq t_n$, both $X_{t_j} - X_{t_{j-1}}$, $j = 1, \ldots, n$, and $Y_{t_j} - Y_{t_{j-1}}$, $j = 1, \ldots, n$, are families of independent Poisson random variables with parameters $\lambda(t_j - t_{j-1})$, $j = 1, \ldots, n$, respectively. □

Transition Probabilities. We will now revert to "omitting the omegas" except where they are needed for clarity.

Definition 7.65. The **transition probabilities** of the Poisson process are $p_t(k) \stackrel{def}{=} P\{X_t = k\}$.

Since X_t is Poisson (λt),

$$p_t(0) = e^{-\lambda t},$$

$$p_t(k) = e^{-\lambda t} \frac{\lambda^k t^k}{k!}.$$

Corollary 7.66 (Markov Property). *Let X be a Poisson process. Let $0 \leq s_1 < \cdots < s_n < s < t$ and let $i, j, i_1, \ldots, i_n$ be positive integers. Then*

$$P\{X_t = j \mid X_s = i, X_{s_1} = i_1, \ldots, X_{s_n} = i_n\} = p_{(t-s)}(j - i).$$

Proof. $X_t = X_s + (X_t - X_s)$. By Theorem 7.64, $X_t - X_s$ is independent of $X_{s_1}, \ldots, X_{s_n}, X_s$, so $P\{X_t = j \mid X_s = i, X_{s_1} = i_1, \ldots, X_{s_n} = i_n\} = P\{X_t - X_s = j - i \mid X_s = i, X_{s_1} = i_1, \ldots, X_{s_n} = i_n\} = P\{X_{s+t} - X_s = k\} = p_t(k)$. □

Remark 7.67. It follows that the Poisson process is a Markov process. and the $p_t(k)$ are its transition probabilities.

Exercise 7.61. Show that for any $s, t \geq 0$, $p_{s+t}(k) = \sum_{i=0}^{k} p_s(i) p_t(k-i)$.

Jump Times and the Strong Markov Property. Let X be a Poisson process of parameter λ. Then $X_0 = 0$, and X_t remains at zero until the time it jumps to one. Let T_1 be the time of the first jump.

Proposition 7.68. T_1 *has an exponential* (λ) *distribution.*

Proof. This is immediate, since $T_1 > t \iff X_t = 0$, so $P\{T_1 > t\} = P\{X_t = 0\} = e^{-\lambda t}$. □

We conclude that the Poisson process starts at zero, remains there for an exponential length of time, and then jumps to one. It seems reasonable to continue the description by saying. "Then it remains at one for an exponential time, then jumps to two, and so on." This is true, but it is less obvious than it seems, because the first jump time is random, so both the time the process reaches one and the time it leaves it are random. We had similar problems with Markov chains. We solved them by introducing stopping times and the strong Markov property. Let us do the same here. We recall the definition of stopping time in a version appropriate for continuous parameter processes[15].

Definition 7.69. Let $\mathcal{F}_t = \sigma\{X_s, \, s \leq t\}$ for each $t \geq 0$. A random variable with values in $[0, \infty]$ is a **stopping time** if, for each $t \geq 0$, $\{T \leq t\} \in \mathcal{F}_t$. The **past before T** is the σ-field $\mathcal{F}_T = \{\Lambda \in \mathcal{F} : \Lambda \cap \{T \leq t\} \in \mathcal{F}_T \, \forall t \geq 0\}$.

Example 7.69.1. Deterministic times are stopping times. So is T_1, the first jump time, for $\{T_1 \leq t\} = \{X_t \geq 1\} \in \mathcal{F}_t$. More generally, the time T_n of the nth jump is a stopping time. Indeed, since the process starts from zero, the nth jump time coincides with the first time that $X_t = n$, so $\{T_n \leq t\} = \{X_t \geq n\} \in \mathcal{F}_t$.

Notice that $\{T = t\} \in \mathcal{F}_t$; this is less obvious than it was in the discrete case: $\{T = t\} = \{T \leq t\} - \bigcup_n \{T \leq t - 1/n\} \in \mathcal{F}_t$. Thus the events $\{T = t\}$, $\{T > t\}$, $\{T \geq t\}$ and $\{T < t\}$ are all in $\mathcal{F}_t$. We leave it to the reader to verify that the conclusions of Proposition 7.26 hold for continuous parameter stopping times, too; in particular, that $S \leq T \implies \mathcal{F}_S \subset \mathcal{F}_T$.

This leads us to the strong Markov property of the Poisson process.

Theorem 7.70 (Strong Markov Property). *Let T be a finite stopping time for the Poisson process. Let $Y_t = X_{T+t} - X_T$. Then the process $\{Y_t, \, t \geq 0\}$*

---

[15]The reason this definition uses $\{T \leq t\}$ instead of $\{T = t\}$ is that it may happen that $P\{T = t\} = 0$ for all t. This can cause problems. The condition "$T \leq t$" is much easier to use.

7.10. The Poisson Process

is a Poisson process, independent of $\mathcal{F}_T$. In particular, if $0 \le i \le j$ are integers and $t > 0$, then

(7.30) $$P\{X_{T+t} = j \mid X_T = i\} = p_t(j - i).$$

Remark 7.71. This says that the restarting property works at stopping times as well as at fixed times. If the stopping time is discrete (i.e. countable-valued) this is similar to the Markov chain case. The new challenge is that in general, it is possible that $P\{T = t\} = 0$ for every t. This is true, for instance, of the jump times. This forces us to use an approximation argument. We first show it for discrete-valued stopping times T_n. Then we approximate the general time by discrete times, and go to the limit. The key facts we use are (a) it is possible to approximate T from above by discrete-valued stopping times, and (b) the sample paths are right-continuous.

Proof. To show (Y_t) is a Poisson process, it is enough to show that its increments are independent and have the proper distributions. The key is the following: for an integer $k \ge 0$, $t > 0$, and $\Lambda \in \mathcal{F}_T$,

(7.31) $$P\{\Lambda;\, X_{T+t} - X_T = k\} = P\{\Lambda\} p_t(k).$$

This is equivalent to $P\{Y_t = k \mid \Lambda\} = p_t(k)$.

Step 1: suppose that T is discrete-valued, with possible values (t_j). Then

$$P\{\Lambda;\, X_{T+t} - X_T = k\} = \sum_j P\{\Lambda \cap \{T = t_j\};\, X_{t_j+t} - X_{t_j} = k\}.$$

Now $\Lambda \in \mathcal{F}_T$, so $\Lambda \cap \{T = t_j\} \in \mathcal{F}_{t_j}$. Apply Theorem 7.64:

$$= \sum_j P\{\Lambda \cap \{T = t_j\}\} p_t(k).$$

Factor out $p_t(k)$ and sum. As claimed, this is

$$= P\{\Lambda\} p_t(k).$$

Step 2: Approximate the original T by discrete times. Define T_n by

$$T_n \stackrel{\text{def}}{=} \frac{k}{2^n} \quad \text{if } \frac{k-1}{2^n} \le T < \frac{k}{2^n}, \quad k = 1, 2, \ldots$$

Now T_n is a stopping time, for $\{T = k2^{-n}\} = \{(k-1)2^{-n} \le T < k2^{-n}\}$. Moreover, $T < T_n$, so $\Lambda \in \mathcal{F}_T \subset \mathcal{F}_{T_n}$. Thus (7.31) holds for T_n.

Moreover, $T < T_n \le T + 2^{-n}$, so that T_n decreases to T as $n \to \infty$. Since X_t is integer-valued and right continuous, X_{T_n} not only converges to X_T, it *equals* X_T for all large enough n; the same is true for X_{T_n+t}. Thus, we can let $n \to \infty$ to get (7.31) for the original T.

Step 3: Now let $0 = t_0 < t_1 < t_2 < \cdots < t_n$, and define increments $I_1, \ldots, I_n$ by $I_j = Y_{t_j} - Y_{t_{j-1}} = X_{T+t_j} - X_{T+t_{j-1}}$. Let $\Lambda \in \mathcal{F}_T$ and let $i_1, \ldots i_n$ be positive integers. We claim that

(7.32) $P\{\Lambda; I_1 = i_1, \ldots, I_n = i_n\} = P\{\Lambda\} p_{t_1}(i_1) \cdots p_{t_n - t_{n-1}}(i_n).$

This will complete the proof, since it identifies (Y_t) as a process of independent Poisson increments—take $\Lambda = \Omega$ to see this—which is independent of each Λ and therefore of $\mathcal{F}_T$.

Let $\Lambda \in \mathcal{F}_T$, and define events Γ_ℓ, for $\ell = 1, \ldots, n$: $\Gamma_\ell \stackrel{\text{def}}{=} \{I_j = i_j,\ j = 1, \ldots, \ell\}$. Note that the left hand side of (7.32) is just $P\{\Lambda \cap \Gamma_n\}$. Now for each ℓ, $T + t_{\ell-1}$ is a stopping time, $\Gamma_\ell = \Gamma_{\ell-1} \cap \{I_\ell = i_\ell\}$, and $\Lambda \cap \Gamma_{\ell-1} \in \mathcal{F}_{T+t_{\ell-1}}$. Apply (7.31) successively for $\ell = n-1, n-2, \ldots, 0$:

$$P\{\Lambda \cap \Gamma_n\} = P\{\Lambda \cap \Gamma_{n-1}\} p_{t_n - t_{n-1}}(i_n)$$
$$= P\{\Lambda \cap \Gamma_{n-2}\} p_{t_{n-1} - t_{n-2}}(i_{n-1}) p_{t_n - t_{n-1}}(i_n)$$
$$\vdots$$
$$= P\{\Lambda\} p_{t_1}(i_1) \ldots p_{t_n - t_{n-1}}(i_n),$$

which establishes (7.32). □

Let us apply this to the jump times $T_1, T_2, T_3, \ldots$. Note that T_1 is exponential with parameter λ. Moreover, T_2 is the first time that $X_t = 2$, and $T_2 - T_1$ is the first jump time of $X_{T_1+t} - X_{T_1}$, which, by the strong Markov property, is a Poisson process independent of $\mathcal{F}_{T_1}$. Thus by Proposition 7.68, $T_2 - T_1$ is exponential (λ), and is independent of $\mathcal{F}_{T_1}$, and therefore of T_1. Thus T_1 and $T_2 - T_1$ are i.i.d. exponentials. Continuing by induction, T_{n+1} is the first time that $X_t = n+1$, and $T_{n+1} - T_n$ is the first jump time of the Poisson process $\{X_{T_n+t} - X_{T_n},\ t \geq 0\}$. By the strong Markov property, this is independent of $\mathcal{F}_{T_n}$, hence so is its first jump time. Therefore $T_{n+1} - T_n$ is exponential (λ) and independent of $\mathcal{F}_{T_n}$. This means it is independent of all the previous T_j. We have proved:

Proposition 7.72. *Let X be a Poisson process with parameter λ, and let $T_1, T_2, \ldots$ be its successive jump times. Then $T_1,\ T_2 - T_1,\ T_3 - T_2, \ldots$ are i.i.d. exponential (λ) random variables.*

Remark 7.73. We now have the desired description of a Poisson process: it starts at zero, stays there an exponential length of time, then jumps to one, stays there an exponential time, then jumps to two, and so on.

Construction of a Poisson Process⋆. We derived the Poisson process from the standpoint of events. The events had to satisfy some properties which assured that they were "uniformly distributed" on the line. It is

7.10. The Poisson Process

legitimate to ask if there really is such a sequence of events. The answer is "Yes," and we can use Proposition 7.72 to construct them. Each event is the jump time of a Poisson process, so the process determines the events, and the events determine the process. So to construct the process, we need merely to construct the jump times.

By Theorem 5.11, there exists a sequence $S_1, S_2, S_3, \ldots$ of i.i.d. exponential (λ) random variables. Set $T_1 = S_1$, $T_2 = S_1 + S_2$, and, in general, $T_n = S_1 + \cdots + S_n$. Then for each n, $T_n - T_{n-1} = S_n$, so $T_1, T_2 - T_1, \ldots$ are i.i.d. exponential (λ) random variables. Then set

$$X_t \stackrel{def}{=} \sum_{n=1}^{\infty} I_{\{t \geq T_n\}}.$$

The reader can verify that $X(0) = 0$, $X_t = n$ on the interval $[T_n, T_{n+1})$, and the T_n are its successive jump times. The inter-jump times are independent exponentials, just as (according to Proposition 7.72) with the Poisson process. It follows that X is, in fact, a Poisson process.

Remark 7.74. The intervals between jumps are i.i.d exponential (λ) random variables, so that the expected inter-jump time is $1/\lambda$. The expected time for the process to reach n is therefore $E\{T_n\} = nE\{T_1\} = n/\lambda$.

A Waiting Time Paradox. The Poisson process is often used for the arrival of events: we say that "the events arrive according to a Poisson process." Customers arrive at a queue, telephone calls arrive at an interchange, Geiger counters click, complicated machines have minor breakdowns, politicians say silly things, and computers surf a popular internet web page, all according to their own Poisson processes. Consider the following observation.

I am very unlucky. Buses arrive at regular intervals, but whenever I take one, I arrive at a bus stop at a time when the interval between buses is twice as long as the average. How can that be?

Suppose that the buses arrive according to a Poisson process. I arrive at some time $t > 0$. I have just missed one bus. How long must I wait for the next? Let X_t be the Poisson process which counts the number of buses arrived, and let T be the arrival time of the next bus. I have to wait a time $T - t$ to catch it. Now $T - t$ is the first jump time of the Poisson process $\{X_{t+s} - X_t,\ s \geq 0\}$, so it is exponential ($\lambda$), and my expected waiting time is $1/\lambda$.

But this already equals the expected interval between buses. I missed the last bus, so the inter-bus interval must be even longer. I missed it by $t - L$. (If there was no previous bus since the day began. set $L = 0$.) What is the distribution of $t - L$?

Let $s < t$. Now $L \leq s$ if $X_s = X_t$, so $P\{L \leq s\} = P\{X_t - X_s = 0\} = e^{-\lambda(t-s)}$. Thus, $P\{t - L \geq s\} = e^{-\lambda s}$ for $0 \leq s \leq t$, so $t - L$ is the minimum of t and an exponential (λ) random variable. If t is large, $E\{t - L\} \sim 1/\lambda$. Thus the expected time between the bus I caught and the bus I missed is $E\{T - L\} = E\{T - t\} + E\{t - L\} \sim 2/\lambda$. But the expected time between any two buses is $E\{T_n - T_{n-1}\} = 1/\lambda$. My interval is twice as long! I am truly unlucky to have chosen that time to arrive at the bus stop!

Question: Is there an intuitive reason that one might expect to arrive at the bus stop between two buses which are farther apart than average?

Problems 7.10

7.62. Show that there are infinitely many Poisson events in $(0, \infty)$.

7.63. Show that for a Poisson process (X_t) with transition function $p_t(k)$ that
$$P\{X_{t_j} = k_j,\ j = 1, \ldots, n\} = p_{t_1}(k_1) p_{t_2 - t_1}(k_2 - k_1) \ldots p_{t_n - t_{n-1}}(k_n - k_{n-1}).$$

7.64. A certain device will fail if it receives k shocks. The shocks arrive according to a Poisson (λ) process. Find the distribution function of its failure time.

7.65. Let $\{X_t,\ t \geq 0\}$ be a Poisson process with rate λ and let T be an independent exponential random variable with mean $1/\mu$. Find the distribution of X_T. [Hint: $\int_0^\infty x^n e^{-x}\, dx = n!$.]

7.66. Messages arrive according to a Poisson process, with an average rate of 3 per hour. Find the probability that no messages arrive from 8 a.m. till noon. Find the distribution of the arrival time of the first message of afternoon.

7.67. Let X_t be a Poisson process with parameter λ. Find the covariance of X_t and X_{t+s}.

7.68. Cars arrive at a gas station according to a Poisson process of rate λ. Trucks arrive as a Poisson process with rate μ.

(a) Find the probability that the first arrival after 10 a.m. is a truck.

(b) If the average number of cars is 10 per hour, the average number of trucks 2 per hour, find the probability that exactly one vehicle arrives between 10:00 and 10:05.

7.69. Let X_t be a Poisson process with parameter λ. Let $r < n$. Given that $X_0 = 0$ and $X_t = n$, find the distribution of the first hitting time T_r of r.

7.70. Passengers arrive at a train station according to a Poisson process with rate λ. Trains arrive according to an independent Poisson process with rate μ. When a train arrives, it picks up all waiting passengers. Find the long-term probability that n passengers are waiting. Compare this with the distribution of the number of passengers when the train first arrives. (See Exercise 7.65.)

7.11. Birth and Death Processes*

We want to describe the behavior of a population of individuals which evolves in continuous time. The situation is similar to a branching process except that births and deaths don't happen simultaneously, so that the

7.11. Birth and Death Processes*

population increases or decreases by one step at a time. The *rates* of births and deaths depend on the size of the population, however.

Represent the population size by a stochastic process $\{X_t,\ t \geq 0\}$ with values in $0, 1, 2, \ldots$. The sample paths of X look roughly—very roughly— like those of the Poisson process: the population is constant between jumps, but now a jump can be negative as well as positive.

We assume that the probabilities of births and deaths at any given time depend only on the state of the process at that time, so that X will be a Markov process. Since the birth and death probabilities depend only on the state, and not on the time, the transition probabilities will be stationary.

Remark 7.75. Birth and death processes are continuous-parameter Markov chains, that is to say, Markov processes on the positive integers which have a continuous time-parameter. A study in any depth of continuous time Markov chains would lead us too far afield, for the class contains processes whose behavior is, well, rather raffish—there was a time when the subject was regarded as the Wild West of probability—and, tempting though it is, we will have to limit our studies to one small subset. However, it is only small by comparison. It is large enough and interesting enough that its study will introduce the standard tools of the subject, and typical enough that the properties we discover will be shared by all well-behaved[16] Markov Chains.

Definition 7.76. The **transition probabilities** of X are
$$p_{ij}(t) \stackrel{\text{def}}{=} P\{X_t = i \mid X_0 = j\}, \quad i, j = 0, 1, 2, \ldots, \ t \geq 0\,.$$

The **transition probability matrix** is
$$\mathbb{P}(t) = \Big(p_{ij}(t)\Big)\,.$$

Note that $\mathbb{P}(0) = I$, the identity matrix. Then $P\{X_{s+t} = j \mid X_s = i\} = p_{ij}(t)$. Thus for all states i and times $t \geq 0$,
$$\sum_{j=0}^{\infty} p_{ij}(t) = 1\,.$$

Since $\{X_t,\ t \geq 0\}$ is a Markov process, then for every $s < t$ and every $s_1 < s_2 < \cdots < s_n < s$ and i, j and $i_1, \ldots, i_n$

(7.33) $\quad P\{X_t = j \mid X_s = i, X_{s_j} = i_j,\ j = 1, \ldots, n\} = p_{ij}(t - s)\,.$

---

[16]As in life, good behavior should be treasured, not assumed. Among the raffish relatives of such solid citizens as the Poisson process are some Markov chains which have infinitely many infinite discontinuities in each finite time interval. That took a while to understand. That is not surprising given that the understanding hinged on questions of topology, not probability.

Consequently (something we leave to the reader to verify), $\{X_t, t \geq s\}$ is conditionally independent of $\{X_u, u \leq s\}$ given X_s.

Hypotheses on the Transition Probabilities and Sample Paths. Let the birth rate and the death rate be λ_i and μ_i, respectively, when the population size is $X_t = i$. The first three of the following properties essentially define what we mean by "rates". The remaining three tell us about the rates themselves. We assume that for $h > 0$ and positive integers i and j,

1° $p_{i\,i+1}(h) = \lambda_i h + o(h)$.
2° $p_{i\,i-1}(h) = \mu_i h + o(h)$.
3° $p_{i\,i}(h) = 1 - (\lambda_i + \mu_i)h + o(h)$.
4° $p_{i,j}(0) = \delta_{ij}$.
5° $\mu_0 = 0$, $\lambda_0 > 0$.
6° If $i \geq 1$, $\lambda_i > 0$ and $\mu_i > 0$.

The parameter λ_i is the birth rate when the population is i: 1° says that the process goes from i to $i+1$—has a birth—at the rate λ_i. Similarly, μ_i is the death rate, the rate of going from i to $i-1$.

Note that 1°, 2° and 3° plus the fact that the transition probabilities sum to one imply that $\sum_{j:|j-i|\geq 2} p_{ij}(h) = o(h)$ as $h \to 0$. For short time periods, the process is unlikely to change by more than one.

Now let us make an assumption on the sample paths, one which hold for the Poisson process[17]. We could derive it from the above hypotheses, but that would take space and time that we wish to use for some new ideas. Therefore we will just accept it.

7° $\{X_t, t \geq 0\}$ *is a pure jump process. That is, for a.e. ω, the sample path $t \mapsto X_t(\omega)$ is right-continuous, positive integer-valued, constant except for jumps, and the jumps occur at discrete times $0 < T_1(\omega) \leq T_2(\omega) \leq T_3(\omega), \ldots$. Furthermore, for a.e. ω. $\lim_{n\to\infty} T_n(\omega) = \infty$.*

This hypothesis tells us that the process is constant between jumps. Furthermore, there are only finitely many jumps in any finite time interval, and therefore $\sup\{X_t : t \leq \tau\} < \infty$ for any $\tau > 0$.

The assumption that the jump times tend to infinity might seem to go without saying, but it is actually a restriction on the μ_i and λ_i. It always holds if they are bounded, but if they are not, the situation is more delicate, as we will see below.

---

[17]Birth and death processes are an intermediate step between the Poisson process and general continuous-time Markov chains, which need not satisfy these assumptions and can be entirely too wild for a sedate first introduction to stochastic processes like this. But they introduce some useful tools, such as infinitesimal generators and Kolmogorov differential equations.

7.11. Birth and Death Processes*

All jumps of the Poisson process are positive, but birth and death processes can have both positive and negative jumps. That makes a difference. For instance, if X is a Poisson process with $X_0 = 0$, then if $X_t = 0$, the first jump must be greater than t. However, if X is a birth and death process, this is no longer true, for there could be two jumps in $(0,t]$, one away from 0, one back, again leading to $X_t = 0$. Let us show that these double jumps do not occur too quickly.

Proposition 7.77. *X has no jumps of size two or greater. For each i, given $X_0 = i$, the probability of having two or more jumps in the interval $[0,h)$ is $O(h^2)$ as $h \to 0$.*

Proof. For the first statement, it is enough to show that there are no jumps of size two or greater in an interval of the form $(n, n+1]$. Since the transition probabilities are stationary, it is enough to show it for $(0,1]$. Let $M = \max\{X_t : 0 \le t \le 1\}$. Then M is finite by 7°, and it is enough to prove that for any $N > 0$,

$$P\{\exists \text{ jump of size} \ge 2 \text{ in } [0,1],\, M \le N\} = 0.$$

(The effect of this is to limit our attention to a finite number of states, on which the λ_i and μ_i are necessarily bounded.)

If there is a jump in $(0,1]$, say at time T, it must occur in some interval $(k/n, (k+1)/n]$. By 7°, the paths are constant for short intervals on either side of a jump, so for large enough n, the jump at T is $X_T(\omega) - X_{T-}(\omega) = X_{(k+1)/n}(\omega) - X_{k/n}(\omega)$. Therefore, it is enough to show that $P\{\exists k \le n : |X_{(k+1)/n} - X_{k/n}| \ge 2,\, M \le N\} \to 0$ as $n \to \infty$.

$$P\{\exists k \le n : |X_{(k+1)/n} - X_{k/n}| \ge 2,\, M \le N\}$$
$$\le \sum_{k=0}^{n-1} P\{|X_{(k+1)/n} - X_{k/n}| \ge 2,\, M \le N\}$$
$$\le \sum_{k=0}^{n-1} \sum_{i=0}^{N} P\{X_k = i\}\left(1 - p_{ii}(1/n) - p_{i,i+1}(1/n) - p_{i\,i-1}(1/n)\right).$$

By 1°–3°, the term in parentheses is $o(1/n)$, so for each i it is bounded by a function, $\rho_i(h)$ such that $\rho_i(h)/h \to 0$ as $h \to 0$. Put $\delta_N(h) = \max_{i \le N} \rho_i(h)$ and note that $\delta_N(h)/h \to 0$ as $h \to 0$. Then the above is

$$\le \sum_{k=0}^{n} \sum_{i=0}^{N} P\{X_{k/n} = i\}\delta_N(1/n) \le n\delta_N(1/n).$$

This tends to zero as $n \to \infty$, which is what we needed to prove.

Now consider the possibility of two or more jumps in $(0,h]$. If $X_0 = i$, then the first jump must be to $i \pm 1$, and the second from there to i or $i \pm 2$.

Divide $(0, h]$ into n equal intervals by $t_1 < t_2 < \cdots < t_n$, where $t_j = jh/n$. If $t \mapsto X_t(\omega)$ has two or more jumps in $(0, h]$, then for large enough n (and "large enough" depends on ω) there will exist $j < k \leq n$ for which

$$X_{t_j}(\omega) = i,\ |X_{t_{j+1}}(\omega) - X_{t_j}(\omega)| = 1,\ X_{t_k}(\omega) = i \pm 1,\ |X_{t_{k+1}}(\omega) - X_{t_k}(\omega)| = 1.$$

There are $n(n-1)/2$ ways to choose the pair (j, k) such that $j < k$, so the probability that this happens is

$$= \frac{n(n-1)}{2} P^i \{|X_{t_1} - X_0| = 1,\ |X_{t_2} - X_{t_1}| = 1\}$$

$$\leq \frac{n(n-1)}{2} (1 - p_{ii}(h/n))(1 - p_{i+1\,i+1}(h/n) + 1 - p_{i-1\,i-1}(h/n))$$

$$\leq \frac{n(n-1)}{2} \left((\lambda_i + \mu_i)\frac{h}{n} + o\!\left(\frac{h}{n}\right)\right)\left((\lambda_{i+1} + \mu_{i+1} + \lambda_{i-1} + \mu_{i-1})\frac{h}{n} + o\!\left(\frac{h}{n}\right)\right)$$

$$= \frac{1}{2}(\lambda_i + \mu_i)(\lambda_{i+1} + \mu_{i+1} + \lambda_{i-1} + \mu_{i-1})h^2 + o(h^2),$$

where we have made free use of 1°–3° and the Markov property. □

The Chapman-Kolmogorov Equations. By the Markov property,

$$P\{X_{t+s} = j \mid X_0 = i\} = \sum_{k=0}^{\infty} P\{X_{t+s} = j,\ X_t = k \mid X_0 = i\}$$

$$= \sum_{k=0}^{\infty} P\{X_{t+s} = j \mid X_t = k\} P\{X_t = k \mid X_0 = i\}.$$

When we write this in terms of the transition probabilities, we get the **Chapman-Kolmogorov equations**: if i and j are positive integers and s and t are positive reals, then

(7.34) $$p_{ij}(t+s) = \sum_{k=0}^{\infty} p_{ik}(t) p_{kj}(s).$$

The Absolute Distribution. Suppose $P\{X_0 = i\} = q_i$. Then $P\{X_t = j\} = \sum_{i=0}^{\infty} P\{X_0 = i\} P\{X_t = j \mid X_0 = i\}$, or

(7.35) $$P\{X_t = j\} = \sum_{i=0}^{\infty} q_i p_{ij}(t).$$

Kolmogorov's Differential Equations. The transition probabilities satisfy a system of differential equations. Let us calculate the derivative of $p_{ij}(t)$

7.11. Birth and Death Processes*

as follows:

$$\begin{aligned}p_{ij}(t+h) &= \sum_{k=0}^{\infty} p_{ik}(h)p_{kj}(t) \\ &= p_{i\,i-1}(h)p_{i-1\,j}(t) + p_{ii}(h)p_{ij}(t) + p_{i\,i+1}(h)p_{i+1\,j}(t) \\ &\quad + \sum_{k \neq i, i\pm 1} p_{ik}(h)p_{kj}(t).\end{aligned}$$

Now $p_{i+1\,j}(t) \leq 1$ so the final term is bounded by $\sum_{\{k:|k-i|\geq 2\}} p_{ik}(h) = o(h)$. Thus we have

$$\begin{aligned}p_{ij}(t+h) - p_{ij}(t) &= (p_{ii}(h) - 1)p_{ij}(t) + p_{i\,i-1}(h)p_{i-1\,j}(t) \\ &\quad + p_{i\,i+1}(h)p_{i+1\,j}(t) + o(h) \\ &= -(\mu_i + \lambda_i)hp_{ij}(t) + \mu_i h p_{i-1\,j}(t) + \lambda_i h p_{i\,i+1}(t) + o(h).\end{aligned}$$

Divide by h and let $h \to 0$ to see:

Theorem 7.78 (Kolmogorov Backward Equations). *For $t > 0$ and states i and j,*

$$(7.36) \qquad \frac{d}{dt}p_{ij}(t) = \mu_i p_{i-1\,j}(t) + \lambda_i p_{i+1\,j}(t) - (\lambda_i + \mu_i)p_{ij}(t).$$

We can write this in terms of matrices. Let $\mathbb{P}(t) = (p_{ij}(t))$ be the transition probability matrix.

Definition 7.79. The **infinitesimal generator** $\mathbb{G}$ of $\mathbb{P}$ is

$$\mathbb{G} \stackrel{\text{def}}{=} \left.\frac{d}{dt}\mathbb{P}(t)\right|_{t=0}.$$

We can compute $\mathbb{G}$ from the properties 1°–6°:

$$(7.37) \qquad \begin{cases} p'_{ii}(0) = -(\lambda_i + \mu_i), \\ p'_{i\,i+1}(0) = \lambda_i, \\ p'_{i\,i-1}(0) = \mu_i, \\ p'_{ij}(0) = 0 \text{ if } |j-i| \geq 2.\end{cases}$$

Then

$$\mathbb{G} = \begin{pmatrix} -\lambda_0 & \lambda_0 & 0 & 0 & 0 & \cdots \\ \mu_1 & -\lambda_1 - \mu_1 & \lambda_1 & 0 & 0 & \cdots \\ 0 & \mu_2 & -\lambda_2 - \mu_2 & \lambda_2 & 0 & \cdots \\ 0 & 0 & \mu_3 & \lambda_3 - \mu_3 & \lambda_3 & \cdots \\ & \vdots & & \ddots & \ddots & \ddots \end{pmatrix}.$$

The infinitesimal generator gives the rates of jumping from one state to another, i.e., the rates of birth and death for each state. The state 0 is special, since $\mu_0 = 0$. Note that the row sums of $\mathbb{G}$ are always zero.

In matrix terms, the Chapman-Kolmogorov equations are
$$\mathbb{P}(s+t) = \mathbb{P}(s)\mathbb{P}(t),$$
and the backward Kolmogorov equations are
$$(7.38) \qquad \frac{d}{dt}\mathbb{P}(t) = \mathbb{G}\mathbb{P}(t).$$

Remark 7.80. (*i*) The formal solution of (7.38) with initial condition $\mathbb{P}(0) = I$ is $\mathbb{P}(t) = e^{t\mathbb{G}}$. (This makes sense if we interpret $e^{t\mathbb{G}} = \sum_{n=0}^{\infty} \frac{t^n}{n!}\mathbb{G}^n$, with $\mathbb{G}^0 = I$.)

(*ii*) The derivation of the backward Kolmogorov equations can be written, formally at least, in terms of matrices:
$\mathbb{P}(t+h) = \mathbb{P}(h)\mathbb{P}(t)$ so that $\mathbb{P}(t+h) - \mathbb{P}(t) = (\mathbb{P}(h) - \mathbb{P}(0))\mathbb{P}(t)$ (using $\mathbb{P}(0) = I$. Divide by h and let $h \to 0$ to see that $\mathbb{P}'(t) = \mathbb{P}'(0)\mathbb{P}(t) = \mathbb{G}\mathbb{P}(t)$.

There is another system of differential equations. Note that $\mathbb{P}(t+h) = \mathbb{P}(t)\mathbb{P}(h)$, leading to $\mathbb{P}(t+h) - \mathbb{P}(t) = \mathbb{P}(t)(\mathbb{P}(h) - \mathbb{P}(0))$. Divide by h and let $h \to 0$ to get
$$\mathbb{P}'(t) = \mathbb{P}(t)\mathbb{G}$$
or
$$(7.39) \qquad \frac{d}{dt}p_{ij}(t) = \lambda_i\, p_{i\,j-1}(t) - (\lambda_i + \mu_i)\, p_{ij}(t) + \mu_i\, p_{i\,j+1}(t),$$
which are[18] the **Kolmogorov forward differential equations.**

Notice that the backward and forward equations are different. Our derivation of the forward equations was purely formal. A careful derivation would reveal that some further assumptions are needed to take the limit, and indeed, unlike the backward equations, the forward equations do not always hold. That is why one usually uses the backward equations.

Example 7.80.1. The infinitesimal generator of the Poisson process is
$$\mathbb{G} = \begin{pmatrix} -\lambda & \lambda & 0 & 0 & 0 & \dots \\ 0 & -\lambda & \lambda & 0 & 0 & \dots \\ 0 & 0 & -\lambda & \lambda & 0 & \dots \\ 0 & 0 & 0 & \lambda & \lambda & \dots \\ & & \vdots & & \ddots & \ddots & \ddots \end{pmatrix}.$$

---

[18] The designations "backward" and "forward" have to do with whether the increment of length h is on the backward or forward side of the interval $(0, t+h)$, i.e., whether we go from 0 to h, and then to $t+h$, or whether we go from 0 to t, and then to $t+h$.

7.11. Birth and Death Processes*

A pure birth process has zero death rates, so its infinitesimal generator is of the form

$$\mathbb{G} = \begin{pmatrix} -\lambda_0 & \lambda_0 & 0 & 0 & 0 & \cdots \\ 0 & -\lambda_1 & \lambda_1 & 0 & 0 & \cdots \\ 0 & 0 & -\lambda_2 & \lambda_2 & 0 & \cdots \\ 0 & 0 & 0 & \lambda_3 & \lambda_3 & \cdots \\ \vdots & & & \ddots & \ddots & \ddots \end{pmatrix}.$$

Here is an example of the use of the Kolmogorov equations. It has a rather surprising consequence: a birth and death process can get from anywhere to anywhere else in an arbitrarily small amount of time.

Proposition 7.81. *For all $t > 0$ and all states i and j, $p_{ij}(t) > 0$.*

Proof. Notice that $p_{ii}(t) \geq P^i\{\text{first jump is } > t\} = e^{-(\lambda_i + \mu_i)t} > 0$. Suppose $i < j$. (The case $i > j$ is similar.) By 1° and 5°,

$$\frac{d}{dt} p_{i\,i+1}(0) = \lambda_i > 0.$$

Thus, $p_{i\,i+1}(h) > 0$ for all small enough h. Choose a strictly positive $h < t/j$ small enough that $p_{k\,k+1}(h) > 0$ for $k = i, i+1, \ldots, j-1$. Then

$$p_{ij}(t) \geq \underbrace{p_{i\,i+1}(h) p_{i+1\,i+2}(h) \ldots p_{j-1,j}(h)}_{>0} \underbrace{p_{jj}(t - (j-1)h)}_{>0}.$$

$\square$

Embedded Chains and Stationary Distributions. Fix $h > 0$. Then $X_0, X_h, X_{2h}, \ldots$ is a discrete time Markov chain with transition probabilities $p_{ij} = p_{ij}(h)$. This is a Markov chain, but it is *not* a birth and death process, since, as all transition probabilities are strictly positive, it can jump more than one step at a time. Since $p_{ij}(h) > 0$ for all $h > 0$, the chain is irreducible and aperiodic.

It follows that the transition probabilities have a limit.

Theorem 7.82. *For all j: $\lim_{t \to \infty} p_{ij}(t) \stackrel{\text{def}}{=} \pi_j$ exists, independent of i.*

Proof. For any $h > 0$, (X_{nh}) is an irreducible, aperiodic chain, so the limits of its transition probabilities, $\lim_{n \to \infty} p_{ij}(nh)$, exist. For each j, define

$$\pi_j = \lim_{n \to \infty} p_{jj}(n)$$

and note that $\lim_{n \to \infty} p_{ij}(n) = \pi_j$ for any state i. We claim that the limit over all t exists. First, take $h = 1/m$, and notice that, $\lim_{n \to \infty} p_{ij}(nh)$ exists for all i, and, since $(p_{ij}(n))$ is a subsequence of $(p_{ij}(nh))$, the limit must equal π_j.

By Kolmogorov's backward differential equations,

$$|p'_{ij}(t)| = |\mu_i p_{i-1\,j}(t) + \lambda_i p_{i+1\,j}(t) - (\lambda_i + \mu_i) p_{ij}(t)| \leq \mu_i + \lambda_i\,.$$

Thus, $p'_{ij}(t)$ is bounded, independent of t. Let $\varepsilon > 0$ and choose h small enough so that $|p_{ij}(t) - p_{ij}(s)| < \varepsilon$ if $|t - s| < h$. Then $\sup_{s \leq h} |p_{ij}(t+s) - p_{ij}(t)| \leq \varepsilon$. Consequently,

$$\pi_j = \lim_{n \to \infty} p_{ij}(nh) \leq \limsup_{t \to \infty} p_{ij}(t) \leq \lim_{n \to \infty} p_{ij}(nh) + \varepsilon = \pi_j + \varepsilon\,.$$

Similarly, $\limsup_{t \to \infty} p_{ij}(t) \geq \pi_j - \varepsilon$. Since ε is arbitrary, we are done.
□

To summarize, all the transition probabilities tend to limits. These must equal the limits along the integers. Since $X_0, X_1, X_2, \ldots$, is a discrete-time Markov chain, the results for Markov chains carry over. For instance, we know that the chain $X_0, X_1, X_2, \ldots$, is irreducible and aperiodic. Therefore all the states are either transient, null-recurrent, or positive recurrent. In the first two cases, all the $p_{ij}(t)$ tend to zero as $t \to \infty$. In the third, the chain has a stationary probability distribution (π_j), and as $t \to \infty$, $P\{X_t = j\} \longrightarrow \pi_j$ for all j and X_t approaches statistical equilibrium.

Let us use this to derive a criterion for the stationary distribution in terms of the infinitesimal generator. We will do this purely formally, and leave it to the interested reader to supply the rigor.

If $\pi = (\pi_0, \pi_1, \ldots)$ is a stationary distribution, then

$$\pi \mathbb{P}(t) = \pi\,, \quad \text{all } t > 0\,.$$

Then

$$\pi \frac{1}{t} \bigl(\mathbb{P}(t) - \mathbb{P}(0) \bigr) = 0\,.$$

Take the limit as $t \to 0$, noting that $\bigl(\mathbb{P}(t) - \mathbb{P}(0) \bigr)/t \longrightarrow \mathbb{G}$. This implies that

(7.40) $$\pi \mathbb{G} = 0\,.$$

Thus a positive-recurrent birth and death process has a stationary probability distribution (π_j), which satisfies the following system of equations:

$$-\lambda_0 \pi_0 + \mu_1 \pi_1 = 0\,,$$
$$\lambda_0 \pi_0 - (\lambda_1 + \mu_1) \pi_1 + \mu_2 \pi_2 = 0\,,$$
$$\lambda_1 \pi_1 - (\lambda_2 + \mu_2) \pi_2 + \mu_2 \pi_3 = 0\,,$$
$$\vdots \qquad \vdots$$

7.11. Birth and Death Processes*

First Jump Time for a Birth and Death Process. Let $X_0 = i$. The first jump time of X is

$$T = \inf\{t > 0 : X_t \neq i\}.$$

Notation. We denote the probabilities and expectations given that $X_0 = i$ by P^i and E^i respectively:

$$P^i\{\Lambda\} = P\{\Lambda \mid X_0 = i\} \quad E^i\{Y\} = E\{Y \mid X_0 = i\}.$$

Proposition 7.83. *Let $X_0 = i$. Then the first jump time from i is exponential with parameter $\lambda_i + \mu_i$.*

Proof. Let $X_0 = i$. Note that $T > t \implies X_{t/n} = X_{2t/n} = \cdots = X_t = i$, and if the equality holds for all n, then $T > t$. (If $X_{t/n} = X_{2t/n}$, for instance, there are either no jumps or at least two jumps in $(t/n, 2t/n)$, but the probability of the latter is $O(t^2/n^2)$, so the probability of that happening in at least one of the n intervals is therefore at most $O(1/n)$.) Thus,

$$
\begin{aligned}
P^i\{T > t\} &= \lim_{n\to\infty} \left(P^i\{X_{t/n} = X_{2t/n} = \cdots = X_t = i\} + O(1/n)\right) \\
&= \lim_{n\to\infty} p_{ii}(t/n)^n \\
&= \lim_{n\to\infty} \left(1 - \frac{(\lambda_i + \mu_i)t + o(1)}{n}\right)^n \\
&= e^{-(\lambda_i + \mu_i)t}.
\end{aligned}
$$

□

The first jump time is exponential. But what is the distribution of the first jump place? The next proposition gives the joint distribution of the first jump T and X_T.

Theorem 7.84. *Let $X_0 = i$ and let T be the first jump time of X. Then T is exponential $(\lambda_i + \mu_i)$, T and X_T are independent and*

$$P^i\{X_T = i+1\} = \frac{\lambda_i}{\lambda_i + \mu_i}, \quad P^i\{X_T = i-1\} = \frac{\mu_i}{\lambda_i + \mu_i}$$

and

$$P^i\{X_T = j\} = 0 \quad \text{if } |j - i| \geq 2.$$

Proof. Let $X_0 = i$ and $j \neq i$. Now $T > t \implies X_t = i$, so for $t, h > 0$,

$$P^i\{T \in (t, t+h], X_T = j\} = P^i\{T > t\}P^i\{T \leq t+h, X_T = j \mid X_t = i, T > t\}.$$

If $T > t$, then T is the first jump time after t, or, for the process restarted at t, the very first jump. Apply the Markov property at time t; this is

$$= P^i\{T > t\}P^i\{T \leq h,\ X_T = j\}.$$

Let $h \to \infty$ to see that $P^i\{T > t,\ X_T = j\} = P^i\{T > t\}P^i\{X_T = j\}$. This proves independence. To find the distribution of X_T, let h be small. Note that if $X_0 = i \neq j$, then the events $\{T \leq h\}$ and $\{X_h = j\}$ only differ when there are two or more jumps in $[0, h]$, which has probability $O(h^2)$. Thus

$$P\{X_T = j\} = \sum_{k=0}^{\infty} P^i\{T \in (kh, (k+1)h],\ X_T = j\}$$

$$= \sum_{k=0}^{\infty} P^i\{T > kh\}(P^i\{X_h = j\} + O(h^2))$$

$$= (p_{ij}(h) + O(h^2)) \sum_{k=0}^{\infty} e^{-(\lambda_i + \mu_i)kh}.$$

Sum the geometric series and take the limit as $h \to 0$, to see that

$$P^i\{X_T = j\} = \frac{p'_{ij}(0)}{\lambda_i + \mu_i}, \quad \text{if } j \neq i.$$

□

When we proved the strong Markov property for the Poisson process, we mentioned that it held for much more general Markov processes. In particular, it holds for birth and death processes. We leave it as an exercise.

Exercise 7.71. Let $\{X_t, t \geq 0\}$ be a birth and death process with transition probability matrix $\mathbb{P}(t)$. Let $\mathcal{F}_t = \sigma\{X_s,\ s \leq t\}$. Let T be a stopping time. Prove the following theorem.

> **Theorem** (Strong Markov Property). Let T be an a.e. finite stopping time. Then the process $\{X_{T+t},\ t \geq 0\}$ is a birth and death process with the same transition probabilities $\mathbb{P}(t)$, and it is independent of $\mathcal{F}_T$.

The Embedded Jump Chain. Let $\{X_t,\ t \geq 0\}$ be a birth and death process, and let $T_1, T_2, \ldots$ be its successive jump times.

Definition 7.85. $X_0, X_{T_1}, X_{T_2}, \ldots$ is the **embedded jump chain**.

The following theorem is a direct consequence of the strong Markov property and Theorem 7.84.

Theorem 7.86. *The embedded jump chain is a discrete-time Markov chain. For each n, given X_{T_n}, $T_{n+1} - T_n$ and $X_{T_{n+1}}$ are conditionally independent. Its transition probability matrix $\mathbb{P}$ is*

$$\mathbb{P} = \begin{pmatrix} 0 & 1 & 0 & 0 & 0 & \cdots \\ \frac{\mu_1}{\lambda_1 + \mu_1} & 0 & \frac{\lambda_1}{\lambda_1 + \mu_1} & 0 & 0 & \cdots \\ 0 & \frac{\mu_2}{\lambda_2 + \mu_2} & 0 & \frac{\lambda_2}{\lambda_2 + \mu_2} & 0 & \cdots \\ & & \vdots & & \vdots & \end{pmatrix}.$$

7.11. Birth and Death Processes*

Proof. If $Y_t \stackrel{def}{=} X_{T_n+t}$, $t \geq 0$, then Y is again a birth and death process with the same transition probabilities, conditionally independent of $\mathcal{F}_T$ given Y_0. Its first jump time is $S \stackrel{def}{=} T_{n+1} - T_n$. Suppose $X_{T_n} = i$, so $Y_0 = i$. By Theorem 7.84, S is exponential $(\lambda_i + \mu_i)$, is (conditionally) independent of $Y_S = X_{T_{n+1}}$, and $P^i\{Y_S = j\} = p'_{ij}(0)/(\lambda_i + \mu_i)$. Thus,

$$P\{X_{T_{n+1}} = j \mid \mathcal{F}_T, X_{T_n} = i\} = P\{X_{T_{n+1}} = i + 1 \mid X_{T_n} = i\}$$
$$= P^i\{Y_S = j\} = \frac{p'_{ij}(0)}{\lambda_i + \mu_i}.$$

Thus, (X_{T_n}) is a Markov chain with the given transition probabilities. □

Remark 7.87. The times spent in a given state waiting for the next jump are called *waiting times*. The waiting time in each state is exponential. Successive waiting times are conditionally independent given the states. We can describe the path informally by saying that the process waits in its present state for an exponential length of time, then jumps to a new state according to the jump chain's transition probabilities. It waits a further exponential time in the new state before jumping to a third state, and so on.

Explosions!*. Consider a pure birth process $\{X_t,\ t \geq 0\}$—a birth and death process, except that there are no deaths: $\mu_i = 0$ for all i. A Poisson process is a pure birth process, for example. Let $\lambda_i > 0$, $i = 0, 1, 2, \ldots$. Let $T_1, T_2, \ldots$ be its successive jump times.

The embedded jump chain is simple: if $X_0 = 0$, then $X_{T_1} = 1$, $X_{T_2} = 2$, $X_{T_3} = 3$, and so on. So clearly, $X_{T_n} \to \infty$. The question is: How long does it take to get there?

Define $S_1 = T_1$, $S_2 = T_2 - T_1$, $S_3 = T_3 - T_2$, and so on. Now $X_{T_n} = n$ so that the post-T_n process $\{X_{T_n+t},\ t \geq 0\}$ is again a pure-birth process, and its first jump time is $S_n = T_{n+1} - T_n$. Now S_n is exponential (λ_n), and, by the strong Markov property, it is independent of $\mathcal{F}_{T_n}$ and hence independent of all the previous T_j, $j \leq n$, and also of all the previous S_j. Thus $S_1, S_2, \ldots$ is a sequence of independent exponential random variables, and $T_n = S_1 + \cdots + S_n$.

Let $T_\infty = \lim_{n \to \infty} T_n$. Apply the monotone convergence theorem:

$$E\{T_\infty\} = \lim_{n \to \infty} E\{T_n\} = \sum_{n=1}^\infty E\{S_n\} = \sum_{n=1}^\infty \frac{1}{\lambda_n}.$$

Suppose that the λ_n grow so fast with n that $\sum_n (1/\lambda_n) < \infty$. Then T_∞ has a finite expectation, and therefore is a.e. finite. But this means that the process blows up at a finite time! That is, if $T_\infty = \lim_n T_n$, and if $T_\infty < \infty$, then the process itself tends to infinity at T_∞: $\lim_{t \to T_\infty} X_t = \infty$.

In other words, if $\sum_n (1/\lambda_n) < \infty$, then X blows up in finite time.

This brings up a few problems, such as, "How is the process defined after T_∞? "Or is it?" "In fact, how is it defined *at T_∞?*"

The answer is that it is *not* defined... yet. Outside of the fact that $\mu_i \equiv 0$—which is not vital to this example—it satisfies all our hypotheses except the requirement in 7° that $\lim_n T_n = \infty$. The hypotheses on the birth and death process tell us what it does when it is finite—but not what it does at T_∞. Something new has to be added to deal with this case. There are a number of possibilities. One is to simply kill the chain at T_∞. This leaves it undefined after T_∞, but this is not a serious problem[19]. In this case, the transition probabilities will not sum to one: $\sum_{j\geq 0} p_{ij}(t) < 1$ in general. These are called the *minimal transition probabilities*.

One can extend the process past T_∞, essentially by introducing a transition probability that kicks in at T_∞. For instance, we can arbitrarily define $X_{T_\infty} = 0$. This means that the process jumps back from infinity to zero at time T_∞, and then recommences. It will eventually explode again, and after the second explosion, it resets to zero once more, and goes on to a third explosion, and so on. Or we could decree that the process jumps back into the state space according to some probability measure. Even this does not exhaust all the possibilities: there are even wilder things the process might do. (Which is a small part of the reason that continuous time Markov chains were considered to be the Wild West of probability... for the time it took them to understand them. Then the Wild West moved on.)

In short we can extend our pure birth process past T_∞ to be a continuous-parameter Markov chain, but the extension is not unique. The transition probabilities of any such extension will satisfy the Kolmogorov backward equations[20]. This means that the Kolmogorov backward differential equations do not always uniquely determine the transition probabilities, for each different way of continuing the chain past T_∞ gives yet another set of transition probabilities which satisfies the backward equations.

Problems 7.11

7.72. A Markov chain has two states, 0 and 1. The waiting times in each are exponential (λ).

(a) Find $P_{00}(t)$.

(b) Let N_t be the number of times the system changes states before time t. Find the distribution of N_t. What is its expectation?

7.73. According to (7.40), if π is a stationary distribution for a birth and death process, then $\pi \mathbb{G} = 0$. Prove it.

---

[19]Indeed, one can take a state not in the original state space, say -1, for instance, and simply set $X_t = -1$ for $t \geq T_\infty$.

[20]...though not the forward equations!

7.11. Birth and Death Processes*

7.74. Consider a pure birth process starting from zero in which all odd states have same parameter, λ_{odd}, and all even states have the same parameter, λ_{even}. If $X_0 = 0$, find $P_1(t) = P\{X_t \text{ odd}\}$, and $P_2(t) = P\{X_t \text{ even}\}$.

[Hint: This can be reduced to a two-state process.]

7.75. Consider branching in continuous time. Let X_t be the number of individuals alive at time t. Let $\lambda > 0$, $\mu > 0$. At time t, each individual has a probability $\lambda h + o(h)$ of giving birth to an additional individual in the interval $(t, t+h)$, and probability $\mu h + o(h)$ of dying. Different individuals are independent. Write this as a birth and death process. What are the λ_i and μ_i?

If $X_0 = 1$, find the probability that the population becomes extinct, i.e., that X_t eventually reaches zero.

[Hint: See Exercise 7.81.]

7.76. Let (X_n) be a birth and death process with $\lambda_0 = \lambda$, $\mu_0 = 0$ and for $n \geq 1$, $\lambda_n = n\lambda$ and $\mu_n = n\mu$, where λ and μ are strictly positive. Prove that X_n is recurrent if $\lambda < \mu$, and find its stationary distribution. What can you say if $\lambda = \mu$?

7.77. Customers arrive in a queue according to a Poisson process of rate λ. The service time for the customer at the head of the queue is exponential (μ). Each customer stays in the queue for an exponential (δ) length of time, after which time the customer gives up and leaves. All these quantities are independent.

(a) Model this by a birth and death process, and show that it is recurrent for any strictly positive λ, μ, and δ.

(b) Show the stationary distribution of the queue length is $\pi_n = \frac{\lambda^n}{\prod_{k=1}^n (\mu k \delta)} \pi_0$. Give the value of π_0 in terms of a series.

(c) Assume that π_0 is known. Find the average length of a busy period.

(d) Find the proportion of customers who are eventually served.

[Hint: Exercises 7.80 and 7.81 will help.]

7.78. Suppose the **conditional rate of failure** of an object is $g(t)$, i.e., that the probability that the first failure occurs in $(t, t+h)$ given no failure occurs before t is $g(t)h + o(h)$. Derive the distribution function of the first failure time.

[Hint: Derive a differential equation.]

7.79. Consider a linear growth birth and death process: $\lambda_n = n\lambda$, $\mu_n = n\mu$. Suppose the initial value is one. Find the distribution of the number living at time of the first death.

Many properties of birth and death processes can be determined by first finding the values of some function, first on the states 0 and 1, and then on all other states by induction. In what follows, (X_t) is a birth and death process with strictly positive rates λ_i and μ_i.

7.80. (Stationary Distribution.) If (π_j) are stationary probabilities for (X_t), then we have seen that $\lambda_0 \pi_0 = \mu \pi_1$, and for $j \geq 2$,

(7.41) $$(\lambda_j + \mu_j)\pi_j = \lambda_{i-1}\pi_{i-1} + \mu_{j+1}\pi_{j+1}.$$

Show that $\pi_k = \frac{\lambda_0 \cdots \lambda_{n-1}}{\mu_1 \cdots \mu_k} \pi_0$. If $C = 1 + \sum_{n=1}^{\infty} \frac{\lambda_0 \cdots \lambda_{n-1}}{\mu_1 \cdots \mu_n} < \infty$, show that the stationary distribution is given by $\pi_0 = 1/C$ and

(7.42) $$\pi_n = \frac{1}{C} \frac{\lambda_0 \cdots \lambda_{n-1}}{\mu_1 \cdots \mu_n}, \quad n \geq 1.$$

7.81. (Recurrence.) Let $(\hat{X}_n)$ be the embedded jump chain, which has $P_{i\,i+1} = \frac{\lambda_i}{\lambda_i + \mu_i}$ and $P_{i\,i-1} = \frac{\mu_i}{\lambda_i + \mu_i}$. Let $v_0(i) = P^i\{\hat{X}_n = 0, \text{ some } n \geq 0\}$.

(a) Show that $v_0(0) = 1$ and for $i \geq 1$,

(7.43) $$v_0(i) = \frac{\lambda_i}{\lambda_i + \mu_i} v_0(i+1) + \frac{\mu_i}{\lambda_i + \mu_i} v_0(i-1).$$

(b) Show that $v_0(i+1) - v_0(i) = \frac{\mu_i}{\lambda_i}(v_0(i) - v_0(i-1))$.

(c) Set $d = v_0(0) - v_0(1)$. Show that for $k \geq 2$,

$$v_0(k) = 1 - d \sum_{j=1}^{k} \frac{\mu_1 \cdots \mu_j}{\lambda_1 \cdots \lambda_j}.$$

(d) Show that a necessary and sufficient condition that $\hat{X}_n$ be recurrent is that

(7.44) $$\sum_{j=1}^{\infty} \frac{\mu_1 \cdots \mu_j}{\lambda_1 \cdots \lambda_j} = \infty.$$

(e) Suppose the sum in (7.44) equals $C < \infty$. Find $v_0(k)$, $k = 0, 1, 2, \ldots$.

[Hint: You may use the fact that the first-hitting probability of 0 is the smallest solution of (7.43) with $v_0(0) = 1$. See Exercise 9.31.]

7.82. There is a timed light switch in the corridor of an apartment building in Paris. If someone enters the corridor when it is dark, they press the switch and the light goes on. If the light is on when they enter, they just go on their way. The light stays on for an exponential time, lasting four minutes on the average. Suppose people arrive according to a Poisson process at the rate of one person every five minutes on the average. Find the stationary probabilities for the light being on and off.

Would it make any difference if each person automatically pressed the button to reset the timer and give a further exponential time, whether or not the light is on?

Chapter 8

Conditional Expectations

8.1. Conditional Expectations

Expectations determine probabilities, for the probability of a set equals the expectation of its indicator function. Once we have the expectations, we also have the probabilities. This is also true for conditional probabilities, so we will focus on conditional expectations, rather than conditional probabilities.

We defined the conditional expectation of a random variable X given the value of another random variable Y in §3.7. We did this for discrete-valued random variables[1]. We would like to be able to define it for arbitrary Y. Even then, there are some disadvantages in restricting our conditioning to random variables. For instance:

• The notation gets clumsy when several random variables are involved:

$$E\{X \mid Y, Z, W, Q, \dots\}.$$

• We are essentially limited to discrete random variables.

• More basically, while additional information can change the probabilities and therefore the expectations, it may not be easily expressible in terms of random variables. (A groom at a race track claims that a certain filly has been held back in her previous races, but they are going to let her run full-out in this race. The groom looks seedy and broke, so his tip may not be too reliable, but still.... This can be important information for a bettor,

---

[1] We also did it for the case in which X and Y have a joint density, but this was by a trick, and is a dead-end: it will not help us here.

and it may well change the conditional probability that the filly wins, but it is not easily put in terms of random variables.)

As we have said, σ-fields provide a very flexible and powerful (if slightly abstract) way of encoding information. We will use them extensively.

It is time for a radical extension of the idea of a conditional expectation, one which will get around the above objections. The key is to concentrate on σ-fields rather than on random variables[2].

Intuitively, σ-fields **are** information. Thus we will make a change of notation, and write $E\{X \mid \sigma(Y)\}$ instead of $E\{X \mid Y\}$: we condition on the σ-field, rather than the random variable. Recall how $E\{X \mid Y\}$ was defined.

If Y is discrete, with possible values $y_1, y_2, \ldots$, define $\phi(y)$ by

$$\phi(y) = \begin{cases} \frac{1}{P\{Y=y\}} \int_{\{Y=y\}} X \, dP & \text{if } P\{Y=y\} > 0, \\ 0 & \text{otherwise}. \end{cases}$$

Then $E\{X \mid \sigma(Y)\} \stackrel{\text{def}}{=} \phi(Y)$.

Notice two things. First, $E\{X \mid \sigma(Y)\}$ is a random variable: conditional expectations are random variables. Second, it is constant on each set $\{Y = y\}$, and its value is the average of X on that set: conditional expectations are averages. Let us pursue these thoughts to derive two fundamental properties. We already know the first.

(i) $E\{X \mid \sigma(Y)\}$ is a $\sigma(Y)$-measurable random variable.

For the second, note that if $\Lambda_i = \{Y = y_i\}$, then

$$\int_{\Lambda_i} E\{X \mid Y\} \, dP = \int_{\Lambda_i} \underbrace{\phi(Y)}_{\phi(y_i)} \, dP = \phi(y_i) P\{\Lambda_i\} = \int_{\Lambda_i} X \, dP.$$

We can extend this to $\sigma(Y)$, since any $\Lambda \in \sigma(Y)$ is a finite or countable disjoint union of the Λ_i, and it follows that

(ii) $\Lambda \in \sigma(Y) \Longrightarrow \int_\Lambda E\{X \mid \sigma(Y)\} \, dP = \int_\Lambda X \, dP.$

It turns out that properties (i) and (ii) characterize $E\{X \mid \sigma(Y)\}$. That leads us to the following definition of conditional expectations with respect to σ-fields.

Definition 8.1. Let $(\Omega, \mathcal{F}, P)$ be a probability space, let X be an integrable random variable and let $\mathcal{G}$ be a sub-σ-field of $\mathcal{F}$. We say that a random variable Z is the **conditional expectation of X given $\mathcal{G}$** if

(CE1) Z is $\mathcal{G}$-measurable, and

---

[2] We are about to introduce the conditional expectations created by Kolmogorov in [21]. We do some boot-strapping to bypass the initial lack of a general existence theorem (it comes at the end, see Theorem 9.37 and Exercise 9.35) but apart from this, our exposition is entirely standard.

8.1. Conditional Expectations

(CE2) $\Lambda \in \mathcal{G} \implies \int_\Lambda Z\,dP = \int_\Lambda X\,dP$.

We denote the conditional expectation by $E\{X \mid \mathcal{G}\}$.

Just as with $E\{X \mid Y\}$, a conditional expectation is a random variable, not just a number. We have defined it by its properties rather than by a formula. This is perfectly legal as long as the properties uniquely determine it. So let us check that the conditional expectation is well-defined.

Proposition 8.2. *If Z and Z' are random variables which satisfy* (CE1) *and* (CE2), *then they are integrable and $Z = Z'$ a.e.*

Proof. Suppose both Z and Z' satisfy (CE1) and (CE2). It is implicit in (CE2) that Z and Z' are integrable. We claim that $P\{Z = Z'\} = 1$. Indeed, the set $\Lambda \stackrel{\text{def}}{=} \{Z' > Z\} \in \mathcal{G}$, so that by (CE2)

$$\int_\Lambda (Z' - Z)\,dP = \int_\Lambda Z'\,dP - \int_\Lambda Z\,dP = \int_\Lambda X\,dP - \int_\Lambda X\,dP = 0.$$

Thus, $P\{Z' > Z\} = 0$, and by symmetry, $P\{Z' < Z\} = 0$, too, which verifies the claim. $\square$

Any two versions of the conditional expectation are a.s. equal. In general, this is the best we can do: the conditional expectation is only defined up to null-sets. We could make the definition unique by defining $E\{X \mid \mathcal{G}\}$ to be an equivalence class, or an element of $L^1(\Omega, \mathcal{F}, P)$. However, it is easier to simply remember that the conditional expectation is defined up to sets of measure zero, and that any equation involving it is only true almost surely[3].

Example 8.2.1. Let X be an integrable random variable. Suppose $\mathcal{G}$ is generated by a countable partition of Ω into disjoint events $\Lambda_1, \Lambda_2, \ldots$. Then $\bigcup_i \Lambda_i = \Omega$, and any set in $\mathcal{G}$ is a finite or countable union of the Λ_i. Note that the Λ_i are atoms of $\mathcal{G}$: no non-empty proper subset of Λ_i can be in $\mathcal{G}$, so that any $\mathcal{G}$-measurable random variable is constant on each Λ_i. Let Z be $\mathcal{G}$-measurable, with $Z = \lambda_i$ on Λ_i. If Z satisfies (ii), then $\lambda_i P\{\Lambda_i\} = \int_{\Lambda_i} Z\,dP = \int_{\Lambda_i} X\,dP$, so if $P\{\Lambda_i\} > 0$, then $\lambda_i = (1/P\{\Lambda_i\}) \int_{\Lambda_i} X\,dP$. (For the sake of completeness, we define $\lambda_i = 0$ if $P\{\Lambda_i\} = 0$.) This leads to the formula for $E\{X \mid \mathcal{G}\}$:

(8.1) $$E\{X \mid \mathcal{G}\}(\omega) = \sum_{i: P\{\Lambda_i\} > 0} \frac{I_{\Lambda_i}(\omega)}{P\{\Lambda_i\}} \int_{\Lambda_i} X\,dP.$$

---

[3] This brings up the curious fact that the equation $E\{X \mid \mathcal{G}\} = E\{X \mid \mathcal{G}\}$ is only true a.e., since the versions of the conditional expectations on the two sides of the equation could be different!

The Λ_i are disjoint, so the series converges for all ω. There is something more to check: that the above formula does indeed satisfy (CE1) and (CE2). But (CE1) is clear, for I_{Λ_i} is $\mathcal{G}$-measurable, hence so is the sum. To check (CE2), notice that it holds for $\Lambda = \Lambda_i$, and that the class of sets for which it holds is closed under countable disjoint unions, and therefore contains $\mathcal{G}$.

Thus $E\{X \mid \mathcal{G}\}$ exists for a σ-field generated by a countable partition. At the minute, that is the *only* case in which we know that the conditional expectation exists. Question: Does it always exist? Answer: Yes, but we cannot prove it... yet. The usual proofs of existence use either the Radon-Nikodym theorem from measure theory, or a Hilbert space projection theorem, neither of which we assume here. We are going to bootstrap our way up and around this impediment. We will state the theorems in their final form, but, for the first time through, we will make an assumption:

> *All necessary conditional expectations exist.*

This is true if all the σ-fields are generated by partitions, but we do not yet know if it is true in general. (For the first reading, the reader should assume that all the σ-fields in the following are partition-generated.) Once we have enough martingale theory we will prove the general existence of conditional expectations, using only partition-generated σ-fields. This will show that the necessary conditional expectations do indeed exist, that the assumption is always satisfied, and that, sometimes, one can indeed pull oneself up by ones bootstraps. Thus, finally, all the theorems—and their proofs—are true exactly as stated, with no restriction on the σ-fields.

8.2. Elementary Properties

It remains to see how to use (CE1) and (CE2). The easiest way to do this is to use them to prove some elementary properties of the conditional expectation. Surprisingly enough, even when we have an explicit formula for the conditional expectation, it is usually easier to use the definition, rather than the formula, to prove things. This is fortunate, since we will soon be dealing with cases where no explicit formula is possible.

Theorem 8.3. *Let X and Y be integrable random variables, a and b real numbers, and $\mathcal{G} \subset \mathcal{F}$ a σ-field. Then*

(i) $E\{E\{X \mid \mathcal{G}\}\} = E\{X\}$.
(ii) $\mathcal{G} = \{\emptyset, \Omega\} \implies E\{X \mid \mathcal{G}\} = E\{X\}$.
(iii) *If X is $\mathcal{G}$-measurable, $E\{X \mid \mathcal{G}\} = X$ a.e.*
(iv) $E\{aX + bY \mid \mathcal{G}\} = aE\{X \mid \mathcal{G}\} + bE\{Y \mid \mathcal{G}\}$ *a.e.*
(v) *If $X \geq 0$ a.e., $E\{X \mid \mathcal{G}\} \geq 0$ a.e.*

8.2. Elementary Properties

(vi) If $X \leq Y$ a.e., $E\{X \mid \mathcal{G}\} \leq E\{Y \mid \mathcal{G}\}$ a.e.

(vii) $|E\{X \mid \mathcal{G}\}| \leq E\{|X| \mid \mathcal{G}\}$ a.e.

(viii) Suppose Y is $\mathcal{G}$-measurable and XY is integrable. Then

(8.2) $$E\{XY \mid \mathcal{G}\} = Y E\{X \mid \mathcal{G}\} \text{ a.e.}$$

(ix) If X and $\mathcal{G}$ are independent, then $E\{X \mid \mathcal{G}\} = E\{X\}$ a.e.

(x) If X_n and X are integrable, and if either $X_n \uparrow X$, or $X_n \downarrow X$, then
$$E\{X_n \mid \mathcal{G}\} \longrightarrow E\{X \mid \mathcal{G}\} \text{ a.e.}$$

Proof. The "a.e.'s" in these formulas reflect the fact that the conditional expectation is only defined up to null sets, so any two versions may differ on a set of probability zero.

The proofs are similar: in each case we have a candidate for the conditional expectation, and confirm it by verifying (CE1) and (CE2).

(i) Just take $\Lambda = \Omega$ in (CE2).

(ii) $E\{X\}$ is constant, hence $\mathcal{G}$-measurable. The only sets on which to check (CE2) are the empty set $\emptyset$, where it is trivial, and the whole space Ω, where it is obvious. This means $Z \stackrel{\text{def}}{=} E\{X\}$ satisfies the properties of the conditional expectation, and hence it *is* the conditional expectation.

(iii) Set $Z \equiv X$. Then (CE1) and (CE2) are immediate.

(iv) Once more we verify that the right-hand side satisfies (CE1) and (CE2). It is clearly $\mathcal{G}$-measurable, and if $\Lambda \in \mathcal{G}$, apply (CE2) to X and Y to see that
$$\int_\Lambda aE\{X \mid \mathcal{G}\} + bE\{Y \mid \mathcal{G}\} \, dP = a \int_\Lambda X \, dP + b \int_\Lambda Y \, dP = \int_\Lambda aX + bY \, dP.$$

(v) Take $\Lambda = \{E\{X \mid \mathcal{G}\} < 0\} \in \mathcal{G}$. Then by (CE2),
$$0 \geq \int_\Lambda E\{X \mid \mathcal{G}\} \, dP = \int_\Lambda X \, dP \geq 0 \Longrightarrow P\{\Lambda\} = 0.$$

(vi) Let $Z = Y - X$ and apply (v).

(vii) This follows from (vi) and (iv) since $X \leq |X|$ and $-X \leq |X|$.

(viii) The right-hand side of (8.2) is $\mathcal{G}$-measurable, so (CE1) holds. Assume first that both X and Y are positive and let $\Lambda \in \mathcal{G}$. First suppose $Y = I_\Gamma$ for some $\Gamma \in \mathcal{G}$. Then
$$\int_\Lambda Y E\{X \mid \mathcal{G}\} \, dP = \int_{\Lambda \cap \Gamma} E\{X \mid \mathcal{G}\} \, dP = \int_{\Lambda \cap \Gamma} X \, dP = \int_\Lambda XY \, dP.$$

Thus (CE2) holds in this case. Now we pull ourselves up by our bootstraps. It follows from (iv) that (CE2) holds if Y is a finite linear combination of indicator functions, i.e., if Y is simple.

If X and Y are arbitrary positive integrable random variables, consider the dyadic approximation of Y: $\underline{Y}_n = k2^{-n}$ on the set $\{k2^{-n} \leq Y < (k+1)2^{-n}\}$, $k = 0, 1, \ldots$. Truncate these: $\underline{Y}'_n \stackrel{def}{=} \min\{\underline{Y}_n, n\}$. Then $\underline{Y}'_n \leq Y$ and $\underline{Y}'_n$ increases to Y as $n \to \infty$. $\underline{Y}'_n$ is positive, simple, and bounded, so

(8.3) $$\int_\Lambda \underline{Y}'_n E\{X \mid \mathcal{G}\} \, dP = \int_\Lambda \underline{Y}'_n X \, dP.$$

Let $n \to \infty$. Since $X \geq 0$, $E\{X|\mathcal{G}\} \geq 0$ by (v) and we can apply the Monotone Convergence Theorem to both sides of (8.3) to see that

(8.4) $$\int_\Lambda Y E\{X \mid \mathcal{G}\} \, dP = \int_\Lambda XY \, dP.$$

In the general case, write X and Y in terms of their positive and negative parts: $X = X^+ - X^-$ and $Y = Y^+ - Y^-$. Note that both Y^+ and Y^- are positive and $\mathcal{G}$-measurable. If X is integrable, so are X^+ and X^-. Moreover, $XY = X^+Y^+ + X^-Y^- - X^+Y^- - X^-Y^+$. Since (8.4) holds for all four products, it holds for XY. Thus, $Y E\{X \mid \mathcal{G}\}$ satisfies (CE1) and (CE2), proving $(viii)$.

(ix) $E\{X\}$ is constant, so it certainly satisfies (CE1). To check (CE2), let $\Lambda \in \mathcal{G}$. Then as X is independent of $\mathcal{G}$, X and I_Λ are independent, so that

$$\int_\Lambda X \, dP = E\{X I_\Lambda\} = E\{X\} P\{\Lambda\} = \int_\Lambda E\{X\} \, dP.$$

proving (CE2).

(x) Assume without loss of generality that X_n increases to X. (Otherwise consider $(-X_n)$.) Let $Z_n = E\{X_n \mid \mathcal{G}\}$. Then $Z_n \leq \{X \mid \mathcal{G}\}$ a.e., and by (vi), Z_n increases to a limit Z. By the Monotone Convergence Theorem 4.12, $E\{Z\} \leq E\{E\{X \mid \mathcal{G}\}\} = E\{X\}$ by (i). X is integrable, so this is finite and Z must be integrable. Each Z_n is $\mathcal{G}$-measurable, hence so is Z. Moreover, again by the Monotone Convergence Theorem, for $\Lambda \in \mathcal{G}$,

$$\int_\Lambda Z \, dP = \lim_n \int_\Lambda Z_n \, dP = \lim_n \int_\Lambda X_n \, dP = \int_\Lambda X \, dP.$$

Thus, Z satisfies both (CE1) and (CE2), and therefore equals $E\{X \mid \mathcal{G}\}$. □

We often need to take a conditional expectation of a conditional expectation. We cannot say much about this in general, but if one σ-field is contained in the other, the result is the conditional expectation with respect to the smaller of the two.

Theorem 8.4. *If X is an integrable random variable, and if $\mathcal{G}_1 \subset \mathcal{G}_2$, then*

(8.5) $$E\{E\{X \mid \mathcal{G}_1\} \mid \mathcal{G}_2\} = E\{E\{X \mid \mathcal{G}_2\} \mid \mathcal{G}_1\} = E\{X \mid \mathcal{G}_1\}.$$

8.2. Elementary Properties

Proof. We will show that each of the first two terms of (8.5) equals the third. First, $E\{X \mid \mathcal{G}_1\}$ is $\mathcal{G}_1$-measurable, and therefore $\mathcal{G}_2$-measurable as well, since $\mathcal{G}_1 \subset \mathcal{G}_2$. By (iii), then, $E\{E\{X \mid \mathcal{G}_1\} \mid \mathcal{G}_2\} = E\{X \mid \mathcal{G}_1\}$.

Consider the second term, $E\{E\{X \mid \mathcal{G}_2\} \mid \mathcal{G}_1\}$. To show this equals $E\{X \mid \mathcal{G}_1\}$, note that it is $\mathcal{G}_1$-measurable, as required. If $\Lambda \in \mathcal{G}_1$, apply (CE2) to $E\{X \mid \mathcal{G}_2\}$ and X successively to see that

$$\int_\Lambda E\{E\{X \mid \mathcal{G}_2\} \mid \mathcal{G}_1\}\, dP = \int_\Lambda E\{X \mid \mathcal{G}_2\}\, dP = \int_\Lambda X\, dP\,.$$

Thus, $E\{E\{X \mid \mathcal{G}_2\} \mid \mathcal{G}_1\}$ satisfies both (CE1) and (CE2) and therefore equals $E\{X \mid \mathcal{G}_1\}$ a.s. This completes the proof. $\square$

Remark 8.5. Note that (i) of Theorem 8.3 is a special case of this, in which $\mathcal{G}_1$ is the trivial σ-field. If we think of the conditional expectation as an average, this is an instance of the principle that the average of averages is itself an average.

Corollary 8.6. *Let X be an integrable random variable and let $\mathcal{G}_1 \subset \mathcal{G}_2$ be σ-fields. Then a necessary and sufficient condition that $E\{X \mid \mathcal{G}_2\} = E\{X \mid \mathcal{G}_1\}$ is that $E\{X \mid \mathcal{G}_2\}$ be $\mathcal{G}_1$-measurable.*

Proof. Suppose $E\{X \mid \mathcal{G}_2\}$ is $\mathcal{G}_1$-measurable. By Theorem 8.3 (iii), $E\{X \mid \mathcal{G}_2\} = E\{E\{X \mid \mathcal{G}_2\} \mid \mathcal{G}_1\}$ and by Theorem 8.4 this equals $E\{X \mid \mathcal{G}_1\}$. The converse is clear. $\square$

Jensen's Inequality. Jensen's inequality (3.21) is true for conditional expectations as well as for ordinary expectations, but its proof, elegant as it is, does not carry over easily.

Theorem 8.7 (Jensen's Conditional Inequality). *Let ϕ be a convex function on an open interval (x_1, x_2) and let X be a random variable whose range is in (x_1, x_2). Suppose both X and $\phi(X)$ are integrable, and that $\mathcal{G} \subset \mathcal{F}$ is a σ-field. Then*

(8.6) $$\phi\bigl(E\{X \mid \mathcal{G}\}\bigr) \leq E\{\phi(X) \mid \mathcal{G}\} \text{ a.e.}$$

Proof. Case 1: Suppose X is discrete, say $X = \sum_i x_i I_{\Gamma_i}$ where the Γ_i are disjoint and $\bigcup_i \Gamma_i = \Omega$.

$$\phi\bigl(E\{X \mid \mathcal{G}\}\bigr) = \phi\Bigl(\sum_i x_i P\{\Gamma_i \mid \mathcal{G}\}\Bigr).$$

For a.e. fixed ω, $\sum_i P\{\Gamma_i \mid \mathcal{G}\} = P\{\bigcup_i \Gamma_i \mid \mathcal{G}\} = 1$ a.e. so by Theorem 3.21, this is

$$\leq \sum_i \phi(x_i) P\{\Gamma_i \mid \mathcal{G}\} = E\{\phi(X) \mid \mathcal{G}\}.$$

In the general case, there exists a sequence of discrete, integrable X_n such that $X_n \uparrow X$ and $|X - X_n| \leq 2^{-n}$. Then $E\{X_n \mid \mathcal{G}\} \uparrow E\{X \mid \mathcal{G}\}$ a.e. by Theorem 8.3 (ix), the monotone convergence theorem for conditional expectations. Now ϕ is convex, therefore continuous, so $\phi(E\{X_n \mid \mathcal{G}\}) \to \phi(E\{X \mid \mathcal{G}\})$. The right-hand derivative $D_+\phi$ of ϕ is increasing. If it never changes sign, ϕ is monotone. Otherwise, there is minimum point a such that ϕ is decreasing on $(-\infty, a]$ and increasing on $[a, \infty)$.

If ϕ is monotone, we can apply the monotone convergence theorem for conditional expectations to $\phi(X_n)$ to finish the proof.

If ϕ is not monotone, let $m < n$ be positive integers and write

$$\phi(X_n) = I_{\{X < a - 2^{-m}\}} \phi(X_n) + I_{\{a - 2^{-m} \leq X \leq a + 2^{-m}\}} \phi(X_n) \\ + I_{\{X > a + 2^{-m}\}} \phi(X_n),$$

so

$$E\{\phi(X_n) \mid \mathcal{G}\} = E\{I_{\{X < a - 2^{-m}\}} \phi(X_n) \mid \mathcal{G}\} \\ + E\{I_{\{a - 2^{-m} \leq X \leq a + 2^{-m}\}} \phi(X_n) \mid \mathcal{G}\} \\ + E\{I_{\{X > a + 2^{-m}\}} \phi(X_n) \mid \mathcal{G}\}.$$

In the first expectation, $X_n < X \leq a$, and ϕ decreases on $(-\infty, a]$, so that $\phi(X_n)$ decreases to $\phi(X)$, and we can go to the limit by Theorem 8.3 (ix). Similarly, in the third expectation, $\phi(X_n)$ decreases to $\phi(X)$. In the middle expectation, ϕ is continuous, therefore bounded, on $[a - 2^{-m+1}, a + 2^{-m+1}]$, and if $n > m$ and $X \in [a - 2^{-m}, a + 2^{-m}]$, then $X_n \in [a - 2^{-m+1}, a + 2^{-m+1}]$, so we can go to the limit by bounded convergence. $\square$

Example 8.7.1. (i) If X is square integrable, then so is $E\{X \mid \mathcal{G}\}$ for any σ-field $\mathcal{G}$. Indeed, by Jensen's inequality, $E\{X \mid \mathcal{G}\}^2 \leq E\{X^2 \mid \mathcal{G}\}$ a.e., and taking expectations of both sides gives $E\{E\{X \mid \mathcal{G}\}^2\} \leq E\{X^2\} < \infty$.

8.3. Approximations and Projections

Here is another interesting characterization of $E\{X \mid \mathcal{G}\}$: it is the best mean-square approximation of X among *all* $\mathcal{G}$-measurable random variables.

Indeed, suppose Y is square-integrable and $\mathcal{G}$-measurable. Then

$$E\{(Y - X)^2\} = E\left\{(Y - E\{X \mid \mathcal{G}\} + E\{X \mid \mathcal{G}\} - X)^2\right\} \\ = E\left\{(Y - E\{X \mid \mathcal{G}\})^2\right\} + E\left\{(E\{X \mid \mathcal{G}\} - X)^2\right\} \\ + 2E\left\{(Y - E\{X \mid \mathcal{G}\})(E\{X \mid \mathcal{G}\} - X)\right\}.$$

8.3. Approximations and Projections

Now $Y - E\{X \mid \mathcal{G}\}$ is $\mathcal{G}$-measurable, so that by Theorem 8.3 (iii) and (i),

$$E\left\{\left(Y - E\{X \mid \mathcal{G}\}\right)\left(E\{X \mid \mathcal{G}\} - X\right)\right\}$$
$$= (Y - E\{X \mid \mathcal{G}\}) \underbrace{E\{E\{X \mid \mathcal{G}\} - X\}}_{=0} = 0,$$

so the cross terms above drop out. Thus we can minimize $E\{(Y - X)^2\}$ by setting $Y = E\{X \mid \mathcal{G}\}$.

In the language of Hilbert spaces, $E\{X \mid \mathcal{G}\}$ is the *projection* of X on the space of square-integrable $\mathcal{G}$-measurable random variables.

The Gaussian Case. Conditional expectations simplify when the random variables are Gaussian. If $X_0, \ldots, X_n$ are jointly Gaussian, then $E\{X_0 \mid X_1, \ldots, X_n\}$ is not just a Borel function of the X_i, it is actually a linear function.

Theorem 8.8. *Let $X_0, \ldots, X_n$ be jointly Gaussian, and let $\hat{X} = E\{X_0 \mid X_1, \ldots, X_n\}$. Then there are constants $a_0, \ldots, a_n$ such that $\hat{X}_0 = a_0 + \sum_{i=1}^n a_i X_i$. Moreover, $X_0 - \hat{X}$ is independent of $X_1, \ldots, X_n$, and the a_i are solutions of*

(8.7)
$$E\{X_0 - \hat{X}\} = 0,$$
$$E\{(X_0 - \hat{X})X_i\} = 0\}, \ i = 1, \ldots, n.$$

Proof. First, (8.7) gives us a system of $n+1$ linear equations in $n+1$ unknowns (a_i), so there is at least one solution. Let $\mathcal{G} = \sigma\{X_1, \ldots, X_n\}$. Suppose $\hat{X} \stackrel{\text{def}}{=} a_0 + \sum_{i=1}^n a_i X_i$ satisfies these equations. First, it is $\mathcal{G}$-measurable. Since $X_0 - \hat{X}$ has mean zero, the remaining equations imply that $X_0 - \hat{X}$ and X_i are uncorrelated, $i = 1, \ldots, n$. But they are jointly Gaussian, so that $X_0 - \hat{X}$ is actually independent of $X_1, \ldots, X_n$, and therefore of $\mathcal{G}$. Thus for any $\Lambda \in \mathcal{G}$,

$$\int_\Lambda X_0 \, dP = E\{(\hat{X} + X_0 - \hat{X}))I_\Lambda\} = E\{\hat{X} I_\Lambda\} + E\{(X_0 - \hat{X})I_\Lambda\}.$$

But $X_0 - \hat{X}$ has mean zero and is independent of $\mathcal{G}$ and therefore of I_Λ, so the last term vanishes. We conclude that for $\Lambda \in \mathcal{G}$, $\int_\Lambda X_0 \, dP = \int_\Lambda \hat{X} \, dP$, hence $\hat{X} = E\{X_0 \mid \mathcal{G}\}$. □

Remark 8.9. This explains why so much of statistics is both more pleasant and more powerful when the underlying distributions are Gaussian. For then the conditional expectation is a projection onto the *linear* space spanned by the X_i, not just onto the much larger space of $\mathcal{G}$-measurable random variables, which includes non-linear functions of the X_i. So that, for instance, least-squares estimates coincide with conditional expectations, and the error is not just orthogonal to, but independent of, the estimate, and, not least, finding conditional expectations reduces to solving linear equation systems such as (8.7).

Problems 8

8.1. Let X and Y be independent random variables and let $f(x,y)$ be a bounded continuous function. Let F_X be the distribution function of X. Show $E\{f(X,Y) \mid Y\} = \int f(x,Y)\,dF_X(x)$. Deduce that $P\{X+Y \leq x \mid Y\} = F_X(x-Y)$.

8.2. Let $\mathcal{G}_1$ and $\mathcal{G}_2$ be independent σ-fields, and let X be an integrable random variable. Show that $E\{E\{X \mid \mathcal{G}_1\} \mid \mathcal{G}_2\} = E\{X\}$.

8.3. Let Y be an integrable random variable and $\mathcal{G}$ a σ-field. Show that if $\mathcal{G}$ is trivial, (i.e., all its sets have probability zero or one), $E\{Y \mid \mathcal{G}\} = E\{Y\}$.

8.4. Let $\mathcal{G}_1$ and $\mathcal{G}_2$ be σ-fields which are conditionally independent relative to a third sigma-field $\mathcal{H}$. If $\mathcal{H}$ is trivial (all events have probability zero or one), then $\mathcal{G}_1$ and $\mathcal{G}_2$ are independent. This is true, in particular, if $\mathcal{H} = \sigma(Y)$ where Y is a.s. constant.

8.5. Let X and Y be integral random variables, and suppose XY is also integrable. Show $E\{XY\} = E\{XE\{Y \mid X\}\}$.

8.6. Let X and Y be integrable random variables and let $\mathcal{G}$ be a σ-field. Suppose that $E\{Y\} = E\{X\}$, that Y is $\mathcal{G}$-measurable, and that $X-Y$ is independent of $\mathcal{G}$. Show that $Y = E\{X \mid \mathcal{G}\}$.

8.7. The Monotone Convergence Theorem, Fatou's Lemma, and the Dominated Convergence cheorem all carry over to conditional expectations. Let X and X_n be integrable random variables. Prove the following.

(a) (Monotone Convergence.) Suppose that $X_n \leq X_{n+1} \leq \ldots$ and $X = \lim_n X_n$. Then $E\{X_n \mid \mathcal{G}\} \uparrow E\{X \mid G\}$.

(b) (Fatou's Lemma.) Suppose the X_n are all positive and $X = \liminf_n X_n$. Then $\liminf_{n \to \infty} E\{X_n \mid \mathcal{G}\} \geq E\{X \mid \mathcal{G}\}$.

(c) (Dominated Convergence.) Suppose that $X = \lim_n X_n$ a.s. and that there exists an integrable random variable Y such that $|X_n| \leq Y$ a.s. for all n. Then $\lim_{n \to \infty} E\{X_n \mid \mathcal{G}\} = E\{X \mid \mathcal{G}\}$ a.s.

8.8. (Gaussian conditional expectations.) Conditional expectations are particularly simple in the Gaussian case: they are linear combinations.

Let $X, Z_1, \ldots, Z_n$ be jointly Gaussian and let $\mathcal{G} = \sigma\{Z_1, \ldots, Z_n\}$. Suppose that $\mathcal{G}$ is not generated by any proper subset of the Z_j. Let $Y = a_0 + \sum_{j=1}^n a_j Z_j$. Show that there exist constants a_j, $j = 0, \ldots, n$ such that $X - Y$ is orthogonal to Z_k for all k, and also to 1. Show that for these a_j,

$$Y = E\{X \mid \mathcal{G}\}.$$

[Hint: Use Exercise 8.6.]

8.9. (Continuation.) Suppose that the Z_j are i.i.d. $N(0,1)$, but X may have a non-zero mean. Find a formula for the a_j, and consequently for $E\{X \mid \mathcal{G}\}$.

Chapter 9

Discrete-Parameter Martingales

9.1. Martingales

A martingale is a mathematical model of a fair game. It takes its name from the French "la grande martingale", the betting strategy in which one doubles the bet after each loss. (If you double the bet with each loss, the first win recoups all previous losses, with a slight profit left over. Since you are bound to win a bet sooner or later, this system guarantees a profit. Its drawbacks have been thoroughly explored.) The term also refers to the back belt of a dress or jacket, a part of a sailing ship's rigging, and a tie-down used on horses' bridles, which might explain some bizarre birthday gifts to prominent probabilists.

In a fair game, the gambler should, on the average, come out even. Mathematically, the gambler's expected winnings and expected losses should cancel out. But it is not enough to have the overall expectation vanish. The expectation must vanish *at the time of the bet.* A simple example will show why. Suppose we choose one of two coins at random, and start flipping it. One of the coins has two heads, and the other has two tails. Bet a dollar on the outcome of each toss at even odds. The probability of heads on each toss is clearly the probability we choose the two-headed coin, one-half, so the bet is fair... for one toss. But if we toss it twice, the bet on the second toss is not fair, for we have already seen the result of the first toss, and know which coin is being flipped, so we can win the second toss by betting on whichever side came up on the first. So it is not the expectation which has to vanish, but the *conditional* expectation, given all that we know at the time we make the bet.

Let us consider a sequence of fair bets, and keep track of the total amount of money the gambler has at each stage. Let X_0 be the gambler's initial fortune, X_1 the fortune after the first bet, X_2 the fortune after the second, and so on. The gambler's winnings on the first bet are $X_1 - X_0$, on the nth, $X_n - X_{n-1}$. For a bet to be fair, the conditional expectation of the winnings, given all the knowledge available at the time of the bet, should be zero. Let us represent the knowledge after the outcome of the nth bet by a σ-field, $\mathcal{F}_n$. Our knowledge when we make the nth bet is $\mathcal{F}_{n-1}$ (for we don't yet know the nth outcome). So for each n, we must have $E\{X_n - X_{n-1} \mid \mathcal{F}_{n-1}\} = 0$.

Martingales[1] can be defined on several different parameter sets, so we will start with the general case, and then specialize. Let $\mathfrak{T} \subset \mathbb{R}$ be a subset of the line. This is the parameter set of the process. Think of $t \in \mathfrak{T}$ as time. (In most cases, $\mathfrak{T}$ will either be the positive integers or the positive half-line. However, the case where $\mathfrak{T}$ is the *negative* integers will turn out to be capital. We will even have use for the case where $\mathfrak{T}$ is just two points, $\mathfrak{T} = \{1, 2\}$.) So we will keep the parameter set general while we investigate a few basic properties. But the reader can safely think of $\mathfrak{T}$ as being $\{0, 1, 2, \dots\}$.

Definition 9.1. A **filtration** on the probability space $(\Omega, \mathcal{F}, P)$ is a family $\{\mathcal{F}_t : t \in \mathfrak{T}\}$ of sub-σ-fields of $\mathcal{F}$ such that for all $s < t \in \mathfrak{T}$, $\mathcal{F}_s \subset \mathcal{F}_t$.

The filtration represents our knowledge at the successive times. This increases with time—in this model we don't forget things—so that the σ-fields increase. Keeping the same parameter set, we define

Definition 9.2. Let $\{\mathcal{F}_t : t \in \mathfrak{T}\}$ be a filtration, and $\{X_t, t \in \mathfrak{T}\}$ be a stochastic process. The process is said to be **adapted** to the filtration if, for all $t \in \mathfrak{T}$, X_t is $\mathcal{F}_t$-measurable.

This is a good example of an apparently abstract definition which is, in fact, highly practical. It simply says that the process—like us—cannot look into the future. For this reason, such processes are called **non-anticipating**. Indeed, $\mathcal{F}_t$ represents what is known at time t—the history up to t—and if X_t is measurable with respect to $\mathcal{F}_t$, that means that its value is known at time t. An adapted process, then, is one that happens in real time.

We can now define a martingale. At the same time, we will define the closely related ideas of submartingale and supermartingale.

---

[1]Martingale theory was developed by J. L. Doob—his book, Stochastic Processes [**12**] summarized his results and remained the state of the art until André Meyer combined martingales and capacity theory in a deep study of probabilistic potential theory. However, with hindsight it is clear that the idea of martingales had been in the air for some time. Paul Lévy had proved results for conditional expectations, like Theorem 9.38, and Doob himself credited the original idea to Jean Ville. See [**29**] for a discussion of martingale pre-history.

9.1. Martingales

Definition 9.3. Let $\mathfrak{T} \subset \mathbb{R}$. A stochastic process $\{X_t, \mathcal{F}_t,\ t \in \mathfrak{T}\}$ is a **martingale** (resp. **submartingale, supermartingale**) if

(i) $\{\mathcal{F}_t,\ t \in \mathfrak{T}\}$ is a filtration;

(ii) $\{X_t,\ t \in \mathfrak{T}\}$ is adapted to the filtration $(\mathcal{F}_t)$;

(iii) for each $t \in \mathfrak{T}$, X_t is integrable;

(iv) For each $s < t \in \mathfrak{T}$, $X_s = E\{X_t \mid \mathcal{F}_s\}$ (resp. $X_s \leq E\{X_t \mid \mathcal{F}_s\}$, $X_s \geq E\{X_t \mid \mathcal{F}_s\}$.)

If a martingale is a fair game, a supermartingale is an unfair game, and a submartingale is a more-than-fair game. In casino gambling, the casino usually has a small edge, so gamblers ordinarily play supermartingales, while the casino itself plays a submartingale. Indeed, the (conditional) expected winnings between times s and t are $E\{X_t - X_s \mid \mathcal{F}_s\}$. This is zero for a martingale, negative for a supermartingale, and positive for a submartingale[2].

Here are some elementary properties.

Proposition 9.4. *Let $\mathfrak{T}$ be a subset of the line, let $\{\mathcal{F}_t,\ t \in \mathfrak{T}\}$ be a filtration, and let $X \equiv \{X_t,\ t \in \mathfrak{T}\}$ be a stochastic process.*

(i) X is a submartingale if and only if $-X$ is a supermartingale. It is a martingale if and only if it is both a sub- and supermartingale.

(ii) Let $s < t \in \mathfrak{T}$. If X is a martingale, $E\{X_s\} = E\{X_t\}$. If X is a submartingale, then $E\{X_s\} \leq E\{X_t\}$. If X is a supermartingale, then $E\{X_s\} \geq E\{X_t\}$.

*(iii) If X is a submartingale relative to a filtration $(\mathcal{F}_t)$, then it is also a submartingale with respect to its **natural filtration** $\mathcal{G}_t \equiv \sigma\{X_s,\ s \leq t\}$.*

Remark 9.5. We will state most of our results for submartingales. The corresponding results for martingales and supermartingales follow immediately from (i).

By (ii), martingales have constant expectations, while the expectations of submartingales increase with time, and the expectations of supermartingales decrease. (iii) shows us that if necessary, we can always use the natural filtration of the processes. However, it is useful to have the flexibility to choose larger filtrations.

---

[2]There are stories about these names. J.L. Doob, who developed the whole theory, originally called the fundamental equality "Property E". He was surprised that it generated little interest. Then he renamed it "the martingale property", and the idea took off. The designations submartingale and supermartingale come from their deep relations with subharmonic and superharmonic functions in potential theory. Although he was aware of the connection when he first defined them, he originally called them upper and lower semimartingales. The reason, he said, was that his daughter was listening to Superman serials on the radio at the time, and he couldn't stand the idea of his supermartingales leaping tall buildings in a single bound. This only lasted for a short time: he changed them to submartingale and supermartingale once her listening habits changed. "Semimartingale" has since resurfaced as the name of a different but related class of processes.

Proof. (*i*) is clear.

(*ii*) If X is a submartingale and $s < t$, $X_s \leq E\{X_t \mid \mathcal{F}_s\}$. Take the expectation of both sides: by Theorem 8.3 (*vi*) and Theorem 8.4, $E\{X_s\} \leq E\{E\{X_t \mid \mathcal{F}_s\}\} = E\{X_t\}$.

(*iii*) By (*i*), it is enough to prove this if X is a submartingale. Then $s < t \in \mathfrak{T} \implies X_s \leq E\{X_t \mid \mathcal{F}_s\}$. Note that X_s is $\mathcal{G}_s$-measurable and $\mathcal{G}_s \subset \mathcal{F}_s$. (Why?) Take the conditional expectation of both sides given $\mathcal{G}_s$ and use Theorem 8.4:

$$X_s = E\{X_s \mid \mathcal{G}_s\} \leq E\{E\{X_t \mid \mathcal{F}_s\} \mid \mathcal{G}_s\} = E\{X_t \mid \mathcal{G}_s\}.$$

□

Proposition 9.6. (*i*) *Suppose* $\{X_t, \mathcal{F}_t, t \in \mathfrak{T}\}$ *is a martingale and* ϕ *is a convex function on some interval. Then, if* $\phi(X_t)$ *is integrable for all* t, $\{\phi(X_t), \mathcal{F}_t, t \in \mathfrak{T}\}$ *is a submartingale.*

(*ii*) *Suppose that* $\{X_t, \mathcal{F}_t, t \in \mathfrak{T}\}$ *is a submartingale and* ϕ *is an increasing convex function. Then, if* $\phi(X_t)$ *is integrable for all* t, $\{\phi(X_t), \mathcal{F}_t, t \in \mathfrak{T}\}$ *is a submartingale.*

Proof. Let $s < t \in \mathfrak{T}$.

(*i*) By Jensen's inequality for conditional expectations,

$$\phi(X_s) = \phi(E\{X_t \mid \mathcal{F}_s\}) \leq E\{\phi(X_t) \mid \mathcal{F}_s\}.$$

The proof for (*ii*) is almost the same: $X_s \leq E\{X_t \mid \mathcal{F}_s\}$ so $\phi(X_s) \leq \phi(E\{X_t \mid \mathcal{F}_s\}) \leq E\{(\phi(X_n) \mid \mathcal{F}_m\}$, where the first inequality holds because ϕ is increasing, and the second because ϕ is convex. □

For example, if X_t is a martingale, then (subject to integrability) $|X_t|$, X_t^2, e^{X_t}, and e^{-X_t} are all submartingales, while if $X_t > 0$, $\sqrt{X_t}$ and $\log(X_t)$ are supermartingales and $1/X_t$ is a submartingale. If X_t is a submartingale and K a constant, then $\max\{X_t, K\}$ is a submartingale, while if X_t is a supermartingale, so is $\min\{X_t, K\}$. (The last two follow because $x \mapsto \max\{x, K\}$ is convex and increasing.)

Much of what follows concerns the parameter set $0, 1, 2, \ldots$. We will make free use of the following exercise, and its obvious extension to martingales and supermartingales.

Exercise 9.1. Let $\{X_n, n = 0, 1, 2, \ldots\}$ be integrable random variables which are adapted to a filtration $\{\mathcal{F}_n, n = 0, 1, 2, \ldots\}$. Show that a necessary and sufficient condition that $\{X_n, \mathcal{F}_n, n = 0, 1, 2, \ldots\}$ be a submartingale is that for each positive integer n, $X_n \leq E\{X_{n+1} \mid \mathcal{F}_n\}$.

9.1. Martingales

Examples. 1° Let $Y_1, Y_2, \ldots$ be a sequence of independent integrable random variables. Set $X_n = \sum_{i=1}^n Y_i$ for $n \geq 1$. Let $\mathcal{F}_n = \sigma\{Y_1, \ldots, Y_n\}$ be the σ-field generated by the first n Y_i's. If the Y_n all have expectation zero, then $\{X_n, \mathcal{F}_n, n = 1, 2, \ldots\}$ is a martingale. It is a submartingale if the $E\{Y_n\}$ are all positive, and a supermartingale if they are all negative. In particular, the simple symmetric random walk is a martingale.

To see this, note that X_n is $\mathcal{F}_n$-measurable, so the process is adapted. Moreover, $X_{n+1} = X_n + Y_{n+1}$. Now Y_{n+1} is independent of $Y_1, \ldots, Y_n$, therefore of $\mathcal{F}_n$, so $E\{Y_{n+1} \mid \mathcal{F}_n\} = E\{Y_{n+1}\}$, and clearly, $E\{X_n \mid \mathcal{F}_n\} = X_n$. Thus $E\{X_{n+1} \mid \mathcal{F}_n\} = X_n + E\{Y_{n+1}\}$. If $E\{Y_n\} = 0$ for all n, $X_n = E\{X_{n+1} \mid \mathcal{F}_n\}$, and (X_n) is a martingale. It is clearly a submartingale if $E\{Y_n\} \geq 0$ for all n, and a supermartingale if $E\{Y_n\} \leq 0$ for all n.

2° Example 1° could describe a sequence of gambles: the gambler bets one dollar each time and Y_j is the gain or loss at the jth play. We would say the game is fair if $E\{Y_j\} = 0$ for all j, that it is favorable if $E\{Y_j\} \geq 0$ for all j, and that it is unfair (or realistic) if $E\{Y_j\} \leq 0$. A more worldly model would allow the gambler to vary the bets. Suppose the gambler stakes K_n dollars on the nth bet; K_n must be positive, and it may well be random—the gambler might vary the bets according to the situation (this is a principal strategy in the system for blackjack) or on a random whim. At the time of the wager, the bettor does not know the result Y_n of the bet, but knows all the previous results $Y_1, \ldots, Y_{n-1}$, so that K_n should be measurable with respect to $\mathcal{F}_{n-1}$. Call the new process $\hat{X}$. Thus the gambler's fortune after the nth wager is $\hat{X}_n = \hat{X}_{n-1} + K_n Y_n$, where K_n is a bounded positive $\mathcal{F}_{n-1}$-measurable random variable. Then $E\{\hat{X}_n \mid \mathcal{F}_{n-1}\} = \hat{X}_{n-1} + E\{K_n Y_n \mid \mathcal{F}_{n-1}\} = \hat{X}_{n-1} + K_n E\{Y_n\}$, since both $\hat{X}_{n-1}$ and K_n are $\mathcal{F}_{n-1}$-measurable. Since K_n is positive, $\hat{X}$ will be a martingale (resp. submartingale, supermartingale) if the fixed-bet process X is a martingale (resp. submartingale, supermartingale).

So, as long as the bets are positive, the gambler cannot change the basic nature of the game by changing the size of the bets. But negative bets would change a submartingale into a supermartingale! So, in a way, for a gambler, the stock market has an advantage over casinos. In casinos, one cannot bet negative dollars. In the stock market one can, by selling short.

3° Here is an important example. Let $(\mathcal{F}_n)$ be a filtration and let X be an integrable random variable. Define $X_n \equiv E\{X \mid \mathcal{F}_n\}$. Then $\{X_n, n = 0, 1, \ldots\}$ is a martingale relative to the filtration $(\mathcal{F}_n)$.

Indeed, $E\{X_{n+m} \mid \mathcal{F}_n\} = E\{E\{X \mid \mathcal{F}_{n+m}\} \mid \mathcal{F}_n\} = E\{X \mid \mathcal{F}_n\} = X_n$ by Theorem 8.4.

4° Let $0 < p < 1$ and $Y_1, Y_2, \ldots$ be a sequence of i.i.d. Bernoulli random variables with $P\{Y_j = 1\} = p$, and $P\{Y_j = 0\} = 1 - p$. Then

$$X_n = \prod_{j=1}^{n} \frac{Y_j}{p}$$

is a martingale relative to its natural filtration. Indeed,

$$E\{X_{n+1} \mid Y_1, \ldots Y_n\} = E\{X_n Y_{n+1}/p \mid Y_1, \ldots, Y_n\} = X_n E\{Y_{n+1}/p\} = X_n.$$

This is an example of a positive martingale whose expectation is always one, and which converges to zero with probability one. (Since $p < 1$, one of the Y_n will eventually vanish, and $X_n = 0$ from then on.)

5° There are numerous examples of martingales in statistics. Here is one which arises in sequential analysis.

Let $X_1, X_2, \ldots$ be a sequence of i.i.d. random variables. Assume that they have probability densities, and that there are two candidates for their common density, which we call (H1) and (H2). The statistical problem is to observe some of the random variables, and then decide whether (H1) or (H2) is correct.

(H1) The probability density of the X_n is $f(x)$.

(H2) The probability density of the X_n is $g(x)$.

Intuitively, we expect X_n to take values where its probability density is relatively large. In particular, if we compare $f(X_n)$ and $g(X_n)$, we expect that on balance, $f(X_n)$ will tend to be larger than $g(X_n)$ if (H1) is true, and smaller if (H2) is true. Thus consider

$$Z_n = \begin{cases} \prod_{i=1}^{n} \frac{g(X_i)}{f(X_i)} & \text{if } f(X_i) > 0,\ i = 1, \ldots, n, \\ 0 & \text{otherwise.} \end{cases}$$

The fraction is called a **likelihood ratio**. Let $\mathcal{F}_n = \sigma\{X_1, \ldots, X_n\}$. Then if (H1) is the correct density, $\{Z_n, \mathcal{F}_n,\ n = 1, 2, \ldots\}$ is a positive supermartingale, and it is a positive martingale if $f(x)$ never vanishes.

To see this note that

(9.1)
$$\begin{aligned} E\{Z_{n+1} \mid \mathcal{F}_n\} &= E\left\{ Z_n \frac{g(X_{n+1})}{f(X_{n+1})} I_{\{f(X_{n+1})>0\}} \,\Big|\, \mathcal{F}_n \right\} \\ &= Z_n E\left\{ \frac{g(X_{n+1})}{f(X_{n+1})} I_{\{f(X_{n+1})>0\}} \,\Big|\, \mathcal{F}_n \right\} \\ &= Z_n E\left\{ \frac{g(X_{n+1})}{f(X_{n+1})} I_{\{f(X_{n+1})>0\}} \right\}, \end{aligned}$$

9.1. Martingales

where we have used Theorem 8.3 and the fact that X_{n+1} and $\mathcal{F}_n$ are independent. But since f is the true density under (H1),

$$E\left\{\frac{g(X_{n+1})}{f(X_{n+1})}I_{\{f(X_{n+1})>0\}}\right\} = \int_{\{x:f(x)>0\}} \frac{g(x)}{f(x)} f(x)\,dx$$
$$= \int_{\{x:f(x)>0\}} g(x)\,dx \leq \int_{-\infty}^{\infty} g(x)\,dx = 1.$$

Thus, $E\{Z_{n+1} \mid \mathcal{F}_n\} \leq Z_n$ for all n, hence Z is a supermartingale. If f never vanishes, there is equality, and Z is a martingale.

We will see in §9.5 that under (H1), $Z_n \to 0$, and under (H2), $Z_n \to \infty$, so that the statistician can eventually tell which of (H1) or (H2) is correct by observing the sequence $Z_1, Z_2, \ldots$.

The Doob Decomposition. The conditions involving supermartingales and submartingales involve inequalities, while the definition of martingales involves an equality. It would appear that sub- and supermartingales are far more general than martingales. But there is a rather remarkable fact: all three behave in much the same way. This is explained by the Doob Decomposition, which tells us that a submartingale can be written as a martingale plus a monotone process. Increasing sequences are relatively easy to handle, so this explains why that, once we know how martingales behave, we know so much about the behavior of sub- and supermartingales.

Theorem 9.7 (Doob Decomposition). *Let $\{X_n, \mathcal{F}_n, n = 0, 1, 2, \ldots\}$ be a submartingale. Then there exists a martingale $\{M_n, \mathcal{F}_n, n = 0, 1, 2, \ldots\}$ and a process $\{A_n, n = 0, 1, 2, \ldots\}$ such that for $n = 0, 1, 2, \ldots$*

(i) $A_n \leq A_{n+1}$ a.e.,

(ii) $A_0 = 0$ and A_n is $\mathcal{F}_{n-1}$-measurable, $n = 1, 2, 3, \ldots$,

(iii) $X_n = M_n + A_n$, $n = 0, 1, 2, \ldots$.

Proof. Let $d_n = E\{X_{n+1} - X_n \mid \mathcal{F}_n\}$. Then $d_n \geq 0$ by the submartingale inequality, and d_n is $\mathcal{F}_n$-measurable. Define $A_0 = 0$, $M_0 = X_0$, and

$$A_n \stackrel{\text{def}}{=} d_1 + \cdots + d_{n-1}, \qquad M_n \stackrel{\text{def}}{=} X_n - A_n, \ n = 1, 2, 3, \ldots.$$

It is clear that (i), (ii) and (iii) hold. We must verify that M_n is a martingale. It is adapted. Calculate

$$E\{M_{n+1} \mid \mathcal{F}_n\} = E\{X_{n+1} - A_{n+1} \mid \mathcal{F}_n\}$$
$$= E\{X_{n+1} - X_n + \underbrace{X_n - A_n}_{M_n} - (A_{n+1} - A_n) \mid \mathcal{F}_n\}$$

which is

$$= d_n + M_n - \underbrace{(A_{n+1} - A_n)}_{d_n}$$
$$= M_n.$$

Thus, M_n is indeed a martingale. □

9.2. System Theorems

The name "system theorem" comes from gambling. There are several such results, and they tell us, roughly, that martingales remain martingales under optional stopping or optional sampling. They were originally used to show that gambling systems do not work. Once proved, they were seen to be powerful, with implications that go beyond gambling, and even beyond probability itself. But let us introduce them as they were originally seen, from the gambling perspective.

It is common knowledge that simple gambling systems, such as the old roulette system, "Wait until one color comes up five times in a row, then bet on the other", do not work. Some people think that the law of averages makes the other color more likely at this time, but the law of averages says no such thing. In fact, a string of five reds in a row has no influence on the color showing up on the next spin of the wheel, for the sixth spin is independent of the last five. But it is easier to debunk the law of averages argument than to prove this fact rigorously. In fact, the proof, while not hard, is surprisingly subtle. (See Corollary 7.29.) The delicate point is this: the time of the fifth same-color spin is random, not fixed, while our hypotheses on the sequence of spins (e.g. that the sequence $X_1, X_2, \ldots$ of outcomes is i.i.d.) only concern fixed times.

Not every random time works. The sequence may not be independent after just any random time. Consider, for example, the first time that the pattern "Five reds followed by a black" occurs, and let T be the time of the fifth red in that sequence. Then T is a random time, X_T is red, and X_{T+1} is black—so the $T+1^{\text{st}}$ spin is not independent of the Tth.

But this is not really a possible gambling system. The gambler cannot use the time T, for it requires knowing the result of the next spin. There is an obvious restriction on the type of random times that gamblers can use in a system: they cannot look into the future. We have seen this before: they must be stopping times. Stopping times are the only random times a gambler can actually use in a gambling system.

Now consider games of chance that might be based on something more general than successive i.i.d. random variables. Follow the fortune of a

9.2. System Theorems

gambler playing such a game. If the game is fair, the gambler's fortune is a martingale. If it is unfair, it is a supermartingale, and if it is super-fair, it is a submartingale. The question is whether a gambler can make decisions at stopping times which change an unfair game into a fair game, or vice-versa. If the gambler's fortune is a supermartingale, can he or she make it into a submartingale by being clever? The system theorems tell us that—modulo some hypotheses—the answer is "No."

Theorem 9.8. *Let $\{X_n, \mathcal{F}_n, n = 0, 1, 2,\}$ be a submartingale and let T be a stopping time. Then $\{X_{n \wedge T}, \mathcal{F}_n, n = 0, 1, 2, \ldots\}$ is a submartingale, as is $\{X_{n \wedge T}, \mathcal{F}_{n \wedge T}, n = 0, 1, 2, \ldots\}$.*

Proof. The properties of stopping times and their associated σ-fields were established in §7.4. We will use them freely.

Note that $X_{n \wedge T}$ is measurable with respect to $\mathcal{F}_{n \wedge T} \subset \mathcal{F}_T$, so $X_{n \wedge T}$ is adapted to both filtrations $(\mathcal{F}_n)$ and $(\mathcal{F}_{n \wedge T})$. Let $\Lambda \in \mathcal{F}_n$.

$$\int_\Lambda X_{(n+1) \wedge T}\, dP = \int_{\Lambda \cap \{T > n\}} X_{n+1}\, dP + \int_{\Lambda \cap \{T \leq n\}} X_{n \wedge T}\, dP,$$

for $X_{(n+1) \wedge T} = X_{n+1}$ if $T > n$. Now $\Lambda \cap \{T > n\} \in \mathcal{F}_n$ so, as (X_n) is a submartingale,

$$\int_{\Lambda \cap \{T > n\}} X_{n+1}\, dP \geq \int_{\Lambda \cap \{T > n\}} X_n\, dP = \int_{\Lambda \cap \{T > n\}} X_{n \wedge T}\, dP,$$

and the above is

$$\geq \int_{\Lambda \cap \{T > n\}} X_{n \wedge T}\, dP + \int_{\Lambda \cap \{T \leq n\}} X_{n \wedge T}\, dP = \int_\Lambda X_{n \wedge T}\, dP,$$

showing that $(X_{n \wedge T})$ is a submartingale relative to the filtration $(\mathcal{F}_n)$. Since $\mathcal{F}_{n \wedge T} \subset \mathcal{F}_n$, this also shows it is a submartingale with respect to the filtration $(\mathcal{F}_{n \wedge T})$. □

This theorem is useful because it has so few hypotheses on X and T. It works for any stopping time and any submartingale. But it doesn't get us directly to X_T, only to $X_{n \wedge T}$. Extensions directly involving X_T require extra hypotheses: either T is bounded in some sense, or X is bounded in some sense. First, we treat the case where the stopping times are bounded.

Theorem 9.9. *Let $\{X_n, \mathcal{F}_n, n = 0, 1, 2, \ldots\}$ be a submartingale. Let $S \leq T$ be bounded stopping times. Then (X_S, X_T) is a submartingale relative to the filtration $(\mathcal{F}_S, \mathcal{F}_T)$.*

Proof. Note that X_S and X_T are measurable with respect to $\mathcal{F}_S$ and $\mathcal{F}_T$ respectively, so we need only show that $X_S \leq E\{X_T \mid \mathcal{F}_S\}$ a.s.

First, there is an N such that $S \leq T \leq N$ a.s. Note that

$$E\{|X_T|\} = \sum_{j=0}^{N} E\{|X_j|, T=j\} \leq \sum_{j=0}^{N} E\{|X_j|\} < \infty,$$

since the X_j are integrable. The same is true for X_S, so both X_S and X_T are integrable. Let $\Lambda \in \mathcal{F}_S$. Then

$$\int_\Lambda X_T - X_S \, dP = \sum_{j=0}^{N} \int_{\Lambda \cap \{S=j\}} X_T - X_j \, dP$$

$$= \sum_{j=0}^{N} \sum_{k=j}^{N} \int_{\Lambda \cap \{S=j\}} \left(X_{T \wedge (k+1)} - X_{T \wedge k}\right) dP,$$

where we have written $X_T - X_j$ as a telescoping series. (You should check that this actually telescopes to the right quantity.) But $\Lambda \cap \{S=j\} \in \mathcal{F}_j$, so by property (CE2) of the conditional expectation, this is

$$= \sum_{j=0}^{N} \sum_{k=j}^{N} \int_{\Lambda \cap \{S=j\}} E\{X_{T \wedge (k+1)} - X_{T \wedge k} \mid \mathcal{F}_j\} \, dP \geq 0,$$

since by Theorem 9.8 $X_{T \wedge k}$ is a submartingale. □

Remark 9.10. The theorem can be restated: "A submartingale remains a submartingale under bounded optional sampling," where "optional sampling" just means sampling the process at a stopping time. It also holds if we replace "submartingale" by "martingale" and "supermartingale", respectively.

Let us move on to the case where the stopping times may be unbounded, but the process itself is bounded: a bounded submartingale (resp. martingale, supermartingale) remains a submartingale (resp. martingale, supermartingale) under arbitrary optional sampling.

Theorem 9.11. *Let $\{X_n, \mathcal{F}_n, n = 0, 1, 2, \ldots\}$ be a bounded submartingale. Let $S \leq T < \infty$ be finite stopping times. Then (X_S, X_T) is a submartingale relative to the filtration $(\mathcal{F}_S, \mathcal{F}_T)$.*

Proof. X_S and X_T are bounded and therefore integrable. They are clearly adapted. We must show that $X_S \leq E\{X_T \mid \mathcal{F}_S\}$ a.s. Now for each n, $S \wedge n \leq T \wedge n$ are bounded stopping times, so by Theorem 9.9, $(X_{S \wedge n}, X_{T \wedge n})$ is a submartingale relative to $(\mathcal{F}_{S \wedge n}, \mathcal{F}_{T \wedge n})$. Let $\Lambda \in \mathcal{F}_S$. Let $m < n$ and note that $\Lambda \cap \{S = m\} \in \mathcal{F}_{S \wedge n}$. (Why?) Thus,

$$\int_{\Lambda \cap \{S=m\}} X_{S \wedge n} \, dP \leq \int_{\Lambda \cap \{S=m\}} X_{T \wedge n} \, dP.$$

Let $n \to \infty$ and use bounded convergence on both sides: $X_{n \wedge T} \to X_T$ and $X_{S \wedge n} \to X_S$, so in the limit, we see

$$\int_{\Lambda \cap \{S=m\}} X_S \, dP \leq \int_{\Lambda \cap \{S=m\}} X_T \, dP.$$

Add this over $M = 1, \ldots, N$ for any large integer N to see that

$$\int_{\Lambda \cap \{S \leq N\}} X_S \, dP \leq \int_{\Lambda \cap \{S \leq N\}} X_T \, dP.$$

Let $N \to \infty$ to see that

$$\int_\Lambda X_S \, dP \leq \int_\Lambda X_T \, dP.$$

Again, the passage to the limit is justified since the integrands are bounded. □

We can extend this to processes which are bounded on only one side.

Theorem 9.12. *Let* $\{X_n, \mathcal{F}_n, n = 0, 1, 2, \ldots\}$ *be a negative submartingale (resp. positive supermartingale). Let* $S \leq T$ *be finite stopping times. Then* (X_S, X_T) *is a submartingale (resp. supermartingale) relative to the filtration* $(\mathcal{F}_S, \mathcal{F}_T)$.

Proof. The function $\phi_N(x) = \max(x, -N)$ is increasing and convex in x. Thus, $\phi_N(X_n)$ is a submartingale by Proposition 9.6. Now $-N \leq \phi_N(X_n) \leq 0$ so $\phi_N(X_n)$ is bounded, hence by Theorem 9.11, $(\phi_N(X_S), \phi_N(X_T))$ is a submartingale. Now X_0 is integrable, and $X_0 \leq \phi_N(X_0) \leq 0$, so $\phi_N(X_0)$ is also integrable and

$$E\{X_0\} \leq E\{\phi_N(X_0)\} \leq E\{\phi_N(X_S)\} \leq 0$$

by Theorem 9.11, using 0 and S as the stopping times. Let $N \to \infty$ and note that $\phi_N(x) \downarrow x$ for all x. By the monotone convergence theorem, $E\{\phi_N(X_S)\} \downarrow E\{X_S\} \geq E\{X_0\}$. Thus, X_S is integrable. Similarly, so is X_T. Now let $\Lambda \in \mathcal{F}_S$ and use the submartingale property:

$$\int_\Lambda \phi_N(X_S) \, dP \leq \int_\Lambda \phi_N(X_T) \, dP.$$

Let $N \to \infty$ and use monotone convergence on both sides to see that $\int_\Lambda X_S \leq \int_\Lambda X_T$. Thus, (X_S, X_T) is a submartingale, as claimed. □

Remark 9.13. It is not always true that a martingale remains a martingale under optional sampling. For example, let (X_n) be a simple symmetric random walk with $X_0 = 1$. It is a martingale. If T is the stopping time $T = \inf\{N \geq 0 : X_n = 0\}$, then $(X_{T \wedge n})$ is a positive martingale, but $(X_0, X_T) = (1, 0)$ is *not* a martingale. On the other hand, $(X_{T \wedge n})$ is a positive supermartingale, and so is (trivially) its limit, so Theorem 9.12 holds, as it must.

Application to the Gambler's Ruin. Let $\{X_n,\ n = 0, 1, 2, \ldots\}$ be a simple symmetric random walk from an integer x. Fix an integer $N > 0$ and let T be the stopping time $T = \inf\{n \geq 0 : X_n \leq 0 \text{ or } X_n \geq N\}$. If $0 < x < N$ (the most interesting case), then $\{X_{T_n},\ n = 0, 1, 2, \ldots\}$ is a gambler's ruin on $[0, N]$. It is bounded, and by the Theorem 9.8, it is a martingale relative to its natural filtration, $\mathcal{F}_n = \sigma(X_0, \ldots, X_n)$. Moreover, X_T equals either zero or N. Let us find the probability p that the gambler goes broke, i.e., that $X_T = 0$. (We derived this earlier in §1.7, but now we will get it by a simple martingale argument.)

By Theorem 9.11, (X_0, X_T) is a martingale, so that $E\{X_0\} = E\{X_T\}$. But $E\{X_0\} = x$ and $P\{X_T = 0\} = p$, so
$$x = E\{X_T\} = NP\{X_T = N\} = N(1-p).$$

Solving for p, the probability that the gambler goes broke is

(9.2) $$p = \frac{N-x}{N}.$$

Let us find the expected duration of the game. Since it ends at T—when either the gambler or the opponent goes broke—we want to find $E\{T\}$.

First, we claim that $\{X_n^2 - n, \mathcal{F}_n,\ n = 0, 1, 2, \ldots\}$ is also a martingale. Indeed,

$$E\{X_{n+1}^2 - (n+1) \mid \mathcal{F}_n\} = E\left\{(X_n + (X_{n+1} - X_n))^2 - n - 1 \mid \mathcal{F}_n\right\}$$
$$= E\{X_n^2 + 2X_n(X_{n+1} - X_n) + \underbrace{(X_{n+1} - X_n)^2}_{1} - n - 1 \mid \mathcal{F}_n\}$$
$$= X_n^2 + 2X_n \underbrace{E\{X_{n+1} - X_n \mid \mathcal{F}_n\}}_{0} - n = X_n^2 - n,$$

proving the claim. Then $X_{n \wedge T}^2 - n \wedge T$ is also a martingale (Theorem 9.8,) hence $E\{X_{n \wedge T}^2 - n \wedge T\} = E\{X_0^2\} = x^2$, or
$$E\{X_{n \wedge T}^2\} = E\{n \wedge T\} + x^2.$$

Let $n \to \infty$. $X_{n \wedge T}$ is bounded by N, and $n \wedge T \uparrow T$, so we can go to the limit on the left-hand side by bounded convergence, and on the right by monotone convergence, to see that
$$E\{X_T^2\} = E\{T\} + x^2.$$

But we know the distribution of X_T: $E\{X_T^2\} = \frac{x}{N} N^2$. Plugging in and solving:

(9.3) $$E\{T\} = x(N-x).$$

There is an easy consequence of this: the simple symmetric random walk is null-recurrent. Indeed, let $x > 0$ and let $T_0 = \inf\{n \geq 0 : X_n = 0\}$ be

9.2. System Theorems

the first hit of the origin. Notice that as $N \to \infty$, the first hitting time of $\{0, N\}$ increases to T_0. Thus,
$$E\{T_0 \mid X_0 = x\} = \lim_{N \to \infty} x(N - x) = \infty \,.$$
It follows that the expected return time to the origin is also infinite, and the random walk must be null-recurrent. Done!

So, there is hope for the gambler. He is bound to go broke, but perhaps not soon: his expected time-to-insolvency is infinite.

Remark 9.14. We know from our work on Markov chains that T is a.s. finite, and we used that above. However, it would be more satisfactory to give a martingale proof of it. We will do just that once we have the martingale convergence theorem. See Example 9.24.1.

The Martingale Maximal Inequality. The maximum of a sequence of random variables, or of functions, for that matter, is notoriously slippery to handle, and results such as maximal inequalities which give control of it are cherished. We will look here at a maximal inequality which falls out quite neatly from the system theorems.

Theorem 9.15 (Maximal Inequality). *Let $\{X_n, \mathcal{F}_n, n = 0, 1, 2 \dots\}$ be a positive submartingale. Then for $\lambda \geq 0$ and $N \geq 1$,*
$$(9.4) \qquad \lambda P\{\max_{n \leq N} X_n \geq \lambda\} \leq E\{X_N\} \,.$$

The following corollary is immediate.

Corollary 9.16. *Let $\{X_n, \mathcal{F}_n, n = 0, 1, 2, \dots\}$ be a positive submartingale. Then for $\lambda \geq 0$,*
$$(9.5) \qquad \lambda P\{\sup_n X_n \geq \lambda\} \leq \sup_n E\{X_n\} \,.$$

Remark 9.17. If (X_n) is a martingale, $(|X_n|)$ is a positive submartingale, and the theorem applies: $\lambda P\{\max_{n \leq N} |X_n| \geq \lambda\} \leq E\{|X_N|\}$.

For comparison, Chebyshev's inequality tells us that $\lambda P\{|X_N| \geq \lambda\} \leq E\{|X_N|\}$. This is the same bound as the martingale inequality, except that it only bounds the one variable $|X_N|$, while the martingale maximum inequality bounds the maximum of the whole sequence $|X_0|, |X_1|, \dots, |X_N|$. Since N could be arbitrarily large—in the millions, for example—this is a far stronger result.

Proof. (of Theorem 9.5.) Let T be the first time the process exceeds λ:
$$T = \begin{cases} \inf\{n \leq N : X_n \geq \lambda\}, & \\ N & \text{if no such } n \text{ exists.} \end{cases}$$

T is a bounded stopping time so Theorem 9.9 applies: (X_T, X_N) is a submartingale. Moreover, as X_T is one of $X_0, \ldots, X_N$, $\max_{n \leq N} X_n \geq \lambda \iff X_T \geq \lambda$. Therefore, applying Chebyshev's inequality and the submartingale property in order,
$$\lambda P\{\max_{n \leq N} X_n \geq \lambda\} = \lambda P\{X_T \geq \lambda\} \leq E\{X_T\} \leq E\{X_N\}.$$
□

There is a companion minimal inequality, which follows from similar considerations.

Proposition 9.18. *Let $\{X_j, \mathcal{F}_j, 1 \leq j \leq N\}$ be a submartingale. Then for $\lambda > 0$,*
$$\lambda P\{\min_{0 \leq j \leq N} X_j \leq -\lambda\} \leq E\{X_N^+\} - E\{X_1\}.$$

Proof. Let $T = \inf\{j : X_j \leq -\lambda\}$, and set $T = N$ if there is no such j. Then $\min_j X_j \leq -\lambda \iff X_T \leq -\lambda$. Now (X_1, X_T) is a submartingale, so
$$E\{X_1\} \leq E\{X_T\} = \int_{\{X_T \leq -\lambda\}} X_T \, dP + \int_{\{X_T \geq -\lambda\}} X_N \, dP,$$
where we used the fact that if $X_T > -\lambda$, then $T = N$ and $X_T = X_N$ in the second integral; but $X_N \leq X_N^+$ so this is
$$\leq -\lambda P\{X_T \leq -\lambda\} + E\{X_N^+\}.$$
This implies the result. □

Problems 9.2

9.2. Prove: to show that an adapted integrable process $\{X_n,\ n = 0, 1, 2, \ldots\}$ is a submartingale, it is enough to show $X_n \leq E\{X_{n+1} \mid \mathcal{F}_n\}$ for all n.

9.3. Let $x_1, x_2, \ldots$ be a sequence of real numbers. Let $\{\mathcal{F}_n,\ n = 1, 2, \ldots\}$ be a filtration. Consider the process $\{x_n, \mathcal{F}_n,\ n = 1, 2, \ldots\}$. Show that this is a submartingale iff the sequence is increasing, a supermartingale iff the sequence is decreasing, and a martingale iff all the x_i are equal[3].

9.4. (Gambling systems don't work.) Let $X_1, X_2, \ldots$ be a sequence of i.i.d. random variables, and let $T_1 < T_2 < \ldots$ be finite stopping times for (X_n). Show that $X_{T_1+1}, X_{T_2+1}, X_{T_3+1}, \ldots$ is a sequence of i.i.d. random variables with the same distribution as the (X_n). (This was the original gambling system theorem, proved by Doob in 1936.)

9.5. A gambler plays the system that the French call "La Grande Martingale". She bets $1 the first bet. If she wins, she stops. If she loses, she doubles the bet, and bets $2 the next time. If she loses this, she redoubles the bet, and continues, doubling the bet at each loss, until she finally wins, and stops. Suppose the bets are independent, and her probability of winning is $p > 0$ each time.

---

[3]Trivial though it may be, this example is sometimes useful to curb over-enthusiastic conjectures.

(a) Show that she is sure to make a profit of $1. What is the expected size of her winning bet?

(b) Suppose that $p = 1/2$. Let S_n be the gambler's fortune after the nth bet. Take $S_0 = 0$, so S_n is 1 if she has won and $-1 - 2 - 4 - \cdots - 2^{n-1}$ if she has not. Show that S_n is a martingale.

(c) Show that $\lim_{n \to \infty} S_n$ exists and equals one, but that $\lim_{n \to \infty} E\{S_n\} = 0$.

(d) Suppose $p < 1/2$, as it will be in casino games such as roulette. Show that, although $S_n \to 1$ a.s., $\lim_{n \to \infty} E\{S_n\} = -\infty$.

9.6. (The von Neumann Extractor) Let $X_1, X_2, \ldots$ be an exchangeable sequence with values in $\{0, 1\}$, such that the probability that $X_n \neq X_{n+1}$ is non-zero. Here is a way to extract a genuinely random sequence from it.

Divide the sequence into non-overlapping pairs $(X_1, X_2), (X_3, X_4), \ldots$. In each pair (X_i, X_{i+1}), if $X_i = X_{i+1}$, throw away the pair. If $X_i \neq X_{i+1}$, keep X_i and throw away X_{i+1}. Let $Y_1, Y_2, \ldots$ be the sequence of random variables kept. Show that the Y_i are i.i.d. Bernoulli $(1/2)$ random variables.

Note: This method, due to John von Neumann, extracts an independent sequence from a sequence that is merely exchangeable, but not necessarily independent.

[Hint: The subtlety—and the reason that it finds its way into this chapter—is that the Y_i are related to the X_j at times determined by the sequence, i.e., random times. Intuitively, Y_i has the same distribution, no matter what has gone before, and, by symmetry, 0 and 1 are equally likely values. Break it up into sets on which $Y_i = X_{2n-1}$, and use the exchangeability of X_{2n-1} and X_{2n} to show that $P\{Y_i = j\} = 1/2$, $j = 0, 1$, and, moreover, that the *conditional* probability is $1/2$, given whatever happens before $2n - 1$. Conclude independence from this.]

The following five exercises concern an optimal stopping problem. You are successively offered a sequence of rewards, $Z_0, \ldots Z_n$. As each is offered, you can either accept it, or reject it and go on to the next, but you can only accept one. You wish to maximize the expected reward. What is your best strategy?

Notice that choosing a reward defines a stopping time, so the problem can be stated formally:

Let $Z_0, \ldots Z_n$ be integrable random variables. Let $\mathcal{F}_j = \sigma\{Z_i : i \leq j\}$. Find a $(\mathcal{F}_j)$-stopping time $\tau \leq n$ which maximizes $E\{X_\tau\}$ over all possible $(\mathcal{F}_j)$-stopping times $T \leq n$.

9.7. Let $X_n = Z_n$ and define X_j by backward induction: $X_j = \max(Z_j, E\{X_{j+1} \mid \mathcal{F}_j\})$. The process $\{X_j,\ j = 0, \ldots, n\}$ is called **Snell's envelope** for the problem. Show that Snell's envelope is the smallest supermartingale which dominates (Z_j). That is, show that

(a) $X_j \geq Z_j$ for all $j \leq n$;

(b) $\{X_j, \mathcal{F}_j,\ j = 1, \ldots, n\}$ is a supermartingale;

(c) if $\{M_j, \mathcal{F}_j,\ j = 1, \ldots, n\}$ satisfies (a) and (b), then $M_j \geq X_j,\ j = 1, \ldots, n$.

9.8. (Continuation.) Let $\tau = \inf\{j : X_j = Z_j\}$.

(a) Show that τ is a stopping time and $\tau \leq n$.

(b) Show that $\{X_{j\wedge\tau}, \mathcal{F}_j, j \leq n\}$ is a martingale, hence $E\{X_0\} = E\{Z_\tau\}$.

9.9. (Continuation.) Show that τ is the optimal stopping time. That is, show that if T is any other stopping time with $T \leq n$, that $E\{Z_T\} \leq E\{Z_\tau\}$.

9.10. You are a contestant at a game show which offers prizes concealed behind three closed doors. The prize behind door i is worth Z_i, $i = 1, 2, 3$. You know that the Z_i are i.i.d. $U(0,1)$ random variables. The host opens each door in turn, and you can either accept the prize or go on to the next door. If you don't accept the prize behind the second door, you go on to the third.

(a) What is Snell's envelope for this, and what is the optimal stopping time τ?

(b) What is the expected prize, i.e., what is $E\{Z_\tau\}$?

9.11. Let (Y_n) be a simple symmetric random walk on states 0–4, stopped the first time it hits 0 or 4. There is a prize $r(i)$ at state i, where $r(0) = 2$, $r(1) = 1$, $r(2) = 3$, $r(3) = 2$ and $r(4) = 2$. You can stop the process at any time j and take the prize at the state you are at, but you can only do this once.

(a) State this as an optimal stopping problem.

(b) Show that there is a concave function f on the state space such that Snell's envelope is $X_j = f(Y_j)$.

(c) Find the optimal stopping time τ.

(d) Show that given $Y_0 = i$, $E\{r(Y_\tau)\} = f(i)$.

9.12. Let $\{X_j^i, \mathcal{F}_j, j = 0, 1, \dots\}$, $i = 1, 2$ be supermartingales. Suppose that there exists a stopping time T such that $X_T^1 \geq X_T^2$ a.s. on $\{T < \infty\}$. Show that (Y_n) is a supermartingale, where

$$Y_n \stackrel{\text{def}}{=} \begin{cases} X_n^1 & \text{if } n < T, \\ X_n^2 & \text{if } n \geq T. \end{cases}$$

9.3. Convergence

The original suspicion that martingales might have interesting convergence properties came from Doob's intuition that information is increasing (that is, the σ-fields increase with n) and, since something—in this case the information—is monotone, if there is any justice, there should be some kind of convergence. And, since sub- and supermartingales are just martingales plus monotone processes, they should converge as well. The intuition was resoundingly correct: given half a chance, a martingale will converge, and so will sub- and supermartingales.

The Upcrossing Inequality. We are going to prove an unusual type of inequality in this section, called the martingale upcrossing inequality[4]. The maximal inequality gave us a bound on the size of the maximum. The

---

[4]This was proved by J.L. Doob for martingales, and by his student, Laurie Snell, for submartingales.

9.3. Convergence

upcrossing inequality gives us a bound on how much the process can "wiggle around."

Let $[a, b]$ be an interval and let $x_0, x_1, x_2, \ldots, x_N$ be a sequence of real numbers. We want to count the number of times the sequence crosses the interval. There are two crossing directions, up and down, but they are so closely related that we only need to keep track of the "upcrossings", the number of times the sequence goes from below a to above b. Let us see how to count them.

Set $\alpha_0 = 0$ and let
$$\alpha_1 = \begin{cases} \inf\{n \leq N : x_n \leq a\}, \\ N+1 & \text{if no such } n. \end{cases}$$

For $k \geq 1$ set
$$\beta_k = \begin{cases} \inf\{n \geq \alpha_k : x_n \geq b\}, \\ N+1 & \text{if no such } n, \end{cases}$$
and
$$\alpha_{k+1} = \begin{cases} \inf\{n \geq \beta_k : x_n \geq b\}, \\ N+1 & \text{if no such } n. \end{cases}$$

Then $\alpha_0 = 0 \leq \alpha_1 \leq \beta_1 \leq \alpha_2 \leq \beta_2 \leq \ldots$ with equality only if the α_k or β_k equals $N+1$. If $\beta_k \leq N$, then $\alpha_k < \beta_k$, $x_{\alpha_k} \leq a$ and $x_{\beta_k} \geq b$, so that there is an upcrossing—and only one—of $[a, b]$ between α_k and β_k, and a downcrossing between β_{k-1} and α_k. So there is one upcrossing for each k for which $\beta_k \leq N$.

Definition 9.19. The **number of upcrossings** of the interval $[a, b]$ by the sequence $x_0, \ldots, x_N$ is
$$\nu_N(a, b) = \sup\{k : \beta_k \leq N\}.$$

This defines the number of upcrossings of $[a, b]$ by any sequence of reals, and by extension, by a stochastic process. Let $\{X_n, n = 0, 1, 2, \ldots\}$ be a stochastic process which is adapted to the filtration $\{\mathcal{F}_n, n = 0, 1, 2, \ldots\}$. Fix $\omega \in \Omega$ and apply the above to the sequence of reals $X_0(\omega), X_1(\omega), X_2(\omega), \ldots$. The α_k and β_k are now random variables, and, in fact, they are stopping times relative to the filtration $(\mathcal{F}_n)$. The number of upcrossings of $[a, b]$ by $X_0(\omega), \ldots, X_N(\omega)$ is $\nu_N(a, b)(\omega)$, so $\nu_N(a, b)$ is also a random variable.

Theorem 9.20 (Upcrossing Inequality). *Let $\{X_n, \mathcal{F}_n, n = 0, 1, 2, \ldots\}$ be a submartingale and let $a < b$ be real numbers. Then the number of upcrossings $\nu_N(a, b)$ of $[a, b]$ by $\{X_n, n = 0, \ldots, N\}$ satisfies*

(9.6) $$E\{\nu_N(a, b)\} \leq \frac{E\{(X_N - a)^+\}}{b - a}.$$

Proof. Fix N and let $\hat{X}_n = a \vee X_n$, $n = 1, \ldots, N$ and let $\hat{X}_{N+1} = \hat{X}_N$. Then $(\hat{X}_n)$ is again a submartingale, and it has the same number of upcrossings as the original sequence (X_n). Now $\alpha_0 = 0$, so

$$\hat{X}_N - \hat{X}_0 = \hat{X}_{\alpha_1} - \hat{X}_{\alpha_0} + \sum_{n=1}^{N}(\hat{X}_{\beta_n} - \hat{X}_{\alpha_n}) + \sum_{n=1}^{N}(\hat{X}_{\alpha_{n+1}} - \hat{X}_{\beta_n}).$$

By the submartingale property,

$$E\{\hat{X}_N - \hat{X}_0\} = \underbrace{E\{\hat{X}_{\alpha_1} - \hat{X}_{\alpha_0}\}}_{\geq 0} + \sum_{n=1}^{N} E\{\hat{X}_{\beta_n} - \hat{X}_{\alpha_n}\} + \sum_{n=1}^{N} \underbrace{E\{\hat{X}_{\alpha_{n+1}} - \hat{X}_{\beta_n}\}}_{\geq 0}$$

$$\geq E\left\{\sum_{n=1}^{N}(\hat{X}_{\beta_n} - \hat{X}_{\alpha_n})\right\},$$

since we have tossed away some positive terms. But note that for all n for which $\beta_n \leq N$, there is an upcrossing between α_n and β_n, so that $\hat{X}_{\beta_n} - \hat{X}_{\alpha_n} \geq b - a$. There are exactly $\nu_N(a,b)$ of these. If $\beta_n = N+1$, all we can say is that $E\{\hat{X}_{\beta_n} - \hat{X}_{\alpha_n}\}$ is positive. Thus the sum is

$$\geq (b-a) E\{\nu_N(a,b)\}.$$

We conclude that

$$E\{\nu_N(a,b)\} \leq \frac{E\{\hat{X}_N - \hat{X}_0\}}{b-a}.$$

Now $\hat{X}_N - \hat{X}_0 = a \vee X_N - a \vee X_0 \leq a \vee X_N - a = (X_N - a)^+$. □

Remark 9.21. We threw away the term $E\{\sum(\hat{X}_{\alpha_{n+1}} - \hat{X}_{\beta_n})\}$, since it is positive. *But the interval from β_n to α_{n+1} is a downcrossing interval. So why is this term not negative?*

The number of upcrossings increases as the length of the sequence increases, so let $\nu_\infty(a,b) = \lim_{N \to \infty} \nu_N(a,b)$. By monotone convergence, $E\{\nu_\infty(a,b)\} = \lim_N E\{\nu_N(a,b)\}$ and we have

Corollary 9.22. *Let $\{X_n, \mathcal{F}_n, n = 0, 1, 2, \ldots\}$ be a submartingale and $a < b$. Then*

$$E\{\nu_\infty(a,b)\} \leq \frac{\sup_N E\{(X_N - a)^+\}}{b-a}.$$

Martingale Convergence. There are numerous proofs of the martingale convergence theorem. We will give Doob's proof here, not only because it is clear, concise, and elegant, but also because it gives insight into convergence itself.

9.3. Convergence

Theorem 9.23 (Martingale Convergence Theorem). *Let $\{X_n, \mathcal{F}_n, n = 0, 1, 2, \ldots\}$ be a submartingale. Suppose that $\sup_n E\{|X_n|\} < \infty$. Then there exists a finite random variable X_∞ such that*
$$\lim_{n \to \infty} X_n = X_\infty \text{ a.s.}$$

Proof. First let us show that $\liminf_{n \to \infty} X_n = \limsup_{n \to \infty} X_n$ a.s. If this is not true, then $\liminf_n X_n < \limsup_n X_n$ with positive probability. Fix ω for which $\liminf_n X_n(\omega) < \limsup_n X_n(\omega)$. (Once we have fixed ω, we are just dealing with a sequence of real numbers.) Then there exist rational $a < b$ for which $\liminf_n X_n(\omega) < a < b < \limsup_n X_n(\omega)$. Then there exist infinitely many n for which $X_n < a$, and infinitely many n for which $X_n > b$. But for this to happen, the process must cross $[a,b]$ infinitely often. Thus, $\nu_\infty(a,b)(\omega) = \infty$.

By the upcrossing inequality,
$$E\{\nu_\infty(a,b)\} \leq \frac{\sup_N E\{(X_N - a)^+\}}{b - a} \leq \frac{\sup_N E\{|X_N|\} + |a|}{b - a},$$
and this is finite by hypothesis. Thus, $P\{\nu_\infty(a,b) = \infty\} = 0$. This is true simultaneously for each of the countable number of rational pairs $a < b$, and it follows that the set of ω for which $\liminf_n X_n(\omega) < \limsup_n X_n(\omega)$ has probability zero. In short, $P\{\liminf_n X_n = \limsup_n X_n\} = 1$, so the limit exists almost surely. Thus, $X_\infty \stackrel{\text{def}}{=} \lim_n X_n = \liminf_n X_n = \limsup_n X_n$ exists. A priori it might be infinite. However, by Fatou's Lemma,
$$E\{|X_\infty|\} \leq \liminf_{n \to \infty} E\{|X_n|\} < \infty.$$
Thus, X_∞ is finite a.s. □

Notice that to show a submartingale converges, it is enough to show that its absolute expectation is bounded.

Corollary 9.24. *Let $\{X_n, \mathcal{F}_n, n = 0, 1, 2, \ldots\}$ be either a positive supermartingale or a negative submartingale. Then X_n converges a.s. as $n \to \infty$.*

Proof. In either case, $|X_n|$ is a positive supermartingale. By definition, X_0 is integrable, so the submartingale inequality implies that for all $n \geq 1$,
$$E\{|X_n|\} \leq E\{|X_0|\} < \infty.$$
Thus the absolute expectation is bounded, and convergence follows from Theorem 9.23. □

Example 9.24.1. We promised to give a martingale proof that the Gambler's Ruin eventually ends. In fact, we can do slightly better, and show that the simple symmetric random walk is recurrent. We showed that in §7.2, but here is another proof.

Let X_n be a simple symmetric random walk. To show it is recurrent, it is enough to show that X_n hits zero a.s., regardless of its initial value. Let $X_0 = k \neq 0$ and let $T_0 = \inf\{n \geq 0 : X_n = 0\}$ be the first hit of zero. Then (X_n) is a martingale, hence so is $(X_{n \wedge T_0})$ (Theorem 9.8.) In fact, if $k > 0$, $(X_{n \wedge T_0})$ is a positive martingale and therefore converges a.s. by the above corollary. Suppose $T_0 = \infty$ and observe:

$$T_0 = \infty \implies |X_{(n+1) \wedge T_0} - X_{n \wedge T_0}| = |X_{n+1} - X_n| \equiv 1 \quad \forall n,$$

so the sequence $(X_{n \wedge T_0})$ cannot converge, a contradiction. Therefore, $T_0 < \infty$ a.s.!

Exercise 9.13. Let $Y_1, Y_2, \ldots$ be bounded i.i.d. random variables with $E\{Y_1\} = 0$. Let $X_n = Y_1 + \cdots + Y_n$. Show that if $P\{Y_1 \neq 0\} > 0$, then with probability one, $\limsup_n X_n = \infty$ and $\liminf_n X_n = -\infty$.

9.3.1. Backward Martingales. We often want to take the limits of martingales as the σ-fields decrease, rather than increase. To handle this, we simply take the parameter set to be $\ldots, -3, -2, -1$ instead of $1, 2, 3, \ldots$, and then ask about the limiting behavior of the martingale X_n as n tends to minus infinity, rather than to infinity. For lack of a better name, such processes are called **backward martingales.**

Let $\{X_n, \mathcal{F}_n, n = \ldots, -3, -2, -1\}$ be a submartingale. Consider the limit as n tends to minus infinity. Since $(\mathcal{F}_n)$ is a filtration, the σ-fields decrease as $n \to -\infty$, so this is the limit as the σ-fields decrease. It turns out that backward martingales converge under even weaker conditions than ordinary martingales.

Theorem 9.25. *Let $X = \{X_n, \mathcal{F}_n, n = \ldots, -3, -2, -1\}$ be a submartingale.*

(i) X_n converges as $n \to -\infty$ to a random variable $X_{-\infty}$, where $-\infty \leq X_{-\infty} < \infty$ a.s. If $E\{X_n\}$ is bounded below, $X_{-\infty}$ is finite and integrable.

(ii) If X is a martingale, then $X_{-\infty}$ exists and is integrable.

Proof. In order to show convergence, we need only show that there are only finitely many upcrossings of any given interval. Define $\nu_{-n}(a,b)$ to be the number of upcrossings of $[a,b]$ by the sequence $X_{-n}, X_{-n+1}, \ldots, X_{-1}$, and let $\nu_{-\infty}(a,b) = \lim_{n \to \infty} \nu_{-n}(a,b)$. The last element of the submartingale is X_{-1}, so by the upcrossing inequality,

$$E\{\nu_{-n}(a,b)\} \leq \frac{E\{(X_{-1} - a)^+\}}{b - a} < \infty.$$

9.4. Uniform Integrability

Notice that the bound is independent of n, so that we can go to the limit as $n \to \infty$ by monotone convergence to see that

$$E\{\nu_{-\infty}(a,b)\} \leq \frac{E\{(X_{-1}-a)^+\}}{b-a} < \infty.$$

Thus, $\nu_{-\infty}(a,b)$ is integrable, hence a.s. finite. This is true simultaneously for all pairs of rational $a < b$, which, as before, implies that as $n \to -\infty$, the lim sup and lim inf of X_n are equal. A priori, the lim inf and lim sup can take on infinite values, so that $X_{-\infty} = \lim_{n \to -\infty} X_n$ exists in the extended reals $[-\infty, \infty]$.

But, in fact, the limit cannot equal plus infinity. Indeed, X_n^+ is a positive submartingale which converges to $X_{-\infty}^+$, and by Fatou's Lemma and the submartingale inequality,

$$E\{X_{-\infty}^+\} \leq \lim_{n \to -\infty} E\{X_{-n}^+\} \leq E\{X_{-1}^+\} < \infty.$$

Thus, $X_{-\infty}^+$ is integrable, hence a.s. finite.

Now suppose the expectations are bounded below, say $E\{X_n\} \geq K$ for all n. Note that $|x| = 2x^+ - x$, so that by the submartingale inequality, $E\{|X_{-n}|\} = 2E\{X_{-n}^+\} - E\{X_{-n}\} \leq 2E\{X_{-1}^+\} + |K|$. It follows from Fatou's Lemma again that $E\{|X_{-\infty}|\} \leq 2E\{X_{-1}^+\} + |K| < \infty$. Thus the limit is not only finite, but integrable.

In case X is a martingale, its expectation is constant, therefore bounded below, so that the limit automatically exists and is integrable. □

Conclusion: Backward martingales always converge to a finite, integrable limit. Backward submartingales and supermartingales also have limits, but they may be infinite. The limits are finite and integrable if the expectations are bounded. However, it is easy to find examples of backward submartingales which converge to minus infinity.

9.4. Uniform Integrability

We will revisit some questions about taking limits under expectations. Up till now we have used the bounded, dominated, or monotone convergence theorems to justify this. We will introduce a new (to us) criterion, called *uniform integrability*. This is particularly well-suited to martingales.

Let $\{X_n, \mathcal{F}_n, n = 0, 1, 2, \dots\}$ be a martingale. Suppose that it converges to X_∞, say. Question: Is $E\{X_\infty\} = \lim_n E\{X_n\}$? Another question: If $m < n$, we know that $X_m = E\{X_n \mid \mathcal{F}_m\}$. But is $X_m = E\{X_\infty \mid \mathcal{F}_m\}$? The first question simply asks if we can go to the limit under the expectation; the bounded and dominated convergence theorems tell us that that the answer is "Yes" if the martingale is bounded or dominated by a random variable Y, but it is "No" in general. (The simple symmetric random walk from 1, stopped when it first hits zero, has the limit $0 \neq 1$.) The second question is more delicate: it asks if we can go to the limit under a conditional expectation. If the answer is yes, then $X_0, X_1, \ldots, X_\infty$ is a martingale. That is, the limit can be added as a final element to the martingale.

It turns out that the hypotheses of the dominated convergence theorem are often difficult to verify in this context. We turn to another criterion, called *uniform integrability*, which can be easier to verify for martingales[5].

Definition 9.26. A family $\{X_\alpha, \alpha \in I\}$ of random variables is **uniformly integrable** if
$$\lim_{N \to \infty} \int_{\{|X_\alpha| > N\}} |X_\alpha| \, dP = 0, \quad \text{uniformly in } \alpha.$$

Remark 9.27. (i) $\{X_\alpha,\ \alpha \in I\}$ is uniformly integrable iff for all $\varepsilon > 0$ there is N_ε such that $\int_{\{|X_\alpha| > N_\varepsilon\}} |X_\alpha| \, dP < \varepsilon$ for all $\alpha \in I$.

(ii) A single integrable random variable is uniformly integrable.

(iii) A finite family of integrable random variables is uniformly integrable.

(iv) A finite union of uniformly integrable families is uniformly integrable.

(v) If there exists an integrable random variable Y such that $|X_\alpha| \leq Y$ a.s. for each $\alpha \in I$, the family $\{X_\alpha,\ \alpha \in I\}$ is uniformly integrable.

(vi) If there exists $K < \infty$ such that $E\{X_\alpha^2\} \leq K$ for all $\alpha \in I$, then $\{X_\alpha,\ \alpha \in I\}$ is uniformly integrable.

(vii) If $\{X_\alpha,\ \alpha \in I\}$ is uniformly integrable, then $E\{|X_\alpha|\}$, $\alpha \in I$ is bounded.

Indeed, (i) just restates the definition, (ii) follows from the dominated convergence theorem, since, as $P\{|X| > N\} \to 0$ as $N \to \infty$, $XI_{\{|X|>N\}}$ converges to zero in measure and is dominated by the integrable random variable $|X|$. (iii) and (iv) are then immediate.

In (v), $|X_\alpha| I_{\{|X_\alpha|>N\}} \leq Y I_{\{Y>N\}}$, which tends to zero since Y is integrable. By dominated convergence, $E\{|X_\alpha| I_{\{|X_\alpha|>N\}}\} \leq E\{Y I_{\{Y>N\}}\} \to 0$, which implies that the X_α are uniformly integrable.

In (vi), if the expectation is bounded by, say, K, write
$$\int_{\{|X|>N\}} |X_\alpha| \, dP \leq \frac{1}{N} \int_{\{|X_\alpha|>N\}} X_\alpha^2 \, dP \leq \frac{1}{N} E\{X_\alpha^2\} \leq \frac{K}{N},$$

---

[5]Uniform integrability is in fact a necessary and sufficient condition for going to the limit under the expectation. It is not widely used in analysis, for it is not easy to verify in most situations. However, as we will see, it is often easy to verify for martingales and submartingales.

9.4. Uniform Integrability

which converges to zero uniformly in α. Finally, for (vii), we just choose N such that $\int_{\{|X_\alpha|>N\}} |X_\alpha|\, dP \leq 1$ for all α. Then $E\{|X_\alpha|\} \leq N+1$. □

Here is an alternate characterization of uniform integrability. We leave the proof as an exercise.

Exercise 9.14. A family $\{X_\alpha,\ \alpha \in I\}$ of random variables is uniformly integrable if and only if $E\{|X_\alpha|\}$ is bounded for $\alpha \in I$, and

$$(9.7) \qquad \lim_{P\{\Lambda\}\to 0} \int_\Lambda |X_\alpha|\, dP = 0 \quad \text{uniformly in } \alpha.$$

The remarks above tell us that the familiar conditions for going to the limit under the expectation—boundedness and domination—imply uniform integrability. Uniform integrability is more general, for it does not imply either. The following theorem shows why the condition is interesting: it is a necessary and sufficient condition for taking the limit under the expectation.

Theorem 9.28. *Let X be a random variable and let $\{X_n,\ n=0,1,2,\dots\}$ be a sequence of integrable random variables such that $X_n \longrightarrow X$ in probability. Then the following conditions are equivalent:*

(i) The sequence is uniformly integrable.

(ii) $X_n \longrightarrow X$ in L^1 as $n \to \infty$;

(iii) $E\{|X_n|\} \longrightarrow E\{|X|\}$ as $n \to \infty$.

Proof. $(i) \implies (ii)$: $X_n \to X \implies |X_n| \to |X|$ in probability. Thus there exists a subsequence (X_{n_k}) which converges a.e. By Fatou's Lemma, $E\{|X|\} \leq \liminf_k E\{|X_{n_k}|\}$. This subsequence is necessarily uniformly integrable, since the original sequence is. By the above remarks, the expectations are bounded, so $E\{|X|\} < \infty$, i.e., $X \in L^1$.

"Convergence in L^1" means that $E\{|X_n - X|\} \longrightarrow 0$. But X is integrable and (X_n) is uniformly integrable, so that the family $|X_n - X|$ is also uniformly integrable. Let $\varepsilon > 0$ and write

$$E\{|X_n - X|\} = \int_{\{|X_n-X|>\varepsilon\}} |X_n - X|\, dP + \int_{\{|X_n-X|\leq\varepsilon\}} |X_n - X|\, dP$$

$$\leq \sup_k \int_{\{|X_n-X|>\varepsilon\}} |X_k - X|\, dP + \varepsilon,$$

for the second integral is bounded by ε. Since $X_n - X$ converges to zero in probability, $P\{|X_n - X| > \varepsilon\}$ tends to zero as $n \to \infty$, and the family $(|X_k - X|)$ is uniformly integrable, so Exercise 9.14 implies that the supremum converges to zero as $n \to \infty$. This proves (ii).

$(ii) \implies (iii)$ is clear. To finish the proof, let us show that $(iii) \implies (i)$, which is the most striking part of the theorem.

Suppose $E\{|X_n|\} \longrightarrow E\{|X|\}$. Choose M for which $P\{|X| = M\} = P\{|X_n| = M\} = 0$ for all n, and truncate X and X_n:

$$X^{(M)} \stackrel{\text{def}}{=} \begin{cases} X & \text{if } |X| \leq M, \\ 0 & \text{if } |X| > M. \end{cases}$$

Then

$$\int_{\{|X_n|>M\}} |X_n|\, dP = E\{|X_n|\} - E\{|X_n^{(M)}|\}$$

$$\longrightarrow \underbrace{E\{|X|\}}_{\text{by hypothesis}} - \underbrace{E\{|X^{(M)}|\}}_{\text{by bounded convergence}}.$$

Now

$$\int_{\{|X_n|>M\}} |X_n|\, dP = E\{|X_n|-|X|\} + E\{|X|-|X^{(M)}|\} + E\{|X^{(M)}|-|X_n^{(M)}|\}.$$

Note that if $\varepsilon > 0$, there exists M_ε such that $E\{|X|\} - E\{|X^{(M_\varepsilon)}|\} < \varepsilon/3$. For this M_ε there exists N_0 such that if $n \geq N_0$, then

$$|E\{|X_n^{(M_\varepsilon)}|\} - E\{|X^{(M_\varepsilon)}|\}| < \varepsilon/3,$$
$$|E\{|X_n|\} - E\{|X|\}| < \varepsilon/3.$$

Thus for $n \geq N_0$,

$$\int_{\{|X_n|>M_\varepsilon\}} |X_n|\, dP \leq \frac{\varepsilon}{3} + \frac{\varepsilon}{3} + \frac{\varepsilon}{3} = \varepsilon.$$

But there are only finitely many $n \leq N_0$, so there exists $M'_\varepsilon > M_\varepsilon$ such that for *all* n,

$$\int_{\{|X_n|>M'_\varepsilon\}} |X_n|\, dP \leq \varepsilon.$$

Thus, (X_n) is uniformly integrable. □

Uniform Integrability and Martingales. Uniform integrability is often easy to check for martingales and submartingales because of its fundamental connection with conditional expectations:

Theorem 9.29. *Let $\{\mathcal{F}_\alpha, \alpha \in I\}$ be a family of sub-σ-fields of $\mathcal{F}$, and let X be an integrable random variable. Let $X_\alpha = E\{X \mid \mathcal{F}_\alpha\}$. Then $\{X_\alpha, \alpha \in I\}$ is a uniformly integrable family.*

Proof. Note that (X_α, X) is a martingale, so $(|X_\alpha|, |X|)$ is a submartingale. Then $P\{|X_\alpha| > N\} \leq \frac{1}{N} E\{|X_\alpha|\} \leq \frac{1}{N} E\{|X|\}$. This tends to zero as N tends to infinity. Thus,

$$\int_{\{|X_\alpha|>N\}} |X_\alpha|\, dP \leq \int_{\{|X_\alpha|>N\}} |X|\, dP.$$

9.4. Uniform Integrability

The right-hand side is the integral of a random variable over sets whose probability goes to zero, so it tends to zero by the dominated convergence theorem. In particular, given $\varepsilon > 0$, there exists $\delta > 0$ such that $P\{\Lambda\} < \delta \implies \int_\Lambda |X|\, dP < \varepsilon$. Choose $N_\varepsilon > E\{|X|\}/\delta$. Then $P\{|X_\alpha| > N_\varepsilon\} < \delta$, and we conclude that $N \geq N_\varepsilon \implies \int_{\{|X_\alpha|>N\}} |X_\alpha|\, dP < \varepsilon$. Thus the family (X_α) is uniformly integrable, as claimed. $\square$

Corollary 9.30. *(i) A positive submartingale with a last element is uniformly integrable.*

(ii) A martingale with a last element is uniformly integrable.

Proof. This is true for any parameter set which has a largest element, so let $\mathfrak{T} \subset \mathbb{R}$ be the parameter set, and τ its largest element. Then $t \in \mathfrak{T} \implies t \leq \tau$.

(i) Suppose $\{X_t, \mathcal{F}_t, t \in \mathfrak{T}\}$ is a positive submartingale. Then $0 \leq X_t \leq E\{X_\tau \mid \mathcal{F}_n\}$. Since the family $E\{X_\tau \mid \mathcal{F}_n\}$ is uniformly integrable, so is the family (X_t) of smaller random variables.

(ii) If (X_t) is a martingale, then $(|X_t|)$ is a positive submartingale, and the conclusion follows from (i). $\square$

Uniform Integrability and Convergence. Martingales, submartingales and supermartingales converge if their absolute expectations are bounded. Uniform integrability guarantees this, so if a submartingale is uniformly integrable, it converges a.e. (Theorem 9.23); but, uniform integrability tells us more. It may even identify the limit.

Theorem 9.31. *Let $X = \{X_n, \mathcal{F}_n, n \geq 0\}$ be a submartingale and let $\mathcal{F}_\infty = \sigma\left(\bigcup_n \mathcal{F}_n\right)$. Then the following are equivalent.*

(i) X is uniformly integrable.

(ii) X converges in L^1.

(iii) X converges a.e. to a limit X_∞, $\{X_n, \mathcal{F}_n, n = 0, 1, \ldots, \infty\}$ is a submartingale, and $E\{X_n\} \longrightarrow E\{X_\infty\}$ as $n \to \infty$.

Remark 9.32. There are two new things here: the convergence is in L^1 as well as a.e. (so that we can go to the limit under the integral whenever necessary) and the limit can be added at the end as the last element of the submartingale.

Proof. (i) $\implies$ (ii): Uniform integrability $\implies E\{X_n^+\} \leq E\{|X_n|\} \leq K$ for some finite K. The martingale convergence theorem implies that X_n converges a.e. to a random variable X_∞. Then uniform integrability implies that it also converges in L^1 (Theorem 9.28) which proves (ii).

$(ii) \implies (iii)$: If X_n converges to X_∞ in L^1, then X_∞ is integrable and $E\{X_n\} \longrightarrow E\{X_\infty\}$. Let $\Lambda \in \mathcal{F}_n$. Then as $k \to \infty$

$$\int_\Lambda X_n \, dP \le \int_\Lambda X_{n+k} \, dP \longrightarrow \int_\Lambda X_\infty \, dP \, .$$

This implies $X_n \le E\{X_\infty \mid \mathcal{F}_n\}$, and that X_∞ can be added to the submartingale as a last element, which proves (iii).

$(iii) \implies (i)$: If $(X_0, \ldots, X_\infty)$ is a submartingale, then $(X_1^+, \ldots, X_\infty^+)$ is a positive submartingale with a last element. It is therefore uniformly integrable by Corollary 9.30, and $E\{X_n^+\} \to E\{X_\infty^+\}$.

But now $E\{X_n^-\} = E\{X_n^+\} - E\{X_n\} \to E\{X_\infty^+\} - E\{X_\infty\}$, so that $E\{X_n^-\} \to E\{X_\infty^-\}$. For positive variables, this implies that (X_n^-) is uniformly integrable by Theorem 9.28. It follows that $(X_n) = (X_n^+ - X_n^-)$ is uniformly integrable. $\square$

Backward martingales and submartingales are even better-behaved than ordinary ones. In fact, a backward martingale is automatically uniformly integrable, and converges a.e. and in L^1. Submartingales and supermartingales are almost as good. They converge a.e. and in L^1 if their expectations are bounded.

Theorem 9.33. *Let* $X = \{X_n, \mathcal{F}_n, n \le -1\}$ *be a backward submartingale. Let* $\mathcal{F}_{-\infty} = \bigcap_n \mathcal{F}_n$. *Then the following statements are equivalent.*

(i) X *is uniformly integrable.*

(ii) X *converges in* L^1 *as* $n \to -\infty$.

(iii) X *converges a.s. to a random variable* $X_{-\infty}$ *and* $\{X_n, \mathcal{F}_n, n = -\infty, \ldots, -2, -1\}$ *is a submartingale.*

(iv) $E\{X_n\}$ *is bounded below.*

Proof. $(i) \implies (ii)$: If X is uniformly integrable, its expectations are bounded, and, in particular, are bounded below, so that by Theorem 9.25, X converges a.s. By Theorem 9.28, it also converges in L^1.

$(ii) \implies (iii)$: L^1-convergence implies the absolute expectations are bounded, so by Theorem 9.25, X also converges a.e. Let $X_{-\infty}$ be the limit. $X_{-\infty}$, being an L^1 limit by assumption, is integrable and $\mathcal{F}_{-\infty}$-measurable. Let $\Lambda \in \mathcal{F}_{-\infty}$. Since $\Lambda \in \mathcal{F}_m$ for all m, if $m < n$,

$$\int_\Lambda X_m \, dP \le \int_\Lambda X_n \, dP \, .$$

Let $m \to -\infty$. By L^1 convergence, we can go to the limit under the left-hand integral to see that

$$\int_\Lambda X_{-\infty} \, dP \le \int_\Lambda X_n \, dP \, ,$$

9.4. Uniform Integrability

which shows $X_{-\infty} \leq E\{X_n \mid \mathcal{F}_{-\infty}\}$ a.s. This implies (iii).

$(iii) \implies (iv)$: $X_{-\infty}$ is part of the submartingale, so it is integrable by definition. Then $E\{X_n\} \geq E\{X_{-\infty}\} > -\infty$.

$(iv) \implies (i)$: This is the only tricky part of the proof. Note first that if X is a backward martingale, it has a last element, and is therefore uniformly integrable by Corollary 9.30.

In general, X is a submartingale. Let $L = \lim_{n \to -\infty} E\{X_n\}$. This is finite by (iv). Let us now tweak the Doob decomposition to cover the parameter set $\{n : n \leq -1\}$. (The steps below just follow the proof of Theorem 9.7, and we refer the reader to that for the details.)

Let

$$d_n = E\{X_n \mid \mathcal{F}_{n-1}\} - X_{n-1},$$
$$A_n = \sum_{k=-\infty}^{n} d_n,$$
$$M_n = X_n - A_n.$$

Then $d_n \geq 0$, and $E\{d_n\} = E\{X_n\} - E\{X_{n-1}\}$. The series converges, for if $m < n$, $E\{\sum_{k=m}^{n} d_n\} = E\{X_n\} - E\{X_m\}$, so by the monotone convergence theorem, $E\{A_n\} = E\{X_n\} - L < \infty$. Thus A_n is finite and even integrable. Furthermore, (M_n) is a backward martingale. Therefore (M_n) is uniformly integrable, and (A_n), being an increasing sequence dominated by the integrable random variable A_{-1}, is also uniformly integrable. Hence so is $(X_n) = (M_n + A_n)$. $\square$

Remark 9.34. The above proof extends the Doob decomposition to include backward submartingales.

Uniform Integrability and System Theorems. If (X_n) is a submartingale and $S \leq T$ are finite stopping times, then we saw that if either the stopping times or the process are bounded, (X_S, X_T) is a submartingale. We will extend this to the case where both the submartingale and the stopping times can be unbounded. Uniform integrability is the key.

Theorem 9.35. *Let $\{X_n, \mathcal{F}_n, n \geq 0\}$ be a uniformly integrable submartingale, and let $S \leq T$ be finite stopping times. Then (X_S, X_T) is a submartingale relative to $(\mathcal{F}_S, \mathcal{F}_T)$.*

Proof. First note that $S \wedge n$ and $T \wedge n$ are bounded stopping times, so that $(X_{S \wedge n}, X_{T \wedge n})$ is a submartingale relative to $(\mathcal{F}_{S \wedge n}, \mathcal{F}_{T \wedge n})$. It converges to (X_S, X_T) as $n \to \infty$. We claim that the submartingale property is maintained under the convergence. The key is to show that $(X_{S \wedge n})$ and $(X_{T \wedge n})$

are uniformly integrable, and the key to that is to show that X_S and X_T are integrable.

By Fatou's Lemma,
$$E\{|X_T|\} \leq \liminf_n E\{|X_{T\wedge n}|\}$$
$$= \liminf_n \left(2E\{X_{T\wedge n}^+\} - E\{X_{T\wedge n}\}\right)$$

since $|x| = 2x^+ - x$. Now $(X_0, X_{T\wedge n})$ is a submartingale, so $E\{X_{T\wedge n}\} \geq E\{X_0\}$. $(X_{T\wedge n}, X_n)$ is also a submartingale, hence so is $(X_{T\wedge n}^+, X_n^+)$, and $E\{X_{T\wedge n}^+\} \leq E\{X_n^+\} \leq E\{|X_n|\} \leq K$, where the last inequality follows because (X_n) is uniformly integrable, and hence has bounded absolute expectations. Thus,
$$E\{|X_T|\} \leq K - E\{X_0\} < \infty,$$
and X_T is integrable.

Now for each ω, $X_{T\wedge n}(\omega)$ is either $X_n(\omega)$ or $X_T(\omega)$, so $|X_{T\wedge n}| \leq |X_T| + |X_n|$. But X_T is integrable and (X_n) is uniformly integrable, so that their sum—and therefore $(X_{T\wedge n})$—is uniformly integrable. So is $(X_{S\wedge n})$.

Let $\Lambda \in \mathcal{F}_S$. If $0 \leq j < n$, then $\Lambda \cap \{S = j\} \in \mathcal{F}_{S\wedge n}$. (Why?) Then, as $(X_{S\wedge n}, X_{T\wedge n})$ is a submartingale,

(9.8) $$\int_{\Lambda \cap \{S=j\}} X_{S\wedge n}\, dP \leq \int_{\Lambda \cap \{S=j\}} X_{T\wedge n}\, dP.$$

Let $n \to \infty$. $X_{S\wedge n} \to X_S$, $X_{T\wedge n} \to X_T$ and both families are uniformly integrable, so we can go to the limit under the integral on both sides of (9.8) to get
$$\int_{\Lambda \cap \{S=j\}} X_S\, dP \leq \int_{\Lambda \cap \{S=j\}} X_T\, dP.$$
Add over j to see that for all $\Lambda \in \mathcal{F}_S$,
$$\int_\Lambda X_S\, dP \leq \int_\Lambda X_T\, dP,$$
or, equivalently, $X_S \leq E\{X_T \mid \mathcal{F}_S\}$. □

If we apply this to an increasing sequence of stopping times, we get the following.

Corollary 9.36. *Let $\{X_n, \mathcal{F}_n,\ n = 0, 1, 2, \ldots\}$ be a submartingale and let $T_1 \leq T_2 \leq \ldots$ be an increasing sequence of finite stopping times. If either each T_n is bounded, or if (X_n) is uniformly integrable, then $\{X_{T_n}, \mathcal{F}_{T_n},\ n \geq 0\}$ is a submartingale.*

Problems 9.4

9.15. The upcrossing inequality can be refined. Prove **Dubin's inequality**: if $\nu_N(a,b)$ is the number of upcrossings of (a,b) by the positive supermartingale $\{X_n, \mathcal{F}_n, \ n = 0, 1, \ldots, N\}$, then

$$P\{\nu_N(a,b) \geq k\} \leq \left(\frac{a}{b}\right)^k \min(X_0, 1).$$

[Hint: Let $\alpha_1 \leq \beta_1 \leq \alpha_2 \leq \beta_2 \leq \alpha_3 \leq \ldots$ be the stopping times defined in Section 9.3 and define (Y_n) by $Y_n = 1$ if $0 \leq n < \alpha_1$, $Y_n = X_n/a$ if $\alpha_1 \leq N < \beta_1$, $Y_n = \frac{b}{a}$ if $\beta_1 \leq n < \alpha_2$, $Y_n = \frac{b}{a}\frac{X_n}{a}$ if $\alpha_2 \leq n < \beta_2$ and, in general, $Y_n = \left(\frac{b}{a}\right)^{k-1}\frac{X_n}{a}$ if $\alpha_k \leq n < \beta_k$, and $Y_n = \left(\frac{b}{a}\right)^k$ if $\beta_k \leq n < \alpha_{k+1}$. Show that (Y_n) is a supermartingale. Exercise 9.12 may help. Note that $Y_0 \leq \min(1, X_0/a)$ and that $Y_N \geq \left(\frac{b}{a}\right)^n I\{\beta_k \leq N\}$. Conclude that $\left(\frac{b}{a}\right)^k P\{\beta_k \leq N\} \leq \min(1, \frac{X_0}{a})$.]

9.16. (Continuation.) Dubin's inequalities cannot be improved, in the sense that there exists a supermartingale for which they are exact. For $0 < a < b$ let (ξ_n) be a sequence of i.i.d. Bernoulli (a/b) random variables. Set $X_0 = a$, $X_1 = b\xi_0$, $X_2 = a\xi_0$, and, in general, $X_{2k+1} = b\xi_0 \cdots \xi_k$, $X_{2k+2} = a\xi_0 \cdots \xi_k$. Let $(\mathcal{F}_k)$ be the natural filtration generated by (X_k).

(a) Show that $\{X_k, \mathcal{F}_k, \ k = 0, 1, 2, \ldots\}$ is a positive supermartingale.

(b) Show that $P\{\nu_\infty \geq k\} = \left(\frac{a}{b}\right)^k \min(1, X_0/a)$.

9.17. Let $\{X_n, \mathcal{F}_n, \ n = 0, 1, 2, \ldots\}$ be a submartingale. Show that if the expected squares of the X_n are bounded, it converges in L^2.

9.18. Let $\{X_n, \mathcal{F}_n, \ n = 0, 1, 2, \ldots\}$ be a submartingale and let $x > 0$. Show that $P\{\max_{0 \leq m \leq n} X_m \geq x\} \leq \frac{1}{x} E\{X_n^+\}$.

9.19. A martingale may converge in probability, but diverge a.e. Let (Y_n) be a sequence of i.i.d. random variables with values ± 1 with probability $1/2$ each. Let $\mathcal{F}_n = \sigma(Y_1, \ldots, Y_n)$. Let B_n be events in $\mathcal{F}_n$ such that $P\{B_n\} \to 0$ and $P\{\limsup_{n \to \infty} B_n = 1\} = 1$. Define $X_0 = 0$, $X_{n+1} = X_n(1 + Y_{n+1}) + I_{B_n}Y_{n+1}$. Show that (X_n) is a martingale which converges to zero in probability, but the set of ω for which $X_n(\omega)$ converges has measure zero.

9.20. Let $\{X_\alpha, \ \alpha \in I\}$ be a family of random variables. Let $\phi > 0$ be an increasing function such that $\phi(x)/x \longrightarrow \infty$ as $x \to \infty$. Show that if $E\{\phi(|X_\alpha|)\}$ is bounded in α, the family is uniformly integrable.

9.21. (Continuation.) Show that if $\{X_\alpha, \ \alpha \in I\}$ is uniformly integrable, there exists a function $\phi > 0$ with $\phi(x)/x \to \infty$ as $|x| \to \infty$, such that $E\{\phi(X_\alpha)\}$ is bounded in α.

9.22. Let $\{X_\alpha, \ \alpha \in I\}$ be a uniformly integrable family of random variables and let $\{\mathcal{G}_\beta, \ \beta \in J\}$ be a family of σ-fields. Let $Z_{\alpha\beta} = E\{X_\alpha \mid \mathcal{G}_\beta\}$. Show that the $Z_{\alpha\beta}$ are uniformly integrable.

9.23. Each day, a certain insurance company has equal probability of paying a claim or of collecting a premium. The probability itself depends on the amount of activity on that day. Suppose the company starts with an initial reserve $X_0 =$

$x \geq 1$, and has reserve X_n on the nth day. On the $n + 1^{\text{st}}$ day, if $X_n \geq 1$, then $X_{n+1} = X_n + Y_n$, where Y_n represents the premium received or claim paid; it is independent of $X_1, \ldots, X_n$, and has values 1, 0, and -1 with probabilities $p_n/2$, $1 - p_n$, and $p_n/2$, respectively. If $X_n = 0$, the company is declared bankrupt, and $X_{n+1} = X_{n+2} = \cdots = 0$.

(a) Prove that $E\{X_n\} = x$, $\forall n$, and that (X_n) converges a.s. to a limit X_∞.

(b) If $\sum_n p_n < \infty$ (i.e., rapidly decreasing activity) show that $E\{X_\infty\} = x$.

(c) Suppose instead that $\sum_n p_n = \infty$. Find $E\{X_\infty\}$.

[Hint: (b): It might help to show that $E\{X_n^2\}$ is bounded. (c): it is not x.]

9.24. Show that any L^2-bounded martingale can be written as the difference of two positive L^2-bounded martingales.

9.25. Say which of the following statements are true, and which are false. If true, give a short proof, or outline of a proof. If false, give a counterexample.

(a) Let T be a stopping time. Then so is $2T$.

(b) If (X_n) is a positive supermartingale, then X_n converges in L^1.

(c) If (X_n) is a positive supermartingale, it is L^1-bounded.

(d) Let (X_n) be a positive supermartingale and let $x > 0$. Then
$$P\{\max_{1 \leq j \leq n} X_j \geq x\} \leq \frac{1}{x} E\{X_n\}.$$

9.26. Let (S_n) be a simple non-symmetric random walk, with probability of a positive step p, and of a negative step, $q = 1 - p$, $p \neq 1/2$.

(a) Find r such that $M_n \stackrel{\text{def}}{=} r^{S_n}$ is a martingale.

(b) Let $a < b$ be integers and suppose $S_0 \in (a, b)$ a.s. $T = \min\{n : S_n = a \text{ or } b\}$. Show that $E\{M_T\} = E\{M_0\}$.

(c) Find $P\{S_T = a\}$. Why doesn't this work if $p = 1/2$?

9.27. (Continuation.) Suppose $p < 1/2$ in the previous problem, and that $X_0 = 0$. Show that $E\{\sup_m S_m\} \leq \frac{p}{1-2p}$.

[Hint: Write the expectation of the max as a sum of $P\{\max \geq n\}$. Then use the previous problem.]

9.5. Applications

Because of their many nice properties, martingales are a useful tool[6]. They occur more often than one might expect, for conditional expectations are averages, and are likely to come up in many situations, even those which might seem to have nothing to do with probability.

In fact, probabilists, confronted with a new problem, will often check to see if there is a martingale or submartingale hidden somewhere in it.

---

[6] Doob and Feller reputedly had an ongoing contest: if Doob proved a result by martingales, Feller would find a non-martingale proof, and vice versa. The final score remains unknown...

9.5. Applications

Application to Conditional Expectations I: Existence. Our very first application will be to justify all that has gone before. ("All" is to be taken in the narrow sense, of course.)

That is, we defined the conditional expectation and derived its properties in §8, but we only knew its existence for discrete σ-fields, i.e., those generated by countable partitions. This is very, very restrictive, and, since we had no means of showing the general existence at that time, we covered ourselves with the blanket assumption that: "All necessary conditional expectations exist." This assumption would be true if we limited ourselves to discrete σ-fields, and it made our proofs rigorous, but it left open the question of what, exactly, we did prove. After all, we have to know when our theorems apply.

In fact, that assumption was safe: the conditional expectation $E\{X \mid \mathcal{G}\}$ exists for any σ-field $\mathcal{G} \subset \mathcal{F}$. Therefore, the assumption is always satisfied, and, happily, the results all hold without any restriction on the σ-fields.

Let $(\Omega, \mathcal{F}, P)$ be a probability space.

Theorem 9.37. *Let $\mathcal{G} \subset \mathcal{F}$ be a σ-field, and let X be an integrable random variable. Then $E\{X \mid \mathcal{G}\}$ exists.*

Proof. We will assume that $\mathcal{G}$ is separable[7], i.e., generated by a countable number of sets $\Lambda_1, \Lambda_2, \ldots$. This does not mean that $\mathcal{G}$ is generated by a partition, for the Λ_i may not be disjoint. For example, for any random variable Y, $\mathcal{G} = \sigma(Y)$ is generated by the sets $\{Y \leq r\}$ for rational r.

We will construct a random variable Z which satisfies (CE1) and (CE2) of the definition. Let $\mathcal{G}_0 = \{\Omega, \emptyset\}$, $\mathcal{G}_1 = \{\Omega, \emptyset, \Lambda_1, \Lambda_1^c\}$, and, in general, let $\mathcal{G}_n$ be the σ-field generated by $\Lambda_1, \ldots, \Lambda_n$. Note that $\mathcal{G}_1 \subset \mathcal{G}_2 \subset \ldots$, so that $(\mathcal{G}_n)$ is a filtration. Note also that all Λ_i are in $\bigcup_n \mathcal{G}_n$, so that $\sigma(\bigcup_n \mathcal{G}_n) = \mathcal{G}$.

Now $\mathcal{G}_n$ is finite, so that $E\{X \mid \mathcal{G}_n\}$ exists—the formula is given in Example 8.1. Thus define $Z_n = E\{X \mid \mathcal{G}_n\}$. Then

(*i*) $\{Z_n, \mathcal{G}_n, \ n \geq 1\}$ is a uniformly-integrable martingale, and therefore converges a.s. and in L^1 as $n \to \infty$ to an integrable random variable Z.

We claim that $Z = E\{X \mid \mathcal{G}\}$. We must verify (CE1) and (CE2).

(*ii*) Z is measurable with respect to $\sigma(\bigcup_n \mathcal{G}_n) = \mathcal{G}$, so Z satisfies (CE1).

(*iii*) Let $m < n$ and let $\Lambda \in \mathcal{G}_m \subset \mathcal{G}_n$. Then, as $Z_n = E\{X \mid \mathcal{G}_n\}$,

$$\int_\Lambda Z_n \, dP = \int_\Lambda X \, dP.$$

---

[7]The extension to the general case is not hard, but it doesn't add any new probabilistic insight, and we will leave it to the experts.

Now $Z_n \to Z$ in L^1, so that we can go to the limit under the first integral to see that

(9.9) $$\int_\Lambda Z\,dP = \int_\Lambda X\,dP, \quad \Lambda \in \mathcal{G}_m.$$

This is true for all $\Lambda \in \mathcal{G}_m$, hence for all $\Lambda \in \bigcup_m \mathcal{G}_m$. But the set of Λ for which (9.9) holds is clearly a monotone class; it contains the field $\bigcup_m \mathcal{G}_m$, and therefore it contains $\sigma(\bigcup_m \mathcal{G}_m) = \mathcal{G}$. Therefore Z satisfies (CE2), hence $Z = E\{X \mid \mathcal{G}\}$. □

Application to Conditional Expectations II: Lévy's Theorem. One of the most immediate martingale applications is to successive conditional expectations. If X is integrable, and if we take a sequence of conditional expectations of it with respect to a monotone sequence of σ-fields, the limit exists, and we can identify it.

Theorem 9.38 (P. Lévy). *Let X be an integrable random variable, let $\mathcal{F}_1 \subset \mathcal{F}_2 \subset \dots$ be an increasing sequence of σ-fields, and let $\mathcal{G}_1 \supset \mathcal{G}_2 \supset \dots$ be a decreasing sequence of σ-fields. Let $\mathcal{F}_\infty = \sigma(\bigcup_n \mathcal{F}_n)$ and $\mathcal{G}_{-\infty} = \bigcap_n \mathcal{G}_n$. Then*

(i) $\lim_{n \to \infty} E\{X \mid \mathcal{F}_n\} = E\{X \mid \mathcal{F}_\infty\}$.

(ii) $\lim_{n \to \infty} E\{X \mid \mathcal{G}_n\} = E\{X \mid \mathcal{G}_{-\infty}\}$.

Proof. *(i)* Let $X_n = E\{X \mid \mathcal{F}_n\}$. Then $\{X_n, \mathcal{F}_n, n \geq 1\}$ is a martingale. The X_n are conditional expectations of X, so it is uniformly integrable. Therefore it converges a.s. to a random variable X_∞. We claim that $X_\infty = E\{X \mid \mathcal{F}_\infty\}$. Being the limit of X_n, X_∞ is $\mathcal{F}_\infty$-measurable. Let $\Lambda \in \bigcup_n \mathcal{F}_n$. Then $\Lambda \in \mathcal{F}_n$ for some n. Thus, as $X_n = E\{X \mid \mathcal{F}_n\}$,

$$\int_\Lambda X_n\,dP = \int_\Lambda X\,dP.$$

We can go to the limit under the integral to see that for each $\Lambda \in \bigcup_n \mathcal{F}_n$,

(9.10) $$\int_\Lambda X_\infty\,dP = \int_\Lambda X\,dP.$$

Now, a technical point: $\bigcup_n \mathcal{F}_n$ is a field, but it may not be a σ-field. However, the class of Λ for which (9.10) holds is clearly a monotone class, and by the monotone class theorem, it contains $\sigma(\bigcup_n \mathcal{F}_n) = \mathcal{F}_\infty$. This proves the claim.

(ii) Set $X_{-n} = E\{X \mid \mathcal{G}_n\}$. Then (X_{-n}) is a uniformly integrable backward martingale which converges to a limit $X_{-\infty}$. For each $m < -n < 0$, X_m is $\mathcal{G}_n$-measurable, hence so is $X_{-\infty} \stackrel{\text{def}}{=} \lim_{m \to -\infty} X_m$. This is true for all n, so $X_{-\infty}$ is measurable with respect to $\bigcap_n \mathcal{G}_n = \mathcal{G}_{-\infty}$.

If $\Lambda \in \mathcal{G}_{-\infty} \subset \mathcal{G}_m$, then
$$\int_\Lambda X_m \, dP = \int_\Lambda X \, dP.$$
Go to the limit under the integral by uniform integrability to see that
$$\int_\Lambda X_{-\infty} \, dP = \int_\Lambda X \, dP, \quad \Lambda \in \mathcal{G}_{-\infty},$$
which implies that $X_{-\infty} = E\{X \mid \mathcal{G}_{-\infty}\}$. $\square$

Application to the Zero-One Law. The proof of the Borel zero-one law (Theorem 5.22) is not hard, but it is instructive to rederive it from martingale considerations.

Let $X_1, X_2, \ldots$ be independent, let $\mathcal{F}_n = \sigma(X_j, j \leq n)$ and set $\mathcal{F}_\infty = \sigma(\bigcup_n \mathcal{F}_n)$. Let $\mathcal{F}_n^* = \sigma(X_j : j \geq n+1)$ and let $\mathcal{F}_\infty^* = \bigcap_n \mathcal{F}_n^*$, be the tail field. Notice that $\mathcal{F}_\infty^* \subset \mathcal{F}_\infty$.

Theorem 9.39. $\Lambda \in \mathcal{F}_\infty^* \implies P\{\Lambda\} = 0 \text{ or } 1$.

Proof. Let $\Lambda \in \mathcal{F}_\infty^*$. Notice that $\mathcal{F}_n$ and $\mathcal{F}_n^*$ are independent, and $\Lambda \in \mathcal{F}_\infty^* \subset \mathcal{F}_n^*$, so Λ is independent of $\mathcal{F}_n$. Therefore, $E\{I_\Lambda \mid \mathcal{F}_n\} = E\{I_\Lambda\} = P\{\Lambda\}$. It follows that $\lim_{n \to \infty} E\{I_\Lambda \mid \mathcal{F}_n\} = P\{\Lambda\}$. But by Theorem 9.38,
$$\lim_{n \to \infty} E\{I_\Lambda \mid \mathcal{F}_n\} = E\{I_\Lambda \mid \mathcal{F}_\infty\} = I_\Lambda \text{ a.e.,}$$
since Λ is certainly in $\mathcal{F}_\infty$. Thus, $P\{\Lambda\} = I_\Lambda$ a.e. But I_Λ takes on only the values zero and one. Therefore, $P\{\Lambda\}$ must be either zero or one. Done! $\square$

The Kolmogorov Strong Law. The strong law of large numbers says that the average of a large number of random variables with the same mean tends to their common expectation. The statement only involves the expectation—the first moment—of the summands, but the theorems we gave all require higher moments. One feels that the "true" theorem should only involve first moments. Kolmogorov's strong law does this. However, it does impose one restriction: the summands must be i.i.d. (and indeed, there are examples of independent but non-identically-distributed sequences which do not satisfy the law of large numbers). The proof is a good example of the "Find the hidden martingale" method.

Theorem 9.40. Let $X_1, X_2, \ldots$ be a sequence of i.i.d. random variables. Suppose that $E\{X_1\} = m$. Then
$$\lim_{n \to \infty} \frac{1}{n} \sum_{k=1}^{n} X_n = m \text{ a.e. and in } L^1.$$

Proof. We will proceed softly, so as not to spoil the surprise. Set

$$S_{-n} = \frac{1}{n} \sum_{k=1}^{n} X_k,$$
$$\mathcal{F}_{-n} = \sigma(S_{-n}, S_{-n-1}, S_{-n-2}, \dots).$$

Note that $\mathcal{F}_{-n-1} \subset \mathcal{F}_{-n}$ so that $(\mathcal{F}_{-n})$ is a filtration. (The reason for the negative indices will soon be clear.) Moreover,

$$\mathcal{F}_{-n} = \sigma(S_{-n}, X_{n+1}, X_{n+2}, \dots)$$
$$= \sigma((X_1 + \dots + X_n), X_{n+1}, X_{n+2} \dots).$$

We claim that $\{S_n, \mathcal{F}_n, n \leq -1\}$ is a backward martingale. Note that

$$nS_n = E\{nS_{-n} \mid \mathcal{F}_{-n}\}$$
$$= E\{X_1 + \dots + X_n \mid \mathcal{F}_{-n}\}$$
$$= E\{X_1 \mid \mathcal{F}_{-n}\} + \dots + E\{X_n \mid \mathcal{F}_{-n}\}.$$

But $X_1, \dots, X_n$ are independent of $X_{n+1}, X_{n+2}, \dots$, so that for each k, $E\{X_k \mid \mathcal{F}_{-n}\} = E\{X_k \mid X_1 + \dots + X_n\}$. The X_k are also identically distributed, so that by symmetry $E\{X_1 \mid \mathcal{F}_{-n}\} = E\{X_2 \mid \mathcal{F}_{-n}\} = E\{X_n \mid \mathcal{F}_{-n}\}$, and therefore, for $k \leq n$,

$$E\{X_k \mid \mathcal{F}_{-n}\} = \frac{1}{n} S_{-n}.$$

Therefore,

$$E\{S_{-n+1} \mid \mathcal{F}_{-n}\} = \frac{1}{n-1} E\{X_1 + \dots + X_{n-1} \mid \mathcal{F}_{-n}\}$$
$$= \frac{1}{n-1}(n-1)S_{-n} = S_{-n}.$$

Therefore, (S_{-n}) is a backward martingale! It is necessarily uniformly integrable and converges a.s and in L^1 to an $\mathcal{F}_{-\infty}$-measurable random variable $S_{-\infty}$, and $(S_{-\infty}, S_{-1})$ is a martingale. It remains to identify the limit. Recall that $S_{-\infty}$ is measurable with respect to the tail field, $\bigcap_n \sigma(X_n, X_{n+1}, \dots)$, which is trivial. Therefore $S_{-\infty}$ is constant, so $S_{-\infty} = E\{S_{-\infty}\} = E\{S_{-1}\} = E\{X_1\} = m$. $\square$

Application to Integration in Infinitely Many Dimensions★. Consider a function $f(x_1, x_2, \dots)$ defined for $0 \leq x_i \leq 1$, $i = 1, 2, 3, \dots$. Assume that it is a nice function. We would like to integrate this with respect to[8]

---

[8]This is really a question about Lebesgue measure on the infinite cube $[0,1] \times [0,1] \times \dots$. Its existence is not hard to show directly, but it can also be derived by noting that it is the distribution of a sequence $X_1, X_2, \dots$ of i.i.d. $U(0,1)$ random variables.

9.5. Applications

$dx_1\, dx_2\, dx_3 \ldots$:

$$\int_0^1 \int_0^1 \cdots\, f(x_1, x_2, \ldots)\, dx_1\, dx_2\, dx_3 \ldots.$$

We have not defined such an integral, so we cannot say exactly what it means, but a first guess would be that it is the limit of iterated integrals:

$$\lim_{n \to \infty} \underbrace{\int_0^1 \cdots \int_0^1}_{n} f(x_1, x_2, \ldots)\, dx_1 \ldots dx_n.$$

The question is, is it?

Let us rephrase this. Let $X_1, X_2, \ldots$ be a sequence of i.i.d. random variables, with a $U(0,1)$ distribution. Let $Z = f(X_1, X_2, \ldots)$. (By "f is a nice function," we mean that Z is an integrable random variable.) Consider $E\{Z \mid X_2, X_3, \ldots\}$. If, say, $X_2 = x_2$, $X_3 = x_3, \ldots$, then, as X_1 is independent of the other X_i, $Z = f(X_1, x_2, x_3, \ldots)$ and $E\{Z \mid X_2 = x_2, X_3 = x_3, \ldots\} = \int f(x_1, x_2, \ldots)\, dx_1$. Thus,

$$E\{Z \mid X_2, X_3, \ldots\} = \int_0^1 f(x_1, X_2, X_3, \ldots)\, dx_1.$$

(We leave it to the reader to check the definition to see that this really is the conditional expectation.)

More generally, let

$$Z_{-n} = E\{Z \mid X_{n+1}, X_{n+2}, \ldots\}$$
$$= \int_0^1 \cdots \int_0^1 f(x_1, \ldots, x_n, X_{n+1}, X_{n+2}, \ldots)\, dx_1 \ldots dx_n,$$

and let $\mathcal{F}_{-n} = \sigma(X_{n+1}, X_{n+2}, \ldots)$. Then $\{Z_{-n}, \mathcal{F}_{-n},\, n \geq 1\}$ is a backward martingale. Therefore it is uniformly integrable and convergent, and

$$\lim_{n \to -\infty} Z_n = E\Big\{Z \mid \bigcap_n \mathcal{F}_{-n}\Big\} = E\{Z\} \text{ a.s.}$$

Therefore, we *define*

$$\int_0^1 \cdots \int_0^1 f(x_1, x_2, \ldots)\, dx_1\, dx_2\, dx_3 \ldots \stackrel{\text{def}}{=} E\{Z\},$$

and note that with this definition,

$$\int_0^1 \cdots \int_0^1 f(x_1, x_2, \ldots)\, dx_1 \ldots dx_n \longrightarrow \int_0^1 \cdots \int_0^1 f(x_1, x_2, \ldots)\, dx_1\, dx_2 \ldots.$$

Application to Sequential Analysis in Statistics. Consider a sequence of i.i.d. random variables, $X_1, X_2, \ldots$. We do not know their exact distribution, but suppose we can narrow it down to two possibilities: their common probability density is either $f(x)$ or $g(x)$. We have no further information to help us choose between them, and will have to rely on experiments.

To avoid trivialities, assume that f and g never vanish. Set $Z_0 = 1$ and let the *likelihood ratio* be

$$Z_n = \prod_{k=1}^{n} \frac{g(X_k)}{f(X_k)}.$$

The statistician must choose between two possibilities:

(H1) The correct density is $f(x)$.

(H2) The correct density is $g(x)$.

We expect that under (H1), the likelihood ratio Z_n will be relatively small, and under (H2), it will be relatively large. The classical likelihood ratio test of these two hypotheses is to take a fixed sample size n, and a well-chosen constant C, and to choose (H1) if Z_n is less than C, and choose (H2) if it is greater.

In practice, the X_i are results of experiments, each of which costs time, money, and/or effort. We do not want to do any more experiments than necessary.

Abraham Wald suggested that it could be more efficient to make the observations one-by-one, wait until there is enough information to make a decision, then decide. This leads to Wald's sequential likelihood ratio test.

In this test, we set $Z_0 = 1$ and observe the sequence $Z_0, Z_1, Z_2, \ldots$. After each observation, we decide if there is enough evidence to choose between (H1) and (H2). If there is, we choose; if there is not, we make another observation. Bearing in mind that we expect Z_n to be small under (H1) and large under (H2), we make the following decision rule.

Decision rule: Take two numbers, $0 < a < 1 < b$. Continue observing as long as $a < Z_n < b$. Stop the first time T_{ab} that Z is not in (a, b):

$$T_{ab} = \inf\{n : Z_n \leq a \text{ or } Z_n \geq b\}.$$

- If $T_{ab} < \infty$ and $Z_{T_{ab}} \leq a$, choose (H1).
- If $T_{ab} < \infty$ and $Z_{T_{ab}} \geq b$, choose (H2).

Question: Does the test stop? That is, is T_{ab} finite?

The key (see §9.1, Example 5°) is this: under (H1), $\{Z_n, \mathcal{F}_n, n \geq 1\}$ is a martingale with mean one.

9.5. Applications

Proposition 9.41. *If (H1) holds, then, unless $f = g$ a.e., $\lim_{n\to\infty} Z_n = 0$.*

Proof. Suppose (H1) holds. Then (Z_n) is a positive martingale, so it converges. Let its limit be Z_∞.

If $f = g$ a.e., then $Z_n = Z_\infty = 1$ a.s. Otherwise, there is an $\varepsilon > 0$ for which $P\{|1 - (g(X_1)/f(X_1))| > \varepsilon\} > 0$. Let Λ_n be the event $\{|1 - (g(X_n)/f(X_n))| > \varepsilon\}$. The Λ_n are independent and have the same strictly positive probability, so that with probability one, Λ_n happens infinitely often. Note that $|Z_n - Z_{n-1}| = |1 - (g(X_n)/f(X_n)|Z_{n-1}$ so that on Λ_n, $|Z_n - Z_{n-1}| \geq \varepsilon Z_{n-1}$, or

$$Z_{n-1} \leq \frac{1}{\varepsilon}|Z_n - Z_{n-1}| \quad \text{on } \Lambda_n.$$

But (Z_n) converges a.s., so that

$$Z_\infty = \lim_{n\to\infty} Z_n \leq \limsup_{n\to\infty} \frac{1}{\varepsilon}|Z_n - Z_{n-1}| = 0.$$

Since Z_∞ is positive, it must vanish a.e. □

Thus, under (H1), Z_n tends to zero, so it eventually gets below a, and the test stops. Under (H2), the martingale is $1/Z_n$, not Z_n, so that $Z_n \to \infty$, and eventually surpasses b. In either case, the test eventually ends.

The numbers a and b are chosen to make the error probabilities small. In this case, there are two kinds of error: one is to choose (H2) when (H1) is true. The other is to choose (H1) when (H2) is true. Given a and b, the probabilities of these errors are:

$$P\{\text{Choose (H2)} \mid (H1)\} = P\{Z_{T_{ab}} \geq b \mid (H1)\},$$
$$P\{\text{Choose (H1)} \mid (H2)\} = P\{Z_{T_{ab}} \leq a \mid (H2)\}.$$

Note that $T_{ab} < \infty$, and either $Z_{T_{ab}} \leq a$ or $Z_{T_{ab}} \geq b$. An exact calculation requires knowing the distribution of the "overshoots", $Z_{T_{ab}} - a$ and $b - Z_{T_{ab}}$, but for the first approximation, we make an assumption: when the test ends, there is no overshoot[9]: $Z_{T_{ab}}$ equals either a or b.

As (Z_n) is a martingale, so is $(Z_{n\wedge T_{ab}})$. Since there is no overshoot, it is bounded, and the system theorem tells us that $(Z_0, Z_{T_{ab}})$ is a martingale. Thus,

$$1 = Z_0 = E\{Z_{T_{ab}} \mid (H1)\}.$$

Let $p_1 = P\{Z_{T_{ab}} = b \mid (H1)\}$, which is the probability of choosing (H2) when (H1) is true. Then, as $P\{Z_{T_{ab}} = a \mid (H1)\} = 1 - p_1$,

$$1 = p_1 b + (1-p_1)a \implies p_1 = \frac{1-a}{b-a}.$$

---

[9] One can make a hand-waving justification: the overshoots are probably small and tend to cancel out anyway!

If (H2) is correct, then $1/Z$ is a martingale. Then $1/Z_{t \wedge T_{ab}}$ is bounded, so, letting $p_2 = P\{Z_{T_{ab}} = a \mid (H2)\}$,

$$1 = E\left\{\frac{1}{Z_{T_{ab}}} \mid (H2)\right\} = p_2 \frac{1}{a} + (1-p_2)\frac{1}{b}.$$

Solving,

$$p_2 = \frac{(b-1)a}{b-a}.$$

There is a balance here: while we can make the errrors as small as we wish by choosing b large enough and a small enough, this comes at the expense of taking more observations. In practice, the statistician first chooses an acceptable probability of error. As the two types of errors may have different consequences, the statistician chooses a maximum permissible error for each, say

$$p_1 \leq \alpha_1 \qquad p_2 \leq \alpha_2.$$

Then the test is constructed to minimize the number of observations. This implies that there will be equality, not inequality in the constraints:

$$p_1 = \alpha_1 \qquad p_2 = \alpha_2.$$

These can then be solved for a and b in terms of the error probabilities α_1 and α_2.

Now the test lasts a random amount of time, namely T_{ab}, so the actual cost is not predictable. However, the *expected cost* is. Therefore, the cost is usually measured by $E\{T_{ab}\}$. We can find this in terms of a and b via a martingale argument. (See Exercise 9.28.) If (H1) is the true density, for example, the expected duration of the test is:

$$E\{T_{ab} \mid (H1)\} = \frac{(1-a)\log b + (b-1)\log a}{(b-a)d}$$

where $d = E\{\log g(X)/\log f(X) \mid (H1)\}$.

It can be shown [**24**] that for this problem, this test minimizes the expected sample size among all tests, sequential or not, which have the same error bounds $p_1 \leq \alpha_1$ and $p_2 \leq \alpha_2$.

Application to Branching Processes. Let $\{Z_n, n = 0, 1, 2, \dots\}$ be a branching process, and let the common family size have the same distribution as N, where N is a square-integrable random variable taking values in $0, 1, 2, \dots$. Let $\mu \stackrel{\text{def}}{=} E\{N\}$ and $\sigma^2 \stackrel{\text{def}}{=} \text{Var}(N)$. Let $(\mathcal{F}_n)$ be the natural filtration of (Z_n). Let $T_0 = \inf\{n \geq 1 : Z_n = 0\}$ be the extinction time of the process. We showed in Theorem 7.59 that if $\mu \leq 1$, then $P\{T_0 < \infty\} = 1$. If $\mu > 1$, then the process may go extinct, but there is a strictly positive probability that it doesn't. The question we would like to answer here is:

9.5. Applications

what happens to Z_n if $\mu > 1$? Intuitively, it must blow up. But can we be more exact than that?

Theorem 9.42. *Let $X_n = Z_n/\mu^n$. Then $\{X_n, \mathcal{F}_n, n \geq 1\}$ is a positive martingale, which therefore converges a.s. to a limit X_∞. If $\mu > 1$ and $\sigma^2 < \infty$, then X_∞ is strictly positive with positive probability.*

Remark 9.43. If $\mu > 1$, the process may not tend to zero. In that case, $Z_n \sim \mu^n Z_0$ as $n \to \infty$ on the set where $X_\infty > 0$. In other words, the branching process either goes to zero or tends to infinity exponentially.

Proof. The population size at the nth stage is Z_n. In the next generation, the ith member is replaced by N_i new individuals, so that $Z_{n+1} = \sum_{i=1}^{Z_n} N_i$. Since the N_i are independent of $\mathcal{F}_n$ and of each other, $E\{Z_{n+1} \mid \mathcal{F}_n\} = Z_n E\{N_i\} = \mu Z_n$. Thus, $E\{X_{n+1} \mid \mathcal{F}_n\} = \mu Z_n/\mu^{n+1} = X_n$. Therefore, (X_n) is a martingale. Since it is positive, it converges a.e. to a random variable X_∞. Notice also that, as $E\{X_n\} = 1$ for all n, $E\{Z_n\} = \mu^n$.

This shows that X_∞ exists, but a priori, it might be identically zero, as indeed it is when $\mu \leq 1$. To show it is strictly positive with positive probability, we will show (X_n) is L^2 bounded, and therefore uniformly integrable, so that $E\{X_\infty\} = 1$.

We turn to the second moments for this. Since the N_i are independent and independent of $\mathcal{F}_n$, $\text{Var}\{Z_{n+1} \mid \mathcal{F}_n\} = Z_n \text{Var}(N_i) = \sigma^2 Z_n$. It follows that $E\{Z_{n+1}^2 \mid \mathcal{F}_n\} = E\{Z_{n+1} \mid \mathcal{F}_n\}^2 + \text{Var}\{Z_{n+1} \mid \mathcal{F}_n\}$ or $E\{Z_n^2 \mid \mathcal{F}_n\} = \sigma^2 Z_n + \mu^2 Z_n^2$. Taking expectations,

$$E\{Z_{n+1}^2\} = \mu^2 E\{Z_n^2\} + \sigma^2 \mu^n.$$

Since $Z_1 = N_1^2$, $E\{Z_1^2\} = \mu^2 + \sigma^2$. By induction,

$$E\{Z_n^2\} = \mu^{2n} + \sigma^2 \mu^{n-1} \frac{\mu^n - 1}{\mu - 1}.$$

Thus,

$$E\{X^2\} = E\left\{\frac{Z_2}{\mu^{2n}}\right\} = 1 + \frac{\sigma^2}{\mu - 1}\left(\frac{1}{\mu} - \frac{1}{\mu^{n+1}}\right)$$

which is bounded if $\mu > 1$. Thus the martingale (X_n) has a bounded second moment and is therefore uniformly integrable. Consequently, it converges in L^1 and

$$E\{X_\infty\} = \lim E\{X_n\} = 1.$$

Thus, $P\{X_\infty > 0\} > 0$, as claimed. $\square$

Problems 9.5

9.28. (Expected Duration of a Sequential Test.) Let $X_1, X_2, \ldots$ be a sequence of i.i.d. random variables with probability density f. Let g be another probability density. Suppose both f and g are strictly positive, and assume $\int |\log g(x) -$

$\log f(x) | f(x) \, dx < \infty$. Define $Z_0 = 1$, $Z_n = \prod_1^n \frac{g(X_i)}{f(X_i)}$. Let $0 < a < 1 < b$ and let T_{ab} be the first exit time of Z from (a,b). Assume there is no overshoot: $X_{T_{ab}}$ equals either a or b.

(a) Show that $(\log Z_n)$ is both a random walk and a supermartingale.

(b) Let $d = E\{\log Z_{n+1} - \log Z_n \mid \mathcal{F}_n\}$. Find the Doob decomposition $\log Z_n = M_n - A_n$, where M_n is a martingale and A_n is an increasing process.

(c) Show that $E\{T_{ab}\} = \dfrac{(1-a)\log b + (b-1)\log a}{(b-a)d}$.

9.29. Let $X_1, X_2, \ldots$ be an irreducible recurrent Markov chain on a state space E. Suppose that $f(i)$ is a positive function on E with the property that $\{f(X_n),\ n = 1, 2, \ldots\}$ is a supermartingale. Prove that f is constant.

9.30. Let (X_n) be a Markov chain with transition probability matrix $\mathbb{P} = (P_{ij})$. A function $v(i)$ on the state space is said to be **excessive** if for all i, $v(i) \geq 0$ and

(9.11) $$v(i) \geq \sum_j P_{ij} v(j).$$

Show that if v is excessive, $\{v(X_n),\ n = 0, 1, 2, \ldots\}$ is a supermartingale, and for any stopping time T, $v(i) \geq E^i\{X_T I_{\{T < \infty\}}\}$.

9.31. (Continuation.) Let A be a subset of the state space and set $T_A = \inf\{n \geq 0 : X_n \in A\}$. Let $v_A(i) = P^i\{T_A < \infty\}$. Show that

(9.12) $$v_A(i) = \begin{cases} 1 & \text{if } i \in A, \\ \sum_j P_{ij} v_A(j) & \text{if } i \in A^c. \end{cases}$$

Conclude that v_A is an excessive function, and, moreover, that it is the smallest positive solution of (9.12).

[Hint: If u is another such, apply Exercise 9.30 to show $u(i) \geq v_A(i)$ for each i.]

9.32. (Continuation: another characterization of v_A.) Show that v_A is the smallest excessive function which dominates I_A.

9.33. Prove that any σ-field $\mathcal{G} \subset \mathcal{F}$, has a separable sub-$\sigma$-field $\mathcal{G}' \subset \mathcal{G}$ such that if $\Lambda \in \mathcal{G}$, there is $\Lambda' \in \mathcal{G}'$ such that $\Lambda \Delta \Lambda'$ is a null set. Use this to extend Theorem 9.37 to arbitrary σ-fields.

Let P and Q be measures on $(\Omega, \mathcal{F})$. We say Q is **absolutely continuous with respect to P**, and write $Q \ll P$ if $\Lambda \in \mathcal{F}$ and $P\{\Lambda\} = 0$ imply that $Q\{\Lambda\} = 0$.

9.34. If $Q \ll P$ and $\varepsilon > 0$, there exists $\delta > 0$ such that $\Lambda \in \mathcal{F}$ and $P\{\Lambda\} < \delta$ imply that $Q\{\Lambda\} < \varepsilon$.

[Hint: If not, there are Λ_n such that $P\{\Lambda_n\} < 2^{-n}$, $Q\{\Lambda_n\} > \varepsilon$. Find $\Gamma_n \downarrow$ for which $P\{\Gamma_n\} \to 0$, $Q\{\Gamma_n\} \geq \varepsilon$.]

9.35. (Radon-Nikodym theorem.) Let P, Q be finite measures on $(\Omega, \mathcal{F})$ such that $Q \ll P$. Then there exists a measurable function X such that for all $\Lambda \in \mathcal{F}$, $Q\{\Lambda\} = \int_\Lambda X \, dP$. X is called the **Radon-Nikodym derivative** of Q with respect to P, and written $X = dQ/dP$.

[Hint: First assume P is a probability measure and F is separable. Find increasing partitions $\mathcal{P}_n \subset \mathcal{P}_{n+1} \subset \ldots$ which generate $\mathcal{F}$, and put $X_n = Q(\Lambda)/P\{\Lambda\}$ if $\Lambda \in \mathcal{P}_n$. Show that (X_n) is a martingale. Use the previous problem to show it is uniformly integrable. Now remove the assumption that P is a probability measure, and use exercise 9.33 to remove the separability assumption.]

9.36. Let (X_n) be a sequence of random variables tending a.s. to a random variable X_∞. Let $\mathcal{F}_n$ be a sequence of σ-fields. Prove the following result, called Hunt's Lemma, which extends Theorem 9.38 to the case where both the random variables and the σ-fields vary.

> **Theorem** (Hunt's Lemma). Suppose that the X_n are dominated by an integrable random variable Z.
> (a) If $\mathcal{F}_1 \subset \mathcal{F}_2 \subset \ldots$, then $\lim_{n\to\infty} E\{X_n \mid \mathcal{F}_n\} = E\{X_\infty \mid \bigvee_n \mathcal{F}_n\}$ a.s.
> (b) If $\mathcal{F}_1 \supset \mathcal{F}_2 \supset \ldots$, then $\lim_{n\to\infty} E\{X_n \mid \mathcal{F}_n\} = E\{X_\infty \mid \bigcap_n \mathcal{F}_n\}$ a.s.

9.6. Financial Mathematics I: The Martingale Connection*

The aim of this section is to show the relation between financial markets and martingales. We will prove a very weak version of what is called the fundamental theorem of asset pricing, namely that in a complete market with no arbitrage, market prices are expectations relative to a certain "martingale measure", and that prices evolve as martingales[10].

Before we get to it, however, we need to learn something of the basic structure of a financial market, of buying, selling, and the idea of arbitrage. This will take time, but once done, the probability enters quickly, and from a surprising direction. The important probabilistic structure is not the one we first think it is.

Let us prepare the path with a quick look at a familiar financial market where the same thing happens, but in a less mysterious way: the race track[11].

A Day at the Races. Race tracks express probabilities in terms of odds: if the odds on a horse are five to two, that means that there are two chances the horse wins, five that it doesn't—seven chances in all—so the probability of winning is, in theory at least, 2/7. More importantly, the odds give the payoff of bets: a bettor who bets $2 at odds of five to two will win $5 if the horse wins and lose the $2 if it doesn't.

But the odds the bettor sees on the track's tote board are not the actual probabilities of winning. They are synthetic probabilities determined by the

---

[10]The stronger version of this theorem for discrete parameters is the Dalang-Morton-Willinger Theorem [10]. It was first discovered by Harrison, Kreps, and Pliska in 1979–1981, and extended by numerous authors. Delbaen and Schachermeyer proved the general continuous-parameter version in 1994. See [11] for a complete account.

[11]Look ahead to Remark 9.61 to see why this is not—quite—a joke.

amount of money bet on each horse. The race track sets the odds. Its aim is to assure that, no matter which horse wins, the track will make a profit of ρ percent (called the *take*) of the total money bet. To see how this works, consider a race between three horses, call them[12] A, B, and C. Suppose the there is a total of \$400 bet on A, \$600 bet on B, and \$1000 bet on C, with a take of fifteen percent, a typical figure. The total betting pool is \$2000, and the track wants fifteen percent of that for a \$300 profit. If A wins at odds of x to one, the track keeps the \$1600 bet on B and C and pays $400x$ to those who bet on A. This leads to $1600 - 400x = 300$, so $x = 13/4$. Thus A starts at odds of $13/4$ to one, or thirteen to four. Similar calculations show that the odds on B are eleven to six and the odds on C are seven to ten. These odds imply that A, B, and C have probabilities of winning of $4/17$, $6/17$, and $10/17$, respectively.

These probabilities are entirely synthetic, of course. They have little to do with the true winning probabilities, for they are derived purely from the betting... and they do not even sum to one.

Let us generalize this.

Exercise 9.37. Horses $H_1, H_2, \ldots, H_n$ are entered in a race. Race goers have bet $\$M_i$ on horse H_i, $i = 1, \ldots, n$.

(a) Find odds on each of the n horses which assure that, after paying all winning bets, the track has $\rho\%$ of the total betting pool left over for profit.

(b) What probabilities p_i are implied by these odds?

(c) Find $\sum_{i=1}^n p_i$. Is it less than, equal to, or larger than one?

Exercise 9.38. (Continuation.) Suppose that the take is $\rho = 0$.

(a) Show that $\sum_{i=1}^n p_i = 1$.

(b) Suppose that for each i, p_i is the true probability that horse H_i wins. Show that all bets are fair.

Exercise 9.39. (Continuation.) Suppose the track runs m races in a day, and that $\rho = 0$ for all of them. Suppose further that the races are independent and the implied probabilities are actually the true probabilities of winning each race. A gambler has an initial fortune F_0. As long as money remains, the gambler will bet on none, one, or several horses to win in each race. The decision of whether or not to bet on a given race, and how much, may depend on the results of the previous races, the remaining fortune, and on the gambler's hunches. Let F_j be the gambler's fortune after the jth race. Show that $F_0, F_1, \ldots, F_m$ is a martingale.

Of course, the true probabilities of winning are not those the bettors see on the tote board. (Those don't even add up to one, since the track has a non-zero take.) However, the important fact for what follows is that if the race track's take is zero, there do exist probabilities—synthetic probabilities,

---

[12]Or Alydar, Bold Ruler, and Citation, if you prefer.

to be sure, but probabilities nonetheless—with respect to which all bets are fair and all gamblers' fortunes are martingales.

9.6.1. The Basic Financial Market. Financial markets in the real world contain multitudes of stocks, bonds, goods, services, contracts, and taxes, which can be bought and sold, delivered or refused, fulfilled, and paid or evaded under the rules, regulations, and customs which apply. They also involve people, which means that any mathematical hypotheses we might make about them will soon be violated.

We will not attempt to analyze anything so complicated. We will look instead at a simple idealized market containing a single stock and a single bond. In the real world, stocks represent shares of a company and have an intrinsic value which may change with the fortunes of the company. In our market, though, a stock is simply something to be bought and sold. Its price varies randomly with time. The bond acts like a bank: buying a bond is equivalent to putting money in a savings account—it earns interest—and selling a bond is equivalent to withdrawing money, or borrowing if there is not enough in the account. (We talk about bonds instead of bank accounts in order to use the metaphor of buying and selling for all items in the market.) The difference between a stock and a bond is that the stock is random, but the bond is not: if we know its price today, we know its price for all future times.

Simple as the market is, we shall have to explore its financial workings fairly deeply before we arrive at the probability. This is because there is another element implicit in the market: contracts or promises which depend on the stock and the bond. These are by no means simple. They are called **derivatives** because their value is derived[13] from the stock and bond values. A typical derivative would be a contract to deliver one share of stock one month hence. This is called a *future*. A *stock option* is another. A typical option would be the promise to sell to the buyer one share of stock one month hence at a fixed price K—*if* the buyer still wishes to buy it; but the buyer is not obligated to buy. If the stock price is less than K, the buyer would refuse the option: the stock would be cheaper at the market price.

The stock option cannot lose money, and may make a profit, so it must be worth something. But what? This is the kind of problem we will address.

Rules of the Market. The market consists of:
 • A stock. The price of one share of the stock at time t is S_t: $\{S_t, t = 0, 1, 2, \dots\}$ is a stochastic process.

---

[13] "Derivative" in this sense has nothing to do with differentiation.

• A bond. Its value at time t is $B_0 e^{rt}$, where $r \geq 0$ is the interest rate[14]. The bond is deterministic and the interest rate is constant.

• Some or possibly all derivatives based on the stock and bond.

Furthermore:

• All items in the market can be traded according to the market trading rules.

The trading rules are simple:

• All items in the market may be freely bought and sold at the market price in any quantity desired.

That is, there is a price. If one wishes to buy at that price, there will be a seller. If one wishes to sell, there will be a buyer, and there is no limit, large or small, to the amount bought and sold.

• Short-selling is allowed.

This means that it is not necessary to own a stock or bond in order to sell it. "Short selling" is the selling of an item that one does not own. This is permitted in real-world markets, though it is normally regulated and subject to credit limits and margin requirements. It is akin to borrowing the stock in order to sell it. As with other borrowing, there is an obligation to return it later.

• There are no transaction costs or hidden expenses.

• Promises will be kept: there is no default, so contracts once made will be honored.

These are the basic rules of trading. There is one more important property, called the "no-arbitrage hypothesis" which has to do with the structure of the market, not the trading rules.

• There is no arbitrage in the market.

An **arbitrage** is a deal which makes a profit with strictly positive probability, but cannot lose money, and which requires zero net investment. When a possible arbitrage presents itself, we speak of an **arbitrage opportunity.** According to the hypothesis, this cannot happen. As they say, "There is no free lunch."

For example, suppose person A offers to sell something for a dollar and person B offers to buy it for two. Then person C could buy it from A and immediately sell it to B, making a sure profit of one dollar. The no-arbitrage hypothesis rules out this kind of transaction[15]. It is basically a hypothesis

---

[14]Note that interest is compounded continuously.

[15]It is also ruled out by the hypotheses, which say that quantities are bought and sold at a unique market price.

on the efficiency of the market. Market prices are supposed to adjust so quickly that the opportunity would disappear before C found out about it.

Remark 9.44. The no-arbitrage requirement actually forces prices to be unique. Note that if an item is sold, the buyer and seller must have agreed on the price. But if one person, say A, is willing to sell an item for the price P, and person B is willing to buy it for the same price, then person C cannot buy or sell it for a different price. Indeed, the reader can verify that if C offers to buy it for a strictly greater amount than P, or sell it for strictly less, this offers person D the possibility of an arbitrage. If C wishes to sell it for less than P, or buy it for more, there will be no takers, since D can deal with A or B at a better price. This means that there is a unique price—which we will call the *market price*—for both buying and selling, and everyone, buyer and seller alike, agrees on it. There are no transaction fees or hidden costs.

Future Values. The no-arbitrage hypothesis determines prices to a great extent, via what are called "hedging arguments", which compare alternate ways of achieving the same aims.

Let us see how it determines two things: the future value of the stock and the future value of a dollar.

Consider a question about stock value: What is the value today of the stock at a later time t? Let us rephrase that. Consider a contract which guarantees to provide one share of stock at time t. How much is that contract worth today, at $t = 0$?

The answer is that the contract is worth S_0, the value of the stock today. Any other value provides an arbitrage opportunity.

Indeed, let P be the contract price. There are two ways to have the stock at time t. One can either buy one share today for S_0 and keep it, or one can buy the contract today for P and collect the stock at time t.

Suppose first that $P < S_0$. We can construct an arbitrage as follows:

(a) Sell one share of stock today for S_0. (We are allowed to sell a stock without owning it—this is "short-selling".)

(b) With the money obtained from the sale, buy the contract for P. This leaves $S_0 - P > 0$. Invest that in a bond.

At this point, the total investment is zero. We own the contract, have a bond worth $S_0 - P > 0$, and we are short one share of stock which we must replace later.

(c) At time t, collect the share of stock from the contract and sell it for S_t, its price at time t. Use that money to buy back the share we sold short, also at S_t. This cancels out, but we still have the bond, which is now worth $(P - S_0)e^{rt} > 0$. When we sell that, we have made a strictly positive profit from zero investment. This is an arbitrage, which contradicts the no-arbitrage hypothesis. Thus the price cannot be strictly less than S_0.

If $P > S_0$, reverse the strategy. Sell the contract for P—that is, agree to provide a share of stock at time t—and buy the stock itself for S_0. Invest the remaining $P - S_0$ in a bond. The total investment is zero. We have one share of stock, a bond, and the obligation to give one share of stock to the contract's buyer when t comes.

At time t, we provide the stock as agreed. Then we sell the bond for a strictly positive profit on zero investment. This is an arbitrage, which cannot happen. Thus P cannot be strictly greater than S_0. The only possible price is S_0.

Even without the no-arbitrage rule, there is a powerful reason to use this price. An investor who doesn't is open to an arbitrage. There is no limit on the amount of a transaction, so being on the wrong side of an arbitrage can cost a *lot* of money.

As the second example, consider the value of a dollar bill: how much is it worth some time in the future? (Cash money does not enter our model except as a way to keep score. We will see why after we answer this question.)

Let us rephrase the question. Consider a contract which guarantees to provide a one-dollar bill at a time t in the future. How much is that contract worth today?

We want to have a dollar at hand at time t. Of course we can do this by buying the dollar today, and keeping it until t. That will cost one dollar. But there is a better way.

We claim that the contract to supply one dollar at time t in the future is worth e^{-rt} dollars today.

Let P be the price of the contract. Suppose $P < e^{-rt}$. Then we do the following.

(a) Sell P worth of bonds. (This is short-selling.)

(b) Use that money to buy the contract for P.

The two transactions cancel, and there is no net investment. But now at time t,

(c) collect the dollar the contract promises and

(d) pay off the bonds at their new price, Pe^{rt}.

We now have cleared the transaction, and have $1 - Pe^{rt}$ left over. Since $P < e^{-rt}$, this is strictly positive. We have made a profit, with no risk whatsoever. This is arbitrage, and cannot happen. Thus, $P \geq e^{-rt}$.

9.6. Financial Mathematics I: The Martingale Connection*

If $P > e^{-rt}$, do the opposite: sell the contract for P at $t = 0$ and use the money to buy P worth of bonds. This is again a zero total investment. Then at time t, sell the bonds for Pe^{rt} and deliver the dollar that the contract promised. This leaves $Pe^{rt} - 1$, which is strictly positive because $P > e^{-rt}$. Another arbitrage! Thus P cannot be strictly greater than e^{-rt}. Since it is neither strictly greater nor strictly smaller than e^{-rt}, it must be equal.

Remark 9.45. This says that the value of a dollar bill some time t in the future is e^{-rt}. Its value decreases with time[16]. The prudent investor will never hold money, but will immediately exchange it for bonds[17]. That is why money is not explicitly in the model.

There is another consequence. We have to take inflation into account when comparing the value of items at different times. For example, salaries in the 1950s seem small by modern standards until we realize that today's dollar is worth about eleven cents in 1950s currency.

So the face value of money in the future must be **discounted** to measure it in terms of todays dollars. It is the discounted dollar, not the face-value dollar, that we must use to measure future value. Thus when we want to discuss future values, we should measure them in discounted dollars, not face-value dollars. Another way to express this is to say that future value should be measured in terms of the bond, not the dollar, since the bond has a constant value in discounted dollars.

Derivatives and Value. From now on, we will measure values in terms of todays dollar, or, as it is usually called, the **discounted dollar**. This allows us to compare values from one time to another. The value of the dollar bill decreases with time, as we saw. At time t, it is only worth e^{-rt} in today's dollars. That means that in terms of today's dollar, the stock and bond at time t are worth

$$\hat{S}_t \stackrel{def}{=} e^{-rt} S_t,$$
$$\hat{B}_t \stackrel{def}{=} e^{-rt} B_t \equiv B_0.$$

This is purely a question of units: S_t and $\hat{S}_t$ are the same object expressed in different units. S_t is its value in nominal dollars, $\hat{S}_t$ in discounted dollars. For the rest of the discussion, we will express values in terms of discounted dollars. We will put hats over some discounted quantities such as the stock price, but this is simply as a reminder.

Let $\{\hat{S}_t,\ t \geq 0\}$ be defined on the probability space $(\Omega, \mathcal{F}, P)$. For $t \geq 0$, define

$$\mathcal{F}_t = \sigma\{\hat{S}_s,\ s \leq t\}.$$

---

[16] This is familiar: it is called inflation.
[17] There are no transaction costs in this model, so the exchange costs nothing.

Suppose for simplicity that $\hat{S}_0$ is constant, so that $\mathcal{F}_0$ is trivial. The bond is deterministic, so all the randomness in the market is supplied by the stock, and $\mathcal{F}_t$ represents the total state of knowledge at time t.

Derivatives get their name because their value is derived from values of the stock and the bond, which means that they are actually functions of the of the stock and bond. But the bond is deterministic, so we can regard a derivative as a function of $(\hat{S}_t)$ alone. In short, it is a random variable. Let us make that into a definition:

Definition 9.46. A **derivative** X on $[0,T]$ is a positive $\mathcal{F}_T$-measurable random variable.

If X is a derivative on $[0,T]$, it represents money, but the amount may depend on the behavior of $\hat{S}_t$ in the interim, i.e., on $[0,T]$. The derivative is positive (i.e., non-negative) by convention. Think of X as a contract that promises to provide X dollars at time T. The contract has the final value X, but this value may not be known before time T itself. So its price at an earlier time t is in some sense an estimate of its final value. We will speak of the price of this contract as "the value of X."

We can only speak of the price of derivatives that are freely traded, i.e., those that are part of the market. In real life, stocks are traded on the stock market, the bond is freely available, and some, but not all, derivatives are traded on exchanges. Other derivatives are sold by banks and other financial firms, but not traded, and still others are not traded at all. However, our market is ideal, not real, and, for the purpose of this section, we will assume that it is complete in the sense that *all* derivatives are freely traded. Therefore, they all have a value.

Definition 9.47. Let X be a derivative. Then $\boldsymbol{V_t(X)}$ denotes the **value of $\boldsymbol{X}$ at time $\boldsymbol{t}$**, measured in discounted dollars.

Notice that we have identified a derivative with its payoff, not with the process involved in arriving at the payoff. So if X is a derivative on $[0,T]$, any two schemes which give at the same payoff X at time T have the same value.

We already know something about V_t: if $t < T$, then
$$V_t(\hat{S}_T) = \hat{S}_t, \quad V_t(\hat{B}_T) = \hat{B}_t \equiv \hat{B}_0.$$

The final value of a derivative X on $[0,T]$ is X itself:
$$V_T(X) = X.$$

Moreover, the time-t value of X is known at time t, so $V_t(X)$ is $\mathcal{F}_t$-measurable. (In fact, it is itself a derivative on $[0,t]$: it is the time-t price of a contract to deliver X at time T.)

9.6. Financial Mathematics I: The Martingale Connection*

The basic problem is easy to state: "Find $V_t(X)$ for all derivatives X and all t."

The arbitrage opportunities discussed above are all elementary, in that they involve a finite number of securities at just two different times. To handle more complex dealings, let us introduce the idea of a portfolio. This is simply a list of the investor's holdings. It could include the stock, the bond, cash, and diverse derivatives. We assume the investor deals with a finite or at most countably infinite number of securities, say $X_1, X_2, X_3, \ldots$. Formally:

Definition 9.48. A **portfolio** is a function $\Phi(t) = (\phi_1(t), \phi_2(t), \ldots)$, $t = 0, 1, 2, \ldots$, where $\phi_k(t)$ is $\mathcal{F}_{t-1}$-measurable, $k = 1, 2, \ldots$. The **value** of the portfolio at time t is

$$M_t \stackrel{def}{=} \sum_{k=1}^{\infty} \phi_k(t) V_t(X_k).$$

Here, $\phi_k(t)$ is the amount of the derivative X_k the investor holds at time t. It can be negative, which corresponds to short-selling. If a security is bought, it is added to the portfolio; if sold, it is subtracted. If the investor buys X_k, for example, then ϕ_k increases. In particular, short-selling something adds a negative amount to the portfolio. The reason that $\Phi(t)$ is $\mathcal{F}_{t-1}$-measurable rather than $\mathcal{F}_t$-measurable is that the investor cannot look into the future: the decision on the portfolio for day t is made after the stock-prices on day $t-1$ are known, but before those on day t are. (Otherwise, the investor could pick only stocks that are sure to go up, an arbitrage.)

If no money is injected into or taken from the portfolio, an increase in some ϕ_k must be balanced by a decrease in other ϕ_j used to pay for it. A portfolio from which no money is injected or extracted is called "self-financing": money used to buy a new security must come from the sale of others, so it does not require an influx of money to run.

The portfolio's value at time $t-1$ is $M_{t-1} = \sum_k \phi_k(t-1) V_{t-1}(X_k)$. Then securities are bought and sold at the old prices to make the new portfolio $\Phi(t)$. Its value—at the old prices—is $\sum_k \phi_k(t) V_{t-1}(X_k)$. No money has been added or removed, so this still equals M_{t-1}. Subtracting, we get

$$(9.13) \qquad \sum_k \big(\phi_k(t) - \phi_k(t-1)\big) V_{t-1}(X_k) = 0, \; t = 1, 2, \ldots.$$

Definition 9.49. A portfolio satisfying (9.13) is called **self-financing**.

We can describe arbitrage opportunities in terms of portfolios.

Definition 9.50. A self-financing portfolio $(\Phi(t))$ is called an **arbitrage opportunity** on $[s, t]$ if $M_t - M_s \geq 0$ a.s. and $P\{M_t - M_s > 0\} > 0$.

Market trading rules allow one to convert the entire portfolio into cash and back at any time, so the owner of the portfolio Φ can have $M(s)$ in cash at time s and $M(t)$ in cash at time t, giving a genuinely riskless profit of $M(t) - M(s)$. This turns an arbitrage opportunity into a true arbitrage. So the no-arbitrage hypothesis is equivalent to saying there are no arbitrage opportunities. The self-financing requirement excludes the possibility that the profit comes from an outside investment.

Remark 9.51. Infinite values are financially impractical but mathematically inescapable. While the stock price is always finite, it may be possible to construct a derivative whose value is infinite. Since nobody could afford it, it should be irrelevant.

The St. Petersburg arbitrage[18] reveals a deeper problem. Suppose an investor invests repeatedly using the strategy "Double up after every loss." Suppose that the probability that the stock price goes up in any interval n to $n+1$ is $p > 0$, and that the different intervals are independent. Let X_n be the derivative which pays \$2 if $S_{n+1} > S_n$, and and which pays zero otherwise. Suppose that X_n costs \$1. The investor buys X_1 for \$1 at time 1. If the stock goes up, she pockets the \$1 profit and stops playing. If not, she buys \$2 worth of X_2. If if the stock goes up, she pockets the \$4, again for a \$1 profit, and stops. Otherwise she doubles the investment again, and so on, until one investment finally pays off, as it must. If the stock falls during the first n intervals and rises on the $n+1^{\text{st}}$, she has paid out $1 + 2 + \cdots + 2^n = 2^{n+1} - 1$, and, as her final investment was \$$2^n$, she gains \$$2^{n+1}$, for a profit of \$1. Since $p > 0$ and the different investments are independent, she is sure to eventually make a one-dollar profit. This is a genuine arbitrage.

Notice that in order to follow this strategy, the investor must short-sell bonds to buy the X_i; but there is no bound on the amount it might take. Indeed, this strategy requires the possibility of an unbounded amount of short-selling. Let us define:

Definition 9.52. A portfolio has **bounded risk** if there exists N such that for all t, $\sum_k V_t(X_k)^- \leq N$.

Here, $V_t(X_k)^-$ is the negative part of $V_t(X_k)$, i.e., the amount that the portfolio is short on X_k. This effectively limits the amount that can be borrowed. This is no restriction in practice, since N could be the total amount of money in circulation in the world, but it does rule out both the sale of infinite-value securities and the St. Petersburg arbitrage.

---

[18]This is a variant of the famous St. Petersburg game proposed by Nicolas Bernoullli in 1738. Its drawbacks in gambling and finance have been thoroughly explored.

9.6. Financial Mathematics I: The Martingale Connection*

Definition 9.53. There is **no-arbitrage with bounded risk** if no self-financing portfolio with bounded risk is an arbitrage opportunity.

Properties of the Value.

Lemma 9.54. *Let X and Y be derivatives on $[0,T]$ of finite value. If $X \leq Y$ a.s., then $V_0(X) \leq V_0(Y)$. If $X = Y$ a.s., then $V_0(X) = V_0(Y)$.*

Proof. Suppose that this is wrong, and that $V_0(X) > V_0(Y)$. Consider the portfolio that has one unit of Y and is short one unit of X. Its value at $t = 0$ is $M_0 \equiv V_0(Y) - V_0(X) < 0$. At time T, the same portfolio is worth $M_T = V_T(Y) - V_T(X) = Y - X > 0$. This is an arbitrage opportunity, which is a contradiction. This proves the first statement, and the second follows immediately. $\square$

Theorem 9.55. *Let $a \geq 0$, let $0 \leq t < T$, and let X and Y be derivatives on $[0,T]$. Then*
   (i) $V_0(0) = 0$, $V_0(1) = 1$, and $V_0(X) \geq 0$.
   (ii) $V_0(X) = 0 \iff X = 0$ a.e.
   (iii) V_0 is linear: $V_0(X+Y) = V_0(X) + V_0(Y)$ and $V_0(aX) = aV_0(X)$.
   (iv) If $t \geq 0$, then $V_0(X) = V_0(V_t(X))$.
   (v) If $A \geq 0$ is bounded and $\mathcal{F}_t$-measurable, then $V_t(AX) = AV_t(X)$.

Proof. These equation are all true for the same reason: their two sides represent two ways to get the same payoff, and therefore have the same value by Lemma 9.54.

(i) $X \equiv 1$ gives the same payoff as a one-dollar bond, and therefore has the same initial value, namely $V_0(1) = 1$. Similarly, $X \equiv 0$ is equivalent to zero bonds, whose value is zero. Finally, derivatives are positive by convention, so by Lemma 9.54, $0 \leq X \implies 0 = V_0(0) \leq V_0(X)$.

(ii) If $X = 0$ a.e. $V_0(X) = 0$ by (i). Conversely, if $V_0(X) = 0$, the portfolio containing X alone has initial value $M_0 = V_0(X) = 0$ and final value $M_T = V_T(X) = X$. This is an arbitrage opportunity unless $X = 0$ a.s.

(iii) Consider two ways to get $X + Y$ at time T: either buy a contract to deliver $X+Y$ at time T, which has value $V_0(X+Y)$, or buy two separate contracts, one to deliver X, the other to deliver Y, which have values $V_0(X)$ and $V_0(Y)$. They have the same payoff at T, and therefore, by Lemma 9.54, the same value. Thus $V_0(X + Y) = V_0(X) + V_0(Y)$. A similar argument shows that $V_0(aX) = aV_0(X)$.

(iv) Consider two ways of getting X at time T: at time zero either (a) buy a contract to deliver X at time T, (value $V_0(X)$) or (b) buy a contract that will deliver $V_t(X)$ at the intermediate time t. (That is buy a contract which will, at time t, deliver a second contract. This second contract will provide X at time T.) If it delivers $V_t(X)$ at time t its value must be $V_0(V_t(X))$. The payoff is X in both cases, so the two values are equal.

(v) Note that A is known at time t, so from then on, it is possible to deal in A units of a security.

Let $\Lambda_1 \stackrel{def}{=} \{V_t(AX) < AV_t(X)\}$ and $\Lambda_2 \stackrel{def}{=} \{V_t(AX) > AV_t(X)\}$. Consider a portfolio Φ which is initially empty and therefore has initial value zero. It remains empty on $\Omega - \Lambda_1 \cup \Lambda_2$, and changes exactly once, at time t, on $\Lambda_1 \cup \Lambda_2$. On Λ_1, (where $V_t(AX) < AV_t(X)$), from time t on Φ contains $AV_t(X) - V_t(AX)$ in bonds, one unit of AX, and minus A units of X (i.e., A units of X were short-sold.) On Λ_2, after time t Φ contains $V_t(AX) - AV_t(X)$ in bonds, A units of X and minus one unit of AX. The value of Φ at day-t prices is $AV_t(X) - V_t(AX) + V_t(AX) - AV_t(X) = 0$ on Λ_1 and $V_t(AX) - AV_t(X) + AV_t(X) - V_t(AX) = 0$ on Λ_2. Therefore it is self-financing.

Both X and AX are derivatives on $[0, T]$ (A is $\mathcal{F}_t$ measurable and $\mathcal{F}_t \subset \mathcal{F}_T$) so at time T, $V_T(X) = X$ and $V_T(AX) = AX$. The bonds still have their original price. Thus on Λ_1, Φ has the value $M_T \equiv AV_t(X) - V_t(AX) + V_T(AX) - AV_T(X) = AV_t(X) - V_t(AX) > 0$. On Λ_2 its value is $M_T \equiv V_t(AX) - AV_t(X) + AV_T(X) - V_T(AX) = V_t(AX) - AV_t(X) > 0$. It is zero elsewhere, so $M_0 = 0$, $M_T \geq 0$, and $M_T > 0$ on $\Lambda_1 \cup \Lambda_2$. This is an arbitrage unless $P\{\Lambda_1 \cup \Lambda_2\} = 0$. Therefore $V_t(AX) = AV_t(X)$ a.s. □

We need another result, one which doesn't come up in real-world finance: only a philosopher would ask[19] whether or not someone can simultaneously make an infinite number of different transactions. But this is mathematics, not the real world. Someone can. However, we impose a (loose) credit limit: we assume a bounded risk.

Proposition 9.56. *Let $X_1, X_2, \ldots$ be a sequence of derivatives on $[0, T]$ with $\sum_i V_0(X_i) < \infty$. Then*

$$V_0\left(\sum_{i=1}^\infty X_i\right) = \sum_{i=1}^\infty V_0(X_i).$$

Proof. Construct $\sum_i X_i$ as follows. At time zero, short sell $V_0(X_i)$ worth of bonds and buy X_i with the proceeds. Do this for every i. The resulting portfolio contains one unit of each X_i and is short $\sum_i V_0(X_i)$ of bonds. This is finite by hypothesis, so that its value is $M_0 \equiv -\sum_i V_0(X_i) + \sum_i V_0(X_i) = 0$, and the portfolio has bounded risk. At time T, take delivery of the X_i and add them together to make $\sum_i X_i$. Let $P = V_0\left(\sum_{i=1}^\infty X_i\right)$. The

---

[19]Practical people wouldn't ask, for the answer is obviously "No." Mathematicians wouldn't ask either, for the answer is obviously "Yes."

portfolio now contains $-\sum_i V_0(X_i)$ of bonds along with $\sum_i X_i$ so its value is $M_T = P - \sum_i V_0(X_i)$, where $P \equiv V_0(\sum_i X_i)$.

Now $V_0(X)$ is increasing in X, so $P \geq V_0(\sum_{i=1}^n X_i) = \sum_{i=1}^n V_0(X_i)$ for each n. Let $n \to \infty$ to see that $P \geq \sum_{i=1}^\infty V_0(X_i)$. If the inequality is strict, $M_T > 0$, which is an arbitrage opportunity[20]. Therefore $P = \sum_i V_0(X_i)$. □

No-Arbitrage and Martingales. Now have seen the basic workings of the market, we can connect it to martingales. The connection is made by two theorems. The first one is this.

Theorem 9.57. *Suppose the market consists of the stock, bond and all derivatives. If the market has no arbitrage with bounded risk, then for each T there exists a probability measure Q on $(\Omega, \mathcal{F}_T)$ such that Q is equivalent[21] to P, and if X is a derivative on $[0, T]$, its value is*

$$V_0(X) = \int X \, dQ. \tag{9.14}$$

Proof. Let $T > 0$. For $\Lambda \in \mathcal{F}_T$, define $Q(\Lambda) = V_0(I_\Lambda)$. This makes sense, since I_Λ is a positive $\mathcal{F}_T$-measurable random variable, and therefore a derivative. By Theorem 9.55 Q is a positive finitely additive set function on $\mathcal{F}_T$, and $Q(\Omega) = V_0(1) = 1$. It is even countably additive: if $\Lambda_1, \Lambda_2, \ldots$ are disjoint sets in $\mathcal{F}_T$, then $\sum_i I_{\Lambda_i} \leq 1$, so that $V_0(\sum_i I_{\Lambda_i}) \leq V_0(1) = 1$. It follows that $\sum_i V_0(\Lambda_i) = V_0(\sum_i I_{\Lambda_i}) \leq 1$. Proposition 9.56 implies

$$Q\left(\bigcup_n \Lambda_n\right) = V_0\left(\sum_n I_{\Lambda_n}\right) = \sum_n V_0(I_{\Lambda_n}) = \sum_n Q(\Lambda_n).$$

Thus Q is a measure. To prove (9.14), just follow the construction of the integral, using the linearity of both the integral and the value function. By definition, $V_0(X) = \int X \, dQ$ if X is the indicator function of a measurable set. Therefore (9.14) holds if X is a positive simple function. In general, X, being positive, is the limit of an increasing sequence $X_0, X_1, X_2, \ldots$ of simple functions, with $X_0 \equiv 0$. Then $X = \sum_{n=1}^\infty (X_n - X_{n-1})$, and

$$\int X \, dQ = \sum_{n=0}^\infty \int (X_n - X_{n-1}) \, dQ = \sum_{n=0}^\infty V_0(X_n - X_{n-1}) = V_0(X),$$

where the next-to-last equality follows because $X_n - X_{n-1}$ is positive and simple, and the last follows by Proposition 9.56. This proves (9.14).

To see that Q and P are equivalent, apply Theorem 9.55 (*ii*) to indicator functions: $P\{\Lambda\} = 0 \iff V_0(I_\Lambda) = 0 \iff Q\{\Lambda\} = 0$. □

---

[20]The no-arbitrage argument fails if $P = \infty$: it is too expensive to sell! However, the rules of the market stipulate that buyers and sellers must agree on prices, and there cannot be agreement on an infinite price, since we would be willing to sell it for a finite amount!

[21]We say that Q and P are *equivalent* if for each $\Lambda \in \mathcal{F}_T$, $P\{\Lambda\} = 0 \iff Q\{\Lambda\} = 0$.

Corollary 9.58. *Under the hypotheses of Theorem 9.57, if X is a Q-integrable derivative on $[0,T]$, then for $t \geq 0$,*

(9.15) $$V_t(X) = E^Q\{X \mid \mathcal{F}_t\},$$

where E^Q is the expectation on $(\Omega, \mathcal{F}, Q)$. Thus $\{V_t(X), \mathcal{F}_t, t \geq 0\}$ is a martingale.

Proof. Let $\Lambda \in \mathcal{F}_t$. $V_t(X)$ is $\mathcal{F}_t$-measurable. It is also a derivative, so by Theorem 9.55 (iv) and (v),

$$\int_\Lambda V_t(X)\,dQ = \int I_\Lambda V_t(X)\,dQ = \int V_t(I_\Lambda X)\,dQ$$
$$= V_0(V_t(I_\Lambda X)) = V_0(I_\Lambda X)$$
$$= \int_\Lambda X\,dQ.$$

But this is true for all $\Lambda \in \mathcal{F}_t$, which identifies $V_t(X)$ as $E^Q\{X \mid \mathcal{F}_t\}$. □

Definition 9.59. A measure Q on $(\Omega, \mathcal{F}_T)$ is an **equivalent martingale measure** if it is equivalent to P (i.e., $\forall \Lambda \in \mathcal{F}_T : P\{\Lambda\} = 0 \iff Q\{\Lambda\} = 0$), and if the discounted stock price $\{\hat{S}_t, \mathcal{F}_t, 0 \leq t \leq T\}$ is a martingale on $(\Omega, \mathcal{F}_T, Q)$.

Let us restate this. If there is no arbitrage in a complete market, then the prices V_0 determine an equivalent martingale measure Q such that, under Q, not only the discounted stock prices $(\hat{S}_t)$, but the discounted prices of all derivatives are martingales. Moreover, once we know the measure Q, we know the prices of all derivatives at all times, for they are just conditional expectations.

Remark 9.60. One might have guessed that prices were expectations, even without the analysis. And it is true. They are. However, they are expectations with respect to **another** probability measure, the martingale measure Q. This is synthetic, derived from the money—the stock and derivative values—not from the actual probability distribution of the stock. Nevertheless, it is fundamental for evaluating worth.

Remark 9.61. It is instructive to revisit the horse-racing example. In one sense, the connection is clear: betting on a horse is equivalent to buying a derivative whose value is zero if the horse loses and is whatever the odds say if it wins. But there is a deeper connection. The tote board quotes odds that insure a profit for the track. In other words, it assures an arbitrage. Setting the track's take to zero removes that, and results in a betting system with no arbitrage. You proved in Exercises 9.37, 9.38 and 9.39 that the quoted odds defined a probability measure which made all gamblers' fortunes into martingales. You produced, in fact, a martingale measure for the race track.

9.6. Financial Mathematics I: The Martingale Connection*

The analogy doesn't end there. In addition to bets on win, place and show, race tracks offer combination bets like the *pick three*, which is a bet on the winners of three successive races. These are clearly derivatives: they depend on the results of the races. Once the martingale measure is established, Theorem 9.62 shows how to price them.

Conversely, a martingale measure generates a system of prices which have no arbitrage.

Theorem 9.62. *Let Q be an equivalent martingale measure and X a Q-integrable derivative. Define*

(9.16) $$V_t(X) \stackrel{\text{def}}{=} E^Q\{X \mid \mathcal{F}_t\},\ t \geq 0\,.$$

This defines a market having no-arbitrage-with-bounded-risk.

Proof. If there is an arbitrage opportunity, there is a self-financing portfolio $\Phi(t)$ whose value M_t satisfies $M_0 = 0$, $M_T \geq 0$, and $P\{M_T > 0\} > 0$. There is a margin limit, so the value of the absolute portfolio must be finite at all times, i.e., $\infty > V_0\big(\sum_k |\phi_k(t)|\, V_t(X_k)\big) = \sum_k \int |\phi_k(t)|\, V_t(X_k)\, dQ$. In particular, M_t is Q-integrable for all t. Now

$$M_T = \sum_{j=1}^T (M_j - M_{j-1}) = \sum_{j=1}^T \sum_{k=1}^\infty \big(\phi_k(j) V_j(X_k) - \phi_k(j-1) V_{j-1}(X_k)\big)$$

$$= \sum_j \sum_k \phi_k(j)\big(V_j(X_k) - V_{j-1}(X_k)\big)$$

$$+ \sum_j \sum_k \big(\phi_k(j) - \phi_k(j-1)\big) V_{j-1}(X_k).$$

The portfolio is self-financing, so the last sum vanishes by (9.13). Take the expectation with respect to the martingale measure Q. As $\phi_k(j)$ is $\mathcal{F}_{j-1}$-measurable,

$$E^Q\{M_T\} = \sum_j \sum_k E^Q\big\{\phi_k(j)\big(V_j(X_k) - V_{j-1}(X_k)\big)\big\}$$

$$= \sum_j \sum_k E^Q\big\{\phi_k(j)\, E^Q\{V_j(X_k) - V_{j-1}(X_k) \mid \mathcal{F}_{j-1}\}\big\}\,.$$

But by (9.16), $j \mapsto V_j(X_k)$ is a Q-martingale, so the inner conditional expectations vanish and $E^Q\{M_T\} = 0$.

Now P and Q are equivalent, and $P\{M_T \geq 0\} = 1$, so $Q\{M_T \geq 0\} = 1$ as well. Then $E^Q\{M_T\} = 0 \Rightarrow Q\{M_T = 0\} = 1$, so $P\{M_T = 0\} = 1$ too. This contradicts the assumption that $P\{M_T > 0\} > 0$. $\square$

Remark 9.63. This settles a problem that has been lurking in the background: does there exist a consistent set of prices, that is, prices for which there is no arbitrage? This is not at all obvious, but we now see that if there exists an equivalent martingale measure, the answer is "Yes."

The problem of finding the value of derivatives comes down to the problem of finding an equivalent martingale measure, that is, a measure which makes the (discounted) stock price into a martingale. The corresponding (discounted) derivative values are given by (9.16). Theorem 9.62 guarantees that these are consistent market prices: there is no arbitrage. To summarize:

• Every consistent set of market prices corresponds to a martingale measure.

• Every martingale measure corresponds to a consistent set of market prices.

If the martingale measure is unique, then the market prices are uniquely determined and we can say that the no-arbitrage hypothesis determines all the prices. If there is more than one martingale measure, as there may be, then there is more than one set of market prices, and the no-arbitrage hypothesis does not uniquely determine the market.

Example 9.63.1. (A Binary Tree model.) The Cox-Ross-Rubinstein model is a simple model of stock prices which is suitable for computation. For simplicity, we will assume that the interest rate is zero.

Let S_n be the stock price for $n = 0, 1, 2, \ldots$, defined on a probability space $(\Omega, \mathcal{F}, P)$, where $\mathcal{F} = \sigma\{S_n, \ n = 0, 1, 2, \ldots\}$ is the natural σ-field. Suppose that at each time n, S_n can either increase by a factor of $b > 1$, or decrease by a factor of $1/b$, independent of the previous S_j. Let $p = P\{\hat{S}_{n+1} = b\hat{S}_n \mid \mathcal{F}_n\}$ be the probability of increasing.

Thus S_n is a Markov chain on $\{b^n,\ n = 0, \pm 1, \pm 2, \ldots\}$. It will be a martingale if $S_n = E\{S_{n+1} \mid S_n\} = pbS_n + (1-p)S_n/b$, or $p = 1/(b+1)$. Under the martingale measure, then, (S_n) is a Markov chain with transition probabilities $P_{b^n b^{n+1}} = 1/(b+1)$ and $P_{b^n b^{n-1}} = b/(b+1)$. If $S_0 = 1$, this uniquely determines the process modulo its initial distribution. We conclude that the martingale measure is unique, and the market prices are uniquely determined.

To get a feeling for how options are used, consider the plight of an investor who, thinking that the price of a stock is about to go down, has short-sold it, and who now hears a rumor that something is about to happen to make the price go way up. When it comes time to replace the short-sold stock—and that time will surely come—the investor will have to do it at the market price at that time, which may be considerably greater than it is now. (The investor has received today's price, but will have to repay the

9.6. Financial Mathematics I: The Martingale Connection*

stock at the future market price, so if the stock's price triples, the investor will lose his or her investment twice over[22].) Therefore, the investor hedges by purchasing an option to buy the stock at time $n = 3$, say, for its present price. If the price goes down, the investor will make a profit from the short-sold stock, and will not exercise the option. But if the price goes up, he will buy the stock at time $n = 3$ for today's price and use it to replace the short-sold stock. There will be no net loss. So the option is insurance: once it is bought, the investor is insured against loss. Question: How much should the option cost?

Suppose for example that $b = 2$, and that today's price is 1. The stock $\hat{S}$ can either double or halve each day, and there are three days. In the worst case, the stock goes up by a factor of eight, and the investor loses seven times the original investment.

The stock's possible values on the third day are $1/8$, $1/2$, 2, and 8. If the investor chooses to exercise the option, he must pay 1 for each share, so his profit will be $\hat{S}_3 - 1$. The investor will only exercise the option if it will profit him, which is if $\hat{S}_3 > 1$. So the payoff is $(S_3 - 1)^+$.

Neither the investor nor the option seller knows the actual probabilities but they can compute the option's price without them. They only have to know the transition probabilities under the martingale measure. With $b = 2$, the doubling probability is $1/3$, the halving probability $2/3$.

Thus the value of the option is $E^Q\{(S_3-1)^+\}$, where the expectation is with respect to the martingale measure Q. To calculate this, use binomial probabilities to find that the price is $E^Q\{(\hat{S}_3-1)^+\} = (8-1)^+(1/27) + (2-1)^+(6/27) = 13/27$. So the option costs $13/27$ dollars per share, so if the investor owns N shares of stock, the option will cost $13N/27$.

It is useful to think of the stock price evolving on a tree: if its value at time n is s, it can take either of two branches, the upper branch, which leads to $2s$, and the lower branch, which leads to $s/2$. Under the martingale measure Q, the upper branch has probability $1/3$ and the lower branch has probability $2/3$. The value V_3 of the option at time 3 is known: $V_3(S_3) = (S_3-1)^+$. We can compute its initial value by using the transition probabilities to successively determine $V_2(S_2) = E^Q\{V_3 \mid S_2\}$, $V_1(S_1) = E^Q\{V_2 \mid S_1\}$, and finally $V_0 = E^Q\{V_1\}$, where Q is the martingale measure. This is straightforward. If $S_j = s$, then S_{j+1} is either $2s$ or $s/2$ with (conditional) probabilities $1/3$ and $2/3$, respectively, giving us

$$V_j(s) = \frac{1}{3}V_{j+1}(2s) + \frac{2}{3}V_{j+1}(s/2).$$

---

[22]Short-selling a stock is riskier than simply buying it. The worst that can happen to an investor who buys a stock is that the stock's value goes to zero, and the investment is lost. In short-selling, however, the investor can lose many times the investment.

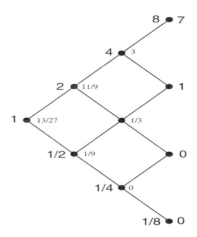

Figure 1. Values of $V_j = E^Q\{(S_3 - 1)^+ \mid \mathcal{F}_j\}$.

If we know the value of a derivative at time n, say, this equation gives us its value at times $n-1$, $n-2$, ..., 1, and 0 successively. See Figure 1.

Example 9.63.2. (Trinomial Tree.) Suppose that there are three possible choices for the stock: $\hat{S}_t$ can either increase by a factor of $b > 1$, decrease by a factor of $1/b$, or stay the same. It is easily seen that there is a whole family of distributions, one for each value of $P\{S_1 = S_0\}$, for which $E\{S_2 \mid \mathcal{F}_1\} = S_1$. Each generates a martingale measure, and each martingale measure leads to a different system of market values. A calculation shows that if $b = 2$, the option in the previous example has values between zero (when $P\{S_1 = S_0\} = 1$) and $13/27$ (when $P\{S_1 = S_0\} = 0$).

Notice that the market values of the stock and bond are the same for all martingale measures. Only the values of the derivatives change.

Problems 9.6

9.40. Consider the Cox-Ross-Rubinstein model on the probablity space $(\Omega, \mathcal{F}, P)$ with step size $b > 1$: $P\{S_{n+1} = bS_n\} = p$, where $0 < p < 1$. Let S_n be the stock price, and assume the interest rate is zero. Let Q be the martingale measure.

(a) Show that $(\log S_n)$ is a simple random walk under either P or Q.

(b) Show that if $P\{S_1 = bS_0\} \neq Q\{S_1 = bS_0\}$, that P and Q are *not* equivalent on $\mathcal{F}_\infty = \sigma\{S_n,\ n \geq 0\}$.

[Hint: Find an event which has probability zero under P, probability 1 under Q.]

9.41. Consider the Cox-Ross-Rubinstein model with step-size $b > 1$. Find the value of the derivative which pays \$100 if $S_1 > S_0$ and zero otherwise.

9.6. Financial Mathematics I: The Martingale Connection*

9.42. In the Cox-Ross-Rubinstein model with step-size $b > 1$ and $S_0 = 1$, it is found that a call option which pays $(S_2 - 1)^+$ at time 2 has value 0.25. Find b.

9.43. Let P be the true measure on $(\Omega, \mathcal{F})$, and let Q be an equivalent martingale measure. By Exercise 9.35, for each t there exists a positive $\mathcal{F}_t$-measurable random variable, ζ_t, called the *Radon-Nikodym derivative of P with respect to Q* such that for each $\Lambda \in \mathcal{F}_t$, $P\{\Lambda\} = \int_\Lambda \zeta_t \, dQ$. Show that $\zeta_t > 0$ a.s., that $\{\zeta_t, t \geq 0\}$ is a Q-martingale of mean 1, and that, if $0 < s < t$, $E^Q\{\zeta_t \mid F_s\} = \zeta_s^{-1} E^P\{X_t \zeta_t \mid \mathcal{F}_s\}$.

The following three problems extend some martingale theorems from discrete to continuous parameter submartingales. The first one extends the upcrossing inequality to a countable dense parameter set $D \subset [0, \infty)$.

9.44. Let $D_1 \subset D_2$ be finite subsets of D. Let $\{x_t, t \in D\}$ be a family of reals and let $a < b$.

(a) Define the number of upcrossings $\nu_{D_i}(a, b)$ of $[a, b]$ by $\{x_t, t \in D_i\}$.

(b) Show that $\nu_{D_1}(a, b) \leq \nu_{D_2}(a, b)$.

[Hint: Show that $\alpha_j^2 \leq \alpha_j^1$ and $\beta_j^2 \leq \beta_j^1$.]

(c) Define $\nu_D(a, b) = \sup\{\nu_{D_n}(a, b) : D_n \text{ finite}, D_n \subset D\}$. Let $\{X_t, \mathcal{F}_t, t \in D\}$ be a submartingale. Show that

$$(9.17) \qquad E\{\nu_D(a, b)\} \leq \sup_{t \in D} \frac{E\{(X_t - a)^+\}}{b - a}.$$

9.45. Let D be a countable dense subset of $[0, \infty)$. Let $\{X_t, \mathcal{F}_t \, t \in D\}$ be a submartingale. Show that with probability one, X_t has both left and right-hand limits along D at each $t > 0$, i.e., that for a.e. ω, $X_{t+} \stackrel{\text{def}}{=} \lim_{s \in D, s \downarrow t, s > t} X_s(\omega)$ and $X_{t-} \stackrel{\text{def}}{=} \lim_{s \in D, s \uparrow t, s < t} X_s(\omega)$ exist for all $t > 0$.

[Hint: If the right-hand limit fails to exist at some t, either X is unbounded there or $\limsup_{s \downarrow t, s > t} X_s(\omega) > \liminf_{s \downarrow t, s > t} X_s(\omega)$. Apply (9.17) to rule this out.]

9.46. Let D be a countable dense subset of $[0, \infty)$. Let $\{X_t, \mathcal{F}_t \, t \in D\}$ be a submartingale. Show that there exists a submartingale $\{\hat{X}_t, \hat{\mathcal{F}}_t, t \geq 0\}$ such that $t \mapsto \hat{X}_t$ is almost surely right-continuous and has left limits at each $t > 0$, $\hat{X}_t = X_{t+}$ for each $t \geq 0$ a.s. What are the σ-fields $\hat{\mathcal{F}}_t$? Show, moreover, that if the original submartingale is right-continuous in probability at each $t \in D$, that $\hat{X}_t = X_t$ a.s. for each $t \in D$.

9.47. Prove Doob's inequality: for $p > 1$, $E\{\max_{j \leq n} |X_j|^p\} \leq \left(\frac{p}{p-1}\right)^p E\{|X_n|^p\}$.

[Hint: First show that if $\xi, \zeta > 0$ satisfy $P\{\xi > a\} \leq \frac{1}{a} E\{\zeta; \xi > a\}$, then $\|\xi\|_p \leq \left(\frac{p}{p-1}\right)^p \|\zeta\|_p$.]

9.48. Let (X_n) be a submartingale with uniformly bounded increments. Show that $\lim_{n \to \infty} X_n$ exists a.s. on $\{\sup_n X_n < \infty\}$.

9.49. Let (ξ_n) be a sequence of i.i.d. Bernoulli (1/2) random variables. Let $X_1 = 1$, and, by induction, $X_n = 2\xi_n X_{n-1}$. Show that X_n is L^1-bounded and converges a.e., but does not converge in L^1.

9.50. Let $(\mathcal{F}_n)$ be a filtration and let $\{M_n, \mathcal{F}_n \ n = 0, 1, 2, \ldots\}$ be an adapted process. Show that M is a martingale if and only if for all bounded stopping times T, $E\{M_T\} = E\{M_0\}$.

Chapter 10

Brownian Motion

Brownian motion is a phenomenon which was observed soon after microscopes were invented. Under high magnification, microscopic particles wiggle around in a totally unpredictable way. While it was observed around 1700, the groundbreaking work on it was by the the botanist Robert Brown, who in 1827 observed pollen grains moving under the microscope with no apparent cause. He did an impressive series of experiments which showed, among other things, that it was not because the particles were alive, as had been thought before: small rock particles exhibited the same motion as pollen grains. He also observed that the motion never ceased, did not appear to have a tangent, and the smaller the particle, the greater the motion. A few years later, when Weierstrass' non-differentiable functions became known, it was remarked that Brownian motion seemed to behave much like one.

The phenomenon was finally explained by Einstein in 1905, who showed that it was caused by molecular motion: the grains of pollen are bombarded on all sides by the molecules of water, and the particles are so small that the imbalance of impacts on one side and the other produces a noticeable motion. In one of the small ironies of science, Einstein had not known of the phenomenon: he actually predicted it as a consequence of his kinetic theory!

The mathematical model goes back to 1900, when Louis Bachelier introduced it in his PhD thesis, primarily as a model for the stock market. Early treatments were light on rigor; Norbert Wiener gave a rigorous construction and proved numerous properties in the early 1920's. The process is often called the Wiener process. (Physicists make a distinction between physical and mathematical phenomena. They call the mathematical description

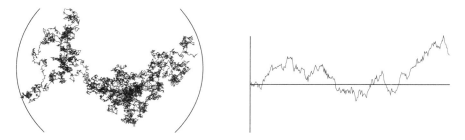

Figure 1. Left: Brownian motion as seen through a microscope. Right: graph of standard Brownian motion.

the Wiener process, and reserve Brownian motion for the physical motion. Probabilists just call both Brownian motion.)

Brownian motion is, in fact, an example of many different types of processes, some of which we have studied, some not. It is

- a Gaussian process;
- a process of independent increments;
- a Markov process;
- a stable process;
- a diffusion process (i.e., a strong Markov process with continuous sample paths);
- and a martingale.

10.1. Standard Brownian Motion

Brownian motion is three-dimensional, but the microscopist sees it in two dimensions. We will just look at its x-component. We would expect it to be continuous—it is the motion of a particle after all, and can't jump—and Gaussian, since it is caused by the sum total of many molecular shocks. The shocks should be symmetric, so it should be centered. And, because the physical environment does not change from second to second, its motion in disjoint time intervals should be similar, and essentially independent. This leads us to propose the following.

Definition 10.1. Standard Brownian motion is a stochastic process $\{B_t,\ t \geq 0\}$ which satisfies the following properties:

(i) $B_0 = 0$;

(ii) $B_{t+s} - B_s$ is $N(0, t)$;

(iii) If $0 \leq t_1 < t_2 < \cdots < t_n$, then the increments B_{t_1}, $B_{t_2} - B_{t_1}$, ..., $B_{t_n} - B_{t_{n-1}}$ are independent.

10.1. Standard Brownian Motion

Then B_t is Gaussian, $E\{B_t\} = 0$, and $E\{B_t^2\} = t$. If $s < t$,
$$E\{B_s B_t\} = E\{B_s^2\} + E\{B_s(B_t - B_s)\}$$
$$= s + E\{B_s\}E\{B_t - B_s\}$$
$$= s.$$

Thus $\{B_t,\ t \geq 0\}$ is a Gaussian process with mean zero and covariance function $\Gamma(s,t) \stackrel{def}{=} E\{B_s B_t\} = s \wedge t$. It is called "standard" because the mean and variance of B_1 are standardized to be zero and one, respectively. By (iii), it is also a process of *independent increments*.

Remark 10.2. The entire distribution of a Gaussian process is determined by its mean and covariance function. Indeed, by (6.7), the joint characteristic function, and therefore the distribution, of any finite number of jointly Gaussian random variables is determined by their means and covariance matrix, (see also Theorem 3.53) and the distribution of any process is determined by its finite-dimensional distributions.

It follows that any Gaussian process with mean zero and covariance function $\Gamma(s,t) = s \wedge t$ is a Brownian motion.

We will deal with the question of existence later. For the moment, just notice that we can define B_t at any *finite* set of times, say $0 = t_0 < t_1 < \cdots < t_n$, by letting $Z_1, Z_2, \ldots Z_n$ be i.i.d. $N(0,1)$ random variables, and setting $B_0 = 0$, $B_{t_k} = \sum_{j=1}^{k} \sqrt{t_j - t_{j-1}}\, Z_j$. This clearly satisfies (i), (ii) and (iii).

The joint distributions are easily calculated. For instance, $B_{t+s} - B_s$ is $N(0,t)$, so it has density
$$p_t(x) \stackrel{def}{=} \frac{1}{\sqrt{2\pi t}} e^{-\frac{x^2}{2t}}.$$

Now $B_s = x$ and $B_{s+t} = y \iff B_s = x$, and $B_{t+s} - B_s = y - x$. Thus, using the fact that B_s and $(B_{s+t} - B_s)$ are independent,
$$P\{B_s \in A, B_{s+t} \in B\} = \int_B \int_A p_s(x_1) p_t(x_2 - x_1)\, dx_1\, dx_2.$$

In general, if $0 < t_1 < \cdots < t_n$ and if $A_1, \ldots A_n$ are Borel sets in $\mathbb{R}$, then

(10.1) $\quad P\{B_{t_i} \in A_i, i = 1, \ldots, n\}$
$$= \int_{A_n} \cdots \int_{A_1} p_{t_1}(x_1) p_{t_2 - t_1}(x_2 - x_1) \ldots p_{t_n - t_{n-1}}(x_n - x_{n-1})\, dx_1 \ldots dx_n.$$

It follows from (10.1) that the joint density of $B_{t_1}, \ldots, B_{t_n}$ is

(10.2) $\quad p_{t_1, \ldots, t_n}(x_1, \ldots, x_n) \stackrel{def}{=} p_{t_1}(x_1) p_{t_2 - t_1}(x_2 - x_1) \ldots p_{t_n - t_{n-1}}(x_n - x_{n-1}).$

Transformations of Brownian Motion. There are a number of transformations of Brownian motion which map it into another, different Brownian motion. For instance, if $s \geq 0$ and $c \neq 0$, let

$$A_t \stackrel{\text{def}}{=} B_{t+s} - B_s \qquad \text{(Translation)},$$

$$C_t \stackrel{\text{def}}{=} cB_{t/c^2} \qquad \text{(Scaling)},$$

$$D_t \stackrel{\text{def}}{=} tB_{1/t} \qquad \text{(Time reversal)}.$$

Proposition 10.3. $\{A_t,\ t \geq 0\}$, $\{C_t,\ t \geq 0\}$, and $\{D_t,\ t \geq 0\}$ are standard Brownian motions.

Proof. All three are Gaussian processes, for they are linear transforms of $\{B_t,\ t \geq 0\}$. They all have mean zero, so we need only check that they have the right covariance function[1]. So let t_1, t_2 be positive.

$$\begin{aligned}
E\{A_{t_1} A_{t_2}\} &= E\{(B_{t_1+s} - B_s)(B_{t_2+s} - B_s)\} \\
&= E\{B_{t_1+s} B_{t_2+s}\} - E\{B_{t_1+s} B_s\} - E\{B_{t_2+s} B_s\} + E\{B_s^2\} \\
&= (t_1 \wedge t_2 + s) - s - s + s \\
&= t_1 \wedge t_2 \,.
\end{aligned}$$

$$\begin{aligned}
E\{C_{t_1} C_{t_2}\} &= E\{cB_{t_1/c^2}\, cB_{t_2/c^2}\} \\
&= c^2 \left(\frac{t_1}{c^2} \wedge \frac{t_2}{c^2} \right) \\
&= t_1 \wedge t_2 \,.
\end{aligned}$$

$$\begin{aligned}
E\{D_{t_1} D_{t_2}\} &= E\{t_1 B_{1/t_1} t_2 B_{1/t_2}\} \\
&= t_1 t_2 \left(\frac{1}{t_1} \wedge \frac{1}{t_2} \right) \\
&= \frac{t_1 t_2}{\max\{t_1, t_2\}} \\
&= t_1 \wedge t_2 \,.
\end{aligned}$$

Thus A, C, and D have the same mean and covariance function as B, and hence are standard Brownian motions. □

Remark 10.4. These simple transformations open up new territory. While the invariance of Brownian motion under translation is familiar—it just says that Brownian motion restarts itself at any time—the other two transformations suggest some novel aspects of the process.

The scaling transformation, for instance, explains why the study of Brownian motion abounds in random fractals[2]. Many classical fractals

---

[1] One can equally well just check the three defining properties.
[2] While Brownian motion has been known for a long time (see [30] for a fascinating account) most of the results in sections 1–9 are either due to, or were known by, Paul Lévy [26]. Lévy

are generated by self-similarity, and Brownian motion is similar to itself under scaling. Much of the seemingly-paradoxical behavior that we will study comes from the fact that, for the first time, we are dealing with genuine fractals. For instance, consider the zero set of Brownian motion: $Z = \{t : B_t = 0\}$. This is a random subset of the half-line. (Think about it!) Magnify it by a factor of $\alpha > 0$: $t \mapsto \alpha t$. Then the new set, αZ, is the zero set of the new Brownian motion $C_t = (1/\sqrt{\alpha})B_{\alpha t}$. So Z and αZ are similar sets. They are not identical, of course, but they have the same properties. We can examine the zero set under a microscope and it will look exactly the same as it does to the naked eye; and if we back off and look at it from farther away, it is still the zero set of a Brownian motion. Many other sets share this property. The reader might want to consider what happens to the graph of B_t when it is magnified, for instance.

Finally, the third transformation introduces a new idea. It says that we can reverse the direction of time in a Brownian motion, and still have a Brownian motion. It is always tempting to speculate about the consequences of going backward in time. With Brownian motion, at least, we know. One notable consequence is that any theorem about the behavior of Brownian motion as $t \to \infty$ can be transformed into another theorem about the behavior of Brownian motion as $t \to 0$.

Brownian Martingales. There are several martingales connected with Brownian motion—including Brownian motion itself—that we will need.

Proposition 10.5. *Let (B_t) be a standard Brownian motion and $\lambda \in \mathbb{R}$. Let $\mathcal{F}_t = \sigma\{B_s : s \leq t\}$. The following are martingales.*
  (i) $\{B_t, \mathcal{F}_t, t \geq 0\}$.
  (ii) $\{B_t^2 - t, \mathcal{F}_t, t \geq 0\}$.
  (iii) $\{e^{\lambda B_t - \frac{1}{2}\lambda^2 t}, \mathcal{F}_t, t \geq 0\}$.

Proof. Let $0 \leq s < t$. All three processes are adapted, so we need only check the conditional expectations. Since the increments are independent, and the increments in $[0, s]$ generate $\mathcal{F}_s$, $B_t - B_s$ is independent of $\mathcal{F}_s$. Thus
  (i) $E\{B_t - B_s \mid \mathcal{F}_s\} = E\{B_t - B_s\} = 0$.

---

certainly knew he was dealing with fractals, but, as the word was not yet invented, he could not say so. (It was his student, Benoît Mandelbrot, who coined the word.)

(*ii*) Note that $B_t^2 = B_s^2 + (B_t - B_s)^2 + 2B_s(B_t - B_s)$, so, as $B_t - B_s$ is independent of $\mathcal{F}_s$,

$$E\{B_t^2 - t \mid \mathcal{F}_s\} = B_s^2 + E\{(B_t - B_s)^2 \mid \mathcal{F}_s\} + 2B_s E\{B_t - B_s \mid \mathcal{F}_s\} - t + s$$

$$= B_s^2 + \underbrace{E\{(B_t - B_s)^2\}}_{t-s} + 2B_s \underbrace{E\{B_t - B_s\}}_{0} - t$$

$$= B_s^2 - s\,.$$

(*iii*) The moment-generating function of a $N(0, \sigma^2)$ random variable is $M(\theta) = e^{\theta^2 \sigma^2/2}$, so that, writing $B_t = B_s + (B_t - B_s)$,

$$E\{e^{\lambda B_t - \frac{1}{2}\lambda^2 t} \mid \mathcal{F}_s\} = e^{\lambda B_s - \frac{1}{2}\lambda^2 t} E\{e^{\lambda(B_t - B_s)} \mid \mathcal{F}_s\}$$

$$= e^{\lambda B_s - \frac{1}{2}\lambda^2 t} \underbrace{E\{e^{\lambda(B_t - B_s)}\}}_{e^{\lambda^2(t-s)/2}}$$

$$= e^{\lambda B_s - \frac{1}{2}\lambda^2 s}\,.$$

□

The Brownian Markov Property. Brownian motion has independent increments, so $A_t = B_{t+s} - B_s$ is not only a Brownian motion, it is independent of B_u for all $u \leq s$. Indeed, if $u_1 < \cdots < u_n \leq s$, $B_{t+s} - B_s$, $B_{u_1} - B_0, \ldots B_{u_n} - B_{u_{n-1}}$ are all independent, so $B_{t+s} - B_s$ is independent of $B_{u_1}, \ldots B_{u_n}$. It follows that $\{A_t,\ t \geq 0\}$ is a Brownian motion, independent of $\{B_u,\ u \leq s\}$. This implies it is a Markov process.

Standard Brownian motion starts from the origin. Brownian motion can also start from other points. This is simple: we get a Brownian motion from x by letting $B_t^x \stackrel{\text{def}}{=} x + B_t$. Then $B_0^x = x$. The density of B_t^x is $P\{B_t^x \in dy\} = p_t(y - x)\,dy$, and its joint density is immediate from (10.2): if $0 < t_1 < \cdots < t_n$, the joint density of $B_{t_1}^x, \ldots, B_{t_n}^x$ is

$$p_{t_1}(x_1 - x) p_{t_2 - t_1}(x_2 - x_1) \ldots p_{t_n - t_{n-1}}(x_n - x_{n-1})\,.$$

By Definition 3.38, the conditional densities of Brownian motion are immediate from the joint density, for the conditional density is the ratio of two densities of the form (10.2). After cancellation, we see that if $0 < s_1 < \cdots < s_n < s$, then

$$P\{B_t \in dx \mid B_{s_1} = y_1, \ldots, B_{s_n} = y_n, B_s = y\} = p_t(x - y)\,dx.$$

Following (3.22), this implies that

$$P\{B_{s+t} \in dx \mid B_{s_1}, \ldots, B_{s_n}, B_s\} = p_t(x - B_s)\,dx\,.$$

10.1. Standard Brownian Motion

This is true for any finite subset $\{s_1, \ldots, s_n\}$ of $[0, s]$, and it follows that if $\mathcal{F}_s \stackrel{\text{def}}{=} \sigma(B_u, u \leq s)$, then

$$P\{B_{s+t} \in dx \mid \mathcal{F}_s\} = P\{B_{s+t} \in dx \mid B_s\} = p_t(x - B_s)\, dx \quad \text{a.s.}$$

This depends only on B_s, which implies that B_{s+t} is conditionally independent of $\mathcal{F}_s$ given B_s. Thus Brownian motion is a Markov process, and the Brownian **transition density** is $p_t(y, x) \stackrel{\text{def}}{=} p_t(y - x)$.

We can do the same for the joint densities and the joint probabilities of the post-s process. We find that each n-tuple $(B_{s+t_1}, \ldots, B_{s+t_n})$ (and therefore the whole post-s process) is conditionally independent of $\mathcal{F}_s$ given B_s. Moreover, it has the same distribution as a Brownian motion starting from B_s. To summarize

Proposition 10.6. *Brownian motion is a Markov process, and for each $s > 0$ we have*

(i) $\{B_{s+t} - B_s,\ t \geq 0\}$ *is a Brownian motion, independent of $\mathcal{F}_s$;*

(ii) $\{B_{s+t},\ t \geq 0\}$ *is a Brownian motion from B_s. Given B_s, it is conditionally independent of $\mathcal{F}_s$.*

Remark 10.7. In fact, Brownian motion is a process of independent increments, and processes of independent increments are Markov processes. The calculations we just made identify its transition probabilities. We are being careful at this point because we will shortly be looking at the strong Markov property, which is deeper. Getting things straight now will make it easier later.

Continuity. Brownian motion represents the motion of a physical particle. Physical particles move continuously—they cannot jump instantaneously. Therefore Brownian motion should have continuous sample paths. It does[3], but it is delicate: continuity is by no means automatic, and we will have to construct a continuous version of the process by hand. The problem is this: a function is continuous if it is continuous at each point. But there are uncountably many points! The related notion of continuity in probability, however, is much simpler.

Definition 10.8. *(i)* A process $X = \{X_t, t \geq 0\}$ is **continuous in probability** if for all s, t and $\varepsilon > 0$,

$$\lim_{s \to t} P\{|X_t - X_s| > \varepsilon\} = 0\,.$$

(ii) X is **almost surely continuous** if

$$P\{\omega : t \mapsto X_t(\omega) \text{ is a continuous function}\} = 1\,.$$

---
[3] See Theorem 10.65 for a different proof.

For example, the Poisson process is not continuous, but it is continuous in probability, for
$$P\{X_s = X_t\} = e^{-\lambda|t-s|} \longrightarrow 1 \quad \text{as } s \to t.$$

Brownian motion is also continuous in probability, for
$$P\{|B_t - B_s| > \varepsilon\} \le \frac{1}{\varepsilon^2} E\{(B_t - B_s)^2\} = \frac{|t-s|}{\varepsilon^2} \longrightarrow 0 \quad \text{as } s \to t.$$

The relevant notion is almost-sure continuity. We watch the particle move. Does it jump around, or does it move continuously? The Brownian sample path is a real-valued function, $t \mapsto B_t(\omega)$. Is it continuous or not?

There are two kinds of discontinuities of a real-valued function $f(t)$ which concern us here. The first is a *jump discontinuity*, in which both the left- and right-hand limits of f exist at a point t, but are not equal. (This includes infinite jumps, such as $1/x$ at the origin.) The Poisson process, for example, has jump discontinuities.

The second kind is an *oscillatory discontinuity*, in which at least one of the two one-sided limits fails to exist. For example, $\sin(1/x)$ has an oscillatory discontinuity at the origin.

Notice that it does little good to prove the process is continuous at each fixed t. For if X_t is the Poisson process, it is certainly not continuous, but it is easy to see that for each $t > 0$, $P\{\lim_{s \to t} X_s = X_t\} = 1$.

The way around this dilemma is to reduce the problem to countable parameter sets, such as the rationals. Let us recall the theorem about submartingales on a countable parameter set proved in Exercise 9.45.

Theorem 10.9. *Let D be a countable dense subset of an open sub-interval $I \subset \mathbb{R}$. Let $X = \{X_t,\ t \in D\}$ be a submartingale. Then*

(i) X has no infinite or oscillatory discontinuities in I.

(ii) With probability one, both

(10.3) $$X_{t+} = \lim_{\substack{s \downarrow\downarrow t \\ s \in D}} X_s \quad \text{and} \quad X_{t-} = \lim_{\substack{s \uparrow\uparrow t \\ s \in D}} X_s$$

exist simultaneously at all $t \in I$.

Note: The limit exists at all $t \in I$, not just for $t \in D$.

Before applying this to Brownian motion, here is a useful estimate for the tails of the Gaussian distribution.

Lemma 10.10. *Let Y be a $N(0, \sigma^2)$ random variable. Then for $\lambda > 0$,*

(10.4) $$P\{|Y| \ge \lambda\} \le \sqrt{\frac{2}{\pi}} \frac{\sigma}{\lambda} e^{-\frac{\lambda^2}{2\sigma^2}}.$$

10.1. Standard Brownian Motion

Proof.
$$P\{|Y| \geq \lambda\} = P\left\{\left|\frac{Y}{\sigma}\right| \geq \frac{\lambda}{\sigma}\right\}$$
$$= 2\int_{\frac{\lambda}{\sigma}}^{\infty} \frac{1}{\sqrt{2\pi}} e^{-\frac{x^2}{2}} dx$$
$$\leq \sqrt{\frac{2}{\pi}} \int_{\frac{\lambda}{\sigma}}^{\infty} e^{-\frac{x^2}{2}} \frac{\sigma x}{\lambda} dx.$$

Change variables: $u = x^2/2$, then integrate to get (10.4). □

Theorem 10.11. *There exists a continuous version of Brownian motion.*

Remark 10.12. The word "version" is a tip-off that we will have to reconstruct the process. We say $(\hat{B}_t)$ is a **version** of (B_t) if for each fixed t, $P\{\hat{B}_t = B_t\} = 1$. Any two versions of a process have the same finite-dimensional distributions. However, there are uncountably many t, so this by no means implies that B and $\hat{B}$ are the same for *all* t. We will have to construct $\hat{B}_t$ carefully to be sure that it is continuous.

Proof. Let us restrict B_t to *rational* values of t. $\{B_t, t \in \mathbb{Q}_+\}$ is a martingale, so by Theorem 10.9, for a.e. ω, $t \mapsto B_t(\omega)$, $t \in \mathbb{Q}_+$ has both left and right limits at all points. Define

$$\hat{B}_t(\omega) = \begin{cases} \lim_{s \downarrow t,\, s \in \mathbb{Q}} & \text{if the limit exists,} \\ 0 & \text{if not.} \end{cases}$$

For a.e. ω the limit exists at all t, not just at rational t, and it follows that $t \mapsto B_t(\omega)$ is a right-continuous function on $[0, \infty)$, and that it also has left limits at each t.

Moreover,
$$E\{(B_t - \hat{B}_t)^2\} = E\{\lim_{s \downarrow t, s \in \mathbb{Q}} (B_t - B_s)^2\}$$
$$= \lim_{s \downarrow t, s \in \mathbb{Q}} \underbrace{E\{(B_t - B_s)^2\}}_{|t-s|} = 0.$$

Thus $\hat{B}_t = B_t$ a.s. for each t, so $(\hat{B}_t)$ is a version of (B_t). It remains to show that it is continuous, not just right-continuous.

Note that $\hat{B}_t$ can only have jump discontinuities, since both left and right limits exist at all points. If there is some t for which $|\hat{B}_t - \hat{B}_{t-}| > \varepsilon > 0$ for some $t \leq 1$, say, then for large enough n there exists $k \leq n$ such that

$|\hat{B}_{k/n} - \hat{B}_{(k-1)/n}| \leq \varepsilon$. Thus,

$$P\{\exists 0 < t \leq 1 : |\hat{B}_t - \hat{B}_{t-}| > \varepsilon\} = \lim_{n \to \infty} P\left\{\exists k \leq n : \left|\hat{B}_{\frac{k}{n}} - \hat{B}_{\frac{k-1}{n}}\right|\right\}$$

$$\leq \lim_{n \to \infty} \sum_{k=0}^{n} P\left\{\left|\hat{B}_{\frac{k}{n}} - \hat{B}_{\frac{k-1}{n}}\right| > \varepsilon\right\}.$$

But $\hat{B}_{k/n} - \hat{B}_{(k-1)/n}$ is $N(0, 1/n)$, so by the lemma, this is

$$\leq \lim_{n \to \infty} n \frac{1}{\varepsilon \sqrt{n}} \sqrt{\frac{2}{\pi}} e^{-\frac{n\varepsilon^2}{2}}$$
$$= 0.$$

Thus, with probability one, there are no jumps of size greater than ε in $(0, 1]$. This is true for a sequence of ε tending to zero, so there are no jumps at all in $(0, 1]$. The same goes for each interval $(k, k+1]$, so there are no jumps at all. □

Remark 10.13. From now on, we will always assume that Brownian motion is continuous.

10.2. Stopping Times and the Strong Markov Property

Brownian motion has the restarting property: it restarts at any t: $\{B_{s+t} - B_t, t \geq 0\}$ is a Brownian motion, independent of $\{B_s, s \leq t\}$. The strong Markov property says that it can restart from any stopping time T. We have seen that the strong Markov property holds for Markov chains and for the Poisson process. The property for Brownian motion is deeper, and the question of stopping times is more delicate, but the final conclusion—the strong Markov property—is the same.

There is a deep connection between the strong Markov property and the right-continuity, not only of the process, but also of the information. Information is coded into the σ-fields, and we can speak of a right-continuous filtration of σ-fields: $(\mathcal{F}_t)$ is said to be **right-continuous** if $\mathcal{F}_t = \bigcap_{s > t} \mathcal{F}_s$. This brings up a problem we did not have to face in the discrete parameter case. The natural filtration $\{\mathcal{F}_t, t \geq 0\}$ defined by $\mathcal{F}_t = \sigma(B_s, s \leq t)$ is not right-continuous, but the filtration $\mathcal{F}_{t+} \stackrel{\text{def}}{=} \bigcap_{\varepsilon > 0} \mathcal{F}_{t+\varepsilon}$ *is*. We are given the natural filtration, but it turns out that we actually need the right-continuous filtration. For instance, Brownian first-hitting times are stopping times relative to the right-continuous filtration, but not always with respect to the natural filtration. On the other hand, we only know the (ordinary) Markov property for the natural filtration. We do not want to brush this difference under the table... well, not yet... for there is a powerful zero-one law (Blumenthal's zero-one law) which depends on the difference between

10.2. Stopping Times and the Strong Markov Property

$\mathcal{F}_t$ and $\mathcal{F}_{t+}$. This means that we will have to keep close track—but luckily, only for a short time—of the particular filtration we use.

Recall the definition of a stopping time relative to a filtration $(\mathcal{F}_t)$.

Definition 10.14. Let $\{\mathcal{F}_t,\ t \geq 0\}$ be a filtration. A random variable with values in $\mathbb{R}^+ \cup \{\infty\}$ is a **stopping time** if, for all $t \geq 0$, $\{T \leq t\} \in \mathcal{F}_t$. The **past before T** is the σ-field

$$\mathcal{F}_T \stackrel{def}{=} \{\Lambda \in \mathcal{F} : \Lambda \cap \{T \leq t\} \in \mathcal{F}_t,\ \forall t\}.$$

Let (F_t) be a filtration and let T be a stopping time relative to either $(\mathcal{F}_t)$ or $(\mathcal{F}_{t+})$. Let T_n be the discrete **dyadic approximation of T**:

$$T_n \stackrel{def}{=} \begin{cases} \frac{k+1}{2^n} & \text{if } \frac{k}{2^n} \leq T < \frac{k+1}{2^n} \\ \infty & \text{if } T = \infty. \end{cases}$$

The reader can verify that in either case, T_n is a stopping time relative to the filtration $(\mathcal{F}_t)$.

Definition 10.15. Let T be a stopping time. Then $\mathcal{F}_{T+} \stackrel{def}{=} \bigcap_{\varepsilon > 0} \mathcal{F}_{T_n}$.

Exercise 10.1. Show the following.

(a) A random variable T with values in the positive extended reals is an $(\mathcal{F}_{t+})$-stopping time iff $\{T < t\} \in \mathcal{F}_t$ for all $t > 0$.

(b) An $(\mathcal{F}_t)$-stopping time is also an $(\mathcal{F}_{t+})$-stopping time. The converse is not true in general.

(c) If T is a $(\mathcal{F}_t)$-stopping time, then $\mathcal{F}_{T+} = \{\Lambda : \Lambda \cap \{T < t\} \in \mathcal{F}_t, \forall\, t\}$. If T is an $(\mathcal{F}_{t+})$-stopping time, $\mathcal{F}_T = \mathcal{F}_{T+}$.

Exercise 10.2. Show that if T is a stopping time relative to either of the filtrations $(\mathcal{F}_t)$ or $(\mathcal{F}_{t+})$, then the dyadic approximations T_n are all discrete-valued stopping times relative to $(\mathcal{F}_t)$, $T < T_n$ on $\{T < \infty\}$, and $T_n \downarrow T$.

Most of the properties of discrete-parameter stopping times carry over to the continuous case. There are some subtleties, though. The reader can check the following.

Exercise 10.3. Let S and T be stopping times, and let (S_n) be a sequence of stopping times. Show the following.

(a) If $S \leq T$, then $\mathcal{F}_S \subset \mathcal{F}_T$.

(b) $S \wedge T$ and $S \vee T$ are stopping times.

(c) If $S_n \leq S_{n+1}$ for all n, $S \stackrel{def}{=} \lim_n S_n$ is a stopping time.

(d) If $S_n \geq S_{n+1}$ for all n, $S \stackrel{def}{=} \lim_n S_n$ is an $(\mathcal{F}_{t+})$-stopping time.
[Hint: Show $\{S < t\} \in \mathcal{F}_t$.]

Proposition 10.16. *Let T be an $(\mathcal{F}_{t+})$-stopping time and let (T_n) be its dyadic approximation. Then, on $\{T < \infty\}$:*

(i) T_n and B_{T_n} are $\mathcal{F}_{T_n}$-measurable;

(ii) T and B_T are $\mathcal{F}_{T+}$-measurable.

Proof. The reason for the restriction to $\{T < \infty\}$ is that neither B_T nor B_{T_n} are defined on $\{T = \infty\}$. Let us just assume T is finite.

(i) T_n can have values $s_k \stackrel{\text{def}}{=} k/2^n$, and $\{T_n = s_k\} \in \mathcal{F}_{s_k}$. For any t, $\{T_n \leq t\} = \bigcup_{k:s_k \leq t}\{T = s_k\} \in \mathcal{F}_t$. Similarly, $\{B_{T_n} \leq x\} = \bigcup_k \{T_n = s_k;\, B_{s_k} \leq x\}$, so $\{B_{T_n} \leq x\} \cap \{T_n \leq t\} = \bigcup_{k:s_k \leq t}\{T_n = s_k;\, B_{s_k} \leq x\} \in \mathcal{F}_t$.

(ii) Let $m < n$. Then $T_n \leq T_m$, so that by (i), both T_n and B_{T_n} are measurable with respect to $\mathcal{F}_{T_n} \subset \mathcal{F}_{T_m}$. Now $T_n \downarrow T$, and $t \mapsto B_t$ is continuous, so that $B_T = \lim_n B_{T_n}$. Thus T and B_T are measurable with respect to $\mathcal{F}_{T_m}$ for all m, and they are therefore measurable with respect to $\bigcap_m \mathcal{F}_{T_m} = \mathcal{F}_{T+}$. $\square$

This brings us to the strong Markov property.

Theorem 10.17. *Let T be a finite stopping time with respect to either filtration $(\mathcal{F}_t)$ or $(\mathcal{F}_{t+})$. Let $\hat{B}_t = B_{T+t} - B_T$. Then $\{\hat{B}_t,\, t \geq 0\}$ is a standard Brownian motion, independent of $\mathcal{F}_{T+}$.*

Proof. It is enough to show that if $f_1, \ldots, f_m$ are bounded and continuous functions, if $t_1 < t_2 < \cdots < t_m$, and if $\Lambda \in \mathcal{F}_{T+}$, then

(10.5) $\qquad E\{f_1(\hat{B}_{t_1}) \cdots f_m(\hat{B}_{t_m});\, \Lambda\} = P\{\Lambda\} E\{f_1(B_{t_1}) \ldots f_m(B_{t_m})\}.$

Indeed, if we take $\Lambda = \Omega$, this tells us that $\hat{B}_{t_1}, \ldots, \hat{B}_{t_m}$ has the same distribution as $B_{t_1}, \ldots, B_{t_m}$. (These determine the distribution. See e.g. Exercise 5.14.) Then, letting Λ be an arbitrary set in $\mathcal{F}_{T+}$, it shows that $\hat{B}$ is independent of $\mathcal{F}_{T+}$.

The plan is to first approximate T by the stopping times T_n. These are discrete-valued stopping times with respect to the original filtration $(\mathcal{F}_t)$, so we can use the ordinary Markov property on each set $\{T_n = s_k\}$ to show that the independence and conditional independence both hold. Then we take the limit as $T_n \downarrow T$ to extend the result to T.

Note that $\hat{B}_t = \lim_{n \to \infty}(B_{T_n+t} - B_{T_n})$. For a fixed n, let $s_k = k 2^{-n}$.

$$E\left\{\prod_{j=1}^m f_j(\hat{B}_{t_j})\,;\, \Lambda\right\} = \lim_{n \to \infty} E\left\{\prod_{j=1}^m f_j(B_{T_n+t_j})\,;\, \Lambda\right\}$$

$$= \lim_{n \to \infty} \sum_{k=1}^{\infty} E\left\{\Lambda \cap \{T \in (s_{k-1}, s_k]\}\,;\, \prod_{j=1}^m f_j(B_{s_k+t_j} - B_{s_k})\right\}.$$

But the set inside the expectation is $\Lambda \cap \{T < s_k\} - \Lambda \cap \{T < s_{k-1}\} \in \mathcal{F}_{s_k}$ since T is an $(\mathcal{F}_{t+})$-stopping time. By the ordinary Markov property applied at s_k, this is

$$= \lim_{n\to\infty} \sum_{k=1}^{\infty} P\{\Lambda \cap \{T \in (s_{k-1}, s_k]\}\} E\left\{\prod_{j=1}^{m} f_j(B_{s_k+t_j} - B_{s_k})\right\}$$

$$= P\{\Lambda\} E\left\{\prod_{j=1}^{m} f_j(B_{t_j})\right\},$$

which verifies (10.5). □

Notice that the post-T process $\{B_{T+t},\ t \geq 0\}$ is a Brownian motion from B_T which is conditionally independent of $\mathcal{F}_{T+}$ given B_T. This follows by writing $B_{T+t} = B_T + (B_{T+t} - B_T) = B_T + \hat{B}_t$, and noting that $B_T \in \mathcal{F}_{T+}$ while $\hat{B}_t$ is independent of $\mathcal{F}_{T+}$ by the theorem.

Two Zero-One Laws. We can get some mileage from the fact that the σ-field in Theorem 10.17 is $\mathcal{F}_{T+}$, not just $\mathcal{F}_T$.

Corollary 10.18 (Blumenthal Zero-One Law). *The σ-field $\mathcal{F}_{0+} = \bigcap_{\varepsilon > 0} \mathcal{F}_{\varepsilon}$ is trivial.*

Proof. Warning: This is yet another proof that revolves in small logical circles.

Let $\Lambda \in \mathcal{F}_{0+}$. By the strong Markov property applied to the stopping time $T \equiv 0$,
$$I_\Lambda = P\{\Lambda \mid \mathcal{F}_{0+}\} = P\{\Lambda \mid B_0\} \text{ a.e.}$$
But $B_0 \equiv 0$, so the right-hand side is constant and equals $P\{\Lambda\}$. The left-hand side is an indicator function and takes on only two values, zero and one, so $P\{\Lambda\}$ must be one of the two. □

Now recall the Brownian motion $D_t = tB_{1/t}$. Let $\mathcal{F}_t^D$ be its natural filtration. Notice that $\mathcal{F}_t^D = \mathcal{F}_{1/t}^*$, where $\mathcal{F}_t^* = \sigma(B_s,\ s \geq t)$. By Corollary 10.18, $\mathcal{F}_{0+}^D = \bigcap_{t>0} \mathcal{F}_t^D$ is trivial; but $\mathcal{F}_{0+}^D = \bigcap_{t>0} \mathcal{F}_t^* = \mathcal{F}_\infty^*$, the tail field of Brownian motion. Thus we have

Corollary 10.19. *The tail field of Brownian motion is trivial.*

Remark 10.20. (*i*) The Blumenthal zero-one law looks remarkably like the Borel zero-one law, but it does not follow from it, or even from a modification of its proof. In fact, this theorem is not about independent random variables per se, or even independent increments. It is about the strong Markov property, and it holds for essentially all continuous-parameter strong Markov processes. Brownian motion is just an interesting special case.

(ii) Notice that while $\mathcal{F}_{0+}$ may be trivial in the sense that its sets have probability zero or one, it is nonetheless rich. In fact, it is the same as the tail field for the reversed Brownian motion C_t, which contains all the information about the limiting behavior of the Brownian paths as $t \to \infty$. So $\mathcal{F}_{0+}$ describes the limiting behavior of the Brownian paths as $t \to 0$.

10.3. The Zero Set of Brownian Motion

Let $A \subset \mathbb{R}$ and define the **first-hitting time of A** by

$$T_A \stackrel{def}{=} \begin{cases} \inf\{t > 0 : B_t \in A\}, \\ \infty & \text{if the above set is empty}. \end{cases}$$

Note the "> 0" in the definition: only strictly positive times count. T_A is defined for all $\omega \in \Omega$, though it may be infinite.

Hitting times are stopping times... with a nuance, which we will discuss later. For the moment, let us just accept the fact that the strong Markov property holds at hitting times. We need one fact:

$1°$ If A is open, then $\{T_A \leq t\} \in \mathcal{F}_{t+}$ for all $t \geq 0$.

Indeed, $t \mapsto B_t$ is continuous so the set of s for which B_s is in A is open. If it is non-empty, it contains rationals. Thus $T_A < t$ iff there are rational r in $(0, t)$ for which $B_r \in A$. In other words, $\{T_A < t\} = \bigcup_{\substack{r<t \\ r \in \mathbb{Q}}} \{B_r \in A\} \in \mathcal{F}_t$. Then $\{T_A \leq t\} = \bigcap_{\varepsilon > 0}\{T_A < t + \varepsilon\}$. This is in $\mathcal{F}_{t+\varepsilon}$ for all $\varepsilon > 0$, and therefore in $\mathcal{F}_{t+}$.

Consider the three hitting times $T_{\mathbb{R}-\{0\}}$, $T_{(0,\infty)}$ and $T_{(-\infty,0)}$. $T_{\mathbb{R}-\{0\}}$ is the first time B_t leaves the origin.

Notice that for any $t > 0$, $P\{B_t \neq 0\} = 1$, so that $T_{\mathbb{R}-\{0\}} < t$ a.s. Therefore $T_{\mathbb{R}-\{0\}} = 0$ a.s. But $T_{\mathbb{R}-\{0\}} = T_{(0,\infty)} \wedge T_{(-\infty,0)}$ so, for a.e. ω, one of the latter two times must be zero. By symmetry, $P\{T_{(0,\infty)} = 0\} = P\{T_{(-\infty,0)} = 0\}$. This probability must be at least one-half, but $\{T_{(0,\infty)} = 0\} \in \mathcal{F}_{0+}$ by $1°$. By the zero-one law, its probability can be only zero or one... and it is not zero. Therefore $P\{T_{(0,\infty)} = 0\} = P\{T_{(-\infty,0)} = 0\} = 1$.

Now $B_0 = 0$, so in order that $T_{(0,\infty)} = 0$, there must be a sequence of points $t_n > 0$, $t_n \downarrow 0$ for which $B_{t_n} > 0$. And if $T_{(-\infty,0)} = 0$ there must be $s_n > 0$, $s_n \downarrow 0$ for which $B_{s_n} < 0$. But the path is continuous, so that between an s_n and a t_m, the path has to cross zero. Therefore there must be a sequence of points $u_n > 0$, $u_n \downarrow 0$ for which $B_{u_n} = 0$. In particular:

Proposition 10.21. *With probability one, (B_t) changes sign infinitely often in any neighborhood of the origin, and 0 is a limit point of the set $\{t > 0 : B_t = 0\}$.*

10.3. The Zero Set of Brownian Motion

This means that the path is bumpy: it goes from positive to negative and back infinitely often in any interval $(0, \varepsilon)$.

Definition 10.22. The **zero set of Brownian motion** is $Z = \{t \geq 0 : B_t = 0\}$.

This is a random set, since it depends on the path: $Z(\omega) = \{t \geq 0 : B_t(\omega) = 0\}$. By the proposition, 0 is a.s an accumulation point of Z.

Consider a typical path. The set of $\{t : B_t > 0\}$ is open, so it is a union of open intervals on which $B_t > 0$. Similarly, the set $\{t : B_t < 0\}$ is also a union of open intervals on which $B_t < 0$. The end points of all these intervals are in Z, so Z is infinite. But it turns out that there are many other points in Z as well.

Theorem 10.23. *Z is a.s. non-empty, nowhere-dense, perfect (i.e., closed with no isolated points) and hence uncountable. Moreover, Z has Lebesgue measure zero.*

Proof. According to a theorem in topology, a non-empty perfect set is uncountable. Therefore we need only prove that Z is a.s. perfect.

Z is non-empty, since it contains the origin. It is closed since $t \mapsto B_t$ is continuous. We must show that it has no isolated points, i.e., no points $z \in Z$ such that z is the only point of Z in some open interval (a, b).

Suppose $B_a \neq 0$ for some a. Let T be the first time after a that B_t hits 0. Then $T \in Z$, and by the strong Markov property, $B_{T+t} - B_T = B_{T+t}$ is a standard Brownian motion. By Proposition 10.21, T must be an accumulation point of Z, for with probability one, B_{T+t} changes sign infinitely often in any neighborhood of T. Thus T is not an isolated point of Z. This is true a.s. simultaneously for all rational $a \geq 0$: the first element of Z after a is not isolated. But any isolated point must be the first point of Z after *some* rational. Therefore there are none.

Finally, let $L(Z)$ be the Lebesgue measure of Z. Then

$$E\{L(Z)\} = E\left\{\int_0^\infty I_{\{B_t=0\}}\, dt\right\} = \int_0^\infty E\{I_{\{B_t=0\}}\, dt$$

$$= \int_0^\infty \underbrace{P\{B_t = 0\}}_{0}\, dt = 0$$

by Fubini's theorem. □

Remark 10.24. The zero set of Brownian motion is essentially a random Cantor set. The Cantor set in $[0, 1]$ is what is left over after successively removing open middle thirds. Similarly, Z is what is left over after successively removing the open intervals in which $B_t > 0$, and the open intervals in which $B_t < 0$. The Cantor set is a closed, nowhere-dense, totally disconnected, uncountable set of zero Lebesgue measure. So is the zero set of Brownian motion.

Hitting Times⋆. First-hitting times are stopping times. This is a deep result in higher dimensions, but it is relatively straightforward for one-dimensional Brownian motion, which is what concerns us here. Even in one dimension, though, there is a nuance: they are stopping times if the underlying filtration is right-continuous. So it may be necessary to use the right-continuous filtration $(\mathcal{F}_{t+})$. As it happens, this changes nothing: the strong Markov property holds for $(\mathcal{F}_{t+})$-stopping times too, so we can simply replace the filtration $(\mathcal{F}_t)$ by $(\mathcal{F}_{t+})$.

Let $T_A = \inf\{t > 0 : B_t \in A\}$ be the first-hitting time of A. Let $\bar{A}$ denote the closure of A.

The following facts are easily verified.

1° If $A \subset B$, then $T_B \leq T_A$.

2° If $A_1 \subset A_2 \subset \ldots$, and $A = \bigcup_n A_n$, then $T_A = \lim_n T_{A_n}$.

Theorem 10.25. (i) For any $A \subset \mathbb{R}$, $B_{T_A} \in \bar{A}$ a.s. on $\{T_A < \infty\}$.

(ii) If A is open, T_A is a stopping time with respect to $(\mathcal{F}_{t+})$.

(iii) If A is compact, T_A is a stopping time with respect to $(\mathcal{F}_t)$.

(iv) For any $A \subset \mathbb{R}$, $T_A = T_{\bar{A}}$ a.s.

Remark 10.26. While (i)–(iii) hold quite generally—they only depend on right-continuity—(iv) also depends on the order properties of the line, and is spectacularly false for Brownian motion in dimensions other than one.

Proof. (i) From the definition, if $T_A(\omega) < \infty$, either $B_{T_A}(\omega) \in A$, or there exist $t_n \downarrow T_A(\omega)$ for which $B_{t_n}(\omega) \in A$. But Brownian motion is continuous, so $B_{t_n}(\omega) \to B_{T_A}(\omega)$, which implies that $B_{T_A}(\omega)$, being a limit point of points of A, is in $\bar{A}$.

(ii) Proved above.

(iii) If A is compact, then there exists a decreasing sequence of open neighborhoods G_n of A such that $G_n \supset \bar{G}_{n+1} \supset G_{n+1}$, and $A = \bigcap_n \bar{G}_n$. Then $T_{G_n} \leq T_{G_{n+1}} \leq \cdots \leq T_A$. Let $\tau \stackrel{\text{def}}{=} \lim_n T_{G_n} \leq T_A$. In fact, $\tau = T_A$, since if τ is finite, then $B_\tau = \lim_n B_{T_{G_n}}$, and, as $B_{T_{G_n}} \in \bar{G}_n$ by (i), $B_\tau \in \bigcap_n \bar{G}_n = A$. Therefore $T_A \leq \tau$, i.e. $\tau = T_A$. But now, $\{T_A \leq t\} = \bigcap_n \{T_{G_n} < t\} \cup \{B_t \in A\}$, which is in $\mathcal{F}_t$.

(iv) Since $A \subset \bar{A}$, certainly $T_{\bar{A}} \leq T_A$, with equality if $T_{\bar{A}} = \infty$. If $T_{\bar{A}} < \infty$, then $B_{T_{\bar{A}}}$ is either in A or is an accumulation point of A. In the first case, $T_{\bar{A}} = T_A$. In the second case, by the strong Markov property, $B_{T_{\bar{A}}+t} - B_{T_{\bar{A}}}$

is a Brownian motion, and therefore it almost surely enters both $(0,\infty)$ and $(-\infty,0)$ immediately, so for any $\varepsilon > 0$, $\{B_{T_{\bar A}+t},\ 0 \leq t < \varepsilon\}$ enters both $(B_{T_{\bar A}},\infty)$ and $(-\infty, B_{T_{\bar A}})$. The paths are continuous, so for a.e. ω, the path $\{B_{T_{\bar A}+t}(\omega),\ 0 \leq t \leq \varepsilon\}$ sweeps out some interval $(B_{T_{\bar A}}(\omega) - \delta, B_{T_{\bar A}}(\omega) + \delta)$. But this interval contains infinitely many points of A. This holds for any $\varepsilon > 0$, so $T_A = T_{\bar A}$. $\square$

Remark 10.27. From this point on, we will not worry about the difference between the filtrations $(\mathcal{F}_t)$ and $(\mathcal{F}_{t+})$. The strong Markov property holds for both. That is enough.

10.4. The Reflection Principle

It is often difficult to handle the maximum of a process, and any theorems which help control it can be useful. The martingale maximal inequality, for instance, gives us an estimate on the maximum of a submartingale. One of the striking things about Brownian motion is that we can get the exact distribution of its maximum. This is called the *reflection principle* of de la Vallée Poussin.

Theorem 10.28 (Reflection Principle). *Let $x > 0$. Then*

(10.6) $$P\{\max_{0 \leq s \leq t} B_s \geq x\} = 2P\{B_t \geq x\}.$$

Proof. Note that $\max_{s \leq t} B_s \geq x \iff T_x \leq t$. So we must show that $P\{T_x \leq t\} = 2P\{B_t \geq x\}$. But

$$P\{B_t \geq x\} = P\{B_t \geq x \mid T_x \leq t\} P\{T_x \leq t\}.$$

By the strong Markov property, $\{B_{T_x+s} - x,\ s \geq 0\}$ is a Brownian motion, independent of $\mathcal{F}_{T_x}$, and therefore of $t - T_x$. Thus it has equal probability of being positive or negative at time $s = t - T_x$, so $P\{B_t \geq x \mid T_x \leq t\} = 1/2$, and the above equals $\frac{1}{2}P\{T_x \leq t\}$. $\square$

Remark 10.29. The intuition behind the reflection principle is one of symmetry: Brownian motion is symmetric. The probability of any given path is the same as the probability of its reflection. In particular, once the process reaches x, it is symmetric about x from then on: the path and its reflection about x have the same probabilities. It is it is exactly as likely to end up at $x - y$ as at $x + y$. That intuition is correct, but, as the time it first reaches x is random, we have to use the strong Markov property to justify it.

Problems 10.4

In the following, (B_t) is a standard Brownian motion, unless explicitly stated otherwise.

10.4. Let $0 < s < t$. What is the distribution of $B_s + B_t$? Show that the conditional distribution of B_s given that $B_t = b$ is $N(\frac{s}{t}b, \frac{s}{t}(t-s))$.

10.5. Let (B_t) be a standard Brownian motion. Show that
$$E\{B_t \mid B_s\} = \begin{cases} B_s & \text{if } s \leq t, \\ \frac{t}{s}B_s & \text{if } s > t. \end{cases}$$

10.6. Find $E\{B_1 B_2 B_3\}$.

10.7. Show that $X_t = B_1 - B_{1-t}$, $0 \leq t \leq 1$ is a Brownian motion.

10.8. Let $\{X_t, t \geq 0\}$ be a uniformly integrable submartingale whose sample paths are right continuous. Extend the system theorem (Theorem 9.35) to apply to (X_t).
[Hint: Use the dyadic approximation. If it helps, assume the paths are continuous. The odds are that when you have finished, you will be able to simply change every occurrence of "continuous" to "right-continuous".]

10.9. Show that the Brownian density $p_t(x) = \frac{1}{\sqrt{2\pi t}} e^{-\frac{x^2}{2t}}$ satisfies the heat equation, $\frac{\partial p}{\partial t} = \frac{1}{2} \frac{\partial^2 p}{\partial x^2}$.

10.10. Fill in the details in the proof of Proposition 10.6.

10.11. Let B_t and W_t be independent standard Brownian motions. Prove that the product $\{B_t W_t, t \geq 0\}$ is a martingale.

10.12. Show that the natural filtration $(\mathcal{F}_t)$ of (B_t) is $\mathcal{F}_t = \sigma(B_r, r \in \mathbb{Q}, r < t)$.

10.13. Let $a < b < c$. Show that $\sigma\{B_t : a < t < b\}$ and $\sigma\{B_s - B_b : b < s < c\}$ are independent.

10.14. Show that with probability one, $B_t/t \longrightarrow 0$ as $t \to \infty$.

10.15. Let $W = \int_0^t B_s \, ds$. Show that $E\{W\} = 0$ and $E\{W^2\} = t^3/3$.

10.16. Let $Z_t = \int_0^t B_s^2 \, ds$. Find $E\{Z_s Z_t\}$ for $s, t \geq 0$.

10.17. Find the probability that (B_t) hits plus one and then hits minus one before time one.

10.5. Recurrence and Hitting Properties

Theorem 10.30. *(i) For all x, $P\{T_x < \infty\} = 1$.*
(ii) With probability one, $\limsup_{t \to \infty} B_t = \infty$ and $\liminf_{t \to \infty} B_t = -\infty$.
(iii) For all $t_0 \geq 0$ and $x \in \mathbb{R}$, $P\{B_t \text{ hits } x \text{ after } t_0\} = 1$.

Proof. (i) $P\{T_x > t\} = 1 - P\{T_x \leq t\} = 1 - 2P\{B_t > x\}$ by the reflection principle. But as $t \to \infty$, $P\{0 < B_t < x\} \to 0$, so that $P\{B_t > x\} \to 1/2$. Thus as $t \to \infty$, $P\{T_x > t\} \longrightarrow 0$.

Both (ii) and (iii) follow easily. By (i), $T_n < \infty$ for $n = 1, 2, \ldots$. The paths are continuous, so $T_n \to \infty$ and $B_{T_n} = n$. Thus $\limsup_{n \to \infty} B_t \geq \lim_n B_{T_n} = \lim_n n = \infty$. By symmetry, $\liminf_{t \to \infty} B_t = -\infty$, proving (ii).

10.5. Recurrence and Hitting Properties

Then (iii) follows by observing that between T_n and T_{-n}, the Brownian path must hit every point between $-n$ and n, which includes x as soon as $n > |x|$. Since both T_n and T_{-n} tend to infinity, the set $\{t : B_t = x\}$ must be unbounded. $\square$

First Exits From Intervals. Let B_t^x be a Brownian motion from x: $B_0^x = x$. Let $a < b$ and put $\tau_{ab} = \inf\{t > 0 : B_t^x \leq a \text{ or } B_t^x \geq b\}$. If $x \in (a, b)$, then $\tau_{ab} = T_a \wedge T_b < \infty$. Let us find the distribution of $B_{\tau_{ab}}^x$.

Note that (B_t^x) is a martingale, so that $(B_{t \wedge \tau_{ab}}^x)$ is a bounded martingale, and hence, by the system theorems, $(B_0^x, B_{\tau_{ab}}^x)$ is also a martingale. Therefore

$$x = B_0^x = E\{B_{\tau_{ab}}^x\} = bP\{B_{T_{ab}}^x = b\} + aP\{B_{T_{ab}}^x = a\}.$$

Let $p = P\{B_{\tau_{ab}}^x = b\}$. Then $x = bp + a(1 - p)$, giving us:

Proposition 10.31. *Let B^x be a Brownian motion from x. If $a < x < b$, then*

$$(10.7) \qquad P\{B_{\tau_{ab}}^x = b\} = \frac{x - a}{b - a} \qquad P\{B_{\tau_{ab}}^x = a\} = \frac{b - x}{b - a}.$$

How long does it take to leave (x, b)? We can find $E\{\tau_{ab}\}$ by using the fact that $\{(B_t^x)^2 - t, \mathcal{F}_t, t \geq 0\}$ is a martingale. (See Proposition 10.5.) Its initial value is x^2, so for any $t > 0$,

$$x^2 = E\{(B_{t \wedge \tau_{ab}}^x)^2 - t \wedge \tau_{ab}\},$$

or

$$E\{t \wedge \tau_{ab}\} = E\{(B_{t \wedge \tau_{ab}}^x)^2\} - x^2.$$

Let $t \to \infty$. We can go to the limit on the left-hand side by the monotone convergence theorem and on the right-hand side by the bounded convergence theorem, for $B_{t \wedge \tau_{ab}}^2$ is bounded. We know the distribution of $B_{\tau_{ab}}$ from (10.7) so we can calculate its expected square to see that

$$E\{\tau_{ab}\} = E\{(B_{\tau_{ab}}^x)^2\} - x^2$$
$$= b^2 \frac{x-a}{b-a} + a^2 \frac{b-x}{b-a} - x^2.$$

After some algebra, we see that:

Proposition 10.32. *Let $x \in (a, b)$. The expected time for (B_t^x) to leave (a, b) is*

$$(10.8) \qquad E\{\tau_{ab} \mid B_0 = x\} = (b-x)(x-a).$$

This result has an interesting corollary. It turns out that the the expected time for Brownian motion to get anywhere is infinite! More specifically:

Corollary 10.33. $E\{T_x \mid B_0 = 0\} = \infty$ for all $x \neq 0$.

Proof. Suppose $x > 0$ (the case $x < 0$ is similar.) Let $a < 0$. Notice that τ_{ax} increases as a decreases, and that $T_x = \lim_{a \to -\infty} \tau_{ax}$. Thus $E\{T_x\} = \lim_{a \to -\infty} E\{\tau_{ax}\} = \lim_{a \to -\infty} (x-0)(x-a) = \infty$. □

10.6. Path Irregularity

Consider the following chain of logic concerning the derivative of Brownian motion at the origin. It must clearly be $\mathcal{F}_{0+}$-measurable, so according to Blumenthal's zero-one law, it is constant. By symmetry, that constant must be zero. Brownian motion restarts at any t, so its derivative at any t is *also* zero. But if its derivative is zero at all points, it must be constant. Conclusion: Brownian motion is identically zero!

There seems to be a problem here. What is wrong? Let's actually calculate the derivative. By definition

$$B'_0 = \lim_{t \downarrow 0} \frac{B_t - B_0}{t} = \lim_{t \downarrow 0} \frac{B_t}{t}.$$

Let $D_s = sB_{1/s}$. Then (D_s) is a Brownian motion, and $B_t/t = D_{1/t}$. As $t \to 0$, $1/t \to \infty$, so that

$$\limsup_{t \to 0} \frac{B_t}{t} = \limsup_{s \to \infty} D_s = \infty$$

by Theorem 10.30. Similarly, $\liminf B_t/t = -\infty$. The limit simply fails to exist: Brownian motion is not differentiable at the origin. So our pretty path of reasoning ends... slightly before it starts. This is not all.

Theorem 10.34. *For a.e. ω, the function $t \mapsto B_t(\omega)$ is nowhere differentiable.*

Proof. See Theorem 10.40 for a complete proof. Let us prove something weaker, but still surprising: the set of points at which the path is differentiable has Lebesgue measure zero.

The derivative of Brownian motion fails to exist at the origin. But $\{B_{t+s} - B_t, \ s \geq 0\}$ is also a Brownian motion, so it has no derivative at $s = 0$, and Brownian motion has no derivative at t. Thus for each t, $P\{B'_t \text{ exists}\} = 0$. The expected Lebesgue measure of the set of points at which the derivative exists is given by

$$E\left\{\int_0^\infty I_{\{B'_t \text{ exists}\}} \, dt\right\} = \int_0^\infty E\{I_{\{B'_t \text{ exists}\}}\} \, dt$$

$$= \int_0^\infty \underbrace{P\{B'_t \text{ exists}\}}_{0} \, dt = 0.$$

10.6. Path Irregularity

Thus the probability that the set of points of differentiability has positive Lebesgue measure is zero. □

Corollary 10.35. *The Brownian sample path is a.s. of unbounded variation on each non-empty interval.*

The corollary follows because a function of bounded variation is differentiable a.e.

Remark 10.36. The Brownian path is non-differentiable, so we cannot measure the speed of the Brownian particle. However, we can determine its *average* speed. Indeed, if the path has unbounded variation in each interval, it has, in fact, infinite variation, which implies that the particle travels an infinite distance in a finite time. So its average speed is evidently infinite. This bothers physicists somewhat...

Quadratic Variation. Theorems on Brownian motion alternate between exhibiting the extreme irregularity of the sample paths and showing how regular that irregularity is. We first showed the paths are continuous, then that they are not differentiable and have infinite variation. Now we will measure this irregularity by the quadratic variation, and show that it is so predictable, and so evenly spread out, that it is deterministic, not random.

Partition $[0, t]$ by $0, \frac{1}{2^n}, \frac{2}{2^n}, \ldots, \frac{[2^n t]}{2^n}, t$, where $[x]$ is the greatest integer in x. Let

$$Q_n(t) \stackrel{\text{def}}{=} \sum_{k=1}^{[2^n t]} \left(B_{\frac{k}{2^n}} - B_{\frac{k-1}{2^n}} \right)^2 ,$$

$$Q(t) = \lim_{n \to \infty} Q_n(t) \quad \text{if the limit exists}.$$

Notice that any function of bounded variation automatically has zero quadratic variation.

Theorem 10.37. *With probability one, $Q(t) = t$ for all $t \geq 0$.*

Proof. To simplify notation, assume that $t = k/2^n$ for some k. Then

$$E\{Q_n(t)\} = \sum_{k=1}^{2^n t} \underbrace{E\left\{ \left(B_{\frac{k}{2^n}} - B_{\frac{k-1}{2^n}} \right)^2 \right\}}_{2^{-n}}$$

$$= t .$$

The increments are independent, so the variances add and we have

$$\operatorname{Var}\{Q_n(t)\} = \sum_{k=1}^{2^n t} \underbrace{\operatorname{Var}\left\{\left(B_{\frac{k}{2^n}} - B_{\frac{k-1}{2^n}}\right)^2\right\}}_{\operatorname{Var}\{B_{1/2^n}^2\}} = 2^n t \operatorname{Var}\{B_{1/2^n}^2\}.$$

If Y is $N(0,1)$, let $C = \operatorname{Var}(Y^2)$. Then $B_{1/2^n}$ has the same distribution as $2^{-n/2} Y$, so

$$\operatorname{Var}\{B_{1/2^n}^2\} = \operatorname{Var}\{2^{-n} Y^2\} = C/4^n.$$

Thus

$$\operatorname{Var}\{Q_n(t)\} = \frac{Ct}{2^n}.$$

We claim that $Q_n(t) \to t$ a.s., uniformly on compact intervals. For $\lambda > 0$,

$$P\{|Q_n(t) - t| > \lambda\} \le \frac{1}{\lambda^2} \operatorname{Var}\{Q_n(t)\} = \frac{Ct}{2^n \lambda^2}.$$

This is summable, so by the Borel-Cantelli lemma, $|Q_n(t) - t| > \lambda$ only finitely often. Thus with probability one, $|Q_n(t) - t| \le \lambda$ for large enough n. This is true simultaneously for a sequence of λ tending to zero, which implies that with probability one, $Q_n(t)$ converges to t for each fixed t. Therefore it converges simultaneously for all rationals.

In fact, it converges for all t, and the convergence is a.s. uniform on compacts. For the Q_n are increasing functions and the limit is continuous, so that by Dini's theorem, pointwise convergence on a dense subset implies uniform convergence on any compact subinterval. $\square$

The Law of the Iterated Logarithm*. The derivative of Brownian motion does not exist: in fact, as $t \to 0$ the lim sup and lim inf of B_t/t are both infinite. If the limit fails to exist, what does happen? We know that B_t tends to zero, but... how fast? The central limit theorem, suggests $B_t/\sqrt{t}$ might be bounded as $t \to 0$, but it is not hard to show that this also has an infinite lim sup and lim inf. Can we tie the behavior of Brownian motion down more tightly? It turns out that we can, and the result is remarkably[4] exact. It is called the Law of the Iterated Logarithm for Brownian motion. It derives from Khintchine's celebrated theorem for limits of sums of random variables, but the proof for Brownian motion is simpler.

Theorem 10.38 (Law of the Iterated Logarithm). *With probability one*

(10.9) $$\limsup_{t \to 0} \frac{B_t}{\sqrt{2t \log\log(1/t)}} = 1, \quad \liminf_{t \to 0} \frac{B_t}{\sqrt{2t \log\log(1/t)}} = -1.$$

---

[4]This is another example of the "regular irregularity" of the Brownian path. While the derivative fails to exist, it fails to exist in a remarkably precise way.

10.6. Path Irregularity

Proof. Brownian motion is symmetric, so that the statement for the lim inf follows from that for the lim sup. Let $\phi(t) = \sqrt{2t \log\log(1/t)}$ and let $\delta > 0$, $\varepsilon > 0$. We will show that with probability one:

(i) for all $\delta > 0$, $\limsup\limits_{t \to 0} \dfrac{B_t}{\phi_t} < 1 + \delta$;

(ii) for all $\varepsilon > 0$, $\limsup\limits_{t \to 0} \dfrac{B_t}{\phi_t} > 1 - \varepsilon$.

Together, these imply the theorem. The key idea is to consider B_t along a sequence of t tending exponentially to zero. So let $\xi \in (0,1)$ and put $t_n = \xi^n$. (The value of ξ will be chosen later.) Let $\delta > 0$ and let

$$A_n \stackrel{\text{def}}{=} \{\exists t \in [t_{n+1}, t_n] : B_t \geq (1+\delta)\phi(t)\}.$$

Now $A_n \subset C_n \stackrel{\text{def}}{=} \{\sup_{t \leq t_n} B_t \geq (1+\delta)\phi(t_{n+1})\}$. Thus $P\{A_n\} \leq P\{C_n\} = 2P\{B_{t_n} \geq (1+\delta)\phi(t_{n+1})\}$ by the reflection principle. By Lemma 10.10, this is

$$\leq \frac{1}{\sqrt{2\pi}} \frac{\sqrt{t_n}}{(1+\delta)\phi(t_{n+1})} e^{-\frac{(1+\delta)^2 \phi^2(t_{n+1})}{2t_n}}.$$

Note that $\phi(t_n) = \sqrt{2\xi^n \log(n\log(1/\xi))}$, and let $\gamma = (1+\delta)^2 \xi$. Then there is a constant c for which this is

$$\leq \frac{c}{(n+1)^\gamma \sqrt{\log(n+1)}}.$$

We can choose ξ close enough to one so that $\gamma > 1$. Then this is the nth term of a convergent series, so by the Borel-Cantelli Lemma, with probability one, A_n happens only finitely often, proving (i).

To prove (ii), let $\varepsilon > 0$. As in (i), let $\delta > 0$ and $\xi \in (0,1)$—values to be chosen later—and put $t_n = \xi^n$. Let $X_n = B_{t_n} - B_{t_{n+1}}$. The X_n are independent and $\text{Var}(X_n) = t_n - t_{n+1} = \xi^n(1-\xi)$. We claim that

(10.10) $$P\{X_n \geq (1-\delta)\phi(t_n) \text{ i.o.}\} = 1.$$

By Lemma 10.10,

$$P\{X_n \geq (1-\delta)\phi(t_n)\} \leq \frac{\sqrt{t_n - t_{n+1}}}{\sqrt{2\pi}\,\phi(t_n)} e^{-\frac{(1-\delta)^2 \phi(t_n)^2}{2(t_n - t_{n+1})}}.$$

Let $\gamma = (1-\delta)^2/(1-\xi)$ and do some algebra to see there is $c > 0$ for which this is

$$\geq \frac{c}{n^\gamma \sqrt{\log n}}.$$

Choose ξ small enough so that $\gamma < 1$. Then this series diverges, so (10.10) holds by the converse Borel-Cantelli Lemma.

Now by (i) and the symmetry of Brownian motion,
$$\liminf_{t\to 0} \frac{B_t}{\phi(t)} > -(1+\delta).$$
Therefore, for large enough n,
$$B_{t_n} = X_n + B_{t_{n+1}} > X_n - (1+\delta)\phi(t_{n+1}).$$
By (10.10) there is a sequence of n tending to infinity for which this is
$$> (1-\delta)\phi(t_n) - (1+\delta)\phi(t_{n+1}).$$
Note that $\phi(t_{n+1})/\phi(t_n) \leq \sqrt{2\xi}$ for all large enough n, so that for these n,
$$B_{t_n} \geq \left(1 - \delta - \sqrt{2\xi}(1+\delta)\right)\phi(t_n).$$
We can choose both δ and ξ small enough so that $\gamma < 1$ (needed for (10.10)) and $\delta + \sqrt{2\xi}(1+\delta) < \varepsilon$. (Indeed, both hold if $\delta < \varepsilon/2$ and $\xi < \varepsilon^2/32$.) Thus there is a.s. a sequence of $n \to \infty$ for which $B_{t_n} \geq (1-\varepsilon)\phi(t_n)$, and (ii) follows. $\square$

There is another law of the iterated logarithm as $t \to \infty$.

Corollary 10.39. *With probability one,*
$$(10.11) \qquad \limsup_{t\to\infty} \frac{B_t}{\sqrt{2t\log\log(t)}} = 1, \quad \liminf_{t\to\infty} \frac{B_t}{\sqrt{2t\log\log(t)}} = -1.$$

Proof. This follows from time-reversal. Let $D_t = tB_{1/t}$. Then $\{D_t, t \geq 0\}$ is a Brownian motion, and
$$\limsup_{t\to\infty} \frac{D_t}{\sqrt{2t\log\log t}} = \limsup_{t\to\infty} \frac{tB_{1/t}}{\sqrt{2t\log\log t}} = \limsup_{s\to 0} \frac{B_s}{\sqrt{2s\log\log(1/s)}} = 1$$
by Theorem 10.38, where $s = 1/t$. $\square$

Proof of Nowhere-Differentiability*. We stated in Section 10.6 that with probability one, the Brownian path is nowhere-differentiable: the set of t for which dB_t/dt exists is empty. The proof we gave only showed that the set of differentiability has Lebesgue measure zero. This is still striking, but considerably weaker. Here is the full proof.

Theorem 10.40. *For a.e. ω, $t \mapsto B_t(\omega)$ is nowhere-differentiable.*

Proof. Let $b > 0$ and put
$$A^b = \{\omega : \exists s \in (0,1) \text{ such that } B'_s(\omega) \text{ exists}, |B'_s(\omega)| \leq b/2\}.$$
If $\omega \in A^b$, then there exists $s \in (0,1)$ and $\delta > 0$ such that $|B_t(\omega) - B_s(\omega)| < b|t-s|$ if $|t-s| < \delta$. Let
$$A_n^b = \{\exists s \in (0,1) : |B_t - B_s| \leq b|t-s| \text{ for all } |t-s| \leq 2/n\}.$$

Then $A_n^b \subset A_{n+1}^b$ and $A^b \subset \bigcup_n A_n^b$. Moreover, $\omega \in A^b$ and $|B_s'(\omega)| \leq b/2$ imply that $\omega \in A_b^n$ for all large enough n. For such an n, there exists k such that $k/n \leq s < (k+1)/n$ and $|B_t - B_s| \leq b|t-s|$ if $t \in [(k-1)/n, (k+2)/n]$. Now if $s \in A_n^b$, then

$$|B_{\frac{k+2}{n}} - B_{\frac{k+1}{n}}| \leq |B_{\frac{k+2}{n}} - B_s| + |B_s - B_{\frac{k+1}{n}}|$$
$$\leq b\frac{2}{n} + b\frac{1}{n} = \frac{3b}{n}.$$

Similarly,

$$|B_{\frac{k}{n}} - B_{\frac{k-1}{n}}| \leq \frac{2b}{n} + \frac{b}{n} = \frac{3b}{n}$$

and

$$|B_{\frac{k+1}{n}} - B_{\frac{k}{n}}| \leq \frac{b}{n} + \frac{b}{n} < \frac{3b}{n}.$$

Thus,

$$P\{A_n^b\} \leq P\left\{\exists k \leq n : \max\{|B_{\frac{k+i}{n}} - B_{\frac{k+i-1}{n}}|, \, i = 0, 1, 2\} \leq \frac{3b}{n}\right\}.$$

The three increments are independent, so the probability that the maximum of the three is less than $3b/n$ equals the cube of the probability that any one of them is less than $3b/n$. Thus

$$P\{A_n^b\} \leq \sum_{k=1}^n P\left\{|B_{\frac{k+1}{n}} - B_{\frac{k}{n}}| \leq \frac{3b}{n}\right\}^3 \leq nP\left\{|B_{\frac{1}{n}}| \leq \frac{3b}{n}\right\}^3.$$

But if $p(x)$ is the density of $B_{1/n}$, then $p(x) \leq \sqrt{n/2\pi}$ for all x, so

$$P\{|B_{1/n}| \leq 3b/n\} \leq \int_{-3b/n}^{3b/n} p(x)\,dx \leq 6b/\sqrt{2n\pi}.$$

Therefore,

$$P\{A_n^b\} \leq n\frac{(6b)^3}{(2\pi n)^{3/2}} \leq \frac{c}{\sqrt{n}}.$$

This goes to zero as $n \to \infty$. Therefore $P\{A^b\} = \lim_n P\{A_n^b\} = 0$. Now A^b increases with b, and the set of ω for which $t \mapsto B_t(\omega)$ is differentiable at at least one point is in $\bigcup_n A^n$. But $P\{\bigcup_n A^n\} \leq \sum_n P\{A^n\} = 0$. □

10.7. The Brownian Infinitesimal Generator*

The transition functions of discrete-parameter Markov chains are all generated by the one-step transition probabilities. With continuous parameter processes, however, there is no minimal step size to use, and one turns instead to the derivative of the transition probabilities. This is called the *infinitesimal generator*.

We defined it for birth and death processes in §7.11. Let us do it for Brownian motion. Since Brownian motion has a continuous state space, as opposed to the discrete state space of birth and death processes, its infinitesimal generator has a different form.

We will show that the infinitesimal generator $\mathbb{G}$ of a standard Brownian motion is
$$\mathbb{G} = \frac{1}{2}\frac{d^2}{dx^2}.$$
More specifically, consider a smooth function f.

Definition 10.41. The **infinitesimal generator** $\mathbb{G}$ of Brownian motion, applied to a function f, is

(10.12) $$\mathbb{G}f(x) \stackrel{def}{=} \lim_{t \downarrow 0} \frac{1}{t} E\{f(B_t^x) - f(x)\},$$

if the limit exists, where $B_t^x = x + B_t$ is Brownian motion from x.

Functions for which the the limit in (10.12) exists for all x are said to be in the **domain of the infinitesimal generator.** It is not always easy to determine whether a given function is in it or not; in fact, it can be quite tricky and also important, since in regions with boundaries, the domain of the infinitesimal generator is closely linked to the boundary behavior of the process. However, for Brownian motion, at least, there is no trouble in showing that smooth functions with compact support are in the domain. The condition of "compact support" is too restrictive in general, but one can get around it by localization, as follows. If f is any smooth function, let g be another smooth function such that $g(x) = 1$ if $|x| \leq R$ for some large R, and $g(x) = 0$ if $|x| \geq R+1$. Then replace f by $\hat{f} \stackrel{def}{=} fg$. Then $\hat{f}$ is smooth, has compact support, and equals f in $[-R, R]$, so that $\mathbb{G}f = \mathbb{G}\hat{f}$ in $[-R, R]$.

Proposition 10.42. *Let f be a smooth function of compact support. Then*
$$\mathbb{G}f(x) = \frac{1}{2}f''(x).$$

Before proving this, recall Taylor's formula: if f is a function with a bounded third derivative, say $|f'''(x)| \leq C$ for all x, then for each x and y,

(10.13) $$f(y) = f(x) + f'(x)(y-x) + \frac{1}{2}f''(x)(y-x)^2 + R_2(x,y),$$

where the remainder satisfies

(10.14) $$|R_2(x,y)| \leq \frac{1}{6}C|y-x|^3.$$

Proof. (of the Proposition). The idea is simple: the first and second order terms in the Taylor expansion are the important ones, so we take the second order expansion with remainder. We handle the first and second order terms directly, and then show that the remainder tends to zero.

Since f has compact support, f and its derivatives are bounded. By (10.13),

$$f(B_t^x) = f(x) + f'(x)(B_t^x - x) + \frac{1}{2}f''(x)(B_t^x - x)^2 + R_2(x, B_t^x).$$

Take the expectation: $B_t^x - x$ has expectation zero and variance t, so $E\{f(B_t^x) - f(x)\} = f''(x)t/2 + E\{R_2(x, B_t^x)\}$. Then

$$E\{|R_2(x, B_t^x)|\} \leq \frac{C}{6} E\{|B_t^x - x|^3\}.$$

$B_t^x - x$ has the same distribution as $\sqrt{t}\xi$, where ξ is $N(0,1)$, so letting $C' = CE\{|\xi|^3\}/6$, this is

$$\leq \frac{C'}{6} t^{\frac{3}{2}}.$$

Thus, as $t \to 0$,

$$\frac{1}{t} E\{f(B_t^x) - f(x)\} = \frac{1}{2} f''(x) + O(\sqrt{t}) \longrightarrow \frac{1}{2} f''(x).$$

□

Proposition 10.43. *Let f be a smooth function of compact support. Then $\{f(B_t) - \int_0^t \mathbb{G}f(B_s)\,ds, \mathcal{F}_t, t \geq 0\}$ is a martingale.*

Proof. Since f has compact support, f and its derivatives are bounded. Let $s < t$, let $h = (t-s)/n$ and set $t_k = s + kh$, $k = 0, \ldots, n$. Using Taylor's expansion,

$$f(B_t) - f(B_s) = \sum_{k=0}^{n-1} f(B_{t_{k+1}}) - f(B_{t_k})$$

$$= \sum_{k=0}^{n-1} f'(B_{t_k})(B_{t_{k+1}} - B_{t_k}) + \frac{1}{2} \sum_{k=0}^{n-1} f'(B_{t_k})(B_{t_{k+1}} - B_{t_k})^2$$

$$+ \sum_{k=0}^{n-1} R_2(B_{t_k}, B_{t_{k+1}}).$$

Take the conditional expectation given $\mathcal{F}_s$ on both sides. Notice that

$$E\{f'(B_{t_k})(B_{t_{k+1}} - B_{t_k}) \mid \mathcal{F}_s\} = E\{f'(B_{t_k}) E\{B_{t_{k+1}} - B_{t_k} \mid \mathcal{F}_{t_k}\} \mid \mathcal{F}_s\} = 0$$

$$E\{f''(B_{t_k})(B_{t_{k+1}} - B_{t_k})^2 \mid \mathcal{F}_s\} = E\{f''(B_{t_k}) \underbrace{E\{(B_{t_{k+1}} - B_{t_k})^2 \mid \mathcal{F}_{t_k}\}}_{h} \mid \mathcal{F}_s\}$$

$$= E\{f''(B_{t_k}) \mid \mathcal{F}_s\} h\,.$$

Thus

$$(10.15) \quad E\{f(B_t) - f(B_s) \mid \mathcal{F}_s\} = \frac{1}{2} E\left\{\sum_{k=0}^{n-1} f''(B_{t_k}) h \mid \mathcal{F}_s\right\}$$

$$+ \sum_{k=0}^{n-1} E\{R_2(B_{t_k}, B_{t_{k+1}}) \mid \mathcal{F}_s\}\,.$$

Now

$$\left|E\{R_2(B_{t_k}, B_{t_{k+1}}) \mid \mathcal{F}_s\}\right| \leq C E\{|B_{t_{k+1}} - B_{t_k}|^3\} = O(h^{\frac{3}{2}})\,.$$

There are $n \sim 1/h$ terms in the last sum of (10.15), so it is $O(\sqrt{h})$. Now $u \mapsto f''(B_u)$ is continuous, so that the first sum on the right in (10.15) is a Riemann sum for the integral $\int_s^t f''(B_u)\,du$, and it converges boundedly to it as $h \to 0$. While we cannot go directly to the limit under the conditional expectation, we can let $\Lambda \in \mathcal{F}_s$ and note that

$$\int_\Lambda f(B_t) - f(B_s)\,dP = \frac{1}{2} \int_\Lambda \sum_{k=0}^{n-1} f''(B_{t_k}) h\,dP + \int_\Lambda O(\sqrt{h})\,dP.$$

Now we can take the limit by bounded convergence as $n \to \infty$ and $h \to 0$ to see that

$$\int_\Lambda f(B_t) - f(B_s)\,dP = \int_\Lambda \left(\int_s^t \mathbb{G}f(B_u)\,du\right) dP,$$

which shows that $E\{f(B_t) - f(B_s) \mid \mathcal{F}_s\} = E\{\int_s^t \mathbb{G}f(B_u)\,du \mid \mathcal{F}_s\}$ a.s. $\square$

Remark 10.44. Notice that Proposition 10.43 tells us that if $\mathbb{G}f \equiv 0$, then $f(B_t)$ is a martingale. Of course, we already knew this, since $\mathbb{G}f = 0 \implies f = ax + b$, and a linear function of Brownian motion is a martingale. Similarly, if $\mathbb{G}f = 1$, then $f(B_t) - t$ is a martingale. But this is also familiar, since we know that $B_t^2 - t$ is a martingale. On the other hand, this does tell us something new: if $\mathbb{G}f$ is *not* identically zero, then $f(B_t)$ is not a martingale. That is, *only* linear functions of Brownian motion are martingales.

Exercise 10.18. Let f be smooth and real-valued. Show that $f(B_t)$, $t \geq 0\}$ is a martingale iff $f(x) = a + bx$ for some constants a and b.

10.8. Related Processes

Reflecting and absorbing Brownian Motions. Brownian motion represents the movement of a particle in a region. Physical regions have boundaries, and it is natural to ask what happens when the particle encounters one. We will discuss two possibilities here, reflecting and absorbing Brownian motion. If the Brownian particle is moving in a glass of water, the glass itself is a boundary of the liquid, and when the particle hits it, it will bounce off—be reflected. If the boundary is sticky, or perhaps electrically charged, the particle may stay there, leading to absorbing Brownian motion.

Let B_t be a standard Brownian motion, with $B_0 = x$. Define $R_t = |B_t|$. Then $\{R_t, t \geq 0\}$ is called **reflecting Brownian motion**. It is called "reflecting" since, when it hits zero, it is reflected back into $[0, \infty)$.

It is a Markov process, and we can get its transition probability by a symmetry argument. Suppose that $p_t(x)$ is the $N(0,t)$ density. Then the Brownian transition probability is $p_t^B(x,y)\,dy = P\{B_t \in dy \mid B_0 = 0\} = p_t(y-x)dy$. Suppose $B_0 = x > 0$. Then $R_0 = x$ as well, and $R_t = y \iff B_t = y$ or $B_t = -y$, so R_t has density $p_t^R(x,y) = p_t^B(x,y) + p^B(x,-y)$, $y > 0$. By symmetry, $p_t^B(x,-y) = p_t^B(-x,y)$ so we have:

$$(10.16) \qquad p_t^R(x,y) = p_t^B(x,y) + p_t^B(-x,y), \quad x,y \geq 0.$$

It is not hard to see that if $s < t$, $P\{R_t \in dy \mid R_s, \mathcal{F}_s\} = p_{t-s}^R(R_s, y)\,dy$, so that reflecting Brownian motion is indeed a Markov process.

Physicists call this the "method of images[5]". Think of Brownian motion from x as being, not one, but a stream of particles issuing from x. They spread out as time goes on, and $p_t^B(x,y)$ is their density at the point y, t instants later. Then (10.16) gets the transition density of reflecting Brownian motion by combining two sources of non-reflecting particles: a source at x, and an "image" source at $-x$.

Absorbing Brownian motion—also called killed Brownian motion—is closely related. In this case, instead of being reflected, the Brownian particle is absorbed when it encounters 0. It then disappears. Call this process A_t:

$$A_t = \begin{cases} B_t & \text{if } t < T_0, \\ \text{undefined} & \text{if } t \geq T_0. \end{cases}$$

Then $\{A_t,\ t \geq 0\}$ is defined until it hits the origin; after, it is undefined. It is a Markov process, and we can get its transition probabilities by noting that if $x, y \geq 0$, $P\{A_t \in dy \mid A_0 = x\} = P^x\{B_t \in dy,\ t < T_0\}$ and this equals $P^x\{B_t \in dy\} - P^x\{B_t \in dy, T_0 \leq t\}$. By the reflection principle,

---

[5]This is due to Lord Kelvin. He was actually describing the electrical field of a charge distribution, not the motion of Brownian particles. But strangely enough, the two are connected.

$P^x\{B_t \in dy, T_0 \leq t\} = P^x\{B_t \in -dy\}$. (For if the particle reaches $-y$, it has already hit 0. By symmetry, $p_t^B(x,-y) = p_t^B(-x,y)$, so we have the transition density for absorbing Brownian motion:

(10.17) $$p_t^A(x,y) = p_t^B(x,y) - p_t^B(-x,y).$$

There is also a physical "image" argument for these transition probabilities. The transition density at the origin must be zero, since the particles disappear when they hit zero. Therefore one needs an image source to cancel the original source at $y=0$. By symmetry, this source must be at $-x$, the mirror-image of x, and must be *negative* in order to cancel the positive source at x. This leads to (10.17).

Absorbing Brownian motion is not defined after its lifetime[6] T_0. In fact,

$$P\{T_0 > t \mid A_0 = x\} = P\{A_t \in [0,\infty) \mid A_0 = x\}$$
$$= \int_0^\infty p_t^B(x,y) - p_t^B(-x,y)\,dy.$$

This equals $\sqrt{1/2\pi t}\,(e^{-(y-x)^2/2t} - e^{-(y+x)^2/2t})$.

We can also treat Brownian motion in a restricted region, say $[0,1]$. Let us consider the case where the particle is reflected whenever it hits either 0 or 1, so that it stays inside the interval. This is trickier, since the particle can be reflected multiple times. Instead of using the reflection principle, let us use the images directly. We want to find a distribution of sources for the unreflected motion which will give the right transition density in $[0,1]$. We argue that, since the particle is reflected at 0, that its derivative in y must be zero—as many particles coming as going. If the initial source is at x, we have to balance it with an image source at $-x$. Then $(\partial/\partial y)(p_t^B(x,y) + p_t^B(-x,y))|_{y=0} = 0$ by symmetry. This fixes the boundary condition at 0. But the particle is reflected from 1 as well, so the derivative must also vanish at 1. This means that we need *two more* images, one to cancel out each of the first two at $y=1$. They will be located symmetrically about 1: one at $2-x$ to balance out the source at x, and one at $2+x$ to balance out the source at $-x$. This balances things at $x=1$, but unbalances them again at the origin. We need another pair of image sources at $-2 \pm x$ to balance the two we just added. But these reunbalance the boundary conditions at $x=1$, so to restore the balance there, we need two more images... As with two facing mirrors, this gives us an infinite sequence of images receding into the distance. If we look at the pattern, we can see that there are sources at $2n \pm x$, $n = 0, \pm 1, \pm 2, \ldots$. We conclude that the

---

[6]If one wishes, one can define it for all t in a slightly artificial way: pick a point ∂ which is *not* in $[0,\infty)$ and set $A_t = \partial$ if $t \geq T_0$. The extended process is still Markovian, and it is defined at all t. The point ∂ is called a "cemetery point" for obvious reasons.

probability density is

(10.18) $$p_t^{01}(x,y) = \sum_{n=-\infty}^{\infty} p_t^B(x, 2n+y) + p_t^B(x, 2n-y).$$

This series converges, since for $t > 0$, $p_t^B(2n+x,y) = (1/\sqrt{2\pi t})\, e^{\frac{(2n+x-y)^2}{2t}}$ and the exponential goes to zero very quickly as n increases.

Now that we have the answer, we can see where the terms come from. They represent multiple reflections, from paths that have hit and been reflected from both 0 and 1 several times.

The Brownian Bridge. Suppose the Brownian path is known from time $t = 1$ onward, but not before. We want to find the best estimate of B_t for $0 \le t \le 1$. By the Markov property, this estimate should depend only on B_1, and be zero at the origin. An easy symmetry argument (each increment of length $1/n$, say, should be treated equally) tells us that the only reasonable choice for the estimate must be linear in t, and therefore equal to tB_1. We want to study the "error", that is, the difference between B_t and its estimate:

$$\mathfrak{B}_t \stackrel{\text{def}}{=} B_t - tB_1, \quad 0 \le t \le 1.$$

The process $\{\mathfrak{B}_t,\ 0 \le t \le 1\}$ is called the **standard Brownian bridge.**

More generally, we can look at the process between any two times, say $0 \le t_1 < t_2$. Given the values outside (t_1, t_2), the expected value of B_t inside is linear, equals B_{t_1} and B_{t_2} at the two endpoints, and so equals $B_{t_1} + ((t-t_1)/(t_2-t_1))(B_{t_2} - B_{t_1})$, $t \in (t_1, t_2)$. The difference between the process and its expected value is then

$$\tilde{B}_t \stackrel{\text{def}}{=} B_{t_1+t} - B_{t_1} - \frac{t}{t_2-t_1}(B_{t_2} - B_{t_1}), \quad t \le t_2 - t_1.$$

We will also call this process a Brownian bridge.

Proposition 10.45. *Let $0 \le t_1 < t_2$, and put $\tau = t_2 - t_1$. Then the process $\{\tilde{B}_t,\ 0 \le t \le \tau\}$ is independent of $\sigma\{B_s, s \notin (t_1, t_2)\}$. Moreover, $\tilde{B}_\tau = \tilde{B}_0 = 0$, and $\tilde{B}$ is a mean zero Gaussian process with covariance function $\Gamma(s,t) = E\{\tilde{B}_s \tilde{B}_t\} = s \wedge t - st/\tau$. In particular, $\operatorname{Var}\{\tilde{B}_t\} = t(\tau - t)/\tau$.*

Proof. Brownian motion is a Gaussian process, so the variables are independent if they are uncorrelated, and we need only check the covariances. If $s \notin (t_1, t_2)$, use the fact that $E\{B_s B_t\} = s \wedge t$ to see that

$$E\{B_s \tilde{B}_t\} = E\left\{B_s\left(B_{t+t_1} - B_{t_1} - \frac{t}{\tau}(B_{t_2} - B_{t_1})\right)\right\}$$
$$= s \wedge (t+t_1) - s \wedge t_1 - \frac{t}{\tau}(s \wedge t_2) + \frac{t}{\tau}(s \wedge t_1).$$

This expression vanishes if either $s \leq t_1$ or $s \geq t_2$. This proves the independence.

It is clear that $\tilde{B}_\tau = \tilde{B}_0 = 0$, and that $\tilde{B}$ has mean zero. Its covariance function is

$$\Gamma(s,t) = E\{\tilde{B}_s \tilde{B}_t\}$$
$$= E\Big\{\big(B_{s+t_1} - B_{t_1} - \frac{s}{\tau}(B_{t_2} - B_{t_1})\big)\big(B_{t+t_1} - B_{t_1} - \frac{t}{\tau}(B_{t_2} - B_{t_1})\big)\Big\}$$
$$= s \wedge t - \frac{st}{\tau},$$

and the rest is immediate. □

Remark 10.46. One consequence of Proposition 10.45 is that $\{B_s, u \leq s \leq v\}$ is conditionally independent of $\sigma\{B_s, s \notin (u,v)\}$ given B_u, B_v. This is a two-point Markov property.

Problems 10.8

In what follows, (B_t) is a standard Brownian motion.

10.19. Show that if a reflecting Brownian motion starts from 0, the first-hitting time of 1 has a finite expectation, but that if it starts from 1, the first-hitting time of 0 has an infinite expectation.

10.20. Let X_0, σ, $\mu \in \mathbb{R}$ and let $X_t = X_0 e^{\sigma B_t + \mu t}$. Then $\{X_t, t \geq 0\}$ is called **logarithmic Brownian motion**. It is often used to model the evolution of stock prices in financial markets. Show that it is a Markov process.

10.21. Let $\sigma \in \mathbb{R}$.

(a) Show that for any $t \geq 0$, $E\{e^{\sigma B_t - \frac{1}{2}\sigma^2 t}\} = 1$.

(b) Conclude that $\{e^{\sigma B_t - \frac{1}{2}\sigma^2 t}, t \geq 0\}$ is a martingale.

[Hint: Moment generating function.]

10.22. Let $a, x \in \mathbb{R}$ and put $X_t = x + B_t + at$. Then $\{X_t, t \geq 0\}$ is a **Brownian motion with drift**. Suppose $0 < x < b$. Find $P\{X_t \text{ hits } b \text{ before } 0\}$. If $a > 0$ and $x > 0$, find the probability that X_t ever hits the origin.

[Hint: Find r for which (e^{rX_t}) is a martingale.]

10.23. Let $a > 0$, $b > 0$. Show that $P\{B_t = a + bt, \text{ some } t \geq 0\} = e^{-2ab}$.

10.24. Let $f(x)$ be a continuous real-valued function on $[0,1]$ with $f(0) = 0$, and let $\varepsilon > 0$. Show that $P\{\sup_{0 \leq t \leq 1} |f(t) - B_t| \leq \varepsilon\} > 0$.

10.25. Let $a < b$, and let $M = \max\{B_t : a \leq t \leq b\}$. Show that for each x, $P\{M = x\} = 0$.

10.26. Let T be a finite stopping time for Brownian motion. Show that with probability one, T is *not* a local maximum of (B_t).

[Hint: Proposition 10.21 may help.]

10.8. Related Processes

10.27. Show that with probability one, the Brownian path attains its maximum on $[0, 1]$ at a unique time t.

10.28. Consider a Brownian motion on $[0, 1]$ which is absorbed when it first hits either 0 or 1. Find a series giving its transition probability.

10.29. Find the probability that a standard Brownian motion hits neither one nor minus one before time $t = 1$.

10.30. Find $A_1 \supset A_2 \supset \ldots$ with $A = \bigcap_n A_n$ such that $\lim T_{A_n} \neq T_A$.

10.31. Let $x \geq 0$. Define $\hat{T}_x = \inf\{t : B_t > x\}$. Note that $\{\hat{T}_x, x \geq 0\}$ is a collection of $(\mathcal{F}_{T+})$-stopping times, so that it is a stochastic process, even though the parameter x represents space, not time.

(a) Show that the density of $\hat{T}_x$ is
$$f_{T_x}(t) = \frac{x}{\sqrt{2\pi}} t^{-\frac{3}{2}} e^{-\frac{x^2}{2t}}.$$

(b) Show that $x \mapsto \hat{T}_x$ is right-continuous and increasing.

(c) Show that it has independent increments: if $x_1 < x_2 < \cdots < x_n$, then $\hat{T}_{x_1}, \hat{T}_{x_2} - \hat{T}_{x_1}, \ldots, \hat{T}_{x_n} - \hat{T}_{x_{n-1}}$ are independent, and that $\hat{T}_{x_n} - \hat{T}_{x_{n-1}}$ has the same distribution as $\hat{T}_{x_2 - x_1}$.

(d) Show that $\hat{T}_2$ has the same distribution as the sum of two independent copies of $\hat{T}_1$.

(e) Use Brownian scaling to show that $\hat{T}_2$ has the same distribution as $4\hat{T}_1$.

(f) Use (c) and (d) to show that the $\hat{T}_x$ are all stable $(1/2)$. (So $\{\hat{T}_x, x \geq 0\}$ is an example of a stable $(1/2)$ process.)

10.32. Let $B_0 = 0$. Show that the probability that (B_t) has at least one zero in the interval (t_0, t_1) is
$$\frac{2}{\pi} \arccos\left(\sqrt{\frac{t_0}{t_1}}\right).$$
[Hint: Use Exercise 10.31 (a) to write the probability as a double integral. One of the two integrals can be done explicitly, and then the second can be done by a change of variables of the form $v = u^2$. Some trig manipulation will change the resulting arctan into an arccos.]

10.33. Let B_t be a standard Brownian motion, let $[x]$ be the greatest integer n such that $n \leq x$, and consider the process (X_t), where $X_t = B_t - [B_t]$. Then $0 \leq X_t < 1$ and X is a Brownian motion away from the boundaries. Discuss its continuity.

10.34. (Convergence of normalized sums to Brownian motion.) Let $X_1, X_2, \ldots$ be a sequence of i.i.d. random variables with mean zero and variance one. Interpolate the partial sums of the (X_i) to make a continuous-parameter process χ_t^n, defined by
$$\chi_t^n = \frac{1}{\sqrt{n}} \sum_{i=1}^{[nt]} X_i,$$

where $[s]$ is the greatest integer $\leq s$.

(a) Show that $t \mapsto \chi_t^n$ is right continuous, and constant on each interval $j/n \leq t < (j+1)/n$.

(b) Show that χ_t^n has mean zero and variance $n[t/n]$.

(c) Show that as $n \to \infty$, the finite-dimensional distributions of the processes χ^n converge to those of a standard Brownian motion. (You may assume that the convergence of joint characteristic functions implies the convergence in distribution of *finite* families of random variables. See Theorem 10.63 for a deeper approach.)

10.35. Let $X_t = e^{\lambda B_t - \frac{1}{2}\lambda^2 t}$. Differentiate with respect to λ and show that for each n, $H_n(B_t, t)$ is a martingale, where $H_0(x,t) \equiv 1$, $H_1(x,t) = x$, $H_2(x,t) = x^2 - t$, and, in general, $H_n(x,t) = xH_{n-1}(x,t) - (n-1)H_{n-2}(x,t)$. (The H_n are called **Hermite polynomials.**)

10.36. Let (B_t) be a standard Brownian motion. Show directly—i.e., without using the law of the iterated logarithm—that $\limsup_{t \to 0} B_t/\sqrt{t} = \infty$ a.s.

10.9. Higher Dimensional Brownian Motion

The motion of a physical particle in a liquid is three-dimensional. We have been studying its x-coordinate. Let's look at the whole package. By symmetry, all coordinates of the motion should have the same distribution, and—given the molecular source of the motion—should be independent. So each coordinate of a three-dimensional Brownian motion should be a one-dimensional Brownian motion. We can model it by $B_t = (B_t^1, B_t^2, B_t^3)$, where the B_t^i are independent standard Brownian motions. We might also want to consider two-dimensional Brownian motion—that is what we observe in a microscope—and—why not?—Brownian motion in four and higher dimensions, too. Without committing ourselves to a favorite dimension, let us consider d-dimensional Brownian motion.

Definition 10.47. A process $B_t = (B_t^1, \ldots, B_t^d)$ is a **standard Brownian motion in $\mathbb{R}^d$** if the coordinate processes (B_t^i) are independent standard Brownian motions on the line. It is a **standard Brownian motion from x** if $(B_t - x)$ is a standard Brownian motion in $\mathbb{R}^d$.

If B_t is a standard Brownian motion, then $x + B_t$ is a standard Brownian motion from x. Let (B_t) be a standard Brownian motion. Then $B_0 = 0$ and B_t has the density in $\mathbb{R}^d$:

$$p_t(x_1, \ldots, x_d) = \frac{1}{(2\pi t)^{\frac{d}{2}}} e^{-\frac{x_1^2 + \cdots + x_d^2}{2t}}.$$

Since B_t is d-dimensional its variance and covariance are given by matrices: $\left(E\{B_t^i B_t^j\}\right) = t I_{d \times d}$, where $I_{d \times d}$ is the $d \times d$ identity matrix. The

10.9. Higher Dimensional Brownian Motion

covariance matrix of B_s and B_t is

$$\Gamma(s,t) = (s \wedge t)\, I_{d \times d}.$$

In terms of the d-tuple $x = (x_1, \ldots, x_d)$, $p_t(x) = (2\pi t)^{-\frac{d}{2}} e^{-\frac{|x|^2}{2t}}$. Note that the distribution of standard Brownian motion is rotation-invariant. In fact, the distribution of the whole process is invariant under rotations. It is a strong Markov process, since each independent coordinate is. Brownian motion from $x \in \mathbb{R}^d$ is denoted B_t^x; it has the same distribution as $x + B_t$. Thus, the distribution of $B_t^x - B_0^x$ is independent of x.

We will not go very deeply into it. We will just look at its hitting and recurrence properties. Since each coordinate is itself a one-dimensional Brownian motion, we already know a fair amount about it. For instance, the Markov property, the translation, scaling, and reversal properties all carry over directly from the one-dimensional case. Its paths are not bounded, since its first coordinate is already unbounded. It is not smooth, since its first coordinate is already non-differentiable. But there are some questions whose answer is not obvious, such as: is the process recurrent? If so, does it return to its exact starting point? Certainly each coordinate is recurrent, since one-dimensional Brownian motion is, but in order to return to the origin, all coordinates must return *at the same time*. Similarly, the process might return to a neighborhood of its initial point, but does it actually hit it?

Let us find its infinitesimal generator. Let f be a smooth function of compact support on $\mathbb{R}^d$. Suppose that f and its partials are bounded. By Taylor's formula,

$$f(y) = f(x) + \sum_{i=1}^{d} \frac{\partial f}{\partial x_i}(x)(y_i - x_i) + \frac{1}{2} \sum_{i,j=1}^{d} \frac{\partial^2 f}{\partial x_i \partial x_j}(x)(y_i - x_i)(y_j - x_j) + R_2(x,y).$$

There is a C such that $|R_2(x,y)| \leq C|y-x|^3$. Let (B_t) be a standard Brownian motion from $x \in \mathbb{R}^d$. Then

$$f(B_t) - f(x) = \sum_{i=1}^{d} \frac{\partial f}{\partial x_i}(x)(B_t^i - x_i) + \frac{1}{2} \sum_{i,j=1}^{d} \frac{\partial^2 f}{\partial x_i \partial x_j}(x)(B_t^i - x_i)(B_t^j - x_j) + R_2(x, B_t).$$

Take the expectation on both sides. Let $\delta_{ij} = I_{i=j}$ be the Kronecker delta. Noting that $E\{B_t^i - x_i\} = 0$ and $E\{(B_t^i - x_i)(B_t^j - x_j)\} = t\delta_{ij}$, we see

$$\mathbb{G}f(x) \stackrel{def}{=} \lim_{t\downarrow 0} \frac{1}{t} E\{f(B_t) - f(x)\}$$

$$= \frac{1}{2} \sum_{i=1}^{d} \frac{\partial^2 f}{\partial x_i^2}(x) + \lim_{t\downarrow 0} \frac{1}{t} E\{R_2(x, B_t)\}.$$

Now $|E\{R_2(x, B_t)\}| \le CE\{|B_t - x|^3\}$. But $B_t - x$ has the same distribution as $\sqrt{t}(B_1 - x)$. If $\sigma_3 = E\{|B_1 - x|^3\}$, then $E\{R_2, B_t^x)\} \le C\sigma_3 t^{3/2}$. Thus the limit above vanishes, and we have

$$\mathbb{G}f(x) = \frac{1}{2} \sum_{i=1}^{d} \frac{\partial^2}{\partial x_i^2} f(x).$$

Remark 10.48. $\Delta f(x) \stackrel{def}{=} \sum_{i=1}^{d} \frac{\partial^2}{\partial x^2} f$ is called the **Laplacian** of f. Thus $\mathbb{G}f = \frac{1}{2}\Delta f$.

Proposition 10.49. *Let f be a smooth function of compact support. Let $X_t = f(B_t) - \int_0^t \mathbb{G}f(B_s)\,ds$. Then $\{X_t, \mathcal{F}_t, t \ge 0\}$ is a martingale, where $\mathcal{F}_t = \sigma(B_s,\ s \le t)$.*

Proof. This is essentially the same as the proof of the one-dimensional case (Theorem 10.43). Let $s < t$, put $h = (t-s)/n$, and set $t_k = s + kh$, $k = 0, \ldots, n$. Then

$$f(B_t) - f(B_s) = \sum_{k=0}^{n-1} \big(f(B_{t_{k+1}}) - f(B_{t_k})\big)$$

$$= \sum_{k=0}^{n-1} \sum_{i=1}^{d} \frac{\partial}{\partial x_i} f(B_{t_k})(B_{t_{k+1}}^i - B_{t_k}^i)$$

$$+ \frac{1}{2} \sum_{k=0}^{n-1} \sum_{i,j=1}^{d} \frac{\partial^2}{\partial x_i x_j} f(B_{t_k}) (B_{t_{k+1}}^i - B_{t_k}^i)(B_{t_{k+1}}^j - B_{t_k}^j)$$

$$+ R_2(B_{t_k}, B_{t_{k+1}}).$$

The derivatives of f are bounded, so that there is a constant C such that $R_2(x, y) \le C|y - x|^3$. The increments are all independent of $\mathcal{F}_s$, so that $E\{f(B_{t_{k+1}}) - f(B_{t_k}) \mid \mathcal{F}_s\} = 0$ and $E\{(B_{t_{k+1}}^i - B_{t_k}^i)(B_{t_{k+1}}^j - B_{t_k}^j) \mid \mathcal{F}_s\}$

vanishes if $i \neq j$, and equals h if $i = j$. Thus,

$$E\{f(B_t) - f(B_s) \mid \mathcal{F}_s\} = \frac{1}{2} \sum_{k=0}^{n-1} \sum_{i=1}^{d} E\left\{\frac{\partial^2}{\partial x_i^2} h f(B_{t_k}) \mid \mathcal{F}_s\right\}$$
$$+ \sum_{k=0}^{n-1} E\{R_2(B_{t_k}, B_{t_{k+1}}) \mid \mathcal{F}_s\}.$$

But $|E\{R_2(B_{t_k}, B_{t_{k+1}}) \mid \mathcal{F}_s\}| \leq CE\{|B_{t_{k+1}} - B_{t_k}|^3\} = Ch^{3/2}E\{|B_1 - x|^3\} \leq C\sigma_3 h^{3/2}$, so the last sum is bounded by $C\sigma_3 n h^{3/2}$, and we have

$$E\{f(B_t) - f(B_s) \mid \mathcal{F}_s\} = \frac{1}{2} \sum_{i=1}^{d} E\left\{\sum_{k=0}^{n-1} \frac{\partial^2}{\partial x_i^2} f(B_{t_k}) \mid \mathcal{F}_s\right\} h + O(nh^{3/2}).$$

Let $n \to \infty$. The sum inside the expectation is a Riemann sum which converges boundedly to $\int_s^t \mathbb{G} f(B_u) \, du$, while the last term goes to zero. Thus $E\{f(B_t) - f(B_s) \mid \mathcal{F}_s\} = \frac{1}{2} E\{\int_s^t \mathbb{G} f(B_u) \, du \mid \mathcal{F}_s\}$. $\square$

Recalling that $\mathbb{G} = \frac{1}{2}\Delta$, we see that if $\Delta f \equiv 0$, then $f(B_t)$ is a martingale. The equation $\Delta f = 0$ is **Laplace's equation**, and its solutions are **harmonic functions**. A harmonic function of a Brownian motion is a martingale. In one dimension, all harmonic functions are linear. The class of harmonic functions is much larger in higher dimensions. In two dimensions, for example, the real and complex parts of holomorphic functions are harmonic. They are important in physics because electrical potentials in regions without charges are harmonic functions.

Notation. Let $r(x) = \sqrt{x_1^2 + \cdots + x_d^2}$. Let $B(x, a) = \{y \in \mathbb{R}^d : |y - x| \leq a\}$ be the closed ball of radius a centered at x. Let $\tau_a = \inf\{t > 0 : |B_t| = a\}$. Let f be the *fundamental potential with pole at the origin*, defined by:

(i) $f(x) \stackrel{\text{def}}{=} \log(1/r(x))$ if $d = 2$;

(ii) $f(x) \stackrel{\text{def}}{=} \frac{1}{r^{d-2}(x)}$ if $d \geq 3$.

Then, for all $d \geq 2$, f is harmonic except at the origin, and $x \mapsto f(x - y)$ is harmonic except at y.

We cannot apply Proposition 10.49 to $f(B_t)$, since f not only fails to have compact support, it blows up at the origin. However, there is a standard way to work around this. Let $g(x)$ be an infinitely-differentiable function which vanishes if $|x| < \varepsilon/2$ or $|x| > R + 1$, and which is identically 1 in $\varepsilon \leq |x| \leq R$. Set $\hat{f}(x) = g(x)f(x)$. Then $\hat{f}$ is smooth, has compact support, and equals $f(x)$ if $\varepsilon \leq |x| \leq R$. If B_t is a Brownian motion from x, and $\varepsilon < |x| < R$, then the theorem applies to $\hat{f}(B_t)$: $\hat{f}(B_t) - \int_0^t \mathbb{G}\hat{f}(B_s) \, ds$ is a martingale. But $\hat{f}(x) = f(x)$ if $\varepsilon \leq |x| \leq R$, so if we define $\tau_a = \inf\{t : |B_t^x| = a\}$, then $\mathbb{G}\hat{f}(B_{t\wedge\tau_\varepsilon \wedge \tau_R}^x) \equiv 0$. Therefore $\hat{f}(B_{t\wedge\tau_\varepsilon \wedge \tau_R}^x) = f(B_{t\wedge\tau_\varepsilon \wedge \tau_R}^x)$ is a martingale[7].

Let $x \in \mathbb{R}$ and let $0 < a < |x| < b$. Let $\tau = \tau_a \wedge \tau_b$. Let B_t^x be Brownian motion from x. The sample paths of B_t^x are unbounded, so that τ_b is a.s. finite, so τ is necessarily finite, too.

Now $\{f(B_{t\wedge\tau}^x), t \geq 0\}$ is a bounded martingale. It converges as $t \to \infty$ to $f(B_\tau^x)$, so $(f(x), f(B_\tau^x))$ is a martingale. Let $p = P\{|B_\tau| = a\}$. Then $f(x) = E\{f(B_\tau^x)\} = pf(a) + (1-p)f(b)$. Solve for p: $P\{|B_\tau^x| = a\} = \frac{f(b)-f(x)}{f(b)-f(a)}$. (Here, "$f(a)$" is shorthand for $f(y)$, $|y| = a$.) Thus, if $0 < a < |x| < b$,

$$(10.19) \qquad P^x\{\tau_a < \tau_b\} = \begin{cases} \frac{\log b - \log |x|}{\log b - \log a} & \text{if } d = 2, \\ \frac{\frac{1}{b^{d-2}} - \frac{1}{|x|^{d-2}}}{\frac{1}{b^{d-2}} - \frac{1}{a^{d-2}}} & \text{if } d \geq 3. \end{cases}$$

Theorem 10.50. *(i) If $d \geq 2$, B_t does not hit points. That is, for any fixed $y \in \mathbb{R}^d$, the probability that $B_t^x = y$ for some $t > 0$ is zero.*

(ii) If $d = 2$, Brownian motion in $\mathbb{R}^2$ is recurrent in the sense that for any $\varepsilon > 0$, the set $\{t : |B_t| < \varepsilon\}$ is a.s. unbounded. Moreover, the Brownian path is a.s. dense in the plane.

Proof. (i) Notice that $f(|x|) \to \infty$ as $|x| \to 0$. Thus, if $x \neq 0$, and $b > |x|$,

$$\lim_{a \to 0} P^x\{\tau_a < \tau_b\} = \lim_{a \to 0} \frac{f(b) - f(x)}{f(b) - f(a)} = 0,$$

Let τ_0 be the first hit of zero. Clearly, $\tau_0 = \lim_{a \to 0} \tau_a$. Thus $P\{\tau_0 < \tau_b\} = 0$. But $\tau_b \to \infty$ as $b \to \infty$, so evidently $\tau_0 = \infty$ a.s.

In particular, if $y \neq x$, B_t^{x-y} never hits the origin. Thus $y + B_t^{x-y}$ never hits y. But $y + B_t^{x-y}$ is a Brownian motion from x. Thus, for any $x, y \in \mathbb{R}^d$, B_t^x never hits y. Even if $y = x$, B_t^x never hits x for strictly positive t.

Indeed, taking $x = 0$, if $\varepsilon > 0$, $B_{t+\varepsilon}$ is a Brownian motion from B_ε by the Markov property, and $B_\varepsilon \neq 0$ a.s. Therefore it never hits zero, so with probability one, there is no $t > \varepsilon$ for which $B_t = 0$. But ε is arbitrary, so B_t is not zero for any strictly positive t.

(ii) Let $a = \varepsilon$ and let $b \to \infty$ in (10.19). If $d = 2$, then $f(b) \to -\infty$ as $b \to \infty$, so that $\lim_{b\to\infty} P\{\tau_\varepsilon < \tau_b\} = 1$. Thus τ_ε is finite with probability one, so B^x eventually hits the ball $B(0, \varepsilon)$. In fact, let $t_0 > 0$. Then $\{B_{t_0+t}^x, t \geq 0\}$ is a Brownian motion from $B_{t_0}^x$. Therefore it must eventually hit $B(0, \varepsilon)$. This means that for any x and any $t_0 > 0$, the set $\{t : B_t^x \in B(0, \varepsilon)\}$ contains points greater than t_0, and is therefore unbounded.

---

[7]The system theorem used here has only been proved for discrete-parameter martingales. We leave it to the reader to make the extension to include this case. See the problems.

10.9. Higher Dimensional Brownian Motion

To see the path is dense, note that, by symmetry, B_t reaches $B(x, \varepsilon)$ with probability one. Choose a sequence $\varepsilon_n \downarrow 0$ and a sequence of points (x_n) which are dense in the plane. Then, for each m and n, B_t hits $B(x_m, \varepsilon_n)$ a.s. There are only a countably many of these balls, so that with probability one, B_t hits *all* of them. Therefore, for a.e. ω, the path $\{B_t(\omega), t \geq 0\}$ is dense in the plane. □

Notice that even if the process starts at the origin, it can never return there, although, in $d = 2$, it is sure to return to any neighborhood of the origin. In order to see the recurrence in higher dimensions, let us first prove a lemma.

Lemma 10.51. *Let $x \neq 0$. Then the process $\{f(B_t^x), \mathcal{F}_t, t \geq 0\}$ is a supermartingale.*

Proof. This is true in all dimensions; we will prove it for $d \geq 3$, and indicate how to modify the proof for $d = 2$ at the end.

For $\varepsilon > 0$, let $T_\varepsilon = \inf\{t : |B_t^x| \leq \varepsilon \text{ or } |B_t^x| > 1/\varepsilon\}$. By Theorem 10.50, $T_\varepsilon \to \infty$ as $\varepsilon \to 0$. First, $f(B_t)$ is integrable, since, as $f(B_{t \wedge T_\varepsilon}^x)$ is a positive martingale, $f(x) = E\{f(B_0^x)\} = E\{f(B_{t \wedge T_\varepsilon}^x)\}$. But as $\varepsilon \to 0$, $f(B_{t \wedge T_\varepsilon}^x) \to f(B_t^x)$ so by Fatou's Lemma, $f(x) \geq E\{f(B_t^x)\}$.

Now let $s \leq t$ and $\delta < \varepsilon < |x|$. $f(B_{t \wedge T_\delta}^x)$ is a martingale, and $T_\varepsilon \leq T_\delta$, so $\left(f(B_{s \wedge T_\varepsilon}), f(B_{t \wedge T_\delta})\right)$ is a positive martingale. Thus for $\Lambda \in \mathcal{F}_s$,

$$\int_{\Lambda \cap \{T_\varepsilon \geq s\}} f(B_s^x) \, dP = \int_{\Lambda \cap \{T_\varepsilon \geq s\}} f(B_{t \wedge T_\delta}^x) \, dP.$$

Let $\delta \to 0$, so $T_\delta \to \infty$ and use Fatou's lemma to see that this is

$$\geq \int_{\Lambda \cap \{T_\varepsilon \geq s\}} f(B_t^x) \, dP.$$

Let $\varepsilon \to 0$ and use dominated convergence on both sides to see that

$$\int_\Lambda f(B_s^x) \, dP \geq \int_\Lambda f(B_t^x) \, dP.$$

This holds for all $0 \leq s < t$ and $\Lambda \in \mathcal{F}_s$, so $(f(B_t^x))$ is a supermartingale.

If $d = 2$, f is no longer positive, but $f(x) + |x|$ is. Thus for $s \leq t$, $f(B_s^x)$ is bounded below by the integral random variable $-\sup\{|B_u| : u \leq t\}$, and we can still apply the Fatou argument. (See Exercise 4.27.) □

Theorem 10.52. *Brownian motion is transient in $d \geq 3$.*

Proof. In fact, $|B_t| \to \infty$ a.s. Indeed, its first coordinate is a Brownian motion, which has an infinite lim sup as $t \to \infty$. Therefore $\limsup_{t \to \infty} |B_t| = \infty$. Now $f(x) = |x|^{2-d}$, so $f(x) \to 0$ as $|x| \to \infty$. Thus $\liminf_{t \to \infty} f(B_t) = 0$;

but $(f(B_t))$ is a positive supermartingale. Therefore it converges [8] as $t \to \infty$, and the limit must be zero. But $f(B_t) \to 0 \Rightarrow |B_t| \to \infty$. □

Problems 10.9

10.37. Show that standard Brownian motion in $\mathbb{R}^d$ is invariant under rotation.

10.38. Let $X_t = (B_t^1, B_t^2)$, be two-dimensional Brownian motion. For $a > 0$, let T_1 be the time that X_t first hits the line $\{(x, y) : x = a\}$ in the plane. Show that $B^2(T_1)$ is a Cauchy random variable with parameter λ.

10.39. (Continuation.) Let $T_t = \inf\{s : B_s^1 = t\}$ be the first hit by X of the line $\{(x, t) : x = t\}$. Let $Y_t = B_{T_t}^2$, $t \geq 0$. Show that $\{Y_t, t \geq 0\}$ is a process of stationary independent increments, and that $Y_{t+s} - Y_t$ is Cauchy (s). The process (Y_t) is called a **Cauchy process**.

10.10. Financial Mathematics II: The Black-Scholes Model*

Bachelier [1] originally introduced Brownian motion to model the stock market[9] market. While it captures the market's unpredictability, it has a drawback: stock prices are always positive, but Brownian motion isn't. To get a more realistic model of the ups and downs of a typical stock, consider **Logarithmic Brownian motion**[10]. It is defined by

$$S_t = S_0 e^{\sigma B_t + \mu t},$$

where (B_t) is a standard Brownian motion, $S_0 \geq 0$, and σ and μ are real numbers. This is closely connected to Brownian motion, but it bypasses the main objection: it cannot take on negative values. The parameter σ is called the **volatility** of the stock, and governs how quickly its price changes.

This is the **Black-Scholes** model. It is considered realistic enough that traders actually use it to value options.

According to Theorem 9.57, if there is no arbitrage in the market, there is an equivalent martingale measure Q, and the prices of all financial derivatives are just expectations relative to it. Our problem is to find Q.

Change of Measure and the Cameron-Martin Theorem. The key to the martingale measure is the Cameron-Martin theorem.

If Q is a measure on $(\Omega, \mathcal{F})$ which is equivalent to P, then $dQ = M\, dP$ for some integrable, a.s. strictly positive random variable M. M is the

---

[8]We have proved this for discrete-parameter martingales. But, as B_t never hits 0, $f(B_t)$ is continuous and we can take the limit for $t = k2^{-n}$ as $k \to \infty$ for every n and conclude the limit over all t exists and is zero.

[9]See Baxter and Rennie [3] for a good overview of the probabilistic approach to financial mathematics. Black and Scholes' original approach [5] was quite different.

[10]It is also called **geometric Brownian motion** and **exponential Brownian motion**, but it is usually "logarithmic Brownian motion" in the financial literature.

Radon-Nikodym density of Q with respect to P. Let us use E^P and E^Q for the expectations with respect to P and Q. For a random variable Y,
$$E^Q\{Y\} = \int Y\, dQ = \int Y M\, dP = E^P\{YM\}.$$
Take $Y \equiv 1$ to see that Q is a probability measure iff $E^P\{M\} = 1$.

Define $M_t = E^P\{M \mid \mathcal{F}_t\}$ for $t \geq 0$. Then (M_t) is a martingale and if Y is $\mathcal{F}_t$-measurable, then
$$E^Q\{Y\} = E^P\{MY\} = E^P\{Y E^P\{M \mid \mathcal{F}_t\}\} = E^P\{YM_t\}.$$
Thus M_t is the Radon-Nikodym density of Q with respect to P on $\mathcal{F}_t$.

Let B_t be a standard Brownian motion on $(\Omega, \mathcal{F}, P)$ and $\mu \in \mathbb{R}$. The following is a special case of the celebrated Cameron-Martin theorem.

Theorem 10.53 (Cameron-Martin). *Let $T > 0$ and let Q be the measure defined by $dQ = e^{-\mu B_T - \frac{1}{2}\mu^2 T} dP$. Then Q is a probability measure, and on the space $(\Omega, \mathcal{F}, Q)$, the process $\{B_t + \mu t, \, 0 \leq t \leq T\}$ is a standard Brownian motion.*

Proof. Let $X_t = B_t + \mu t$, and let $M_t = e^{-\mu B_t - \frac{1}{2}\mu^2 t}$. Then $dQ = M_T dP$. By Proposition 10.5, M_t is a mean-one martingale (under P), so Q has mass one. We claim
$$(10.20) \qquad E^Q\{e^{i\theta(X_t - X_s)} \mid \mathcal{F}_s\} = e^{-\frac{1}{2}\theta^2(t-s)}, \quad 0 \leq s < t, \, \theta \in \mathbb{R}.$$
Now the right-hand side of (10.20) is a $N(0, t-s)$ characteristic function, so this implies that $X_t - X_s$ is $N(0, t-s)$, independent of $\mathcal{F}_s$. In other words, (X_t) has independent Gaussian increments of the right mean and variance, and is therefore a standard Brownian motion. The theorem will be proved once we show (10.20).

Let $s < t$, $\theta \in \mathbb{R}$, and let Y be a bounded $\mathcal{F}_s$-measurable random variable.
$$E^Q\{Y e^{i\theta(X_t - X_s)}\} = E^P\{Y e^{i\theta(X_t - X_s)} M_T\} = E^P\{Y e^{i\theta(X_t - X_s)} M_t\}$$
since Y and $X_t - X_s$ are $\mathcal{F}_t$-measurable and (M_t) is a P-martingale. Now $M_t = M_s e^{-\mu(B_t - B_s) - \frac{1}{2}\mu^2(t-s)}$, so, expanding $X_t - X_s$, this is
$$= E^P\{M_s Y e^{(i\theta - \mu)(B_t - B_s)}\} e^{-\frac{1}{2}\mu^2(t-s) + i\theta\mu(t-s)}$$
$$= E^P\{M_s Y E^P\{e^{(i\theta - \mu)(B_t - B_s)} \mid \mathcal{F}_s\}\} e^{-\frac{1}{2}\mu^2(t-s) + i\theta\mu(t-s)},$$
for $M_s Y$ is $\mathcal{F}_s$-measurable. Now (B_t) is a standard Brownian motion under P, so $B_t - B_s$ is $N(0, t-s)$, independent of $\mathcal{F}_s$. Thus the conditional expectation is just the ordinary expectation. By Exercise 6.10[11] with $z =$

---
[11]This exercise just extends the Gaussian moment generating function to complex arguments.

$i\theta - \mu$, the above is

$$= E^P\{YM_s\}\, e^{\frac{1}{2}(i\theta-\mu)^2(t-s)}\, e^{-\frac{1}{2}\mu^2(t-s)+i\theta\mu(t-s)} = E^Q\{Y\}\, e^{-\frac{1}{2}\theta^2(t-s)},$$

where we have used the fact that Y is M_s-measurable and M_s is the Radon-Nikodym derivative of Q on $\mathcal{F}_s$. But this is true for all bounded $\mathcal{F}_s$-measurable Y, which proves (10.20). □

Let us use the Cameron-Martin theorem to find *a* martingale measure. It turns out to be unique, so that it is in fact *the* martingale measure, although we will not prove uniqueness here.

Let $S_t = S_0 e^{\sigma B_t + \mu t}$ be the stock price. If the interest rate is $r \geq 0$, the discounted stock price is $\hat{S}_t = e^{-rt}S_t = S_0 e^{\sigma B_t + (\mu - r)t}$. We want to find a Q such that $\{\hat{S}_t, \mathcal{F}_t, 0 \leq t \leq T\}$ is a martingale on $(\Omega, \mathcal{F}, Q)$.

In order to do that, let $W_t \stackrel{\text{def}}{=} B_t + \alpha t$ for some α. In terms of W, $\hat{S}_t = S_0 e^{\sigma W_t - (\alpha\sigma + \mu - r)t}$. According to the Cameron-Martin theorem, there is a Q such that W is a Q-Brownian motion, in which case $\hat{S}$ is a Q-martingale iff $\alpha\sigma + \mu - r = \sigma^2/2$, or $\alpha = \sigma/2 - (\mu - r)/\sigma$. Thus if $T > 0$, Q is given explicitly on $\mathcal{F}_T$ by

$$dQ \stackrel{\text{def}}{=} e^{-(\frac{\mu-r}{\sigma} - \frac{1}{2}\sigma)B_T - \frac{1}{2}(\frac{\mu-r}{\sigma} - \frac{1}{2}\sigma)^2 T}\, dP.$$

Thus we have

Theorem 10.54. *For $T > 0$ there exists a unique equivalent martingale measure Q on $\mathcal{F}_T$, such that on $(\Omega, \mathcal{F}_T, Q)$, $\hat{S}_t = S_0 e^{\sigma W_t - \frac{1}{2}\sigma^2 t}$, where $\{W_t,\ 0 \leq t \leq T\}$ is a standard Brownian motion. If X is a derivative on $[0, T]$ expressed in discounted dollars, then the value of X at time zero is $V_0(X) = E^Q\{X\}$, and the value at time $t \leq T$ is $V_t(X) = E^Q\{X \mid \mathcal{F}_t\}$.*

Remark 10.55. (i) We do not need Q explicitly. It is enough to know that it exists, and that the stock price is given in terms of a Brownian motion.

(ii) We have only shown that there is such a Q. The fact that Q is unique is deeper; we cannot prove it here. We will just accept it.

Example 10.55.1. The European call option with maturity T gives the right, but not the obligation, to buy a stock at time T at a fixed price K, called the **strike price**. Its payoff is $(S_T - K)^+$. Let us find its value at time zero.

The payoff is given in time-T dollars. Its value in discounted dollars is $e^{-rT}(S_T - K)^+ = (\hat{S}_T - e^{-rT}K)^+$. Thus its value at time zero is $V_0 = E^Q\{(\hat{S}_T - e^{-rT}K)^+\}$.

This only depends on the value of $\hat{S}_T$, whose distribution is known. It has a closed-form expression in terms of the standard normal distribution function $\Phi(t) = \frac{1}{\sqrt{2\pi}} \int_{-\infty}^{t} e^{\frac{-x^2}{2}}\, dx$.

In terms of the $N(0,1)$ random variable $\xi \stackrel{\text{def}}{=} W_T/\sqrt{T}$, $\hat{S}_T = e^{\sigma\sqrt{T}\xi - \frac{1}{2}\sigma^2 T}$. Then $\hat{S}_T > e^{-rT}K$ if $\xi > (\log \frac{K}{S_0} - rT + \frac{1}{2}\sigma^2 T)/\sqrt{\sigma^2 T} \stackrel{\text{def}}{=} \nu$. Thus,

$$V_0 = \int_\nu^\infty \left(S_0 e^{\sqrt{\sigma^2 T} x - \frac{1}{2}\sigma^2 T} - e^{rT} K\right) \frac{e^{-\frac{x^2}{2}}}{\sqrt{2\pi}} dx$$

$$= \frac{S_0}{\sqrt{2\pi}} \int_\nu^\infty e^{-\frac{x^2 - 2\sqrt{\sigma^2 T} x + \sigma^2 T}{2}} dx - \frac{e^{-rT}K}{\sqrt{2\pi}} \int_\nu^\infty e^{-\frac{x^2}{2}} dx$$

$$= \frac{S_0}{\sqrt{2\pi}} \int_{\nu - \sqrt{\sigma^2 T}}^\infty e^{-\frac{x^2}{2}} dx - \frac{e^{-rT}K}{\sqrt{2\pi}} \int_\nu^\infty e^{-\frac{x^2}{2}} dx.$$

Noting that $1 - \Phi(t) = \Phi(-t)$, this is

$$= S_0 \Phi(\sqrt{\sigma^2 t} - \nu) - e^{-rT} K \Phi(-\nu).$$

This leads to the Black-Scholes formula:

Proposition 10.56. *The value at time zero of a European call option of strike price K and maturity T with interest rate r is*

$$V_0 = S_0 \Phi\left(\frac{\log \frac{S_0}{K} + rT + \frac{1}{2}\sigma^2 T}{\sqrt{\sigma^2 T}}\right) - e^{-rT} K \Phi\left(\frac{\log \frac{S_0}{K} + rT - \frac{1}{2}\sigma^2 T}{\sqrt{\sigma^2 T}}\right).$$

Remark 10.57. Note that the original drift μ has dropped out, and only the volatility σ enters. The price of an option depends only on the volatility. This is striking: options on two different stocks with the same volatility will have the same price, even if the value of one is rapidly rising and the other is deeply declining.

In fact, this is true of general derivatives, not just options, for μ has already dropped out in Theorem 10.54: $\hat{S}_t$ only depends on σ.

10.11. Skorokhod Embedding*

A simple symmetric random walk can be embedded in a Brownian motion[12]. Let B_t be a standard Brownian motion. Define stopping times (T_n) by:

$$T_0 = 0,$$
$$T_1 = \inf\{T > 0 : |B_t| = 1\},$$

and, by induction,

$$T_{n+1} = \inf\{t > T_n : |B_t - B_{T_n}| = 1\}, \ n = 2, 3, \dots.$$

Notice that the T_n are all finite, they increase with n, and that B_{T_n} and $B_{T_{n+1}}$ differ by one. Now $B_{T_0} = B_0 = 0$, so the process $\{B_{T_n}, n = 0, 1, 2, \dots\}$ has integer values and moves up and down one integer at a time.

---

[12]This section is taken from Chacon and Walsh [**8**].

Moreover, it is a martingale, so that $B_{T_{n+1}} - B_{T_n} = \pm 1$ with probability 1/2 each. This implies the following remarkable fact.

Exercise 10.40. Prove that

(a) The random variables B_{T_1}, $B_{T_2} - B_{T_1}$, $B_{T_3} - B_{T_2}, \ldots$ are i.i.d.

(b) The random variables T_1, $T_2 - T_1$, $T_3 - T_2, \ldots$ are i.i.d.

(c) $B_{T_{n+1}} - B_{T_n}$ and $T_{n+1} - T_n$ are independent. Consequently, all random variables in (a) and (b) are independent.

It follows that $\{B_{T_n},\ n = 0, 1, 2, \ldots\}$ is a simple symmetric random walk.

The key step was to embed a plus-minus one random variable by means of a stopping time. We just did that repeatedly to get the random walk. This has a far-reaching generalization due to Skorokhod: one can embed essentially any random variable by means of stopping times.

This provides an excuse to introduce potential theory. In higher dimensions, potential theory describes the behavior of electrical and gravitational potentials, and, more generally, studies the solutions of elliptic and parabolic partial differential equations. We only need its poor cousin, one-dimensional potential theory, which, although it still describes the behavior of electrical potentials, is quite simple, even though it shares the same basic ideas.

The nice thing about this is that once we have established a few basic properties, the proof of the Skorokhod embedding theorem itself can be done with four pictures.

Some Basic Facts About Potentials. The first thing we need is the **potential kernel**, which is the function $U(x,y) = -|y - x|$. If μ is a measure on R, the **potential of μ** is the function[13]

$$U\mu(x) \stackrel{\text{def}}{=} \int_{-\infty}^{\infty} U(x,y)\, \mu(dy) = -\int |y - x|\, \mu(dy).$$

This is finite if x is μ-integrable. If μ is a probability measure, there is a random variable X with distribution μ, and we can write the potential in terms of X:

$$U\mu(x) = -E\{|X - x|\}.$$

Here are the basic facts we will need.

---

[13] In electromagnetism, the measure μ represents a distribution of electrical charges, and $U\mu(x)$ is the electrical potential field arising from it.

10.11. Skorokhod Embedding★

1° If $\int |y|\,\mu(dy) < \infty$, $U\mu(x)$ is a concave, continuous function.

2° If μ is a probability measure with mean zero, and if δ_0 is the unit mass at zero, then $U\mu \leq \mu\delta_0$. Furthermore, $U\delta_0(x) - U\mu(x) \to 0$ as $|x| \to \infty$.

3° If $\mu, \mu_1, \mu_2, \ldots$ are probability measures such that $U\mu(x) \to U\mu(x)$ for all x, then $\mu_n \to \mu$ weakly.

Notice that 3° implies that the potential determines the measure. Indeed, if μ and ν are two measures with finite potentials and $U\mu = U\nu$, apply 3° with $\mu_n \equiv \mu$ to conclude that $\mu \Longrightarrow \nu$, which implies that $\mu = \nu$.

Proof. Let us give quick proofs of these. For 1°, $x \mapsto U(x, y)$ is clearly concave, so for any $\lambda \in [0, 1]$ and $a < x < b$, $U(x, y) \geq \lambda U(a, y) + (1-\lambda)U(b, y)$. Integrate the inequality with respect to μ to see $U\mu(x) \geq \lambda U\mu(a) + (1-\lambda)U\mu(b)$. Thus $U\mu$ is concave on $\mathbb{R}$, and therefore continuous.

For 2°, first suppose $x \leq 0$ and let X have distribution μ:

$$\begin{aligned} U\delta_0(x) - U\mu(x) &= -|x| + E\{|X - x|\} \\ &= x + E\{X - x; X > x\} + E\{x - X; X \leq x\} \\ &= x + E\{X - x\} + 2E\{x - X; X \leq x\}. \end{aligned}$$

But X has mean zero so this becomes $2E\{x - X; X \leq x\} \geq 0$. The case $x \geq 0$ is similar, which gives a formula worth recording for future reference:

$$(10.21) \qquad U\delta_0(x) - U\mu(x) = \begin{cases} 2E\{x - X; X \leq x\} & \text{if } x \leq 0, \\ 2E\{X - x; X \geq x\} & \text{if } x > 0. \end{cases}$$

For 3°, define

$$f_{ax}(y) = 2U(x, y) - U(x-a, y) - U(x+a, y) = |y-x+a| - 2|y-x| + |y-x-a|.$$

If we take some care with signs, we see that f_{ax} is a tent function: it is continuous, it vanishes outside of $(x - a, x + a)$, it equals $2a$ if $y = x$ and it is linear in each of the intervals $(x - a, x)$ and $(x, x + a)$. (Its graph looks like a tent.)

Suppose that $\mu, \mu_1, \mu_2, \ldots$ are probability measures. If $U\mu_n(x) \to U\mu(x)$ for all x, then for each a and x, $\int f_{ax}(y)\,\mu_n(dy) \longrightarrow \int f_{ax}\mu(dy)$. By Exercise 6.17[14], $\mu_n \Longrightarrow \mu$, proving 3°. □

We need one more fact about what is called the balayage[15] of potentials.

---

[14]That exercise is a consequence of the fact that any continuous function of compact support is the uniform limit of finite linear combinations of tent functions.

[15]Balayage is French for "sweeping out". It describes taking the mass of a measure in an interval and "sweeping it out" to the endpoints.

4° Let ν be a probability measure with finite mean and let $[a,b]$ be a finite interval. Let B be Brownian motion with initial distribution ν and define
$$T_{ab} = \inf\{t \geq 0 : B_t \leq a \text{ or } B_t \geq b\}.$$
Let ν' be the distribution of $B_{T_{ab}}$. Then $U\nu'$ is linear in $[a,b]$ and equals $U\mu$ outside $[a,b]$.

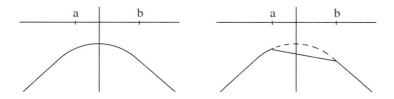

Figure 2. The potentials $U\nu$ (left) and $U\nu'$ (right)

Proof. (of 4°.) To see this, write it in terms of expectations:
$$U\nu'(x) = -E\{|B_{T_{ab}} - x|\} = -E\{E\{|B_{T_{ab}} - x| \mid B_0\}\}.$$

First suppose $x \notin [a,b]$. We claim that $\{|B_{t \wedge T_{ab}} - x|,\ t \geq 0\}$ is a martingale. Indeed, it is constant if $B_0 \notin (a,b)$, and, if $B_0 \in (a,b)$, then the process can never hit x, so that $B_{t \wedge T_{ab}} - x$ never changes sign. Thus, if $x < a$, then $|B_{t \wedge T_{ab}} - x| = B_{t \wedge T_{ab}} - x$, which is indeed a martingale. Similarly, it is a martingale if $x > b$. Therefore $U\nu'(x) = -E\{|B_{T_{ab}} - x|\} = -E\{|B_0 - x|\} = U\nu(x)$, $x \notin [a,b]$.

It remains to show that $x \mapsto U\nu'(x)$ is linear in $[a,b]$, and for this, it is sufficient to show that $x \mapsto E\{|B_{T_{ab}} - x| \mid B_0\}$ is linear. (Indeed, $U\nu'(x)$ is just the expectation of this.) But we can calculate this explicitly: if $B_0 \notin [a,b]$, $B_{T_{ab}} = B_0$, and if $B_0 \in [a,b]$, we know the exit probabilities from the interval. Thus, for $x \in [a,b]$,
$$E\{|B_{T_{ab}} - x| \mid B_0\} = (B_0 - x)I_{\{B_0 > b\}} + (x - B_0)I_{\{B_0 < a\}}$$
$$+ \left(\frac{b - B_0}{b - a}(x - a) + \frac{B_0 - a}{b - a}(b - x)\right) I_{\{B_0 \in [a,b]\}}.$$
This is indeed linear in x. □

Skorokhod's Theorem.

Theorem 10.58 (Skorokhod's Embedding Theorem). *Let B_t be a standard Brownian motion from the origin and let μ be a probability measure on $\mathbb{R}$ with mean zero and a finite second moment. Then there exists a stopping time T such that B_T has distribution μ, and $E\{T\} = \int x^2\, \mu(dx)$.*

10.11. Skorokhod Embedding*

To prove this we will construct a sequence $T_0 \leq T_1 \leq T_2 \leq \ldots$ of stopping times[16] which increase to a limit T, such that B_T has the distribution μ. Let μ_n be the distribution of $\mathcal{B}_{T_n}$. The following pictures will explain our construction... and the following words will explain the pictures.

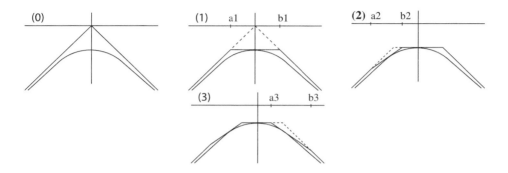

Figure 3. The potentials **(0)** $U\delta_0$ and $U\mu$, **(1)** $U\mu_1$, **(2)** $U\mu_2$ and **(3)** $U\mu_3$.

Choose $T_0 = 0$, so $\mu_0 = \delta_0$. Then the potentials $U\mu_0 = -|x|$ and $U\mu$ are as in **(0)** of Figure 3. (See 2°) Next, choose x such that $U\mu(x) < U\mu_0$, draw a tangent to the graph of $U\mu$ at x, and let a_1 and b_1 be as in **(1)**. (The fact that a_1 and b_1 are finite follows from 2°.)

Now let $T_1 = T_{a_1 b_1}$, the first exit from (a_1, b_1). Then by 4°, B_{T_1} has the distribution μ_1 whose potential is graphed in **(1)**. Continuing in the same vein, we choose another x for which $U\mu(x) < U\mu_1(x)$ and draw the tangent to the graph of $U\mu$ at x. The tangent will intersect the graph of $U\mu_1$ at points a_2 and b_2. Then we let T_2 be the first exit of (a_2, b_2) after T_1. The strong Markov property assures us that B_t restarts at T_1, which is equivalent to starting Brownian motion with initial distribution μ_1 and letting it go until it first exits from (a_2, b_2). Let μ_2 be the distribution of B_{T_2}. Then the potentials $U\mu_2$ and $U\mu$ are as shown in **(2)**. At the next step, we choose another point x, find the corresponding interval (a_3, b_3), and let T_3 be the first exit from that interval after T_2, and so on. At each stage, $U\mu_n$ is piecewise linear and $U\mu_n \geq U\mu$. The stopping times increase, and their limit, T is the stopping time we are after.

We have not been specific about the exact choice of the functions $U\mu_n$ and, in fact, it doesn't much matter. What is important is that we can choose them so that they decrease to the function $U\mu$. Indeed, any concave function is the infimum of a countable family of linear functions, and each $U\mu_n$ is the infimum of finitely many.

---

[16] For those who have seen proofs which rely on randomized stopping times, these are *non-randomized* times.

The important point is that $U\mu_n \downarrow U\mu$, so that by 3°, $\mu_n \Longrightarrow \mu$. At the same time, if T is finite, $B_{T_n} \to B_T$ by continuity, so B_T has distribution μ.

It remains to show that $T < \infty$.

Lemma 10.59. *Let ν be a measure on $\mathbb{R}$ with potential $U\nu$. Then $\int x^2 \, d\nu$ is equal to the area between the curves $y = -|x|$ and $y = U\nu(x)$.*

Proof. Write (10.21) in terms of the measures:

$$(10.22) \qquad U\delta_0(x) - U\nu(x) = \begin{cases} 2\int_{-\infty}^{x}(x-y)\,\nu(dy) & \text{if } x \leq 0, \\ 2\int_{x}^{\infty}(y-x)\,\nu(dy) & \text{if } x > 0. \end{cases}$$

so that the area between the two curves is

$$\int_{-\infty}^{\infty} \bigl(-|y| - U\nu(y)\bigr)\,dy$$

$$= 2\int_{-\infty}^{0} dy \int_{-\infty}^{y} (y-x)\nu(dx) + 2\int_{0}^{\infty} dy \int_{0}^{\infty} (x-y)\nu(dx)$$

$$= 2\int_{-\infty}^{0} \nu(dx) \int_{x}^{0} (y-x)\,dy + 2\int_{0}^{\infty} \nu(dx) \int_{0}^{x} (x-y)\,dy$$

$$= \int_{-\infty}^{\infty} x^2 \, \nu(dx).$$

$\square$

Returning to the proof of the embedding theorem, note that since $B_{t \wedge T_n}^2 - t \wedge T_n$ is a mean-zero martingale and $B_{t \wedge T_n}$ is bounded, we can let $t \to \infty$ to conclude that $E\{T_n\} = E\{B_{T_n}^2\}$. But the right-hand side equals the area between $-|x|$ and $U\mu_n$ which is bounded by the area between $-|x|$ and $U\mu$, and this equals $\int x^2 d\mu < \infty$ by Lemma 10.59. Thus the expectations $E\{T_n\}$ are bounded, and we conclude both that $T < \infty$ a.s. and $E\{T\} = \lim E\{T_n\} = \int x^2 \, d\mu$.

This finishes the proof of Skorokhod's Embedding Theorem! $\square$

Remark 10.60. In applications, the fact that $E\{T\}$ equals the variance of μ gives some useful control over T. If the target distribution only has a pth moment for some $p > 1$, this method gives a stopping time with a $p/2^{\text{th}}$ moment, though the proof is more complicated. In fact, if we are willing to lose all control over the time T, we can embed *any* probability distribution in a Brownian motion. See the problems.

10.11.1. Embedding Sums of Random Variables. Let $X_1, X_2, \ldots$ be i.i.d. random variables with mean zero and variance σ^2. Let $S_0 = 0$ and $S_n = X_1 + \cdots + X_n$, $n = 1, 2, \ldots$. Then S_n is a random walk. Let μ be the distribution of X_1. Then μ has mean zero, so there is a stopping time τ_1 with mean σ^2 such that B_{τ_1} has distribution μ. Now $B_{\tau_1+t} - B_{\tau_1}$ is a Brownian motion independent of B_{τ_1}, so there is a stopping time τ_2 for it such that $B_{\tau_1+\tau_2} - B_{\tau_1}$ also has the distribution μ. Moreover, it is independent of B_{τ_1}. Let $T_1 = \tau_1$ and $T_2 = \tau_1 + \tau_2$. We continue, by embedding μ again in $(B_{T_2+t} - B_{T_2})$ with a time τ_3, and defining $T_3 = T_2 + \tau_2$, and so on. We end up with two sequences:

- stopping times $0 < T_1 \leq T_2 \leq \ldots$ such that $T_1, T_2 - T_1, T_3 - T_2, \ldots$ are i.i.d. and have mean σ^2; and
- random variables $B_0 = 0, B_{T_1}, B_{T_2}, \ldots$ such that $B_{T_1}, B_{T_2} - B_{T_1}, B_{T_3} - B_{T_2}, \ldots$ are i.i.d. with the same distribution as Y_1.
- Let $\xi_n = B_{T_n}$, $n = 0, 1, 2, \ldots$. The process $\{\xi_n, n = 0, 1, 2, \ldots\}$ has exactly the same distribution as $\{S_n, n = 0, 1, 2, \ldots\}$, and is therefore a random walk.

The stopping times T_n are sums of i.i.d. integrable random variables, so by Kolmogorov's strong law of large numbers,

$$(10.23) \quad \lim_{n\to\infty} \frac{T_n}{n} = E\{T_1\}; \quad \text{and} \quad \lim_{n\to\infty} \frac{T_{mn}}{n} = mE\{T_1\}.$$

both a.s. and in L^1.

In other words, the random walk (S_n) can be embedded in Brownian motion, and the stopping times are not too large. If the X_i have a non-zero mean, say λ, then one can embed the $X_i - \lambda$, such that $\{B_{T_n} + n\lambda, n = 0, 1, 2, \ldots\}$ has the same distribution as the random walk (X_n).

We saw at the top of the section that the simple symmetric random walk could be embedded in Brownian motion. Now we see that this was not special: any square-integrable random walk can be embedded in Brownian motion.

Application to the Central Limit Theorem. Let $X_1, X_2, \ldots$ be i.i.d. random variables with mean zero and variance one. Let $S_n = X_1 + \cdots + X_n$. We have just seen that (S_n) can be embedded in (B_t) by means of stopping times T_n, $n = 1, 2, \ldots$. By the bullets above, $T_{n+1} - T_n$, $n = 1, 2, \ldots$ are i.i.d. with mean one, and $(B_{T_n}) \stackrel{d}{\sim} (S_n)$.

Let us see why the basic central limit theorem (Theorem 6.37) is a direct consequence of this.

Theorem 10.61 (Central Limit Theorem). $\dfrac{1}{\sqrt{n}} S_n \Longrightarrow N(0, 1)$.

Before proving this let us note that if T_1 and T_2 are integrable stopping times for standard Brownian motion, then

$$(10.24) \quad E\left\{(B_{T_2} - B_{T_1})^2\right\} = E\{|T_2 - T_1|\}.$$

Indeed, if $T_1 \le T_2$, this is a consequence of the fact that $B_t^2 - t$ is a martingale. In the general case, apply this to the ordered stopping times $\tau \stackrel{def}{=} T_1 \wedge T_2$ and $\tau' \stackrel{def}{=} T_1 \vee T_2$; the result follows since $|T_2 - T_1| = \tau' - \tau$.

Proof. Embed (S_n) in (B_t) by stopping times T_n, so that (S_n) has the same distribution as $\xi_n \stackrel{def}{=} B_{T_n}$.

Rescale: let $\hat{B}_t^n = \frac{1}{\sqrt{n}} B_{nt}$. Then $(\hat{B}_t^n)$ is a standard Brownian motion and $\hat{B}_{\frac{T_n}{n}}^n \stackrel{d}{\sim} \frac{1}{\sqrt{n}} S_n$. We claim that this converges in distribution to $N(0,1)$. Since $\hat{B}_1^n$ is already $N(0,1)$, it is enough to show that $\hat{B}_{\frac{T_n}{n}}^n - \hat{B}_1^n \longrightarrow 0$ in probability. (See Exercise 6.38.) By Chebyshev's inequality and (10.24)

$$P\{|\hat{B}_{\frac{T_n}{n}}^n - \hat{B}_1^n| > \varepsilon\} \le \frac{1}{\varepsilon^2} E\left\{(\hat{B}_{\frac{T_n}{n}}^n - \hat{B}_1^n)^2\right\} = \frac{1}{\varepsilon^2} E\left\{\left|\frac{T_n}{n} - 1\right|\right\} \longrightarrow 0.$$

$\square$

Application to the Law of the Iterated Logarithm. We proved the law of the iterated logarithm for Brownian motion, but Khintchine proved it for sums of independent random variables. We can use embedding to derive Khintchine's original theorem.

In order to keep the notation under control, let $\phi(t) = \sqrt{2t \log \log t}$ for $t > e$.

Theorem 10.62 (Khintchine's Law of the Iterated Log). *Let $X_1, X_2, \ldots$ be i.i.d random variables with mean zero and variance one, and put $S_n = X_1 + \cdots + X_n$. Then as $n \to \infty$,*

$$\limsup_{n \to \infty} \frac{S_n}{\sqrt{2n \log \log n}} = 1, \quad \liminf_{n \to \infty} \frac{S_n}{\sqrt{2n \log \log n}} = -1.$$

Proof. Embed (S_n) in a standard Brownian motion by stopping times (T_n). We must show that $\limsup_{n \to \infty} \frac{B_{T_n}}{\phi(n)} = 1$. By (10.23), $T_n/n \to 1 \implies \phi(T_n)/\phi(n) \to 1$, so it is enough to show that

(10.25) $$\limsup_{n \to \infty} \frac{B_{T_n}}{\phi(T_n)} = 1.$$

There is an immediate upper bound: (T_n) is a subsequence so its lim sup cannot be larger than that of the whole process:

$$\limsup_{n \to \infty} \frac{B_{T_n}}{\phi(T_n)} \le \limsup_{t \to \infty} \frac{B_t}{\phi(t)} = 1$$

by Corollary 10.39. To show the lim sup over T_n equals one, it is enough to show that it is greater than $1 - \eta$ for any $\eta > 0$. The main step is to show

10.11. Skorokhod Embedding*

that if $B_t/\phi(t)$ is close to one, it remains there for a reasonable interval of time. The following estimate does that. Let $\varepsilon > 0$, $\delta > 0$, $t > 0$ and consider

$$P\{\inf_{s \leq \varepsilon t}(B_{t+s} - B_t) \leq -\delta\phi(t)\} = P\{\sup_{s \leq \varepsilon t} B_s \geq \delta\phi(t)\} = P\{\sup_{s \leq \varepsilon} B_s \geq \frac{\delta}{\sqrt{t}}\phi(t)\}$$

where the last equality follows from Brownian scaling. By the reflection principle, this equals $2P\{B_\varepsilon \geq \delta\phi(t)/\sqrt{t}\}$. Since $\phi(t)/\sqrt{t}$ is increasing, this probability decreases with t, and it follows from Lemma 10.10 that

$$(10.26) \quad t \geq e^n \implies P^x\{\inf_{s \leq \varepsilon t}(B_{t+s} - B_t) \leq -\delta\phi(t)\} \leq \sqrt{\frac{\varepsilon}{\pi\delta^2}}\frac{1}{n^{\frac{\delta^2}{\varepsilon}}\log n}.$$

With this established, let $\eta \in (0,1)$ and put $\delta = \eta/2$, $\varepsilon = \eta^2/8$. Define stopping times τ_n by

$$\tau_n = \inf\{t > e^n : B_t > (1-\varepsilon)\phi(t)\}, \quad n = 1, 2, \ldots.$$

Then τ_n is finite, $\tau_n \geq e^n \to \infty$, and $B_{\tau_n} \geq (1-\varepsilon)\phi(\tau_n)$. Let I_n be the (random) interval $I_n = [\tau_n, (1+\varepsilon)\tau_n]$. Let A_n be the event

$$A_n = \left\{\inf_{0 \leq s \leq \varepsilon\tau_n}(B_{\tau_n+s} - B_{\tau_n}) \geq -\delta\phi(\tau_n)\right\}.$$

We claim that on A_n, $B_t/\phi(t) \geq (1-\eta)$ for $t \in I_n$. Indeed, $B_{\tau_n} \geq (1-\varepsilon)\phi(\tau_n)$, so

$$B_{\tau_n+s} = B_{\tau_n} + B_{\tau_n+s} - B_{\tau_n} \geq (1-\varepsilon)\phi(\tau_n) - \delta\phi(\tau_n)$$
$$= (1-\varepsilon-\delta)\phi(\tau_n+s)\frac{\phi(\tau_n)}{\phi(\tau_n+s)}$$
$$\geq (1-\varepsilon-\delta)(1-\varepsilon)\phi(\tau_n+s),$$

for large t, where we have used the facts that $s < \varepsilon\tau_n$ and $\phi(\tau_n)/\phi((1+\varepsilon)\tau_n) \longrightarrow 1/\sqrt{1+\varepsilon} > 1-\varepsilon$ as $n \to \infty$. Substituting for ε and δ, we see that

$$\omega \in A_n, \ t \in I_n \implies B_t(\omega) \geq (1-\eta)\phi(t).$$

But $(B_{\tau_n+s} - B_{\tau_n})$ is a Brownian motion independent of $\mathcal{F}_{\tau_n}$, and hence of τ_n; apply[17] (10.26) at τ_n to conclude that $P\{A_n^c\} \leq C/n^{\frac{\delta^2}{\varepsilon}}\sqrt{\log n}$. Since $\delta^2 > \varepsilon$, this is summable, so by the Borel-Cantelli lemma, A_n happens for all large enough n with probability one. Thus, for a.e. ω,

(a) $T_n(\omega)/n \to 1$.

(b) For all large enough n, $t \in I_n \implies B_t(\omega) \geq (1-\eta)\phi(t)$.

By (a), $|m - T_m(\omega)| < m\varepsilon/3$ for all large enough m. If m is in the middle third of I_n—and for large n, there will be such m—then $T_m(\omega) \in I_n$ and

---

[17] Note that (10.26) concerns fixed times, and we apply it to a case where both the stopping time and the length of the interval are random. The reader should check that the strong Markov property does, indeed, justify this.

$B_{T_m}(\omega)/T_m(\omega) \geq (1-\eta)\phi(T_m(\omega))$. Thus for a.e. ω there exists a sequence of m tending to infinity such that $B_{T_m}/\phi(T_m) \geq 1 - \eta$. Since η is arbitrary, this proves (10.25). □

Application to Donsker's Invariance Principle★. There are a number of limit theorems for random walks—the central limit theorem is a prime example—which are almost invariant under changes of underlying distributions. Moreover, they coincide with known limit theorems for Brownian motion. The intriguing question "why?" is answered by Donsker's Invariance Principle. Donsker discovered a type of central limit theorem for processes: rescaled random walks converge in distribution to a Brownian motion. At the same time, it explains why Brownian motion turns up so often in practice: so many physical processes have distributions close to it.

The exact sense of convergence involves weak convergence in Banach spaces, which is outside the scope of this volume. However, we can prove a result which easily implies the invariance principle[18].

Let $X_1, X_2, \ldots$ be i.i.d. random variables of mean zero and variance one, and let $S_k = X_1 + \cdots + X_k$. As in §10.11.1, embed the random walk in a Brownian motion via Skorokhod embedding times $T_1 < T_2 < \ldots$. Let $\xi_n \stackrel{def}{=} B_{T_n}$. Then (ξ_n) is a copy of the random walk (S_n). We can extend both (S_n) and (ξ^n) to continuous parameter processes by interpolating linearly in between the integers: if $[t]$ is the greatest integer in t, put $S_t = S_{[t]} + (t-[t])(S_{[t]+1} - S_{[t]})$, and do the same for ξ. Then $t \mapsto \xi_t$ and $t \mapsto S_t$ are defined for all $t \geq 0$, are continuous, and are linear in between the integers.

Now rescale (B_t), (S_t) and (ξ_t): let $\hat{B}^n_t = \frac{1}{\sqrt{n}}B_{nt}$, $\hat{S}^n_t = \frac{1}{\sqrt{n}}S_{nt}$, and $\hat{\xi}^n_t = \xi_{nt}$. Note that $\hat{B}^n$ is again a standard Brownian motion, and that $(\hat{S}^n_t)$ is embedded in it by $\hat{S}^n_t \stackrel{d}{\sim} \hat{\xi}^n_t = \hat{B}^n_{T_k/n}$. Both $\hat{S}^n_t$ and $\hat{\xi}^n_t$ are continuous, and linear on each $[k/n, (k+1)/n]$. (It is no coincidence that this scaling gives $\hat{S}^n_t$ and $\hat{\xi}^n_t$ the same mean and variance as Brownian motion.)

Here is the basic result.

Theorem 10.63. *As $n \to \infty$, $\sup_{t \leq 1} |\xi^n_t - B^n_t| \to 0$ in probability.*

Since the rescaled random walks $\hat{S}^n$ and $\hat{\xi}^n$ have the same distribution, this implies that $\hat{S}^n$ converges weakly to a Brownian motion[19] in $C[0,1]$.

---

[18]This puts us in an interesting position: we can prove a theorem that we cannot state.

[19]For those familiar with weak convergence in the general setting, this also says that if we regard them as random variables with values in $C[0,1]$, then $(\hat{S}^n_t)$ converges weakly to (B_t) on $C[0,1]$. This is, in fact, the usual statement of the invariance principle.

10.11. Skorokhod Embedding*

Proof. This has been done for a single value of t in the proof of Theorem 10.61. Now we must show that the convergence is uniform, which is more delicate.

By (10.23), $T_{[nt]}/n \to t$ a.s. for each t. This is true simultaneously for all rational t. The $T_{[nt]}/n$ are increasing functions of t, and the limit is continuous, so Dini's theorem implies that with probability one, $T_{[nt]}/n \to t$, for all $t \geq 0$, and the convergence is uniform on compact sets.

Let $\varepsilon > 0$ and $\gamma > 0$. Let $\eta(\delta)$ be the Brownian modulus of continuity: $\eta(\delta) = \sup\{|B_t - B_s| : |t - s| \leq \delta, s, t \leq 2\}$. Brownian motion is continuous, so $\eta(\delta) \to 0$ as $\delta \to 0$ a.s. Thus we can choose $\delta > 0$ small enough so that

$$P\left\{\eta(\delta) > \frac{\varepsilon}{2}\right\} < \frac{1}{2}\gamma.$$

Since $T_{[nt]}/n \to t$ uniformly, there exists $n_0 > 3/\delta$ such that for $n > n_0$,

$$P\left\{\sup_{t \leq 1}\left|\frac{1}{n}T_{[nt]} - t\right| > \frac{\delta}{3}\right\} < \frac{1}{2}\gamma.$$

Thus with probability at least $1 - \gamma$:

(10.27) $\qquad \eta(\delta) \leq \dfrac{1}{2}\varepsilon \quad \text{and} \quad \sup_{t \leq 1}\left|\dfrac{1}{n}T_{[nt]} - t\right| \leq \dfrac{1}{3}\delta.$

Suppose (10.27) holds. Let $t \in [0,1]$, say $t \in [j/n, (j+1)/n)$, and

$$|\hat{\xi}^n_t - \hat{B}^n_t| \leq |\hat{\xi}^n_t - \hat{\xi}^n_{j/n}| + \underbrace{|\hat{\xi}^n_{j/n} - \hat{B}^n_{T_j/n}|}_{=0} + |\hat{B}^n_{T_j/n} - \hat{B}^n_t| \stackrel{\text{def}}{=} \Delta_1 + \Delta_2.$$

Now $\hat{\xi}^n$ is linear on $[j/n, (j+1)/n]$, so $\Delta_1 \leq |\hat{\xi}^n_{(j+1)/n} - \hat{\xi}^n_{j/n}| = |\hat{B}^n_{T_{j+1}/n} - \hat{B}^n_{T_j/n}|$. But

$$\left|\frac{T_{j+1}}{n} - \frac{T_j}{n}\right| \leq \left|\frac{T_{j+1}}{n} - \frac{j+1}{n}\right| + \frac{1}{n} + \left|\frac{j}{n} - \frac{T_j}{n}\right| \leq \frac{\delta}{3} + \frac{\delta}{3} + \frac{\delta}{3} = \delta.$$

Thus $\Delta_1 \leq \eta(\delta) \leq \varepsilon/2$. Similarly, as $n > n_0$,

$$\left|\frac{T_j}{n} - t\right| \leq \left|\frac{T_j}{n} - \frac{j}{n}\right| + \left|\frac{j}{n} - t\right| \leq \frac{\delta}{3} + \frac{1}{n} < \delta,$$

and $\Delta_2 \leq \eta(\delta) \leq \varepsilon/2$. These both hold with probability at least $1 - \gamma$, so

$$n \geq n_0 \implies P\left\{\sup_{t \leq 1}\left|\hat{\xi}^n_t - \hat{B}^n_t\right| > \varepsilon\right\} \leq \gamma.$$

ε and γ are arbitrary, so this implies convergence in probability. $\square$

Remark 10.64. The uniformity of convergence in Proposition 10.63 implies that many functions of the random walk also converge in distribution. Here is an example involving the maximum. We will pose it as a problem.

Exercise 10.41. Show that for $x > 0$, $\lim_{n\to\infty} P\{\max\{S_j : j \leq n\} \geq x\sqrt{n}\} = \sqrt{\frac{2}{\pi}} \int_x^\infty e^{-\frac{x^2}{2}}\, dx$.

10.12. Lévy's Construction of Brownian Motion★

We promised to prove that Brownian motion actually exists, and that any fears that we are talking about an empty subject are baseless. The problem is this: we must find a probability space $(\Omega, \mathcal{F}, P)$ and a family of random variables $\{B_t,\ t \geq 0\}$ on it which satisfy the hypotheses of Brownian motion. While we do not have general existence theorems for stochastic processes at hand, we can construct a sequence of i.i.d. $N(0,1)$ random variables (Theorem 5.11.) This is sufficient: we can construct Brownian motion from this. We will follow the original construction of Paul Lévy. It has the virtue of proving both the existence and the continuity at the same time.

Note that it is not hard to define Brownian motion at any finite number of points: just add independent mean zero Gaussian random variables of the correct variance. The problem is to interpolate. Suppose, for example, that we know the value of Brownian motion at, say, $k2^{-n}$ and $(k+1)2^{-n}$. We can use the Brownian bridge to interpolate. At the point $t = (k+1/2)2^{-n}$, we know that $B_t - (B_{k2^{-n}} + B_{(k+1)2^{-n}})/2$ has mean zero and variance $2^{-(n+2)}$, and, moreover, it is independent of B_t for t outside this interval. So, if we know the values of B_t at multiples of $1/2^n$, we can interpolate to get its values at multiples of $1/2^{n+1}$.

Let us reverse-engineer this to construct the process. If we have constructed the values at $t = k2^{-n}$, $k = 0, 1, 2, \ldots$ we interpolate by taking an independent $N(0,1)$ random variable X, and setting the value of B at $(k+1)2^{-n}$ equal to $(B_{k2^{-n}} + B_{(k+1)2^{-n}})/2 + 2^{-(n+1)/2}X$—here $2^{-(n+2)/2}X$ has the same distribution as the Brownian bridge at that point.

The construction can be written neatly in terms of the Schauder tent-functions f_{nk}, where

$$f_{nk}(t) = \begin{cases} 2^{-(n+2)/2} & \text{if } t = (k+\frac{1}{2})2^{-n} \\ 0 & \text{if } t \notin (k2^{-n}, (k+1)2^{-n}) \\ \text{linear} & \text{in between}. \end{cases}$$

The graph of f_{nk} is triangular. It has support $(k2^{-n}, (k+1)2^{-n})$, so for each n the f_{nk}, $k = 0, 1, \ldots$ have disjoint support. It is only necessary to define the process on $0 \leq t \leq 1$, since we can glue independent copies of this together to get Brownian motion on $[0, \infty)$.

10.12. Lévy's Construction of Brownian Motion*

Theorem 10.65. *Let X and X_{nk}, $n = 0, 1, 2, \ldots, k = 0, 1, 2, \ldots$ be i.i.d $N(0, 1)$ random variables. Then, with probability one, the series*

$$(10.28) \qquad W_t \stackrel{\text{def}}{=} tX + \sum_{n=0}^{\infty} \sum_{k=0}^{2^n - 1} X_{nk}\, f_{nk}(t)$$

converges uniformly for $0 \le t \le 1$. Moreover, $\{W_t,\ 0 \le t \le 1\}$ is a standard Brownian motion.

Proof. Let us first prove uniform convergence. For $m = 0, 1, 2, \ldots$, define

$$(10.29) \qquad W_t^m = tX + \sum_{n=0}^{m} \sum_{k=0}^{2^n - 1} X_{nk} f_{nk}(t)\,.$$

Notice that for $n > m$, that each f_{nk} vanishes at all multiples of 2^{-m}. Thus $W_t^m = W_t$ for all multiples of 2^{-m}. The f_{nk} are piecewise-linear, hence so is their sum, so W^m is a polygonal approximation of W: it equals W at multiples of 2^{-m}, and is linear in between. Now $W^m - W^{m-1} = \sum_{k=0}^{2^m - 1} X_{m,k} f_{mk}$. Since the f_{mk} have disjoint support,

$$\sup_{t \in [0,1]} |W_t^m - W_t^{m-1}| = 2^{-(m+2)/2} \max\{|X_{mk}|,\ k = 0, \ldots, 2^m - 1\}\,.$$

The probability that the max exceeds $1/m^2$ is smaller than the sum of the probabilities, so

$$P\Big\{2^{-\frac{m+2}{2}} \max\{|X_{mk}| : k = 0, \ldots, 2^m - 1\} > \frac{1}{m^2}\Big\} \le 2^m P\Big\{|X_{m1}| \ge \frac{2^{\frac{m+2}{2}}}{m^2}\Big\}$$

$$\le \frac{1}{\sqrt{2\pi}} 2^{\frac{m}{2} - 1} m^2 e^{-\frac{2^{m+1}}{m^4}}\,.$$

where we have used Lemma 10.10. This is summable, so by the Borel-Cantelli Lemma, $\sup_{t \in [0,1]} |W_t^{m+1} - W_t^m| \le 1/m^2$ for all large enough m, a.s. But $1/m^2$ is itself summable, so this implies that with probability one, (W_t^m) converges uniformly on $[0, 1]$.

Thus W_t is a well-defined mean-zero Gaussian process. It is a uniform limit of continuous functions, so it has continuous sample paths. To show that it is a Brownian motion, it suffices to show that its increments are independent and have the correct variance.

Fix m. Suppose that the increments $\Delta_{mk} \stackrel{\text{def}}{=} W_{(k+1)2^{-m}}^m - W_{k2^{-m}}^m$ are i.i.d. $N(0, 2^{-m})$. This is true initially, since there is only one increment of length 2^0, and it is $N(0, 1)$. Since $W = W^m$ at multiples of 2^{-m}, Δ_{mk} is also the increment of the limit, W. At stage $m+1$, this interval is cut in two, creating two new increments, which we denote by Δ'_{mk} and Δ''_{mk}, respectively: $\Delta'_{mk} = \frac{1}{2}\Delta_{mk} + 2^{-(m+2)/2} X_{m+1,k}$ and $\Delta''_{mk} = \frac{1}{2}\Delta_{mk} - 2^{-(m+2)/2} X_{m+1,k}$. They are Gaussian with mean zero; and $\text{Var}(\Delta'_{mk}) = \text{Var}(\Delta_{mk})/4 + 2^{-(m+2)} = 2^{-(m+2)} + 2^{-(m+2)} = 2^{-(m+1)}$, and the same is true for Δ''_{mk}. To see that the $m+1^{\text{st}}$-stage increments are independent, note that Δ'_{mk} and Δ''_{mk} are independent of Δ_{mj}, $j \neq k$, and also of the $X_{m+1,j}$, $j \neq k$, so they are independent of the other new increments Δ'_{mj} and Δ''_{mj}, for $j \neq k$. It remains to see that the two are independent of each other. They are Gaussian, so we need only check the covariance: $E\{\Delta'_{mj}\Delta''_{mj}\} = E\{(\frac{1}{2}\Delta_{mk} + 2^{-(m+2)/2}X_{m+1,k})(\frac{1}{2}\Delta_{mk} - 2^{-(m+2)/2}X_{m+1,k})\} = 2^{-m}/4 - 2^{-(m+2)} = 0$. Thus, the increments are independent at the $m+1^{\text{st}}$ stage. By induction, the increments Δ_{mk} are i.i.d. $N(0, 2^{-m})$.

It follows that increments of W over disjoint intervals with dyadic rational endpoints are independent. (Such an increment is a sum of Δ_{nk} for large enough n.) Finally, any increment is a limit of increments with dyadic rational end points, hence W_t has independent increments, and for $s < t$, $W_t - W_s$ is $N(0, t-s)$. □

Remark 10.66. Note that Theorem 10.65 gives an independent proof of the continuity of the Brownian paths.

10.13. The Ornstein-Uhlenbeck Process*

Consider the process $Y_t = e^{-t}B_{e^{2t}}$, $-\infty < t < \infty$. Note that the parameter set is the whole real line and that each Y_t has mean zero and variance one. More generally, let $\sigma, \lambda > 0$ and consider the process defined by

(10.30) $$X_t = \sigma e^{-\lambda t} B_{\frac{1}{2\lambda}e^{2\lambda t}}, \quad -\infty < t < \infty.$$

Again, the parameter set is the whole line, not just $[0, \infty)$. This is called an **Ornstein-Uhlenbeck process**. (The particular choice of parameters is linked to the Langevin equation, (10.39).)

Definition 10.67. A process $\{Z_t, t \in T\}$ is **stationary** if, for any n, h and $t_1, \ldots, t_n \in T$ such that $t_1 + h, \ldots, t_n + h$ are also in T, $(Z_{t_1}, \ldots, Z_{t_n})$ and $(Z_{t_1+h}, \ldots, Z_{t_n+h})$ have the same distribution.

The distributions of a stationary process are independent of translation: $\{Z_{t+h}, t \in T\}$ has exactly the same distribution as the original process $\{Z_t, t \in T\}$. Brownian motion is not stationary, but the Ornstein-Uhlenbeck process is.

Let us give some of its elementary properties.

1° It is a mean-zero Gaussian process with $E\{X_t^2\} = \sigma^2/\lambda$ and covariance function

$$\Gamma(s,t) \stackrel{\text{def}}{=} E\{X_s X_t\} = \frac{\sigma^2}{2\lambda}e^{-\lambda|t-s|}.$$

10.13. The Ornstein-Uhlenbeck Process*

2° It is a stationary process.

3° It is reversible: $\{X_{-t}, -\infty < t < \infty\}$ and $\{X_t, -\infty < t < \infty\}$ have the same distribution.

4° $E\{X_t \mid X_s\} = e^{-\lambda|t-s|} X_s$.

5° It is a strong Markov process.

These are easily seen. For instance, to see 1°, note that X is clearly a mean-zero Gaussian process, and write Γ in terms of B:

$$\Gamma(s,t) = \sigma^2 e^{-\lambda(s+t)} E\{B_{\frac{1}{2\lambda}e^{2\lambda s}} B_{\frac{1}{2\lambda}e^{2\lambda t}}\} = \frac{\sigma^2}{2\lambda} e^{-\lambda(s+t)} e^{-2\lambda s} \wedge e^{-2\lambda t}.$$

Note that $|t-s| = s+t - 2(s \wedge t)$, so this equals $\sigma^2 e^{-\lambda|t-s|}/2\lambda$, as claimed.

For 2°, note that the distribution of an n-tuple of jointly Gaussian random variables is determined by their mean and covariance matrix. In this case, the means are zero and, since the covariance function $\Gamma(s,t) = \sigma^2 e^{-\lambda|t-s|}$ is translation-invariant, so is the covariance matrix.

Then 3° follows since the covariance function is also invariant under a change of sign of s and t: $\Gamma(s,t) = \Gamma(-s,-t)$, so (X_{-t}) and (X_t) have the same distribution.

The conditional expectation in 4° equals $\sigma e^{-\lambda t} E\{B_{\frac{1}{2\lambda}e^{2\lambda t}} \mid B_{\frac{1}{2\lambda}e^{2\lambda s}}\}$. If $s < t$, 4° follows from the fact that B_t is a martingale, and it holds if $t \leq s$ by reversal (interchange s and t and use 3°).

Finally, 5° follows from the strong Markov property of Brownian motion.

The transition probabilities of an Ornstein-Uhlenbeck process are readily found from the Brownian transition probabilities. In particular, we can talk of the Ornstein-Uhlenbeck process starting from $x = 0$ at time $t = 0$. A moments thought reveals, given that $X_0 = 0$, that X_t for $t \geq 0$ has the same distribution as $\sigma e^{-\lambda t} B_{\frac{1}{2\lambda}e^{2\lambda t}} - B_{\frac{1}{2\lambda}}$. A direct computation leads to

Proposition 10.68. *Given that $X_0 = 0$, the Ornstein-Uhlenbeck process $\{X_t, t \geq 0\}$ is a mean-zero Gaussian Markov process with covariance function*

$$\Gamma(s,t) = \frac{\sigma^2}{2\lambda}\left(e^{-\lambda|t-s|} - e^{-\lambda(s+t)}\right).$$

Note that as s and t tend to infinity the covariance function tends to the stationary covariance $\frac{\sigma^2}{2\lambda} e^{-\lambda|t-s|}$. Consequently, the process itself tends to a stationary process.

The Ornstein-Uhlenbeck Infinitesimal Generator*. Let $\{X_t^x, t \geq 0\}$ be an Ornstein-Uhlenbeck process with initial value x and let f be a smooth function of compact support. We want to find

$$\mathbb{G}f(x) \stackrel{def}{=} \lim_{t\downarrow 0} \frac{1}{t} E\{f(X_t^x) - f(x)\}.$$

In terms of a Brownian motion, $X_t = \sigma e^{-\lambda t} B_{e^{2\lambda t}}$, so

$$\mathbb{G}f(x) = \lim_{t\downarrow 0} \frac{1}{t} E\{f(\sigma e^{-\lambda t} B_{\frac{1}{2\lambda} e^{2\lambda t}}) - f(x) \mid \sigma B_{\frac{1}{2\lambda}} = x\}$$

$$= \lim_{t\downarrow 0} \frac{1}{t} E\{f(\sigma e^{-\lambda t} B_{\frac{1}{2\lambda} e^{2\lambda t}}) - f(\sigma B_{\frac{1}{2\lambda} e^{2\lambda t}}) \mid \sigma B_{\frac{1}{2\lambda}} = x\}$$

$$+ \lim_{t\downarrow 0} \frac{1}{t} E\{f(\sigma B_{\frac{1}{2\lambda} e^{2\lambda t}}) - f(x) \mid \sigma B_{\frac{1}{2\lambda}} = x\}$$

$$\stackrel{def}{=} \Upsilon_1 + \Upsilon_2.$$

By Taylor's formula,

$$f(\sigma e^{-\lambda t} B_{\frac{e^{2\lambda t}}{2\lambda}}) - f(\sigma B_{\frac{e^{2\lambda t}}{2\lambda}}) = f'(\sigma B_{\frac{e^{2\lambda t}}{2\lambda}})\sigma B_{\frac{e^{2\lambda t}}{2\lambda}}(e^{-\lambda t} - 1) + O\big((e^{-\lambda t} - 1)^2\big).$$

The last term is $O(t^2)$, so

$$\Upsilon_1 = \lim_{t\downarrow 0} E\{f'(\sigma B_{\frac{e^{2\lambda t}}{2\lambda}})\sigma B_{\frac{e^{2\lambda t}}{2\lambda}} \mid \sigma B_{\frac{1}{2\lambda}} = x\}\frac{e^{-\lambda t} - 1}{t} + \lim_{t\downarrow 0} \frac{1}{t} O\big((e^{-\lambda t} - 1)^2\big)$$

$$= -\lambda x f'(x).$$

We can evaluate Υ_2 from the Brownian infinitesimal generator $\mathbb{G}_B$:

$$\Upsilon_2 = \lim_{t\downarrow 0} \frac{1}{t} E\{f(\sigma B_{\frac{e^{2\lambda t}}{2\lambda}}) - f(x) \mid \sigma B_{\frac{1}{2\lambda}} = x\}$$

$$= \lim_{t\downarrow 0} \frac{(e^{2\lambda t} - 1)}{2\lambda t} \underbrace{\lim_{t\downarrow 0} \frac{E\{f(\sigma B_{\frac{e^{2\lambda t}}{2\lambda}}) - f(\sigma B_{\frac{1}{2\lambda}}) \mid \sigma B_{\frac{1}{2\lambda}} = x\}}{\frac{1}{2\lambda}(e^{2\lambda t} - 1)}}_{\sigma^2 \mathbb{G}_B f}$$

$$= \frac{\sigma^2}{2} f''(x).$$

Therefore,

(10.31) $$\mathbb{G}f(x) = \frac{\sigma^2}{2} f''(x) - \lambda x f'(x).$$

Recurrence Properties.

As in Proposition 10.43 we can show that for a bounded smooth f,

(10.32) $$f(X_t) - \int_0^t \mathbb{G}f(X_s)\,ds, \quad t \geq 0 \quad \text{is a martingale}.$$

10.13. The Ornstein-Uhlenbeck Process*

We can use this to find first exit probabilities and expected first exit times. First solve the differential equations $\mathbb{G}F = 0$ and $\mathbb{G}G = 1$. (We do not need the general solution—particular solutions will do.) We find

$$F(x) = \int_0^x e^{\frac{\lambda y^2}{\sigma^2}}\, dy\,, \qquad G(x) = \frac{2}{\sigma^2}\int_0^x e^{\frac{\lambda y^2}{\sigma^2}} \left(\int_0^y e^{-\frac{\lambda z^2}{\sigma^2}}\, dz\right) dy\,.$$

The formulas are not particularly enlightening, but we can extract some useful facts. For instance, G is an even function, F is odd, both tend to infinity quickly, roughly as $e^{\lambda x^2/\sigma^2}$, and, above all, the integral over z is bounded, so that $|G(x)| \le C \int_0^x e^{\frac{\lambda y^2}{\sigma^2}}\, dy = C|F(x)|$ for some C. Thus the ratio G/F is bounded.

Let $a < x < b$, let $X_0 = x$, and let T_{ab} be the first time that (X_t) leaves (a, b). By (10.32), $F(X_{t \wedge T_{ab}})$ and $G(X_{t \wedge T_{ab}}) - t \wedge T_{ab}$ are martingales. (Neither F nor G is bounded, so this is not immediate. See the exercise below.) Then let $t \to \infty$ to see that if $X_0 = x$,

$$F(x) = E^x\{F(X_{T_{ab}})\}\,.$$

If $p = P^x\{X_{T_{ab}} = a\}$, rewrite this equation as $F(x) = pF(a) + (1-p)F(b)\,.$ Solving for p,

$$P^x\{X_{T_{ab}} = a\} = \frac{F(b) - F(x)}{F(b) - F(a)}\,, \qquad P^x\{X_{T_{ab}} = b\} = \frac{F(x) - F(a)}{F(b) - F(a)}\,.$$

Exercise 10.42. Justify the above by letting g be a continuous function of compact support which is identically one on $[a, b]$, and applying (10.32) to gF and gG.

Now let $a \to -\infty$. Then $F(a) \to -\infty$ and $T_{ab} \to T_b$, so

$$P^x\{T_b < \infty\} = \lim_{a \to -\infty} \frac{F(x) - F(a)}{F(b) - F(a)} = 1\,.$$

Similarly, let $b \to \infty$ to see that $P^x\{T_a < \infty\} = 1$. Thus the process is recurrent, and, like Brownian motion, hits all points. Let us see how long it takes.

Now $G(X_{t \wedge T_{ab}}) - t \wedge T_{ab}$ is a martingale, so $E^x\{G(X_{t \wedge T_{ab}})\} - E^x\{G(X_0)\} = E^x\{t \wedge T_{ab}\}$. We can go to the limit on each side as $t \to \infty$, on the left by bounded convergence and on the right by monotone convergence to see that

$$E^x\{T_{ab}\} = E^x\{G(X_{T_{ab}})\} - E^x\{G(X_0)\}\,.$$

We know the distribution of $X_{T_{ab}}$:

$$E^x\{T_{ab}\} = G(a)\frac{F(b) - F(x)}{F(b) - F(a)} + G(b)\frac{F(x) - F(a)}{F(b) - F(a)} - G(x)\,.$$

Now let $a \to -\infty$ to find the expectation of T_b:
$$E^x\{T_b\} = \lim_{a \to -\infty} E^x\{T_{ab}\} = (F(b) - F(x)) \lim_{a \to -\infty} \frac{G(a)}{|F(a)|} + G(b) - G(x).$$
This is finite, for the ratio G/F is bounded. Thus we have proved:

Proposition 10.69. *The Ornstein-Uhlenbeck process is positive-recurrent: it reaches all points of the line, and the expected time to reach any given point is finite.*

Remark 10.70. We assumed above that T_{ab} is a.s. finite. But is it?

Question: Brownian motion has no drift pushing it back towards the origin, so it should be easier for it to reach far-away points. Yet the expected time for a Brownian motion to reach a given point is infinite, that for an Ornstein-Uhlenbeck process, finite. Why?

10.14. White Noise and the Wiener Integral*

To an electrical engineer, noise is an unwanted random disturbance of delicate electronics. It is likely to be called "white noise" if it contains a lot of high-frequencies; more technically, white noise, like white light, is supposed to contain all frequencies equally. A second, shorter description is that it is the derivative of Brownian motion. Both descriptions make more sense as metaphors than as mathematics. A noise with a perfectly flat frequency spectrum would have infinite energy, and the derivative of Brownian motion quite simply does not exist.

Indeed, it turns out that white noise cannot be defined as a stochastic process[20] $\{\dot{W}_t,\ t \geq 0\}$. However, it can be defined rigorously in a more global way. For instance, it can be defined as a Schwartz distribution, or—our choice—as a stochastic measure.

Let us first state some preliminaries. Let $L^2 = L^2(\Omega, \mathcal{F}, P)$ be the set of all square-integrable random variables, with norm $\|X\|_2 \stackrel{\text{def}}{=} E\{X^2\}^{1/2}$.

$1°$ L^2 is a metric space with metric[21] $d(X, Y) = \|Y - X\|_2$. In particular, $\|X - Z\| \leq \|X - Y\|_2 + \|Y - Z\|_2$.

$2°$ (Riesz-Fischer Theorem) L^2 is complete: if (X_n) is a sequence in L^2 such that $\|X_n - X_m\|_2 \longrightarrow 0$ as $m, n \to \infty$, there is a random variable $X_\infty \in L^2$ such that $\|X_n - X_\infty\|_2 \to 0$.

---

[20]This does not stop people from talking about $\dot{W}_t$, or writing integrals like $\int f(t)\dot{W}_t\, dt$. In fact, that last is an intriguing idea. See the next section...

[21]There is a technicality here: $\|X - Y\|_2 = 0 \implies X = Y$ a.s., so elements of L^2 are equivalence classes of random variables: two random variables are in the same equivalence class if they are equal a.e. However, we will abuse terminology, and speak of random variables as if they were elements of L^2.

10.14. White Noise and the Wiener Integral*

$3°$ L^2 is equipped with an inner product: $\langle X, Y \rangle = E\{XY\}$, and $\langle X, X \rangle = \|X\|_2^2$.

$4°$ $\|X - X_n\|_2 \to 0 \Longrightarrow \|X_n\|_2 \longrightarrow \|X\|_2$.

We will accept these. Let $\mathbb{R}_+ \stackrel{def}{=} [0, \infty)$

Let $\mathcal{B} = \mathcal{B}(\mathbb{R}_+)$ be the class of Borel sets on the half-line. For $A \in \mathcal{B}$, let $|A|$ denote the Lebesgue measure of A.

Definition 10.71. White noise on $\mathbb{R}_+$ is a Gaussian process $\{\dot{W}(A), A \in \mathcal{B}, |A| < \infty\}$ such that

(i) $\dot{W}(A)$ is $N(0, |A|)$;

(ii) $E\{\dot{W}(A)\dot{W}(B)\} = |A \cap B|$.

Proposition 10.72. *Let* $\{\dot{W}(A), A \in \mathcal{B}\}$ *be a white noise. Then* $(\dot{W})$ *is a Gaussian L^2-valued measure. That is*

(i) $\dot{W}(\emptyset) = 0$ *a.s.;*

(ii) $A \cap B = \emptyset \Longrightarrow A$ *and* B *are independent;*

(iii) *Let* $A_1, A_2, \ldots$ *be disjoint Borel sets whose union has finite Lebesgue measure. Then* $\dot{W}(\bigcup_n A_n) = \sum_n \dot{W}(A_n)$ *a.s., where the sum converges in the L^2 norm..*

Proof. Note that if $|A| < \infty$, that $\dot{W}(A)$ is Gaussian, and therefore square-integrable, so that $\dot{W}$ takes values in L^2.

(i) Clear.

(ii) $\dot{W}(A)$ and $\dot{W}(B)$ are jointly Gaussian, and if $A \cap B = \emptyset$, then $E\{\dot{W}(A)\dot{W}(B)\} = |A \cap B| = 0$, so they are uncorrelated, therefore independent.

(iii) This proof has two stages. First, we show white noise is finitely additive. Then we show it is countably additive.

Let A and B be disjoint. We claim that $\dot{W}(A \cup B) = \dot{W}(A) + \dot{W}(B)$ a.s. Indeed,

$$E\{(\dot{W}(A \cup B) - \dot{W}(A) - \dot{W}(B))^2\} = E\{\dot{W}(A \cup B)^2\} + E\{\dot{W}(A)^2\}$$
$$+ E\{\dot{W}(B)^2\} - 2E\{\dot{W}(A \cup B)\dot{W}(A)\} - 2E\{\dot{W}(A \cup B)\dot{W}(B)\}$$
$$= |A| + |B| + |A| + |B| - 2|A| - 2|B|$$
$$= 0.$$

To see countable additivity, let $A_1, A_2, \ldots$ be disjoint sets such that $A \stackrel{def}{=} \bigcup_n A_n$ has finite Lebesgue measure. By additivity, $\dot{W}(A) - \sum_{m=1}^n \dot{W}(A_m) =$

$W(A) - \dot{W}\left(\bigcup_{m=1}^n A_m\right) = \dot{W}\left(A - \bigcup_{m=1}^n A_m\right)$. Thus,

$$E\left\{\left(\dot{W}(A) - \sum_{m=1}^n \dot{W}(A_m)\right)^2\right\} = E\left\{\dot{W}(A - \bigcup_{m=1}^n A_m)^2\right\} = \left|A - \bigcup_{m=1}^n A_m\right|.$$

But as $n \to \infty$, $A - \bigcup_{m=1}^n A_m \longrightarrow \emptyset$, so its Lebesgue measure goes to zero. Therefore the sum converges in L^2 to $\dot{W}(A)$. □

Remark 10.73. (a) $\dot{W}(A)$ is only defined for sets of finite Lebesgue measure.

(b) Note that (i) and (iii) imply that white noise is a measure on $\mathcal{B}$; but it is not a real-valued measure. Rather, it is an L^2-valued measure.

(c) This is the first time we have used a stochastic process with a parameter set which is not a subset of the reals. But we should think of white noise as a measure rather than a process, and, of course, we can expect to see an integral which uses it.

(d) Given a white noise $\dot{W}$, we can define a process (W_t) by $W_t \stackrel{\text{def}}{=} \dot{W}([0,t])$. Then $\{W_t, t \geq 0\}$ is a standard Brownian motion, as one can see by calculating its covariance function.

Does white noise actually exist? Yes, and we can construct it from a standard Brownian motion. In order to avoid worrying about sets of infinite measure, let us just construct it on subsets of $(0,1]$. It is then a short step to define it on the line.

Let $\{W_t, t \geq 0\}$ be a standard Brownian motion. Let $(a,b] \subset (0,1]$ be an interval. Define $\dot{W}((a,b]) \stackrel{\text{def}}{=} W_b - W_a$. Note that $E\{W((a,b])^2\} = b-a = |(a,b]|$. If $(a_i, b_i]$ $i=1,2,\ldots,n$ are disjoint intervals, let $A = \bigcup_i (a_i, b_i]$ and define

$$\dot{W}(A) = \sum_{i=1}^n (W_{b_i} - W_{a_i}).$$

The Brownian increments are disjoint, and therefore independent. Then $\dot{W}(A)$ is Gaussian, mean zero, and has variance $\sum_{i=1}^n (b_i - a_i) = |A|$. There may be several ways of writing A as a finite union of disjoint right-semiclosed intervals, but all lead to the same $\dot{W}(A)$. For example, if $a < b < c$, $\dot{W}((a,c]) = W_c - W_a$. But $(a,c] = (a,b] \cup (b,c]$ so the definition also gives $\dot{W}((a,c]) = W_b - W_a + W_c - W_b$, which also equals $W_c - W_a$.

The set $\mathcal{B}^0$ of all finite unions of right semi-closed sub-intervals of $(0,1]$ is closed under finite unions, finite intersections, and complements, so it is a field.

We can summarize this: on the field $\mathcal{B}^0$, $\dot{W}$ satisfies:

(i) $\dot{W}(A)$ is a $N(0, |A|)$ random variable.

(ii) $E\{\dot{W}(A)\dot{W}(B)\} = |A \cap B|$.

As in Proposition 10.72, it follows from (i) and (ii) that $\dot{W}$ is finitely additive on $\mathcal{B}^0$ and, in particular, for $A, B \in \mathcal{B}^0$,

$$\dot{W}(A \cup B) = \dot{W}(A) + \dot{W}(B) - \dot{W}(A \cap B).$$

We cannot talk about countable additivity yet, for a countable union of sets in $\mathcal{B}^0$ is not necessarily in $\mathcal{B}^0$. First we need to extend $\dot{W}$ from $\mathcal{B}^0$ to the Borel sets $\mathcal{B}$.

Recall the symmetric difference of sets: $A \triangle B \stackrel{\text{def}}{=} (A - B) \cup (B - A)$. Notice that the indicator function $I_{A \triangle B}$ of $A \triangle B$ equals $|I_A - I_B|$. Let us measure the difference between two sets A and B by $d(A, B) \stackrel{\text{def}}{=} |A \triangle B|$. Then $d(A, B) = \int |I_B(x) - I_A(x)| \, dx$. Note that d satisfies the triangle inequality, for

$$d(A, C) = \int |I_A(x) - I_C(x)| \, dx \leq \int |I_A(x) - I_B(x)| + |I_B(x) - I_C(x)| \, dx$$
$$= d(A, B) + d(B, C).$$

In terms of the symmetric difference, we have

$$E\{(\dot{W}(B) - \dot{W}(A))^2\} = |A \triangle B|, \quad A, B \in \mathcal{B}^0.$$

Indeed, $\dot{W}(B) = \dot{W}(B - A) + \dot{W}(B \cap A)$ so $\dot{W}(B) - \dot{W}(A) = \dot{W}(B - A) - \dot{W}(A - B)$. These two sets are disjoint, so

$$E\{(\dot{W}(B) - \dot{W}(A))^2\} = E\{\dot{W}(B - A)^2\} + E\{\dot{W}(A - B)^2\}$$
$$= |A - B| + |B - A|$$
$$= |A \triangle B|.$$

If $A \in \mathcal{B}$, by Theorem 1.7 there exist sets $A_n \in \mathcal{B}^0$ such that $A_n \triangle A \to 0$. Then $E\{\dot{W}(A_m) - \dot{W}(A_n))^2\} = |A_n \triangle A_n| \leq |A_n \triangle A| + |A \triangle A_n|$. This tends to zero as m and n tend to infinity, so that $\dot{W}(A_n)$, $n = 1, 2, 3, \ldots$ is a Cauchy sequence. By the Riesz-Fischer theorem, it converges in L^2 to a limit that we will call $\dot{W}(A)$. Then $\|\dot{W}(A_n) - \dot{W}(A)\|_2 \to 0$.

Now $\dot{W}(A_n)$ converges to $\dot{W}(A)$ in L^2, so that it also converges in distribution. Thus $\dot{W}(A)$ is also Gaussian, and has mean zero and variance equal to $\lim_n E\{\dot{W}(A_n)^2\} = \lim_n |A_n| = |A|$. It is easily seen that this limit is unique, i.e., it does not depend on the particular sequence (A_n).

This defines $\dot{W}(A)$ for all $A \in \mathcal{B}$. Let $A, B \in \mathcal{B}$. Choose A_n and B_n in $\mathcal{B}^0$ such that $|A_n \triangle A| \to 0$ and $|B_n \triangle B| \to 0$. Then $\dot{W}(A_n)$ and $\dot{W}(B_n)$

converge in L^2 to $\dot{W}(A)$ and $\dot{W}(B)$, respectively, and

$$\begin{aligned} E\{\dot{W}(A)\,\dot{W}(B)\} &= \lim_n E\{\dot{W}(A_n)\,\dot{W}(B_n)\} \\ &= \lim_n |A_n \cap B_n| \\ &= |A \cap B|\,. \end{aligned}$$

To summarize: we have defined $\dot{W}(A)$ for all $A \in \mathcal{B}$, and shown that it satisfies both (i) and (ii) of the definition. Thus $\{\dot{W}(A),\ A \in \mathcal{B}\}$ is a white noise. We can easily extend it to a white noise on all of $(0,\infty)$ by adding independent white noises on each interval $(n, n+1]$.

Stochastic Integrals★. If white noise is a measure, there should be an integral associated with it, and, since white noise is a basic process, this integral should be interesting. Very interesting. We will only consider the most basic case here, which integrates deterministic, not random, functions. This is called the **Wiener integral**[22]. (The next step would be the Ito integral, which can handle random integrands. We will not take that step, not because it is too difficult—in fact we already have almost all the necessary technical tools—but because it is too interesting. We would have to add another hundred pages to this book simply to study its first implications in probability, analysis, ecology, biology, chemistry, physics and finance.)

The integral with respect to white noise is called a **stochastic integral**, and the usual notation for it is $\int f(t)\,dW_t$. But in order to emphasize the fact that it is really an integral with respect to a measure, much like those we have already dealt with, we will (temporarily) call it $\int f(t)\,\dot{W}(dt)$.

Lebesgue pioneered a straightforward procedure to define such integrals. First find elementary functions whose integral is obvious. Integrate those, then extend the integral by linearity to finite linear combinations. Then extend it by continuity to the limits of such functions. This last step requires cleverness—though often the cleverness is hidden in the choice of elementary functions—but in this case it, too, is straightforward[23].

The elementary functions are indicator functions: $I_A(t)$, where $A \in \mathcal{B}$ has finite Lebesgue measure. Then

$$\int I_A(t)\,\dot{W}(dt) \stackrel{\text{def}}{=} \dot{W}(A)\,.$$

---

[22]Norbert Wiener constructed his stochastic integral for use in his prediction theory [33]. It can only handle deterministic integrands, so it has been supplanted by the more powerful Ito integral, but it is quick to construct, and is just right for our purpose. The basic ideas in the construction of the two integrals are similar.

[23]Perhaps it is straightforward because the cleverness was hidden even earlier, when the measure $\dot{W}$ was chosen to be L^2-valued. More on this later.

Extend this by linearity to *simple functions*, which are finite linear combinations of elementary functions. If $A_1, \ldots, A_n$ are Borel sets of finite Lebesgue measure, and if $a_1, \ldots, a_n$ are reals, put $f(t) = \sum_{i=1}^n a_i I_{A_i}(t)$. Then

$$(10.33) \qquad \int f(t)\, \dot{W}(dt) = \sum_{i=1}^n a_i\, \dot{W}(A_i).$$

Let us calculate the expected square of the integral of f:

$$E\left\{\left(\int f(t)\, \dot{W}(dt)\right)^2\right\} = E\left\{\left(\sum_{i=1}^n a_i \dot{W}\{A_i\}\right)^2\right\}$$

$$= \sum_{i=1}^n \sum_{j=1}^n a_i a_j E\{\dot{W}(A_i)\, \dot{W}(A_j)\}$$

$$= \sum_{i=1}^n \sum_{j=1}^n a_i a_j |A_i \cap A_j|$$

$$= \sum_{i=1}^n \sum_{j=1}^n a_i a_j \int I_{A_i}(t) I_{A_j}(t)\, dt$$

$$= \int \left(\sum_{i=1}^n a_i I_{A_i}(t)\right)^2 dt$$

$$= \int f^2(t)\, dt.$$

First, is the integral well-defined? There may be more than one way to write f as a simple function, which could lead to different values for the integral. However, if there are b_j and B_j such that we can also write $f = \sum_{j=1}^m b_j I_{B_j}$, then (10.33) gives a different formula for the integral, but the result must be the same, for

$$E\left\{\left(\sum_{i=1}^n a_i\, \dot{W}(A_i) - \sum_{j=1}^m b_j \dot{W}(B_j)\right)^2\right\} = \int (f(t) - f(t))^2 dt = 0.$$

Thus the integral is well-defined for simple functions. It is easily seen that it satisfies the following:

(i) If f and g are simple, a and b real, then

$$\int (af(t) + bg(t))\, \dot{W}(dt) = a \int f(t)\, \dot{W}(dt) + b \int f(t)\, \dot{W}(dt).$$

(ii) $E\{(\int f(t)\, \dot{W}(dt))^2\} = \int f^2(t)\, dt$.

We need a result on the approximation by simple functions.

Proposition 10.74. *Let f be a Borel function on $\mathbb{R}_+$ such that $\int f^2(t)\,dt < \infty$. Then there exists a sequence of simple functions (f_n) such that $\int (f(t) - f_n(t))^2\,dt \to 0$ as $n \to \infty$. In addition, $\int f_n^2(t)\,dt \longrightarrow \int f^2(t)\,dt$.*

Proof. Define

$$f_n(t) = \begin{cases} \frac{k}{n}\operatorname{sgn}(f(t)) & \text{if } t < n,\ k \le n^2 - 1,\ \frac{k}{n} \le |f(t)| < \frac{k+1}{n}, \\ 0 & \text{if } t \ge n \text{ or } |f(t)| \ge n. \end{cases}$$

Then $f_n^2(t) \le f^2(t)$. If $t < n$ and $|f(t)| < n$, then $|f(t) - f_n(t)| \le 1/n$. Thus as $n \to \infty$, $f_n(t) \to f(t)$ pointwise. Now $(f(t) - f_n(t))^2 \le 4f^2(t)$, so it follows by the dominated convergence theorem that $\int f_n^2(t)\,dt \to \int f^2(t)\,dt$ and $\int (f(t) - f_n(t))^2\,dt \to 0$. □

The integral is now defined for simple functions. The next step is to extend it to square-integrable Borel functions. Let f be square-integrable, and let (f_n) be a sequence of approximating simple functions guaranteed by Proposition 10.74. Then $\int f_n(t)\dot{W}(dt)$, $n = 1, 2, \ldots$ is a Cauchy sequence in L_2, for

$$E\left\{\left(\int f_n(t)\,\dot{W}(dt) - \int f_n(t)\,\dot{W}(dt)\right)^2\right\} = \int (f_n(t) - f_m(t))^2\,dt$$

$$\le 2\int (f_n(t) - f(t))^2 + (f(t) - f_m(t))^2\,dt\,,$$

using $(a+b)^2 \le 2a^2 + 2b^2$. But this goes to zero as $m, n \to \infty$, so the sequence is Cauchy and therefore converges in L^2. Define

$$\int f(t)\,\dot{W}(dt) \stackrel{\text{def}}{=} \lim_{n \to \infty} \int f_n(t)\,\dot{W}(dt)\,,$$

where the limit is in L^2.

Check the expected square: let $I_n = \int f_n(t)\,\dot{W}(dt)$, $I_\infty = \int f(t)\,\dot{W}(dt)$. Then $E\{I_n^2\} = \int f_n^2(t)\,dt$, and this converges to $\int f^2(t)\,dt$ by the proposition. Moreover, since $\|I_n - I_\infty\|_2 \to 0$, $E\{I_n^2\} \to E\{I_\infty^2\}$. Thus,

$$E\left\{\int f(t)\,\dot{W}(dt)\right\} = \int f^2(t)\,dt\,,$$

for all square-integrable f. It follows from this that the integral is well-defined. We can summarize this in the following theorem.

Theorem 10.75. *$\int f(t)\,\dot{W}(dt)$ is defined for all square-integrable Borel functions f on $\mathbb{R}_+$. If a, b are real and f, g are square-integrable Borel functions, then*

(i) $\int (af(t) + bg(t))\,\dot{W}(dt) = a\int f(t)\,\dot{W}(dt) + b\int g(t)\,\dot{W}(dt)$;

(ii) $\int f(t)\,\dot{W}(dt)$ *is Gaussian with mean zero and variance* $\int f^2(t)\,dt$;

10.14. White Noise and the Wiener Integral* 401

(iii) $E\{(\int f(t)\dot{W}(dt))(\int g(t)\dot{W}(dt))\} = \int f(t)g(t)\,dt$.

Proof. (i) is true for simple functions, and follows in general by taking limits in L^2. To see (ii), note first that it is true for simple functions. Let (f_n) be the simple functions guaranteed by Proposition 10.74. Then $\int f_n^2(t)\,dt \to \int f^2(t)\,dt$, while the L^2 limit of Gaussian random variables is necessarily Gaussian, and when Gaussian variables converge, so do their means and variances.

(iii) follows from (ii) by polarization. Let $\dot{W}(f) = \int f(t)\dot{W}(dt)$, $\dot{W}(g) = \int g(t)\dot{W}(dt)$. Then $E\{\dot{W}(f)\dot{W}(g)\} = (E\{(\dot{W}(f) + \dot{W}(g))^2\} - E\{(\dot{W}(f) - \dot{W}(g))^2\})/4$; but this equals $(\int (f+g)^2\,dt - \int (f-g)^2\,dt)/4 = \int fg\,dt$. □

Remark 10.76. 1° When people write "$\int f(t)\dot{W}_t\,dt$," this is the integral they mean.

2° We asked at the end of Chapter 3 if there was anything interesting about vector spaces generated by infinitely many independent random variables. We hinted that this led to Hilbert spaces. It has taken us a while, but we have finally arrived. Both $L^2(\Omega, \mathcal{F}, P)$ and the space $L^2(\mathbb{R}_+, \mathcal{B}, dx)$ of square-integrable Borel functions are Hilbert spaces. They are vector spaces: functions and random variables can be added, subtracted, and multiplied by constants. Both spaces have inner products: $\langle X, Y \rangle = E\{XY\}$, and $\langle f, g \rangle = \int f(t)g(t)\,dt$, respectively. They are infinite-dimensional, of course, and they also have the notion of convergence.

We have used as little of L^2 theory as we could to get this far, but now that we have the result, it is interesting to step back and describe it from Hilbert space point of view. Let $\dot{W}(f) = \int f(t)\,dW_t$. The stochastic integral takes a square-integrable function—an element of $L^2(\mathbb{R}_+, \mathcal{B}, dx)$—and maps it to its integral, $\dot{W}(f)$, which is a square-integrable random variable, an element of $L^2(\Omega, \mathcal{F}, P)$. Its expected square equals the integral of the square of the function, which says that the norms of the integrand and its integral in their respective spaces are equal.

Thus the stochastic integral maps $L^2(\mathbb{R}_+, \mathcal{B}, dx)$ into $L^2(\Omega, \mathcal{F}, P)$. This map is continuous, even norm-preserving. Proposition 10.74 tells us that the simple functions are dense in $L^2(\mathbb{R}_+, \mathcal{B}, dt)$. Thus we can extend the integral from the simple functions to all of $L^2(\mathbb{R}_+, \mathcal{B}, dt)$ by continuity. Putting this together, we have

Corollary 10.77. *The Wiener integral is a linear map $f \mapsto \dot{W}(f)$ of $L^2(\mathbb{R}_+, \mathcal{B}, dt)$ into $L^2(\Omega, \mathcal{F}, P)$. It is norm-preserving, and therefore preserves inner products. Moreover, it is the unique continuous linear map from $L^2(\mathbb{R}_+, \mathcal{B}, dt)$ into $L^2(\Omega, \mathcal{F}, P)$ which maps I_A to $\dot{W}(A)$, $A \in \mathcal{B}$, $|A| < \infty$.*

Let us now change to standard notation, and denote the integral by $\dot W(f) = \int f(t)\, dW_t$ instead of $\int f(t)\, \dot W(dt)$.

We have defined the integral over $\mathbb{R}_+$. To integrate between limits, define
$$\int_a^b f(t)\, dW_t \stackrel{\text{def}}{=} \int f(t) I_{(a,b]}(t)\, dW_t\,.$$

Consider the integral as a function of its upper limit: $t \mapsto \int_0^t f(s)\, dW_s$ is a stochastic process. Let $\mathcal{F}_t = \sigma\{W_s,\ s \le t\}$.

Exercise 10.43. Let $f \in L^2(\mathbb{R}_+, \mathcal{B}, dx)$ and let $X_t = \int_0^t f(s)\, dW_s$. Show that
(a) $\{X_t, \mathcal{F}_t,\ t \ge 0\}$ is a martingale.
(b) $\{X_t^2 - \int_0^t f^2(s)\, ds, \mathcal{F}_t,\ t \ge 0\}$ is a martingale.
(c) It is possible to define X_t such that $t \mapsto X_t$ is continuous a.s.

[Hint: Follow the proof of the continuity of Brownian motion.]

Notation. If $\sigma(t)$ is a square-integrable deterministic function and $g(t)$ is a (possibly random) integrable function, then $dX_t = \sigma(t)\, dW_t + g(t)\, dt$ means that $X_t = X_0 + \int_0^t f(s)\, dW_s + \int_0^t g(s)\, ds$. Further, if $f(t)$ is a deterministic function such that $f(s)\sigma(s)$ is square integrable and $f(s)g(s)$ is integrable, then
$$\int_0^t f(s)\, dX_s \stackrel{\text{def}}{=} \int_0^t f(s)\sigma(s)\, dW_s + \int_0^t f(s)g(s)\, ds\,.$$

The stochastic calculus of Wiener integrals is somewhat limited. There is no satisfactory chain rule, for example. We need the Ito integral for that. However, there is an integration-by-parts formula. In differential form, it is $F\, dX = FX - X\, dF$. To be precise:

Proposition 10.78. *Let f be a deterministic, integrable function, g a (possibly random) integrable function, and let F be an anti-derivative of f: for all $0 \le s < t$, $\int_s^t f(u)\, du = F(t) - F(s)$. Let $\sigma(t)$ be a deterministic square-integrable function, and let $dX_t = \sigma(t)\, dW_t + g(t)\, dt$. Then for $0 \le a < b < \infty$,*

(10.34) $$\int_a^b F(t)\, dX_t = F(b)X_b - F(a)X_a - \int_a^b f(t)\, X_t\, dt\,.$$

Proof. There are two integrals, a stochastic integral and a Lebesgue integral:
$$\int_a^b F(t)\, dX_t = \int_a^b F(t)\sigma(t)\, dW_t + \int_a^b F(t)g(t)\, dt\,.$$
The integration-by-parts formula for the Lebesgue integral is well-known, so we can concentrate on the other. Let $Y_t = \int_0^t \sigma(s)\, dW_s$, so that $dY_t = \sigma(t)\, dW_t$. Let G be an anti-derivative of g and notice that $X_t - X_0 =$

$Y_t + G(t) - G(0)$. By Exercise 10.43 we can assume that Y is continuous. Let $h = (b-a)/n$ and write the stochastic integral as:

$$F(b)Y_b - F(a)Y_a = \sum_{k=0}^{n-1} \left(F((k+1)h)Y_{(k+1)h} - F(kh)Y_{kh} \right)$$

$$= \sum_{k=0}^{n-1} (F((k+1)h) - F(kh)) Y_{(k+1)h} + \sum_{k=0}^{n-1} F(kh)(Y_{(k+1)h} - Y_{kh})$$

$$\stackrel{\text{def}}{=} S_1 + S_2.$$

Now a trick: Let $\Delta_k = (kh, (k+1)h]$ and write the sums S_1 and S_2 as integrals of step functions. For S_1, approximate Y by a step-function Y^n, constant on the intervals Δ_k: $Y_t^n = \sum_k Y_{(k+1)h} I_{\Delta_k}(t)$. Now $\left(F((k+1)h) - F(kh) \right) Y_{(k+1)h} = Y_{(k+1)h} \int_{\Delta_k} f(t) dt = \int_{\Delta_k} f(t) Y_t^n dt$. Sum over k to see:

$$S_1 = \int_a^b Y_t^n f(t)\, dt,$$

which is a Lebesgue integral. Y is continuous, so that as $n \to \infty$, $Y^n \to Y$ uniformly on $[a,b]$, and

$$S_1 \to \int_a^b Y_t f(t)\, dt.$$

For S_2, approximate F by the function $F^n(t) = \sum_k F(kh) I_{\Delta_k}(t)$, and write the increments of Y as stochastic integrals: $F(kh)(Y_{(k+1)h} - Y_{kh}) = F(kh) \int_{\Delta_k} \sigma(t)\, dW_t = \int_{\Delta_k} F^n(t)\sigma(t)\, dW_t$. Sum over k to see that $S_2 = \int_a^b F^n(t)\sigma(t)\, dW_t$. F is continuous, so F^n converges uniformly to F on $[a,b]$, and $\int |F^n(t) - F(t)|^2 \sigma^2(t)\, dt \to 0$. Thus,

$$S_2 \to \int_a^b F(t) \underbrace{\sigma(t)\, dW_t}_{dY_t},$$

where the convergence is in L^2. Putting these together,

(10.35) $$\int_a^b F(t)\, dY_t = F(b)Y_b - F(a)Y_a - \int_a^b f(t)\, Y_t\, dt.$$

By the classical integration by parts,

(10.36) $$\int_a^b F(t) g(t)\, dt = F(b)G(b) - F(a)G(a) - \int_a^b f(t) G(t)\, dt.$$

Add (10.35) and (10.36) and use the fact that $X_t = Y_t + G(t)$ to get (10.34). □

Problems 10.14

10.44. Let $\{V_t,\ t \geq 0\}$ be an Ornstein-Uhlenbeck process from zero. Show that as $s \to \infty$, $\{V_{s+t},\ t \geq 0\}$ tends in distribution to stationarity.

10.45. Show that the stochastic integral is a homeomorphism between the two Hilbert spaces $L^2(\mathbb{R}_+, \mathcal{B}, dx)$ and the subspace of $L^2(\Omega, \mathcal{F}, P)$ which is generated by $\{\dot{W}(A), A \in \mathcal{B}\}$.

10.46. Let (B_t) be a standard Brownian motion. Expand the function $t \mapsto B_t$ in a Fourier cosine series on $[0, \pi]$. The coefficients are random. What are their distributions? Show that the series converges with probability one for each $t \in [0, \pi]$.

10.47. Let $(\phi_n(t))$ be a complete sequence of orthonormal functions on the interval $[0, 1]$. Expand a standard white noise $\dot{W}$ in the ϕ_n.

(a) Interpret the usual formula for the coefficients in terms of stochastic integrals. What is the (joint) distribution of the coefficients?

(b) Show that with probability one, the series fails to converge in L^2. Show, moreover, that in the case where the ϕ_n are the usual sines or cosines, the series diverges a.s. at each $t \in (0, 1)$.

(c) White light contains all frequencies equally. Do you think the name "white noise" is appropriate?

10.15. Physical Brownian Motion*

Brownian motion is supposed to describe a physical particle[24]. Physicists have a problem with this: the particle's average speed is infinite. If it has a non-zero mass this means that its average energy is also infinite. But the physical particles observed in the microscope have a small but definitely non-zero mass, and they just as definitely have a finite energy. So Brownian motion has some drawbacks as a physical model for particle motion.

Ornstein and Uhlenbeck proposed a different theory of Brownian motion. Their reasoning is roughly this: consider a particle of mass m immersed in a liquid. It is bombarded on all sides by the molecules of the liquid, and, while it is heavy enough that a single molecular impact produces a negligible motion, there are a large number of impacts each second. The impacts from different sides tend to cancel out, but there are fluctuations, and at any instant, there will be a net imbalance, which produces a visible motion.

Let us look more closely at these forces. Let V_t be the velocity of the particle at time t. The particle is subject to two forces from its surroundings. First, there is the net force on the stationary particle coming from the bombardment by the fluid molecules. The different molecules are independent, and the individual impacts are small. The central limit theorem

---

[24] This section was inspired by Nelson's lovely book [30]. It gives a fascinating account of the development, from Thomas Brown's original experiments, to the Ornstein-Uhlenbeck theory and Einstein's discoveries.

guarantees that the net force will be Gaussian and symmetric. Since the forces at different times are caused by different molecules, they should be independent. Moreover, the statistical properties of the force should be the same from time to time and place to place. But this describes white noise! We conclude that the total force in a time interval dt is $\hat{\sigma}\dot{W}(dt)$ where $\dot{W}$ is a standard white noise and $\hat{\sigma}$ will be determined later.

The second force comes directly from the particle's motion. When it moves, it tends to encounter molecules coming towards it more quickly than it does molecules coming from behind it. This leads to a slight imbalance of forces; the molecules coming towards it provide a slightly larger force which should be proportional to the particle's velocity and in the opposite direction. We call this a *viscosity force*, $-\lambda V(t)$. It tends to push the velocity back to zero.

Thus the net force in a short time interval $(t, t+dt)$ is $\hat{\sigma}\dot{W}(dt) - \lambda V_t\, dt$. By Newton's law, $F = mA$, so the net acceleration—that is, the net change in velocity—in an interval is the net force divided by the mass. Call the net change of velocity dV. Then we have Langevin's equation for V:

$$(10.37) \qquad dV_t = \frac{\hat{\sigma}}{m}\dot{W}(dt) - \frac{\lambda}{m}V_t\, dt\,.$$

This makes sense if we integrate it from 0 to t, letting $W_t = \dot{W}([0,t])$ be a standard Brownian motion:

$$(10.38) \qquad V_t - V_0 = \frac{\hat{\sigma}}{m}W_t - \frac{\lambda}{m}\int_0^t V_s\, ds\,.$$

Langevin's equation was the first example of what is now called a stochastic differential equation. It can actually be solved, and the solution is called the **Ornstein-Uhlenbeck velocity process.**

Technically, it is the integral equation (10.38) which is rigorously defined. Its differential form (10.37) is best thought of as shorthand for the integral equation. However, the differential form is so intuitive, and the passage from one to the other so easy, that one usually writes the differential form, and talks of "stochastic differential equations" rather than "stochastic integral equations".

In any case, Langevin's equation is quite intuitive: it says that the particle is being knocked around by a chaotic background force (represented by $\dot{W}$) and a restoring force, $-\lambda V_t$ which returns it to equilibrium. In fact, it is a perturbation of the simple harmonic motion familiar from elementary physics courses. Indeed, if there is no white noise ($\hat{\sigma} = 0$), (10.37) becomes an ordinary differential equation whose solution is sines and cosines of λt, i.e., simple harmonic motion. The additional random force perturbs the motion. So the Ornstein-Uhlenbeck process can be thought of as a randomly perturbed harmonic oscillator, which performs a stochastic no-longer-quite-so-simple harmonic motion.

Solving the Langevin Equation. Consider the Langevin equation with initial condition:

(10.39) $$\begin{cases} dV_t = \hat{\sigma} dW_t - \lambda V_t\, dt\,, \\ V_0 = v_0\,, \end{cases}$$

where we have absorbed the mass m of (10.37) into $\hat{\sigma}$ and λ.

Theorem 10.79. *The Langevin initial-value problem (10.39) has the unique solution*

(10.40) $$V_t = v_0 e^{-\lambda t} + \hat{\sigma} \int_0^t e^{-\lambda(t-s)}\, dW_s\,.$$

Proof. We will say nothing about the reason behind the following calculations lest we spoil the surprise.

Let f be a continuously differentiable function. Multiply both sides of Langevin's equation by $f'(t)$ and integrate with respect to dV:

$$\int_0^t f'(s)\, dV_s = \hat{\sigma} \int_0^t f'(s)\, dW_s - \lambda \int_0^t f'(s) V_s\, ds\,.$$

Apply the integration-by-parts formula to the Riemann integral:

$$= \hat{\sigma} \int_0^t f'(s)\, dW_s - \lambda f(t) V_t + \lambda f(0) V_0 + \lambda \int_0^t f(s)\, dV_s\,.$$

Thus,

$$\int_0^t (f'(s) - \lambda f(s))\, dV_s = \hat{\sigma} \int_0^t f'(s)\, dW_s - \lambda f(t) V_t - \lambda f(0) V_0\,.$$

Choose f such that $f'(s) - \lambda f(s) \equiv 0$ and $\lambda f(t) = 1$. There is only one f that satisfies this: $f(s) = e^{-\lambda(t-s)}/\lambda$. The left-hand side of this equation vanishes and $V_0 = v_0$, leaving $0 = \hat{\sigma} \int_0^t e^{-\lambda(t-s)}\, dW_s - V_t + e^{-\lambda t} v_0$. This gives (10.40). □

V_t is a Gaussian process with mean zero and covariance function

$$\Gamma(s,t) = E\{(V_s - v_0 e^{-\lambda s})(V_t - v_0 e^{-\lambda t})\}$$

$$= \hat{\sigma}^2 E\left\{ \int_0^s e^{-\lambda(s-u)}\, dW_u \int_0^t e^{-\lambda(t-u)}\, dW_u \right\}$$

$$= \frac{\hat{\sigma}^2}{2\lambda} \int_0^{s \wedge t} e^{-\lambda(s+t-2u)}\, du$$

$$= \frac{\hat{\sigma}^2}{2\lambda}\left(e^{-\lambda|t-s|} - e^{-\lambda(t+s)}\right), \quad s,t \geq 0\,.$$

Compare this with Proposition 10.68: V_t is an Ornstein-Uhlenbeck process. If the initial value V_0 is a $N(0, \hat\sigma^2/2\lambda)$ random variable independent of the white noise, then a similar calculation shows that its covariance function is $\Gamma(s,t) = (\hat\sigma^2/2\lambda)e^{-\lambda|t-s|}$, so that V_t is stationary. In particular, its equilibrium variance is $\hat\sigma^2/2\lambda$.

Notice that the covariance drops off exponentially. If $|t-s|$ is large, V_s and V_t are nearly independent. The larger the λ, the more nearly independent the values are, which suggests that, if λ tends to infinity, (V_t) might tend in some sense to a white noise.

In order to find $\hat\sigma$ for our physical model, we turn to a result of statistical mechanics: the average energy of a particle in equilibrium is $kT/2$, where T is the temperature and k is Boltzmann's constant. (Properly speaking, this is the energy per degree of freedom, but in one dimension, there is only one degree of freedom.) In equilibrium, the average energy equals the expected energy, so that $E\{mV_t^2/2\} = kT/2$ or

$$E\{V_t^2\} = \frac{kT}{m}.$$

Thus $\hat\sigma^2 = 2\lambda kT/m$. Let

(10.41) $$D \stackrel{\text{def}}{=} \frac{kT}{\lambda m}.$$

Then $\hat\sigma^2 = 2\lambda^2 D$ and Langevin's equation becomes

(10.42) $$dV_t = \lambda\sqrt{2D}\, dW_t - \lambda V_t\, dt,$$

with solution

(10.43) $$V_t^\lambda = v_0 e^{-\lambda t} + \lambda\sqrt{2D}\int_0^t e^{-\lambda(t-s)} dW_s.$$

We found the infinitesimal generator for the process with drift λ and equilibrium variance σ^2 in (10.31). The equilibrium variance of V^λ is $2\lambda D$ so that its infinitesimal generator is

(10.44) $$\mathbb{G}f(x) = \lambda^2 D f''(x) - \lambda x f'(x).$$

The Physical Particle. Now that we have the velocity V_t^λ from (10.43), we can find the position X_t^λ by integrating:

$$X_t^\lambda = X_0^\lambda + \int_0^t V_s^\lambda\, ds.$$

Since V^λ is continuous, X_t^λ has a continuous derivative, and, in particular, it has finite energy. Its acceleration is not finite, but physicists are used to dealing with rigid body impacts involving infinite accelerations, so this is not a problem. What does appear to be a problem, however, is that it is not a Brownian motion, at least, not of the type we have been studying. The question arises: What is the relation between the two? The answer is that for large λ, (X_t^λ) approximates a Brownian motion. This because if λ is large, the velocity process at any two different times s and t becomes nearly independent, that is, very like a white noise. In the limit as $\lambda \to \infty$, it is one, and its its integral tends to a true Brownian motion.

We should be more precise. We have studied convergence in distribution for random variables in §6.2, but we have not studied convergence in distribution for processes.

Definition 10.80. A sequence of processes $\{X_t^n,\ t \geq 0\}$, $n = 1, 2, \ldots$ **converges in distribution** to a process $\{X_t^\infty,\ t \geq 0\}$ if all the finite-dimensional distributions of X^n converge to the corresponding distributions of X^∞. We write $X^n \Longrightarrow X^\infty$.

Remark 10.81. There are numerous ways of defining convergence in distribution for stochastic processes, some of which give powerful results but require deeper mathematics. This may be the weakest useful definition, but it is simple, it needs nothing we have not already seen, and it shows why the physical particle motion approximates Brownian motion.

For $A \in \mathcal{B}$, define
$$V^\lambda(A) \stackrel{\text{def}}{=} \int_A V_t^\lambda \, dt.$$

Theorem 10.82. *Let $\lambda \to \infty$ while holding D constant. Then:*

(i) $\{X_t^\lambda,\ t \geq 0\}$ converges in distribution to a Brownian motion with infinitesimal generator
$$\mathbb{G} = D \frac{d^2}{dx^2}.$$

(ii) $\{V^\lambda(A),\ A \in \mathcal{B}\}$ converges in distribution to the white noise $\sqrt{2D}\,\dot{W}$.

Remark 10.83. Einstein, using completely different reasoning, came up with the same infinitesimal generator. In terms of the physical constants k, T, m and λ, $D = kT/\lambda m$ and X satisfies the stochastic differential equation
$$dX_t = \sqrt{\frac{2kT}{\lambda m}}\, dW_t,$$
for a standard Brownian motion W. This confirms two observations made long before Einstein's theory: (a) smaller particles move faster, and (b) particles move faster at higher temperatures.

Proof. We can define all the V^λ as stochastic integrals with respect to the same white noise. Since L^2 convergence implies convergence in distribution,

10.15. Physical Brownian Motion*

it is enough to prove pointwise convergence in L^2 for $\{X_t^\lambda,\ t \geq 0\}$ and $\{V^\lambda(A),\ A \in \mathcal{B}\}$.

Let $\varphi_\lambda(t) = \lambda e^{-\lambda t},\ t \geq 0$. Then $V_t^\lambda = \sqrt{2D}c \int_0^t \lambda e^{-\lambda(t-s)}\, dW_s = \sqrt{2D} \int_0^t \varphi_\lambda(s)\, dW_s$. Consider the integral

$$\int_0^\infty f(t) V_t^\lambda dt = \lambda\sqrt{2D} \int_0^\infty f(t) \int_0^t e^{-\lambda(t-s)}\, dW_s$$

$$= \lambda\sqrt{2D} \int_0^\infty f(t) e^{-\lambda t} \underbrace{\int_0^t e^{\lambda s}\, dW_s}_{\stackrel{\text{def}}{=} X_t}\, dt$$

$$= \lambda\sqrt{2D} \int_0^\infty f(t) e^{-\lambda t} X_t\, dt.$$

Let $F(t) = -\int_t^\infty f(s) e^{-\lambda s}\, ds$ and integrate this by parts. Note that $F(t) = 0$ for large t since f has compact support, and $X_0 = 0$ so this is

$$= \lambda\sqrt{2D} \int_0^\infty F(t) e^{\lambda t}\, dW_t$$

$$= \sqrt{2D} \int_0^\infty \int_t^\infty \lambda e^{-\lambda u}\, du\, e^{\lambda s}\, dW_s$$

$$= \sqrt{2D} \int_0^\infty \int_s^\infty \lambda e^{-\lambda(u-s)}\, du\, dW_s.$$

We can identify the inner integral as a convolution. If $\varphi_\lambda(t) = \lambda e^{-\lambda t},\ t \geq 0$, the inner integral is $f * \varphi_\lambda(s)$. Thus,

(10.45) $$\int_0^\infty f(g) V_t^\lambda\, dt = \sqrt{2D} \int_0^\infty f * \phi_\lambda(s)\, dW_s.$$

But φ_λ is an **approximate identity** ($\varphi \geq 0$, $\int \varphi(t)\, dt = 1$ and for all $\varepsilon > 0$, $\int_{-\varepsilon}^{\varepsilon} \varphi_\lambda(t)\, dt \to 1$ as $\lambda \to \infty$) so that $f * \varphi_\lambda \to f$ in the mean square as $\lambda \to \infty$: $\int |f(t) - f * \varphi_\lambda(t)|^2\, dt \to 0$. Thus the integral converges in L^2:

(10.46) $$\int_0^\infty f(t) V_t^\lambda dt \longrightarrow \sqrt{2D} \int_0^\infty f(t)\, dW_t,$$

Let us suppose that $X_0^\lambda = 0$, so that $X_t^\lambda = \int_0^t V_s^\lambda\, ds$. To prove (i), take $f(s) = I_{[0,t]}(s)$ in (10.46) to see that for each t, $X_t^\lambda \to \sqrt{2D}\, W_t$ in L^2. Therefore the process (X_t^λ) converges in distribution to a Brownian motion.

(ii) Take $f = I_A$ for a Borel set $A \subset [0, \infty)$ with compact closure to see that $V^\lambda(A) \to \sqrt{2D} \int_A dW = \sqrt{2D}\, \dot{W}(A)$. Then extend this to A of finite Lebesgue measure. This implies that $\{V^\lambda(A),\ A \in \mathcal{B}\}$ converges in distribution to $\{\dot{W}(A),\ A \in \mathcal{B}\}$. □

10.16. What Brownian Motion Really Does

> *This is an homage to Willy Feller. It is taken from a lecture he gave to cap off his probability course. It is hard to catch his style on paper, but he was as brilliant a lecturer as mathematician, and, whenever he needed an arbitrary number, it was* always *seventeen.*

You may have wondered: If planar Brownian motion doesn't hit points, what *does* it do? The answer is easy. It spends all its time writing "seventeen".

What is "seventeen"? Write it in script in pen or pencil on a piece of paper—or in chalk on a blackboard—and you will see: it is a connected open set:

To write it, the Brownian motion must trace it out, from the up-stroke at the start of the "s" to the flourish on the last "n" without leaving the open set. This is possible: one can bound the probability of doing it by the probability that the first coordinate goes from zero to some value a while the maximum modulus of the second coordinate remains less than some ϵ. So, while its handwriting might be spidery, Brownian motion can write out "*seventeen*" in the time-period $[0, 1]$. Let

$$p \stackrel{\text{def}}{=} P\{B_t \text{ writes "}seventeen\text{" in } 0 \leq t \leq 1\} > 0.$$

By scaling, $B_t = (B_t^1, B_t^2) \stackrel{d}{\sim} c(B_{t/c^2}^1, B_{t/c^2}^2) = cB_{t/c^2}$. Thus,

$$P\{\sqrt{2}B_{t/2} \text{ writes "}seventeen\text{" in } 0 \leq t \leq 1\} = p$$

or, if $s = t/2$,

$$P\{B_s \text{ writes "}seventeen\text{" in } 1/\sqrt{2} \text{ scale in } [0, 1/2]\} = p.$$

B_s has the same probability of writing "*seventeen*" in $[1/2, 1]$, and the two are independent, so that

$$P\{B_s \text{ does } not \text{ write "}seventeen\text{" in } 1/\sqrt{2} \text{ scale in } [0, 1]\} \leq (1 - p)^2.$$

Similarly, for any n,
$$p = P\{nB_{t/n^2} \text{ writes "{\it seventeen}" in } 0 \le t \le 1\}$$
$$= P\{B_s \text{ writes "{\it seventeen}" in } 1/n \text{ scale in } [0, 1/n^2]\}.$$

But it has the same probability of writing it in the intervals $[(k-1)/n^2, k/n^2]$, $k = 1, \ldots, n^2$ independently, so
$$P\{B_s \text{ does } not \text{ write "{\it seventeen}" in } 1/n \text{ scale in } [0,1]\} \le (1-p)^{n^2}.$$

Let $n \to \infty$. This probability goes to zero, and we conclude that
$$P\{B_s \text{ writes "{\it seventeen}" in } some \text{ scale in } [0,1]\} = 1.$$

The same is true in the time intervals $[1,2]$, $[2,3]$ and so on: the probability of writing it out in $[k, k+1]$ is one. Of course, there is nothing magic about starting with the time interval $[0,1]$. We could have started with any interval $[0, \varepsilon]$. Thus we see

Theorem 10.84. *With probability one, Brownian motion writes {\it seventeen} infinitely often in all non-empty open time intervals.*

Remark 10.85. That's what it does! Note that "seventeen" was not arbitrary. Brownian motion cannot write "fifteen" "sixteen" "eighteen" or "nineteen". (Why not?)

Problems 10.16

10.48. Let B_t^i, $i = 1, 2$ be two independent standard Brownian motions, and let $0 < \rho < 1$.

(a) Let $X_t \stackrel{\text{def}}{=} \rho B_t^1 + \sqrt{1-\rho^2} B_t^2$. Is $\{X_t, 0 \le t \ge 0\}$ a standard Brownian motion?

(b) Let Z be $N(0,1)$, and let $X_t \stackrel{\text{def}}{=} \sqrt{t} Z$, $0 \le t \le 1$. Then X_t is $N(0,t)$. Is (X_t) a standard Brownian motion?

10.49. Show that if a process is stationary, Gaussian, Markov, and continuous in probability, then it is of the form $X_t + c$, where X_t is an Ornstein-Uhlenbeck process and c is a constant.

10.50. Let B_t be a standard Brownian motion. Define a process X_t by $X_t = |B_t - 2n|$ if $|B_t - 2n| \le 1$. Show that X_t is a Brownian motion on $[0,1]$ with reflecting boundaries at 0 and 1.

10.51. There are stopping times T for which $P\{B_T = 0\} > 0$, but none for which B_T is a local maximum with positive probability.

10.52. Let (B_t) be a standard Brownian motion, and, for each x, let $T_x^0 = \inf\{t : B_t = x\}$, $T_x = \inf\{t : B_t > x\}$. Show that for each x, $P\{T_x^0 = T_x\} = 1$, but that for a.e. ω, there exist x for which $T_x(\omega) > T_x^0(\omega)$.

Bibliography

[1] Bachelier, Louis, 1900. Théorie de la spéculation, Ann. Sci. École Norm. Sup. 17, 21-86.

[2] Bachelier, Louis, 1964. Theory of speculation, The Random Character of Stock Market Prices (P. Cootner, ed.), MIT Press, pp. 17-78, Translated from French by A. James Boness.

[3] Baxter, M. and A. Rennie, Financial Calculus: An Introduction to Derivative Pricing, Cambridge University Press, Cambridge,1996.

[4] Billingsley, Patrick, 1999. Convergence of Probability Measures, 2nd ed., John Wiley & Sons, New York.

[5] Black, F. and M. Scholes, 1973. The pricing of options and corporate liabilities, Journal of Political Economy, 81, 637-654.

[6] Borel, E., 1909. "Les probabilités dénombrables et leurs applications arithmétiques", Rendiconti del Circolo Matematico di Palermo 27, 247-271.

[7] Bayer, D. and P. Diaconis, 1992. "Trailing the Dovetail Shuffle to its Lair", Annals of Applied Probability 2, 294.

[8] Chacon, R. and J. B. Walsh, 1976. One-dimensional potential embeddings, Séminaire de Probabilités X de L'Univ. de Strasbourg, Lecture Notes in Math, vol. 511, 19-23.

[9] Chung, K. L., A Course in Probability Theory, Academic Press, 1974.

[10] Dalang, R. C., A. Morton, and W. Willinger, 1990. Equivalent martingale measures and no-arbitrage in stochastic securities market models, Stochastics and Stochastics Reports 29, 185–201.

[11] Delbaen, F. and W. Schachermayer, The Mathematics of Arbitrage, Springer, 2006.

[12] Doob, J. L., 1953. Stochastic Processes, John Wiley & Sons Inc., New York.

[13] Devlin, Keith, The Unfinished Game, Basic Books, 2008.

[14] Durrett, R., Probability: Theory and Examples, Second edition, Duxbury Press, Belmont, CA., 1996.

[15] Dudley, R. M., Real Analysis and Probability, Cambridge University Press, Cambridge, 1989.

[16] Feller, William, An Introduction to Probability Theory and its Applications, Vol. 1, Third edition, John Wiley and Sons, New York, 1968.

[17] Feller, William, An Introduction to Probability Theory and its Applications, Vol. 2, Third edition, John Wiley and Sons, New York, 1971.

[18] Keynes, John Maynard, A Treatise on Probability, Macmillan Co., London, 1921.

[19] Karlin, Samuel, A First Course in Stochastic Processes, Academic press, New York, 1966.

[20] Khoshnevisan, Davar. Probability, American Mathematical Society, Graduate Studies in Mathematics 80, Providence RI, 2007.

[21] Kolmogorov, A. N., Grundbegriffe der Wahrscheinlichkeitsrechnung, Springer, Berlin, 1933.

[22] Kolmogorov, A. N., Foundations of Probability, Chelsea Publishing Company, New York, 1950. [Translation edited by Nathan Morrison].

[23] Laplace, P. S., Théorie Analytique des Probabilités, Vol. I and II, Paris, Courcier, 1812.

[24] Lehmann, E. L., Testing Statistical Hypotheses, Wiley, New York, 1959.

[25] Lévy, P., Calcul des Probabilités, Gauthier-Villars, Paris, 1925.

[26] Lévy, P., Processus stochastiques et mouvement brownien, Gauthier-Villars, 1948. Réédition en 1992 par Jacques Gabay.

[27] Lindeberg, J. W., 1922. Eine neue Herleitung des Exponentialgesetzes in der Wahrscheinlichkeitsrechnung, Math. Z. 15, 211-225.

[28] Markov, A. A., 1910. Recherches sur un cas remarquable d'épreuves dépendantes, Acta Math. 33, 87-104.

[29] Mazliak, Laurant, 2009. How Paul Lévy saw Jean Ville and Martingales, Electronic Journal for History of Probability and Statistics, Vol. 5, no. 1.

[30] Nelson, Edward, Dynamical Theories of Brownian Motion, Princeton University Press, Princeton, NJ. 1967.

[31] Royden, H. L., Real Analysis, 3rd edition, Macmillan, New York, 1988.

[32] Scheffé, H., The Analysis of Variance, Wiley, New York, 1959.

[33] Wiener, Norbert, Extrapolation, Interpolation, and Smoothing of Stationary Time Series, Wiley, New York, 1949.

[34] Williams, D., Probability with Martingales, Cambridge University Press, Cambridge, 1991.

Index

L^2, 394
σ-field, 1
σ-field, xv, 1–3, 8, 43, 192, 204, 205, 207, 266

absorption, 228
accessible, 201, 213
algebra, 1
aperiodic, 203, 218, 223, 226
arbitrage, 318–320, 327–330
ascii, 23

Bachelier, Louis, 335
bad luck, 249
balayage, 379
Bayes rule, 18
Bernstein, S., 136
Bernstrein polynomials, 137
bi-stochastic, 228
binary tree, 330
binomial coefficient, 25
binomial theorem, 66
Black-Scholes, 374, 377
bond, 317–319
Borel field, 1
Borel measurable, 44
Borel sets, 3, 207, 395
Borel, Émile, 138
Borel-Cantelli, 135, 356
bounded convergence theorem, 126
branching process, 312
Brown, Robert, 335
Brownian
    bridge, 365
    Markov property, 340
    self-similarity, 339
    time-reversal, 339
    transformations, 338
    zero set, 348
Brownian density, 337
Brownian motion, xv, 346, 349, 353, 354, 356, 374–377, 380, 381, 383, 384, 388, 392–394, 396, 402, 404, 405, 408, 409
    absorbing, 363
    construction, 388
    in [0,1], 364
    killed, 363
    logarithmic, 374
    nowhere-differentiability, 354
    quadratic variation, 355
    reflecting, 363
    standard, 336
    unbounded variation, 355
Buffon's needle problem, 51

Cantelli strong law, 134
Cantor set, 349
capture-recapture, 57
card shuffling, 227
central limit theorem, 383
characteristic function, 337
combination, 22, 25
communicate, 201, 202, 215
conditional expectation
    existence, 305

continuity
  almost sure, 342
  Brownian, 341
continuous
  version, 343
continuous parameter, 191
convergence
  almost everywhere, 117
  in distribution, 118, 409
    for processes, 408
  in probability, 117
  in the mean, 117
  pointwise, 117
  vague, 168
  weak, 168, 169, 191
convex function, 86
convolution, 107
correlation, 99, 100, 107, 112
countable additivity, 4, 7, 395
counting principle, 22
covariance, 99, 100, 108, 112, 407
  function, 391
  matrix, 108, 391
covariance function, 337, 366
covariance matrix, 158
craps, 39
cylinder sets, 192

density, 50, 83
  bivariate normal, 112
  Cauchy, 92, 93
  conditional, 104–106
  joint, 97, 99, 105, 111
  marginal, 98, 99
  multivariate Gaussian, 110
  normal, 90
  transformation, 94, 101
  uniform, 89
derivative, 317, 322, 323, 327, 329, 330,
    332, 376, 377
  Brownian motion, 354
  Radon-Nikodym, 376
difference equation, 32, 33
discontinuity
  jump, 41, 342, 343
  oscillatory, 342
discrete parameter, 191
distribution, 43, 195, 200, 337, 391
  absolutely continuous, 50, 97
  Bernoulli, 65, 70
  binomial, 65
  Cauchy, 92

  conditional, 103
  continuous, 49
  discrete, 49
  exponential, 90
  finite dimensional, 192
  Gamma, 93
  Gaussian, 90
    tails, 342
  geometric, 68
  hypergeometric, 69, 70, 72
  initial, 194, 195, 223, 226
  joint, 96
  negative binomial, 69
  normal, 90
  Poisson, 66, 67
  standard normal, 90
  stationary, 223, 225, 226, 228
  uniform, 51, 89, 140
distribution function, 40–42, 44, 45, 50
  conditional, 103, 104
  joint, 96–98
  marginal, 98
dollar
  discounted, 321, 376
  today's, 321
dominated convergence theorem, 125
Donsker, M., 386
Doob decomposition, 281, 301
Doob, J. L., 37, 85, 277, 290, 292
Dow-Jones, 192
drunkard's walk, 196
dyadic approximation, 75, 346

Einstein, Albert, 335, 408
embedded chain, 257
embedding, 377, 382
  random walk, 382
equally likely, 10
equation
  backward differential, 255
  Chapman-Kolmogorov, 195, 254
  difference, 33
  forward differential, 256
  Kolmogorov, 254
  Langevin, 405, 406
  renewal, 219
  stochastic differential, 405
  stochastic integral, 405
equilibrium, 196, 198, 218, 223, 225
equivalence class, 202, 213, 394
equivalence relation, 201
ergodic, 218

Index

events, 8
exchangeability, 70
exchangeable, 70, 71
expectation, 54, 81
   as integral, 81
   conditional, xv, 103, 105, 265, 266, 268, 270, 298, 305, 306, 375
   discrete, 55, 76
   general, 75
   linearity, 57
   matrix, 108
   non-existence, Cauchy, 93
   properties, 56
   vector, 108
explosion, 261
extinction, 237–239

fair game, 275, 283
false positives, 17
Fatou's Lemma, 124, 297, 302
Feller, W., 410
Fermat, P., xiii
field, 1, 2, 192, 396
filtration, 277, 278, 291, 345
   right continuous, 350
finitely additive, 395, 397
first exits, 353
first hitting time, 206, 211
flush, 25
Fourier inversion, 152
full house, 26
function
   Borel, 44, 49, 400
   characteristic, xiv, 79, 151–153, 155, 337
      Bernoulli, 155
      binomial, 155
      Cauchy, 156
      exponential, 156
      Gaussian, 156
      joint, 155
      joint Gaussian, 158
      Poisson, 156
   compact support, 360
   concave, 86, 379, 381
   convex, 86, 278
   covariance, 337, 390, 391
   distribution, 40, 42, 152
      normal, 376
   elementary, 398
   excessive, 314
   gamma, 93
   generating, 79
   indicator, 43
   moment generating, 91, 151
   non-differentiable, 335
   probability generating, 235, 236, 238
   probability mass, 50
      joint, 97
   Rademacher, 141
   right continuous, 343
   Schauder, 388
   simple, 399, 401
   subharmonic, 277
   superharmonic, 277
   tent, 379
future, 317

Galton, Francis, 234
gambler's ruin, 30, 32, 191, 196, 286, 293
   duration, 286
gambling systems, 210
Gaussian
   bivariate, 107
   bound, 342
   existence, 110
   joint, 109
   linear combinations, 109
   multivariate, 108, 109, 157
   multivariate density, 110
generating function, 79
geometry, 112

hand, 25
   bridge, 25
   poker, 25
Hilbert space, 113, 401
hitting time, 344, 350
horse-racing, 328

increments
   independent, 389
independence, xiv, 18, 113
   conditional, 199, 200, 347, 366
   events, 19
   of random variables, 48
   pairwise, 20
independent increments, 243, 337, 341
inequality
   Chebyshev, 85, 134, 135, 287
   Jensen, 88, 89, 271, 278
   Lyapounov, 89
   martingale maximal, 287

martingale minimal, 288
  Schwarz, 85, 100
  upcrossing, 290, 291
infinitesimal generator, 255, 359
  Brownian, 360, 392, 408
  domain, 360
  Ornstein-Uhlenbeck, 392, 407
  Poisson, 256
inflation, 321
inner product, 395, 401
integrability
  uniform, 295, 299, 301, 307
integrable, 55, 56, 76
integral
  expectation, 81
  Ito, 398, 402
  Lebesgue, xiv, 82, 402
  Lebesgue-Stieltjes, 82
  Riemann, 82, 106
  stochastic, xvi, 398, 401, 402, 408
  Wiener, xvi, 398, 401, 402
integration
  Monte Carlo, 141
integration by parts, 402
invariance principle, 386
irreducible, 202, 215, 218, 221, 223, 226, 228

Jacobian, 101, 102, 107, 110
joint density, 97
joint distribution
  Brownian, 337
joint distribution function, 98
jump chain, 260
jump time, 259
jump times, 246

Khintchine, A., 356, 384
Kolmogorov differential equations, 254
Kolmogorov strong law, 307
Kolmogorov, A. N., xiii

Lévy, Paul, 152, 388
Laplace transform, 84
law of large numbers, 133, 161
  Cantelli, 134, 136, 139
  Kolmogorov, 135, 383
  strong, 134
  weak, 134
law of the iterated logarithm, 356, 358, 384

law of the unconscious statistician, 56, 82
Lazzerini, M., 52
Lebesgue measure, 350, 354, 396
Lebesgue sets, 3
Lebesgue, Henri, 398
lemma
  Kronecker, 135
likelihood ratio, 310
limit
  Cèsaro, 221
linear function of Gaussian, 91

market, 317, 318, 322, 329, 374
  stock, 279
market price, 330, 332
Markov chain, xv, 192–194, 196, 200, 207–209, 211, 218, 223, 225, 226, 228
  finite state, 226
Markov chains, 359
  periodic, 226
Markov property, 346
  Brownian, 340
  Poisson process, 245
  strong, 246, 251, 260, 344, 346, 349–351, 381, 385
Markov time, 205
Markov, A.A., 205
martingale, xv, 211, 276–278, 281, 284, 285, 287, 294, 298, 304, 311, 313, 315, 316, 328, 336, 351, 353, 362, 375, 378
  backward, 294, 295, 300, 308
  Brownian, 361
martingale measure, 328, 330, 332, 376
matrix
  covariance, 108, 110, 337, 391
  transition probability, 193, 194, 200, 228
  n-step, 195
maximal inequality, 351
mean, 60, 61, 63, 64, 112
  geometric, 69
  hypergeometric, 72
  Poisson, 67
  uniform, 89
measure
  L^2-valued, 395
  change of, 374
  equivalent, 374
  martingale, 328, 330, 332

Index

probability, 4
synthetic, 328
measure theory, xiii
median, 63, 64
  Bernoulli, 65
  exponential, 90
  geometric, 69
  uniform, 89
memoryless, 90
method of images, 363
mode, 63
  Bernoulli, 65
  geometric, 69
moment, 61
  absolute, 61
  central, 61
moment generating function, 62
  Bernoulli, 65
  binomial, 66
  Gaussian, 91
  generates moments, 62
  geometric, 68
  Poisson, 66
moments, 85
monotone class, 1, 2
monotone class theorem, 4
monotone convergence theorem, 123
Monte Carlo method, 52
mutation, 198

normal
  bivariate, 107
  multivariate, 108
normal numbers, 137, 138, 140
nowhere-differentiability
  of Brownian motion, 354
null recurrent, 220, 223, 225, 226

odds, 315, 316
option, 317, 330, 376, 377
  European, 376, 377
optional stopping, 282
optional time, 205
orthogonal, 100, 113

partition, 55
Pascal's triangle, 26, 27
Pascal, Blaise, xiii
past before T, 207
perfect set, 349
period, 203, 215, 221
periodic, 221

permutation, 22, 23
Poisson process, 243
  distribution, 245
portfolio, 323
  self-financing, 323, 329
positive definite, 108, 109
positive recurrent, 220, 223, 225, 226, 394
potential, 378, 381, 382
  of a measure, 378
potential field, 378
potential theory, xvi, 277, 378
power series, 59
principle
  duck, 106
probability
  absorption, 228
  conditional, xiv, 15, 103, 106
  density, 50
  equivalent, 327
  measure, 4, 52
  space, 8, 11, 39
  synthetic, 315
probability generating function, 59
probability integral transformation, 46
probability space, 6
process
  birth and death, xv, 251, 360
  branching, 197, 234, 235, 312
  continuous parameter, 191
  diffusion, 336
  Gaussian, 336, 337, 390, 395
  Markov, 192, 336, 341, 363, 391
  Ornstein-Uhlenbeck, 390, 391, 394, 405, 407
  Poisson, xv, 243, 342
  stable, 336
  stationary, 390, 407
  stochastic, xiv, 191
  velocity, 405, 408
  Wiener, 335
projection, 272
property
  Markov, 192, 194, 200, 208, 229
  memoryless, 90
  strong Markov, 210, 214, 391
    discrete case, 209
Pyncheon, T., 67

quadratic form, 111
quadratic variation, 355
queuing, 197

race track, 316
radius of convergence, 59
random variable, 39, 42, 75
    Bernoulli, 140
    Cauchy, 160
    discrete, 54
    existence, 45
    extended real values, 81
    Gaussian, 91, 391
    independent, 78
    uniform, 136, 140
random walk, xv, 196, 203, 215, 286, 377
    embedded, 378
    null recurrent, 286
    reflecting, 196
    three dimensions, 216
    transient in three dimensions, 217
    two dimensions, 215
recurrence, 211, 215, 352
recurrence class, 213, 226
recurrent, 212, 213, 218, 393
reflection principle, 351
restarting, 194
reversible, 391
Riemann integral, 80
roulette, 204, 282

sample path, 341
    irregularity, 354
sample space, 8
sampling
    with replacement, 24
    without replacement, 24
semimartingale, 277
sequential analysis, 280, 310
short-sell, 279, 318
Skorokhod embedding, 377, 378, 380, 382–384, 386
Skorokhod, A.V., 377
Snell, Laurie, 290
standard deviation, 60
state
    recurrent, 212
    transient, 212
state space, 193
Stirling's formula, 214–216
stochastic differential equation, 405, 408
stock, 317, 322, 331, 374
stopping time, 204–208, 210, 211, 283, 284, 302, 344, 350
strike price, 377

strong Markov property, 209, 391
submartingale, 276, 277, 284, 287, 288, 293, 294, 298
    backward, 300
success runs, 199
supermartingale, 276, 284
supporting line, 88
symmetric difference, 6
synthetic probabilities, 315, 316

tail field, 347
Taylor's formula, 360, 392
    remainder, 360
tent function, 379
    Schauder, 388
theorem
    bounded convergence, 126
    Cameron-Martin, 374–376
    central limit, 90, 135, 356, 383
    Dini's, 356
    dominated convergence, 125, 127, 220, 296
    ergodic, 225
    Fourier inversion, 159
    Fubini, 106
    Glivenko-Cantelli, 136
    Helly-Bray, 169
    Lévy inversion, 159
    martingale convergence, 292, 294
    monotone class, 4, 49
    monotone convergence, 123, 155, 353
    normal number, 138, 140
    Poisson limit, 68
    Radon-Nikodym, 268
    Riesz-Fischer, 394
    Skorokhod, 380
    system, 282, 301
    Weierstrass, 136
tight, 169
tightness, 169
tote board, 315, 328
trading rules, 318
transformation
    scaling, 338
transient, 211–213, 223, 225, 228
transition probability, 193, 341, 359
    n-step, 194
    Poisson process, 245
    stationary, 193
trinomial tree, 332

uniform integrability, 295–297

Index

upcrossing inequality, 291
upcrossings, 291

vague convergence, 168
de la Vallée Poussin, 351
value, 322
variance, 60, 407
   geometric, 69
   hypergeometric, 72
   Poisson, 67
viscosity, 405
volatility, 374, 377

Watson, Henry William, 234
weak convergence, 168
Weierstrass, 335
white noise, 394, 395, 405, 408
Wiener, Norbert, 335
Wright, Sewall, 198

zero set, 348
zero-one law, 307
   Blumenthal, 347
   Borel, 347

DATE DUE

PRINTED IN U.S.A.